Actin-Binding Proteins and Disease

PROTEIN REVIEWS

Recent Volumes in this Series

Cristobal G. dos Remedios • Deepak Chhabra

Editors

Actin-Binding Proteins and Disease

 Springer

Cristobal G. dos Remedios
Muscle Research Unit
Bosch Institute, School of Medical
    Sciences
The University of Sydney
Sydney 2006
Australia

Deepak Chhabra
Discipline of Anatomy and Histology
Bosch Institute, School of Medical
    Sciences
The University of Sydney
Sydney 2006
Australia

ISBN 978-0-387-71747-0   e-ISBN 978-0-387-71749-4

Library of Congress Control Number: 2007926248

springer.com

Foreword

The idea for this book arose from a lengthy review on actin and actin-binding proteins (ABPs) written by the editors in 2004 for Physiological Reviews (dos Remedios et al. 2003 Physiol. Rev. 83,433-473). At that time it emerged that there were a number of human diseases and disorders that involved either actin by itself, or was the result of altered assembly of the actin cytoskeleton produced by ABPs.

This book is assembled into three sections. We begin with an overview of actin where we review the structure of this ubiquitous protein, the biophysics of actin polymerisation, and set the groundwork for a detailed discussion on how actin comes to play a role in human disease.

Actin and Human Disease

We cover the history of its discovery, the seminal work of Fumio Oosawa in describing the thermodynamics of actin assembly up to the relatively recent X-ray crystallography of the actin monomer described by the Dominguez laboratory. We also introduce the biophysics of actin polymerisation and the concept of critical concentration needed to understand how the ABPs function.

Sparrow and Laing describe the increasing number of mutations in three human actin genes that cause specific disease phenotypes. Mutations in the actin sequence alter specific sites on the monomer that are required for the interaction of actin with itself (actin polymerisation) as well as with ABPs. Mutations that affect assembly/disassembly of thin filaments will have a major impact on the development of force by actomyosin and they discuss potential effects of mutations on skeletal and cardiac muscle force development as well as their effects on non-muscle motile structures such as the stereocilia of hair cells. The recently-developed ability to express mutant actins will no doubt accelerate investigations of the effects of these mutations.

In the next chapter, Filip Braet and his colleagues describe the discovery of actin-binding drugs that have wide-ranging effects on cells and provide a unique opportunity to gain new insights into actin-mediated processes. New compounds are increasingly being isolated, identified and synthesized and are likely to become important, not only by providing a tool to further advance our understanding of the actin cytoskeleton, but also as well as potential new therapeutic compounds.

Despite its highly conserved primary structure, actin is involved in diseases such as autoimmune hepatitis (AIH). Actin and ABPs are a major target for autoimmune reactions. AIH and other diseases appear to result, at least in part, from autoantibodies to F-actin. Ian Mackay and his colleagues discuss how various technical obstacles have precluded collection of good data and then provide preliminary data based on functional effects of isolated human autoantibodies on the in vitro motility of actin in an actin-myosin motility assay. They discuss the nature of epitopes engaged by T and B lymphocytes and the circumstances in which human anti-F-actin may cause pathogenic effects in vivo.

Actin-Binding Proteins

Next we move on to the biology and diseases associated with abnormalities in ABPs. Mira Krendel and Enrique De La Cruz provide a lucid overview of the field and refer to a number of ABPs that perhaps should, but will not be discussed in detail in this volume.

A major feature of this volume is the detailed and scholarly review of the role of ADF/cofilin in non-muscle cells. Cofilin was first identified over 25 years ago by the senior author (Jim Bamberg) and his colleagues. They provide details of the biochemistry, structure and biology of study of the actin-depolymerizing factor/cofilin (ADP/cofilin) family that are needed to define the roles of ADF/cofilin in disease. They then describe their roles in cornea defects, immune disease, musculoskeletal diseases, gametogenesis and in central nervous disorders such as Alzheimer's disease, mental retardation, dementias, as well as aging and senescence.

Separate chapters are devoted to gelsolin, profilin and thymosin β4. These were selected because they essentially cover the most abundant and thus biologically important ABPs in cells. All authors here have diligently struck to their brief of focussing on the roles of ABPs in disease.

ABPs and Human Disease

This final section of the volume deals with defects in ABPs that are known to cause disorders and disease in man. A probable reason why this has not been a focus in the past is that actin and its binding proteins play essential roles in the biology of cells and any major defect is therefore likely to be lethal. However, some defects are clearly not lethal. In an elegant treatise on the role of ABPs in cancer, Leen Van Troys, Joel Vandekerckhove and Christophe Amp describe the role of ABPs (and actin) in tumor cell migration, the role of the extracellular matrix in this process.

Each chapter is copiously reference to enable to reader to source the original article and we have edited the entire book so that the relationships between the chapters is highlighted. We hope the book will pioneer a new focus on the roles of actin and ABPs in human disease.

Cris dos Remedios
Deepak Chhabra

Contents

Contributors

Paul D. Allen
Brigham and Women's Hospital, 5 Francis Street, Boston, MA 02115, USA

Christophe Ampe
Department of Biochemistry, Faculty of Medicine and Health Sciences, Ghent University and Medical Protein Research, Flanders Interuniversity Institute for Biotechnology (VIB), A. Baertsoenkaai 3, B–9000 Gent, Belgium.

James R. Bamburg
Department of Biochemistry and Molecular Biology and Molecular, Cellular and Integrative Neurosciences Program, Colorado State University, Fort Collins, CO 80523-1870, USA

Elaine L. Bearer
Department of Pathology and Laboratory Medicine, Brown University Brown Medical School, Providence, RI 02912, USA

Filip Braet
Australian Key Centre for Microscopy and Microanalysis (AKCMM), Electron Microscope Unit, The University of Sydney, NSW 2006, Australia

Leslie D. Burtnick
Chemistry Department and Centre for Blood Research, Life Sciences Institute, University of British Columbia, Vancouver, V6T 1Z1, Canada

Deepak Chhabra,
Discipline of Anatomy and Histology, Bosch Institute, School of Medical Sciences, The University of Sydney, Sydney 2006, Australia

Enrique M. De La Cruz
Molecular Biophysics and Biochemistry, Yale University, 266 Whitney Avenue, New Haven, CT 06520, USA

Cristobal G. dos Remedios
Muscle Research Unit, Bosch Institute, School of Medical Sciences,
The University of Sydney, Sydney 2006, Australia

Brett D. Hambly
Department of Pathology, Bosch Institute, The University of Sydney,
Sydney, 2006, Australia

Andrew W. Kinley
Department of Biochemistry and Molecular Biology and Molecular, Cellular
and Integrative Neurosciences Program, Colorado State University, Fort
Collins, CO 80523-1870, USA

Mira Krendel
Department of Cell and Developmental Biology, SUNY Upstate Medical
University, Syracuse, NY 13210, USA.

Nigel G. Laing
Centre for Medical Research, University of Western Australia, Western
Australian Institute for Medical Research, QEII Medical Centre, Nedlands,
Western Australia 6009, Australia

Choon C. Liew
Brigham and Women's Hospital, 5 Francis Street, Boston, MA 02115, USA

Peter S. Macdonald
Victor Chang Cardiac Research Institute, and Cardiopulmonary Transplant
Unit, St Vincent's Hospital, Victoria Street, Darlinghurst NSW, Australia

Ian R. Mackay
Department of Biochemistry and Molecular Biology, Monash University,
Clayton Victoria 3800, Australia

Michael T. Maloney
Department of Biochemistry and Molecular Biology and Molecular, Cellular
and Integrative Neurosciences Program, Colorado State University, Fort
Collins, CO 80523-1870, USA

Roberto Martinez-Neira
Muscle Research Unit, Bosch Institute, School of Medical Sciences,
The University of Sydney, Sydney 2006, Australia

Pierre D. J. Moens
School of Biological, Biomedical and Molecular Sciences, The University
of New England, Armidale, NSW 2351, Australia

Dan Nicolau
Department of Electrical Engineering and Electronics, Brownlow Hill,
Liverpool, L69 3GJ, United Kingdom

Neil J. Nosworthy
Muscle Research Unit, Bosch Institute, School of Medical Sciences,
The University of Sydney, Sydney 2006, Australia

Hiroyuki Nunoi
Division of Pediatrics, Department of Reproductive and Developmental
Medicine Faculty of Medicine, University of Miyazaki, 5200 Kihara,
Kiyotake-chou, Miyazaki, 889-1692, Japan

Chi W. Pak
Department of Biochemistry and Molecular Biology and Molecular, Cellular
and Integrative Neurosciences Program, Colorado State University, Fort
Collins, CO 80523-1870, USA

Robert C. Robinson
Institute of Molecular and Cell Biology Proteos, 61 Biopolis Drive, Singapore
138673

Lilian Soon
Australian Key Centre for Microscopy and Microanalysis (AKCMM),
Electron Microscope Unit, The University of Sydney, NSW 2006, Australia

John C. Sparrow
Department of Biology, University of York, York, YO10 5DD, United Kingdom

Ilan Spector
Department of Physiology and Biophysics, Health Science Center,
State University of New York at Stony Brook (SUNY), Stony Brook,
NY 11794-8661, New York, USA

Maurizio Stefani
Muscle Research Unit, Bosch Institute, School of Medical Sciences,
The University of Sydney, Sydney 2006, Australia

Pall Thordarson
School of Chemistry, The University of Sydney, Sydney 2006, Australia

Ban-Hock Toh
Centre for Inflammatory Diseases, Monash Institute of Medical Research,
Monash University, Clayton Victoria 3800, Australia

Masako Tsubakihara Muscle Research Unit
Bosch Institute, School of Medical Sciences, The University of Sydney,
Sydney 2006, Australia

Joel Vandekerckhove
Department of Biochemistry, Faculty of Medicine and Health Sciences,
Ghent University and Medical Protein Research, Flanders Interuniversity
Institute for Biotechnology (VIB), A. Baertsoenkaai 3, B–9000 Gent, Belgium

M. Van Troys
Department of Biochemistry, Faculty of Medicine and Health Sciences,
Ghent University and Medical Protein Research, Flanders Interuniversity
Institute for Biotechnology (VIB), A. Baertsoenkaai 3, B–9000 Gent, Belgium

Katrien Vekemans
Abdominal Transplant Surgery, Department of Surgery, Catholic University
of Leuven, Leuven 3000, Belgium

Senga Whittingham
Department of Biochemistry and Molecular Biology, Monash University,
Clayton Victoria 3800, Australia

1
Actin: An Overview of Its Structure and Function

Deepak Chhabra and Cristobal G. dos Remedios

Introduction

Actin forms the principal components of cytoskeletal microfilaments. The assembly of globular actin monomers (G-actin) to form filaments (F-actin) is essential in allowing the cytoskeleton to carry out its various functions including, but not limited to, cell division and locomotion.

Actin, 42 kDa, was discovered from muscle extracts (Straub 1942), and is now known to be abundant in all eukaryotic cells (Poglazov 1983). α-actin consists of a single polypeptide chain of 375 residues with an amino acid sequence that is so conserved that actin was erroneously thought to have the same sequence in all tissues and species (Bray 1972).

Actin is now classified into six species-independent isoforms (Vandekerckhove and Weber 1978) that in vertebrates are the products of different genes (Vandekerckhove and Weber 1979a,b). There are two smooth muscle actins (α-smooth and γ-smooth muscle), two striated muscle actins (α-skeletal and α-cardiac), and two cytoplasmic actins (β-cytoplasmic and γ-cytoplasmic). Unusually, chicken is an exception having three cytoplasmic actins (Bergsma et al. 1985). Figure 1 displays sequence alignments of the six isoforms of actin. The α-actins are mainly expressed in all types of muscle (skeletal, cardiac and smooth); β-actin is primarily expressed in nonmuscle cells; γ-actins are expressed mainly in nonmuscle and smooth muscle cells.

Structure of G-Actin

The structure of G-actin has been extensively studied from cocrystals of actin complexed with bovine pancreatic DNase I (Kabsch et al. 1990), profilin (Schutt et al. 1993; Chik, Lindberg and Schutt 1996), and gelsolin segment 1 (McLaughlin et al. 1993). These actin-binding proteins (ABPs) stabilized the actin monomer and reduced its propensity to self-assemble. In 2001, the Dominguez group (Otterbein, Graceffa and Dominguez 2001) finally solved

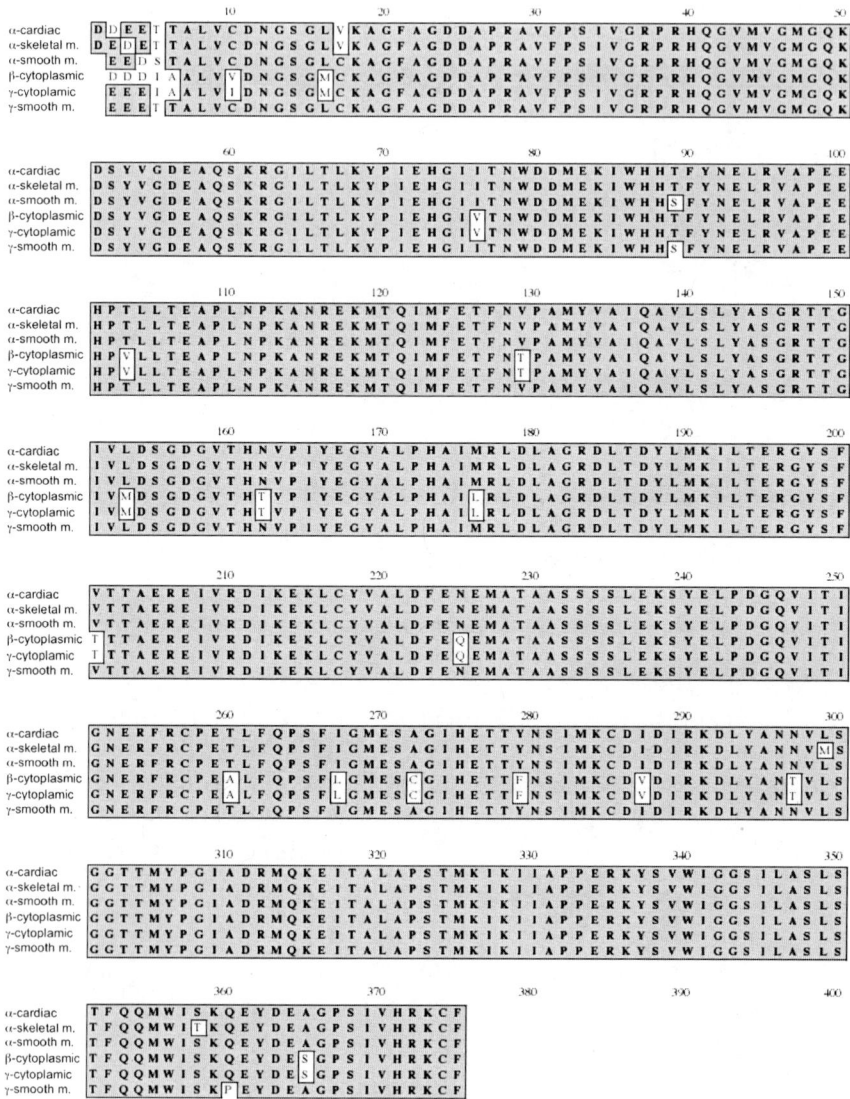

FIG. 1. Sequence comparison of the six isoforms of human actin. Identical residues are shaded in gray

the structure of uncomplexed actin containing bound ADP (Fig. 2). They were successful where others had failed by suppressing actin polymerization by covalently modifying cys-374 with a rhodamine conjugate.

Actin is a globular protein with overall dimensions $67 \times 40 \times 37$ Å. It has large (residues 145–337) and small (residues 1–144 and 338–375) domains that, despite their names, are of approximately equal size. By convention, actin is normally oriented such that the small domain lies to the right and the

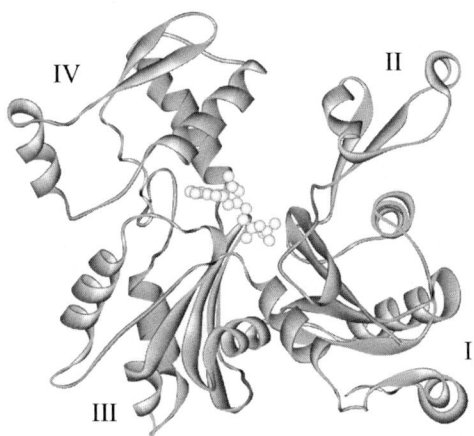

FIG. 2. Atomic structure of uncomplexed actin (gray) bound to ADP (white) (Otterbein, Graceffa and Dominguez 2001). Subdomains are labeled I, II, III, and IV

large domain to the left. The small domain is further divided into subdomain 1 (bottom right, residues 1–32, 70–144 and 338–375) and subdomain 2 (top right, residues 33–69). The large domain is further divided into subdomain 3 (bottom left, residues 145–180 and 270–337) and subdomain 4 (top left, residues 181–269). The N- and C-termini are both present in subdomain 1. The top of subdomain 2 (the DNase I binding loop) appears to be influenced by the binding of DNase I. In the uncomplexed state (Fig. 2) reported by Otterbein, Graceffa and Dominguez (2001), it appears as an α-helix. Conversely, in the original structure of actin bound DNase I (Kabsch et al. 1990) it appears as a coil. Exposed subdomains 1 and 3, and subdomains 2 and 4 are, respectively, named the barbed and pointed ends (Huxley 1963; Moore, Huxley and DeRosier 1970). The same name is applied to the exposed ends of F-actin.

High-affinity nucleotide and cation binding sites are located in the deep interdomain cleft of actin. Due to differences in affinity, the nucleotide-binding site is usually occupied by ATP ($K_{eq} = 10^{10}$ M^{-1} (Engel et al. 1977)) or ADP-Pi rather than ADP ($K_{eq} = 10^8$ M^{-1} (Neidl and Engel 1979)). The adjacent high-affinity metal binding site is occupied by Mg^{2+} *in vivo* (K_{eq} for $Mg^{2+} = 10^8$ M^{-1} (Gershman, Selden and Estes 1986)) due to its high intracellular concentration (Weber, Herz and Reiss 1969; Kitazawa, Shuman and Somlyo 1982). However, due to the high concentration of $CaCl_2$ in buffers used in actin preparation, Ca^{2+} (K_{eq} for $Ca^{2+} = 5 \times 10^8$ M^{-1}, Gershman, Selden and Estes 1986) is normally the bound cation *in vitro*.

The nucleotide–cation complex interacts with both domains of actin restricting their motion at the flexible hinge region formed by α-helix 137–144 (Tirion and Benavraham 1993) thereby maintaining the native conformation of the protein (Barany et al. 1961; Tonomura and Yoshimura 1961). A change in the relative orientations of the domains (Tirion et al. 1995; Page, Lindberg and Schutt 1998) or subdomains (Lorenz, Popp and Holmes 1993) results in the narrowing of the inter-domain cleft, as observed in the open and closed states of the actin monomer (Schutt et al. 1993; Chik, Lindberg and Schutt 1996). Removal of either the nucleotide or cation denatures the actin monomer (West 1971; Kinosian et al. 1993).

Actin contains multiple intermediate affinity (K_d for $Mg^{2+}/Ca^{2+} \approx 10^{-4}$ M; K_d for $K^+ \approx 10^{-6}$ M) and low affinity (K_d for $Mg^{2+}/Ca^{2+} \approx 10^{-2}$ M) cation binding sites that when saturated initiate polymerization or paracrystallization, respectively (Strzelecka-Golaszewska, Prochniewicz and Drabikowski 1978; Carlier, Pantaloni and Korn 1986a; Estes et al. 1992).

Structure of F-Actin

Our current understanding of the structure of F-actin is limited to various proposed atomic models based on low-resolution (25–30 Å) electron microscope studies, X-ray diffraction data (Holmes et al. 1990), molecular simulation (Tirion et al. 1995), directed mutation algorithms (Lorenz, Popp and Holmes 1993), and its involvement in the mechanism of muscle contraction (Schutt et al. 1993).

The actin filament appears to be a dynamic, responsive element. Its atomic structure (Dedova et al. 2002) and state of assembly (see chapter by Krendel and De La Cruz) is influenced by ABPs. The actin filament can exist in multiple conformations depending on the type of bound cation and nucleotide, the isoform of actin (Orlova et al. 1997; Orlova and Egelman 1995), and the presence of other proteins bound to actin (McGough et al. 1997; Owen and DeRosier 1993).

F-actin is a helical polymer with a maximum diameter of 90–95 Å. There are 13 molecules per six left-handed turns with a repeat of 360 Å along the axis. The rotation per subunit is −166°. This rotation allows F-actin to appear morphologically as either a single-start left-handed (genetic) helix or two right-handed steep helices that twine slowly around each other.

The first and most accepted model of the structure is F-actin (Fig. 3). Lorenz et al. (1993) proposes that subunits are aligned tangentially to the filament axis such that the large domain is at a small radius and the small domain is at a large radius. Each monomer in the filament makes contact with four others. Because of this helical symmetry, there are extensive contacts between the two-start helices involving residues 322–325 with 243–245; 286–289 with 202–204; 166–169 and 375 with 41–45. Contacts between the single-start genetic helix are between residues 110–112 and 195–197. Within the helix, residues 40–45 and 63–64 of the "lower" monomer and residues 166, 169, 171, 173, 285 and 289 of the upper monomer form a longitudinal hydrophobic pocket holding the helix together. Residues 264–273 of G-actin form a finger like loop that has been proposed to plug the hydrophobic pocket between adjacent monomers.

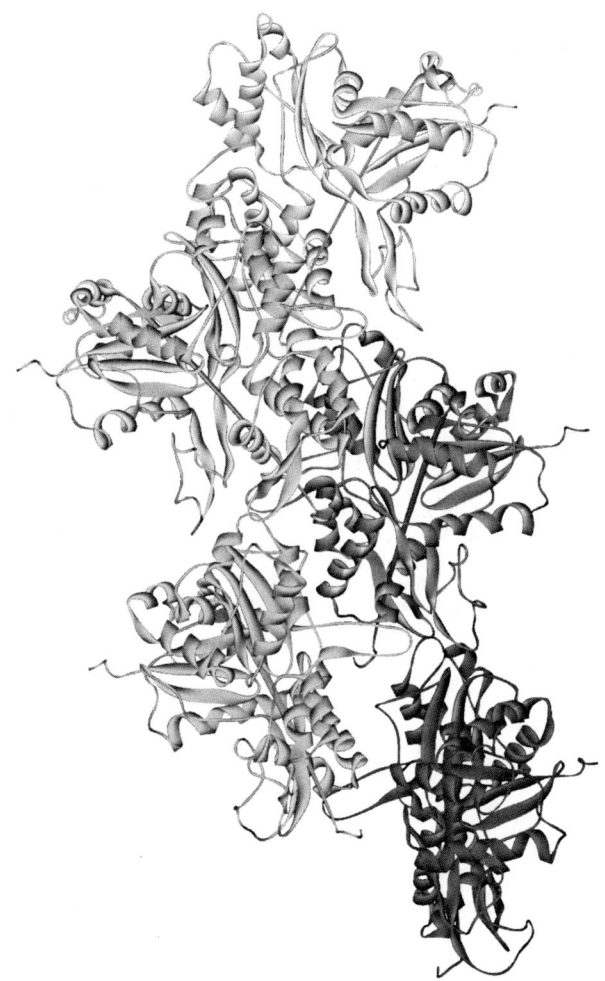

FIG. 3. A model of the actin filament (Lorenz, Popp and Holmes 1993). The single-start left-handed helix starts from the monomer in red going to green, orange, cyan then gray. The two right-handed helices start for the first helix from the monomer in red going to orange then gray, while the second helix is from the monomer in green going to cyan. The gray and cyan monomers are localized at the barbed end of the filament; the red and green monomers are localized at the pointed end of the actin filament. Reproduced from dos Remedios et al. 2003 with permission from the American Physiological Society (*See Color Plates*)

Assembly of Actin

Critical Concentration

Under appropriate conditions, G-actin monomers self-assemble to form F-actin polymers/filaments of indefinite length. Polymerization only occurs if the concentration of G-actin exceeds the *critical concentration* (C_C).

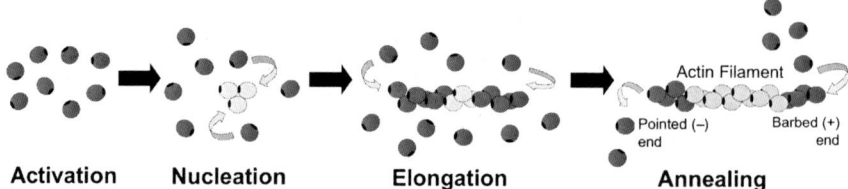

Fig. 4. The polymerization of actin consists of four steps – activation, nucleation, elongation, and annealing. See text for details. Image courtesy of Ms. Lan Kim Nguyen (University of Sydney) (*See Color Plates*)

The C_C is defined as the concentration of monomers coexisting with polymer at the steady state of polymerization (Oosawa et al. 1959; Kasai, Asakura and Oosawa 1962; Oosawa 1983). In other words, at levels of actin below the C_C, actin will fail to polymerize and remain as a monomer. The C_C is a function of the rate constants for monomer association and dissociation and is dependent on solvent conditions including type and concentrations of salts, ionic strength of buffer ($C_C \downarrow$ as ionic strength $\uparrow$), pH ($C_C \downarrow$ as pH $\downarrow$), and temperature ($C_C \downarrow$ as temperature $\uparrow$) (Asakura, Kasai and Oosawa 1960; Grazi and Trombetta, 1985; Zimmerle and Frieden 1988; Wang, Sampogna and Ware 1989). Within cells, a large number of ABPs sequester actin and allow monomeric actin to exist at concentrations 100–1,000 times the C_C *in vitro* (Pollard, Blanchoin and Mullins 2000).

The polymerization of actin is essentially a condensation reaction and usually consists of four steps, namely, activation, nucleation, elongation, and annealing (Gaszner et al. 1999) as illustrated in Fig. 4.

Activation

Activation is due to the binding of Mg^{2+}, K^+ or Ca^{2+} to the intermediate-affinity cation binding sites, and *in vitro*, the exchange of Ca^{2+} for Mg^{2+} at the high-affinity cation binding site (Carlier, Pantaloni and Korn 1986a). Not only does activation neutralize negative charges on actin to reduce electrostatic repulsion between monomers (Frieden 1983; Carlier, Pantaloni and Korn 1986b), but it also induces a conformational change that facilitates the nucleation process (Gaszner et al. 1999). This conformationally unique monomer has been named the F-actin monomer (Rich and Estes 1976), G*-actin (Rouayrenc and Travers 1981), or activated G-actin (Shu, Wang and Stracher 1992).

Despite great interest in the process of divalent cation exchange as part of the activation process, it is most likely an artifact due to the artificial association of Ca^{2+} actin during *in vitro* protein preparation. In fact, the conformation of Mg-ATP-G-actin is close to that of F-actin protomers, but fairly dissimilar to that of Ca-G-actin (Strzelecka-Golaszewska et al. 1993, 1996; Mejean et al. 1988; Adams and Reisler 1994; Chik, Lindberg and Schutt 1996;

Miki and Kouyama 1994; Roustan et al. 1985). Consequently, activation of actin is most probably more rapid *in vivo* where actin is already bound to Mg^{2+}. The conformational similarity between Mg-ATP-actin and the F-actin subunit may also partly explain the more polymeric nature of Mg-actin compared to Ca-actin (Selden, Estes and Gershman 1983).

Nucleation

Nucleation is the energetically unfavorable process of aggregating actin monomers to form a relatively stable oligomer, or nucleus, which can support the addition of monomeric actin at both ends leading to the formation of F-actin filaments (Wegner and Engel 1975; Oosawa 1983). Nucleation is the rate-limiting step of polymerization and in the case of actin the nucleus is a trimer (Barden, Grant and dos Remedios 1982). Nucleation is enhanced by the interaction of ATP-actin with ADP-actin (Pantaloni, Carlier and Korn 1985).

The lower free energy of nucleation for Mg-ATP-G-actin than for Ca-ATP-G-actin (Oosawa and Asakura 1975; Oosawa 1983; Tobacman and Korn 1983; Newman et al. 1985), coupled with the more polymeric conformation of Mg-ATP-G-actin compared to Ca-ATP-G-actin results in a lower C_C for actin with bound Mg^{2+} than bound Ca^{2+}. Thus, in the absence of ABPs, actin inside cells is probably more polymeric than *in vitro*.

Elongation and Annealing

Elongation refers to the association and dissociation of actin monomers at both ends of the actin filament. These newly added protomers are "locked" into position by conformational changes or annealing. Association predominantly occurs at the barbed end of the filament and dissociation at the pointed end.

Despite earlier theories (Oosawa and Asakura 1975; Wegner 1976) hydrolysis of the bound ATP does not appear to be tightly coupled to the polymerization process. In fact, actin can polymerize if it contains bound ADP (Pollard 1984) or a nonhydrolyzable analog of ATP (Cooke 1975a,b), or no bound nucleotide at all (Kasai, Nakano and Oosawa 1965). A short delay exists between the incorporation of ATP-G-actin onto a filament end and hydrolysis of the bound nucleotide (Pardee and Spudich 1982; Pollard and Weeds 1984; Carlier, Pantaloni and Korn 1984; Carlier 1990). A longer delay exists for release of the generated inorganic phosphate (Carlier and Pantaloni 1986). ATP hydrolysis primarily occurs at the barbed (growing) end of filaments (Carlier et al. 1997). The bound ATP of F-actin protomers is essentially non-exchangeable (Hegyi, Szilagyi and Belagyi 1988). However, once disassociated, ADP-actin subunits rapidly exchange their bound ADP for ATP in solution (Neidl and Engel 1979).

The structural polarity of F-actin and the irreversible nucleotide hydrolysis during actin assembly has implications for the rate and direction of filament growth at opposite ends of the actin filament. The C_C for the pointed end is 12–15-fold higher than for the barbed end under physiological conditions (Wegner and Isenberg 1983). This difference in C_C may result in a continual flux of F-actin subunits from the pointed to the barbed end producing unidirectional growth of the actin filament. This head-to-tail polymerization or *treadmilling* of actin (Wegner 1976) is modulated by a number of ABPs within cells (dos Remedios et al. 2003).

Actin: Not Only a Structural Protein

It was previously thought that actin filaments acted purely as structural scaffolding within the cytoplasm of cells. Reports of nuclear actin were questioned by researchers who cited the lack of a nuclear localization sequence within actin (at 42 kDa actin is too large to diffuse across the nuclear membrane). Conversely, actin possesses two functional leucine-rich nuclear export sequences (Wada et al. 1998). Furthermore actin was not visualized in the nucleus upon labeling with fluorescent phalloidin (Rando, Zhao and Crabtree 2000).

Despite the controversy, actin is now recognized to reside within the nuclear compartment. It appears that actin primarily exists as monomers or short oligomers (Gonsior et al. 1999) that are extremely dynamic (McDonald et al. 2006). It is likely that actin enters the nucleus by binding cofilin, which contains an SV-40-type nuclear translocation signal (Nishida et al. 1987; Iida, Matsumoto and Yahara 1992) with which much of it remains associated in the nucleoplasm (Chhabra and dos Remedios 2005). A number of ABPs, including cofilin (Chhabra, Bao and dos Remedios 2002), profilin (Stuven, Hartmann and Gorlich 2003), and members of the gelsolin/villin family (CapG and supervillin) have been found in the nuclear compartment. Presumably, profilin and thymosin β4 are also small enough to enter the nucleus by passive diffusion. Consequently, there are a number of emerging roles of actin in various non-structural processes, including regulation of cell turnover and gene expression. For further details on these actin-binding proteins, we refer the reader to the chapters on cofilin (Maloney et al.), profilin (Moens), gelsolin (Burtnick and Robinson), and thymosin β4 (Au et al).

Cell Turnover

During times of cell stress actin has been reported to move into the nucleus, possibly in association with cofilin (Iida, Matsumoto and Yahara 1992; Nishida et al. 1987; Pendleton et al. 2003). While this initially seems a consequence of cell-stress induced disorganization, it is emerging that nuclear translocation of actin may act as a protective mechanism by decreasing the proliferation potential of these cells (Wada et al. 1998). It is likely that various actin regulatory proteins may also play a role in this actin-induced cell arrest. For example both cofilin and CapG have been reported in the nucleus and are regulated by phosphorylation. During ATP depletion, phosphorylation of both these ABPs is reduced in an attempt to preserve ATP concentrations. Dephosphorylated cofilin preferentially accumulates in the nucleoplasm (Nebl, Meuer and Samstag 1996; Ohta et al. 1989), whereas dephosphorylated CapG leaves the nucleus to enter the cytoplasm (Onoda and Yin 1992). Thus cell regulation of nuclear actin may switch from a CapG to a cofilin-mediated system during ATP depletion. It is likely that a similar mechanism operates for a number of ABPs that can be phosphorylated.

There is a growing body of evidence that actin is indeed a regulator of cell turnover rather than acting as a purely dynamic structural element. A mutant of actin (G244D) is associated with cellular transformation and cancer (Leavitt et al. 1982, 1984). Interestingly, this mutation causes increased proliferation rather than arrest of cell growth that would be expected due to impaired cytokinesis (see chapter by Sparrow and Laing for a review of actin mutations).

Actin is implicated in the morphological changes in cells undergoing apoptosis (voluntary cell suicide). During apoptosis, cytoskeletal microfilaments are disrupted by cleavage of actin by the interleukin-1β-converting enzyme (ICE) family of proteases (Kayalar et al. 1996) and irreversible activation of gelsolin, an actin severing protein, by cleavage with caspase-3 (Geng et al. 1998; Kamada et al. 1998; Kothakota et al. 1997). However, it has recently been reported that gelsolin, along with cofilin, modulates interaction of actin and DNase I *in vitro* (Chhabra, Nosworthy and dos Remedios 2005). It is therefore plausible that cleavage of actin and its regulatory proteins may disinhibit DNase I whose activity is normally abolished by binding to actin (Lazarides and Lindberg 1974).

Collectively these reports suggest that actin is a key player in controlling cell turnover by decreasing proliferation during times of stress while being involved in apoptotic progression once irreversible cell damage has occurred. A breakdown in this process may lead to cell transformation and cancer.

Gene Expression

Actin has been identified as a component of mammalian SWI/SNF-like BAF chromatin remodeling complexes (Olave, Reck-Peterson and Crabtree 2002) and core histone complexes (Covelo et al. 2006) that influence the packaging of chromatin. Consequently the state of actin assembly may expose promoter regions of genes permitting the binding of transcriptional machinery. Recently, the association of actin with RNA polymerase I (Fomproix and Percipalle 2004; Philimonenko et al. 2004), II (Hofmann et al. 2004; Kukalev et al. 2005), and III (Hu, Wu and Hernandez 2004) has been shown to be essential for their transcriptional activity. Depletion of actin or its sequestration in the nucleus results attenuates the nuclear export of RNA and protein (Hofmann et al. 2001) possibly in association with exportin-6 (Stuven, Hartmann and Gorlich 2003). Thus actin is involved in many aspects of gene expression including chromatin remodeling, transcription, and nuclear export of RNA.

Concluding Remarks

Actin is a ubiquitous protein in all eukaryotic cells. The structure of G-actin is known in atomic detail and several plausible models of the actin filament have been proposed. While the role of actin and ABPs are well understood in terms of cell motility and locomotion, it is becoming clear that actin performs many roles in the cell independent of its structural functions. It is these relatively new insights into the cellular function of actin that must be the focus of future research in order to understand the role of actin and its regulatory proteins in disease.

References

Adams, S. B. and Reisler, E. 1994. Sequence 18–29 on actin: Antibody and spectroscopic probing of conformational changes. Biochemistry 33, 14426–14433.

Asakura, S., Kasai, M. and Oosawa, F. 1960. The effects of temperature on the equilibrium state of actin solutions. J. Polym. Sci. 44, 35–49.

Barany, M., Nagy, F., Finkelman, F. and Chrambach, A. 1961. Studies on the removal of the bound nucleotide on actin. J. Biol. Chem. 236, 2917–2925.

Barden, J. A., Grant, N. J. and dos Remedios, C. G. 1982. Identification of the nucleus of actin polymerization. Biochem. Int. 5, 685–692.

Bergsma, D. J., Chang, K. S. and Schwartz, R. J. 1985. Novel chicken actin gene: Third cytoplasmic isoform. Mol. Cell. Biol. 5, 1151–1162.

Bray, D. 1972. Cytoplasmic actin: A comparative study. Cold Spring Harb. Symp. Quant. Biol. 37, 567–571.

Carlier, M. F. 1990. Actin polymerization and ATP hydrolysis. Adv. Biophys. 26, 51–73.

Carlier, M. F., Laurent, V., Santolini, J., Melki, R., Didry, D., Xia, G. X., Hong, Y., Chua, N. H. and Pantaloni, D. 1997. Actin depolymerizing factor (ADF/cofilin) enhances the rate of filament turnover: Implication in actin-based motility. J. Cell Biol. 136, 1307–1322.

Carlier, M. F. and Pantaloni, D. 1986. Direct evidence for ADP-Pi-F-actin as the major intermediate in ATP-actin polymerization. Rate of dissociation of P_i from actin filaments. Biochemistry 25, 7789–7792.

Carlier, M. F., Pantaloni, D. and Korn, E. D. 1984. Evidence for an ATP cap at the ends of actin filaments and its regulation of the F-actin steady state. J. Biol. Chem. 259, 9983–9986.

Carlier, M. F., Pantaloni, D. and Korn, E. D. 1986a. Fluorescence measurements of the binding of cations to high-affinity and low-affinity sites on ATP-G-actin. J. Biol. Chem. 261, 10778–10784.

Carlier, M. F., Pantaloni, D. and Korn, E. D. 1986b. The effects of Mg^{2+} the high-affinity and low-affinity sites on the polymerization of actin and associated ATP hydrolysis. J. Biol. Chem. 261, 10785–10792.

Chhabra, D., Bao, S. and dos Remedios, C. G. 2002. The distribution of cofilin and DNase I *in vivo*. Cell Res. 12, 207–214.

Chhabra, D., Nosworthy, N. J. and dos Remedios, C. G. 2005. The N-terminal fragment of gelsolin inhibits the interaction of DNase I with isolated actin, but not with the cofilin–actin complex. Proteomics 5, 3131–3136.

Chhabra, D. and dos Remedios, C. G. 2005. Cofilin, actin and their complex observed *in vivo* using fluorescence resonance energy transfer. Biophys. J. 89, 1902–1908.

Chik, J. K., Lindberg, U. and Schutt, C. E. 1996. The structure of an open state of β-actin at 2.65 Å resolution. J. Mol. Biol. 263, 607–623.

Cooke, R. 1975a. The role of the bound nucleotide in the polymerization of actin. Biochemistry 14, 3250–3256.

Cooke, R. 1975b. The bound nucleotide of actin. J. Supramol. Struct. 3, 146–153.

Covelo, G., Sarandeses, C. S., Diaz–Jullien, C. and Freire, M. 2006. Prothymosin-α interacts with free core histones in the nucleus of dividing cells. J. Biochem. 140, 627–637.

Dedova, I. V., Dedov, V., Nosworthy, N. J., Hambly, B. D. and dos Remedios C. G. 2002. Cofilin and DNase I affect the conformation of the small domain of actin. Biophys. J. 82, 3134–3143.

Engel, J., Fasold, H., Hulla, F. W., Waechter, F. and Wegner, A. 1977. The polymerization of muscle actin. Mol. Cell. Biochem. 18, 3–13.

Estes, J. E., Selden, L. A., Kinosian, H. J. and Gershman, L. C. 1992. Tightly-bound divalent cation of actin. J. Muscle Res. Cell Motil. 13, 272–284.

Frieden, C. 1983. Polymerization of actin: Mechanism of the Mg^{2+}-induced process at pH 8 and 20 degrees C. Proc. Natl Acad. Sci. USA 80, 6513–6517.

Fomproix, N. and Percipalle, P. 2004. An actin–myosin complex on actively transcribing genes. Exp. Cell Res. 294, 140–148.

Gaszner, B., Nyitrai, M., Hartvig, N., Koszegi, T., Somogyi, B. and Belagyi, J. 1999. Replacement of ATP with ADP affects the dynamic and conformational properties of actin monomer. Biochemistry 38, 12885–12892.

Geng, Y. J., Azuma, T., Tang, J. X., Hartwig, J. H., Muszynski, M., Wu, Q., Libby, P. and Kwiatkowski, D. J. 1998. Caspase-3-induced gelsolin fragmentation contributes to actin cytoskeletal collapse, nucleolysis, and apoptosis of vascular smooth muscle cells exposed to proinflammatory cytokines. Eur. J. Cell Biol. 77, 294–302.

Gershman, L. C., Selden, L. A. and Estes, J. E. 1986. High affinity binding of divalent cation to actin monomer is much stronger than previously reported. Biochem. Biophys. Res. Commun. 135, 607–614.

Gonsior, S. M., Platz, S., Buchmeier, S., Scheer, U., Jockusch, B. N. and Hinssen, H. 1999. Conformational difference between nuclear and cytoplasmic actin as detected by a monoclonal antibody. J. Cell Sci.112, 797–809.

Grazi, E. and Trombetta, G. 1985. Effects of temperature on actin polymerized by Ca^{2+}. Biochem. J. 232, 297–300.

Hegyi, G., Szilagyi, L. and Belagyi, J. 1988. Influence of the bound nucleotide on the molecular dynamics of actin. Eur. J. Biochem. 175, 271–274.

Hofmann, W., Reichart, B., Ewald, A., Muller, E., Schmitt, I., Stauber, R. H., Lottspeich, F., Jockusch, B. M., Scheer, U., Hauber, J. and Dabauvalle, M. C. 2001. Cofactor requirements for nuclear export of Rev response element (RRE) – and constitutive transport element (CTE) – containing retroviral RNAs. An unexpected role for actin. J. Cell Biol. 152, 895–910.

Hofmann, W. A., Stojiljkovic, L., Fuchsova, B., Vatgas, G. M., Mavrommatis, E., Philimonenko, V., Kysela, K., Goodrich, J. A., Lessard, J. L., Hope, T. J., Hozak, P. and de Lanerolle, P. 2004. Actin is part of pre-initiation complexes and is necessary for transcription by RNA polymerase II. Nat. Cell Biol. 6, 1094–1101.

Holmes, K. C., Popp, D., Gebhard, W. and Kabsch, W. 1990. Atomic model of the actin filament. Nature 347, 44–49.

Hu, P., Wu, S. and Hernandez, N. 2004. A role for beta-actin in RNA polymerase III transcription. Genes Dev. 18, 3010–3015.

Huxley, A. F. 1963. Electron microscope studies on the structure of natural and synthetic protein filaments from striated muscle. J. Mol. Biol. 7, 281–308.

Iida, K., Matsumoto, S. and Yahara, I. 1992. The KKRKK sequence is involved in heat shock-induced nuclear translocation of the 18-kDa actin-binding protein, cofilin. Cell Struct. Funct. 17, 39–46.

Kabsch, W., Mannherz, H. G., Suck, D., Pai, E. F. and Holmes, K. C. 1990. Atomic structure of the actin:DNase I complex. Nature 347, 37–44.

Kamada, S., Kusano, H., Fujita, H., Ohtsu, M., Koya, R. C., Kuzumaki, N. and Yoshihide, T. 1998. A cloning method for caspase substrates that uses the yeast two-hybrid system: Cloning of the antiapoptotic gene gelsolin. Proc. Natl Acad. Sci. USA 95, 8532–8537.

Kasai, M., Asakura, S. and Oosawa, F. 1962. The cooperative nature of G–F transformation of actin. Biochim. Biophys. Acta 57, 22–31.

Kasai, M., Nakano, E. and Oosawa, F. 1965. Polymerization of actin free from nucleotide and divalent cation. Biochim. Biophys. Acta 94, 494–503.

Kayalar, C., Ord, T., Testa, M. P., Zhong, L. T. and Bredesen, D. E. 1996. Cleavage of actin by interleukin 1β-converting enzyme to reverse DNase I inhibition. Proc. Natl Acad. Sci. USA 93, 2234–2238.

Kinosian, H. J., Selden, L. A., Estes, J. E. and Gershman, L. C. 1993. Nucleotide binding to actin-cation dependence of nucleotide dissociation and exchange rates. J. Biol. Chem. 268, 8683–8691.

Kitazawa, T., Shuman, H. and Somlyo, A. P. 1982. Calcium and magnesium binding to thin and thick filaments in skinned muscle fibres: Electron probe analysis. J. Muscle Res. Cell Motil. 3, 437–454.

Kothakota, S., Azuma, T., Reinhard, C., Klippel, A., Tang, J., Chu, K., McGarry, T. J., Kirschner, M. W., Koths, K., Kwiatkowski, D. J. and Williams, L. T. 1997. Caspase-3-generated fragment of gelsolin: Effector of morphological change in apoptosis. Science 278, 294–298.

Kukalev, A., Nord, Y., Palmberg, C., Bergman, T. and Percipalle, P. 2005. Actin and hnRNP U cooperate for productive transcription by RNA polymerase II. 2005. Nat. Struct. Mol. Biol. 12, 238–244.

Lazarides, E. and Lindberg, U. 1974. Actin is the naturally occurring inhibitor of deoxyribonuclease I. Proc Natl Acad. Sci. USA 71, 4742–4746.

Leavitt, J., Bushar, G., Kakunaga, T., Hamada, H., Hirakawa, T., Goldman, D. and Merril, C. 1982. Variations in expression of mutant β actin accompanying incremental increases in human fibroblast tumorigenicity. Cell 28, 259–268.

Leavitt, J., Gunning, P., Porreca, P., Ng, S. Y., Lin, C. S. and Kedes, L. 1984. Molecular cloning and characterization of mutant and wild-type human beta-actin genes. Mol. Cell. Biol. 4, 1961–1969.

Lorenz, M., Popp, D. and Holmes, K. C. 1993. Refinement of the F-actin model against X-ray fiber diffraction data by the use of a direct mutation algorithm. J. Mol. Biol. 234, 826–836.

McDonald, D., Carrero, G., Andrin, C., de Vries, G. and Hendzel, M. J. 2006. Nucleoplasmic β-actin exists in a dynamic equilibrium between low-mobility polymeric species and rapidly diffusing populations. J. Cell Biol. 172, 541–552.

McGough, A., Pope, B., Chiu, W. and Weeds, A. 1997. Cofilin changes the twist of F-actin: Implications for actin filament dynamics and cellular function. J. Cell Biol. 138, 771–781.

McLaughlin, P. J., Gooch, J. T., Mannherz, H. G. and Weeds, A. G. 1993. Structure of gelsolin segment-1–actin complex and the mechanism of filament severing. Nature 364, 685–692.

Mejean, C., Hue, H. K., Pons, F., Roustan, C. and Benyamin, Y. 1988. Cation binding sites on G-actin: A structural relationship between antigenic epitopes and cation exchange. Biochem. Biophys. Res. Commun. 152, 368–375.

Miki, H. and Kouyama, T. 1994. Domain motion in actin observed by fluorescence resonance energy transfer. Biochemistry 33, 10171–10177.

Moore, P. B., Huxley, H. E. and DeRosier, D. J. 1970. Three-dimensional reconstruction of F-actin, thin filaments and decorated thin filaments. J. Mol. Biol. 50, 279–295.

Nebl, G., Meuer, S. C. and Samstag, Y. 1996. Dephosphorylation of serine 3 regulates nuclear translocation of cofilin. J. Biol. Chem. 271, 26276–26280.

Neidl, C. and Engel, J. 1979. Exchange of ADP, ATP and 1,N–ethenoadenosine 5′-triphosphate on G-actin. Equilibrium and kinetics. Eur. J. Biochem. 101, 163–169.

Newman, J., Estes, J. E., Selden, L. A. and Gershman, L. C. 1985. Presence of oligomers at subcritical actin concentrations. Biochemistry 24, 1538–1544.

Nishida, E., Iida, K., Yonezawa, N., Koyasu, S., Yahara, I. and Sakai, H. 1987. Cofilin is a component of intranuclear and cytoplasmic actin rods induced in cultured cells. Proc. Natl Acad. Sci. USA 84, 5262–5266.

Ohta, Y., Nishida, E., Sakai, H. and Miyamoto, E. 1989. Dephosphorylation of cofilin accompanies heat shock-induced nuclear accumulation of cofilin. J. Biol. Chem. 267, 16143–16148.

Olave, I. A., Reck-Peterson, S. L. and Crabtree, G. R. 2002. Nuclear actin and actin-related proteins in chromatin remodeling. Annu. Rev. Biochem. 71, 755–781.

Onoda, K. and Yin, H. L. 1993. gCap39 is phosphorylated. Stimulation by okadaic acid and preferential association with nuclei. J. Biol. Chem. 268, 4106–4112.

Oosawa, F. 1983. Macromolecular assembly of actin. In Muscle and Nonmuscle Motility. A. Stracher (Editor). Academic Press, New York. pp. 151–216.

Oosawa, F. and Asakura, S. 1975. Thermodynamics of the Polymerization of Protein. Academic Press, New York.

Oosawa, F., Asakura, S., Hotta, K., Imai, N. and Ooi, T. 1959. G–F transformations of actin as a fibrous condensation. J. Polym. Sci. 37, 323–326.

Orlova, A., Chen, X., Rubenstein, P. A. and Egelman, E. H. 1997. Modulation of the yeast F-actin structure by a mutation in the nucleotide-binding cleft. J. Mol. Biol. 271, 235–243.

Orlova, A. and Egelman, E. H. 1995. Structural dynamics of F-actin. I. Changes in the C terminus. J. Mol. Biol. 245, 582–597.

Otterbein, L., Graceffa, P. and Dominguez, R. 2001. The crystal structure of uncomplexed actin in the ADP state. Science 293, 708–711.

Owen, C. and DeRosier, D. J. 1993. A 13 Å map of the actin–scruin filament from the *Limulus* acrosomal process. J. Cell Biol. 123, 337–344.

Page, R., Lindberg, U. and Schutt, C. E. 1998. Domain motions in actin. J. Mol. Biol. 280, 463–474.

Pantaloni, D., Carlier, M. F. and Korn, E. D. 1985. The interaction between ATP-actin and ADP actin. J. Biol. Chem. 266, 6572–6578.

Pardee, J. D. and Spudich, J. A. 1982. Mechanism of K+-induced actin assembly. J. Cell Biol. 93, 648–654.

Pendleton, A., Pope, B., Weeds, A. and Koffer, A. 2003. Latrunculin B or ATP depletion induces cofilin-dependent translocation of actin into nuclei of mast cells. J. Biol. Chem. 278, 14394–14400.

Philimonenko, V. V., Zhao, J., Iben, S., Dingova, H., Kysela, K., Kahle, M., Zentgraf, H., Hofmann, W. A., de Lanerolle, P., Hozak, P. and Grummt, I. 2004. Nuclear actin and myosin I are required for RNA polymerase I transcription. Nat. Cell Biol. 6, 1165–1172.

Poglazov, B. F. 1983. Actin and coordination of metabolic processes. Biochem. Int. 6, 757–765.

Pollard, T. D. 1984. Polymerization of ADP-actin. J. Cell Biol. 99, 769–777.

Pollard, T. D. Blanchoin, L. and Mullins, R. D. 2000. Molecular mechanisms controlling actin filament dynamics in nonmuscle cells. Annu. Rev. Biophys. Biomol. Struct. 29, 545–576.

Pollard, T. D. and Weeds, A. G. 1984. The rate constant for ATP hydrolysis by polymerized actin. FEBS Lett. 170, 94–98.

Rando, O. J., Zhao, K. and Crabtree, G. R. 2000. Searching for a function for nuclear actin. Trends Cell Biol. 10, 92–97.

dos Remedios, C. G., Chhabra, D., Kekic, M., Dedova, I., Tsubakihara, M., Berry, D. and Nosworthy, N. J. 2003. Actin binding proteins: Regulation of cytoskeletal microfilaments. Physiol. Rev. 83, 433–473.

Rich, S. A. and Estes, J. E. 1976. Detection of conformational changes in actin by proteolytic digestion: Evidence for a new monomeric species. J. Mol. Biol. 104, 777–792.

Rouayrenc, J. F. and Travers, F. 1981. The first step in the polymerization of actin. Eur. J. Biochem. 116, 73–77.

Roustan, C., Benyamin, Y., Boyer, M., Bertrand, R., Audermard, E. and Jauregui-Adell, J. 1985. Conformational changes induced by Mg^{2+} on actin monomers. FEBS Lett. 181, 119–123.

Schutt, C. E., Myslik, J. C., Rozycki, M. D., Goonesekere, N. C. and Lindberg, U. 1993. The structure of crystalline profilin–beta-actin. Nature 365, 810–816.

Selden, L. A., Estes, J. E. and Gershman, L. C. 1983. The tightly bound divalent cation regulates actin polymerization. Biochem. Biophys. Res. Commun. 116, 478–485.

Shu, W. P., Wang, D. and Stracher, A. 1992. Chemical evidence for the existence of activated G-actin. Biochem. J. 283, 567–573.

Straub, F. B. 1942. Actin. Stud. Med. Inst. Szeged. 2, 3–15.

Strzelecka-Golaszewska, H., Moraczewska, J., Khaitlina, S. Y. and Mossakowska, M. 1993. Localization of the tightly bound divalent-cation-dependent and nucleotide-dependent conformation changes in G-actin using limited proteolytic digestion. Eur. J. Biochem. 211, 731–742.

Strzelecka-Golaszewska, H., Prochniewicz, E. and Drabikowski, W. 1978. Interaction of actin with divalent cations. I. The effects of various cations on the physical state of actin. Eur. J. Biochem. 88, 219–227.

Strzelecka-Golaszewska, H., Wozniak, A., Hult, T. and Lindberg, U. 1996. Effects of the type of divalent cation, Ca^{2+} or Mg^{2+}, bound at the high-affinity site and of the ionic composition of the solution on the structure of F-actin. Biochem. J. 316, 713–721.

Stuven, T., Hartmann, E. and Gorlich, D. 2003. Exportin 6: A novel nuclear export receptor that is specific for profilin–actin complexes. EMBO J. 22, 5928–5940.

Tirion, M. M. and Benavraham, D. 1993. Normal mode analysis of G-actin. J. Mol. Biol. 230, 186–195.

Tirion, M. M., Benavraham, D., Lorenz, M. and Holmes, K. C. 1995. Normal modes as refinement parameters for the F-actin model. Biophys. J. 68, 5–12.

Tobacman, L. S. and Korn, E. D. 1983. The kinetics of actin nucleation and polymerization. J. Biol. Chem. 258, 3207–3214.

Tonomura, Y. and Yoshimura, J. 1961. Removal of bound nucleotide and calcium G-actin by treatment with EDTA. J. Biochem. 50, 79–80.

Vandekerckhove, J. and Weber, K. 1978. At least six different actins are expressed in a higher a mammal: An analysis based on the amino acid sequence of the amino-terminal tryptic peptide. J. Mol. Biol. 126, 783–802.

Vandekerckhove, J. and Weber, K. 1979a. The amino acid sequence of actin from chicken skeletal muscle actin and chicken gizzard smooth muscle actin. FEBS Lett. 102, 219–222.

Vandekerckhove, J. and Weber, K. 1979b. The complete amino acid sequence of actins from bovine aorta, bovine heart, bovine fast skeletal muscle, and rabbit slow skeletal muscle. Differentiation 14, 123–133.

Wada, A., Fukuda, M., Mishima, M. and Nishida, E. 1998. Nuclear export of actin: A novel mechanism regulating the subcellular localization of a major cytoskeletal protein. EMBO J. 17, 1635–1641.

Wang, F., Sampogna, R. V. and Ware, B. R. 1989. pH dependence of actin self-assembly. Biophys. J. 55, 293–298.

Weber, A., Herz, R. and Reiss, I. 1969. The role of magnesium in the relaxation of myofibrils. Biochemistry 8, 2266–2271.

Wegner, A. 1976. Head to tail polymerization of actin. J. Mol. Biol. 108, 139–150.

Wegner, A. and Engel, J. 1975. Kinetics of the cooperative association of actin to actin filaments. Biophys. Chem. 3, 215–225.

Wegner, A. and Isenberg, G. 1983. 12–Fold difference between the critical monomer concentrations of the two ends of actin filaments in physiological salt conditions. Proc. Natl Acad. Sci. USA 80, 4922–4925.

West, J. J. 1971. Binding of nucleotide to cation-free G-actin. Biochemistry 10, 3547–3553.

Zimmerle, C. T. and Frieden, C. 1988. Effect of pH on the mechanism of actin polymerization. Biochemistry 27, 7766–7772.

2
Actin Genetic Diseases

John C. Sparrow and Nigel G. Laing

Introduction

The human genome, as for other mammals, contains six actin genes, ACTA1, ACTA2, ACTB, ACTC, ACTG1, and ACTG2. Four of these genes are differentially expressed in cardiac (ACTC), smooth (ACTA2), enteric (ACTG2), and skeletal muscles (ACTA1); two are described as cytoplasmic actin genes (ACTB and ACTG1) and are expressed in all cells.

Actin mutations in human populations were unknown until relatively recently. However, in the past 6–10 years mutations that cause human disease have been found in the ACTC cardiac muscle gene which cause either dilated (Olson et al. 1998) or hypertrophic (Mogensen et al. 1999; Olson et al. 2001) cardiomyopathies and in the ACTA1 gene causing congenital myopathies (Nowak et al. 1999; see Sparrow et al. 2003 for review; Kaindl et al. 2004) including congenital fiber-type disproportion (CFTD) (Laing et al. 2004). Most recently mutations in the ACTG1 gene have been associated with autosomal dominant deafness (Zhu et al. 2003; van Wijk et al. 2003).

The vast majority of actin disease mutations are dominant. This is true of all the cardiomyopathy, CFTD, and deafness mutations so far described. In the case of nemaline myopathy, where more than 100 different mutations are now known, only six recessive mutations have been reported (Sparrow et al. 2003; Nowak et al. 2007). The preponderance of dominant mutant alleles appears not to be unusual for actin mutations. It has also been the case for all mutations of the Drosophila flight muscle-specific actin gene, Act88F (Hiromi and Hotta 1985; Sparrow et al. 1992; Cripps et al. 1994; An and Mogami 1996; Sheterline, Clayton and Sparrow 1998). This common dominance of actin mutations is almost certainly due to the fact that within cells the major functions of actin are as F-actin, usually associated with other proteins in complexes. In striated muscle along the length of the thin filaments the F-actin core is in intimate contact with nebulin, tropomyosin, and the regulatory troponin complex. The "barbed" ends are embedded in the Z-disc binding especially with α-actinin, CapZ, and titin while the pointed ends are capped by tropomodulin. The dominant actin mutations that cause disease are mostly missense mutations.

Some may produce defective actin monomers which could interfere with actin polymerization (defective actin–actin binding or conformational changes affecting F-actin stability), perhaps acting as "capping" proteins causing premature termination of F-actin filaments (Hennessey et al. 1992) and reducing the capacity for the wild-type actins to produce normal function. In fact, few F-actin binding proteins bind to single monomers within F-actin filaments – a mixture of wild type and mutant monomers are thus most likely to generate F-actin regions with altered affinity for specific actin-binding proteins.

One might wonder why human actin mutants were not described before 1998. They have been found previously in model organisms – yeast, nematodes, and flies. However, these genomes are more readily accessible to mutagenesis and although actin mutants have been recovered in large numbers in the single yeast ACT1 actin gene (Wertman, Drubin and Botstein 1992), in *Caenorhabditis elegans* (Waterston, Hirsh and Lane 1984) and *Drosophila melanogaster* (Sparrow et al. 1992) the known actin mutants are restricted to muscle actin genes. In *Drosophila*, expression of the Act88F actin gene, in which all known fly actin mutants occurs, is restricted largely to the indirect flight muscles (Nongthomba et al. 2001), where it encodes all the sarcomeric actin (Ball et al. 1987). Thus even severe actin mutations that cause a dominant flightless phenotype have no effect on fly viability and only a small effect on fertility (Nongthomba et al. 2001).

Three major reasons why actin mutants are generally difficult to recover and are likely to be at very low frequencies in human populations are (1) that actin is a ubiquitously expressed protein; (2) it is a highly conserved protein with many binding partners, so mutations in a large fraction of actin residues appear to cause a severe phenotype in humans and other organisms (see Sheterline, Clayton and Sparrow 1998; Sparrow et al. 2003); and (3) that most actin mutations are dominant and, given the usually severe effects, are not passed on to offspring. Thus, familial actin mutants are likely to be relatively mild dominant alleles or recessive. Severe dominant alleles (see ACTA1) are likely to be *de novo* mutations.

For the ACTA1 gene mutations there is a range of severity from death shortly after birth through to survival in some cases into old age. Survival to birth may be explained by the pre-parturition isoform switch from cardiac to skeletal actin during normal skeletal muscle development (Ilkovski et al. 2005). In the case of cytoplasmic actin, five ACT1G disease-causing mutant alleles are known so far and no disease-causing ACTB alleles have yet been described. The ACT1G mutations may be phenotypically mild alleles since despite the ubiquity of γ-actin in all human tissues the clinical phenotype is apparently restricted to hearing loss.

Cardiomyopathy (ACTC)

Familial cardiomyopathy is a frequent cause of sudden death in young people (Maron 1997). Two related genetic diseases result from mutations of the ACTC gene – hypertrophic cardiomyopathy (HCM) and dilated cardiomyopathy (DCM). A detailed exposition on the ACTC mutations and these diseases will be found in the chapter by Stefani et al. in this volume.

ACTA1 Congenital Skeletal Myopathies

The congenital myopathies are a group of human skeletal muscle diseases in which patients show varying degrees of muscle weakness, often with a tall thin face and facial myopathy. On muscle biopsy the disease shows the presence of characteristic abnormalities. Five different muscle phenotypes, each of which will predominate in patients with specific ACTA1 mutations, are seen, but they are not mutually exclusive (Sparrow et al. 2003; Schröder, Durling and Laing 2004). These phenotypes are known as actinopathy (where homogeneously stained areas contain almost exclusively thin filament materials), sarcoplasmic (nemaline) rods, intranuclear rods, core-like areas (where there is myofibrillar disruption and absence of mitochondria), and congenital fiber-type disproportion. The commonest of these is nemaline myopathy.

The first ACTA1 mutants causing human congenital myopathies – actin myopathy, nemaline myopathy and intranuclear rod myopathy – were described by Nowak et al. (1999). Since then an increasingly large number of ACTA1 mutants causing these diseases has been identified. The count was 67 in 2003 (Sparrow et al. 2003), over 100 in 2005 (Laing and Nowak 2005) and currently stands at 116 in the report of the ENMC workshop 2006 (Wallgren-Pettersson and Laing 2006). Mutations causing nemaline myopathy have been described in a number of other genes encoding sarcomeric proteins, i.e., nebulin (NEB), tropomyosin (TPM2 and TPM3), troponin T (TNNT1), and cofilin (Agrawal et al. 2007).

The majority of all the known dominant ACTA1 mutations lead to amino acid substitutions i.e., missense mutants. Since heterozygous parents of ACTA1 null homozygous children are without clinical symptoms (Sparrow et al. 2003; Nowak et al. 2007) the conclusion must be that most of the dominant mutations must cause dominant negative effects, that is, they interfere with the functions of wild-type actins and other actin-binding proteins. The exceptions to dominance in ACTA1 are currently very interesting. Seven ACTA1 actin null homozygous human patients have now been identified, four in a subset of the Pakistani community in the UK, where the carrier frequency must be relatively high; two in Gypsy patients and one in a UK Gujarati family (Nowak et al. 2007).

These homozygotes for null mutations in the ACTA1 gene show a nemaline phenotype (Sparrow et al. 2003; Nowak et al. 2007). A few individuals survive for a few months, one patient even longer, and appear to do so by the upregulation of cardiac, ACTC, gene expression. Similarly, transgenic knockouts of the mouse ACTA1 gene homologue lead to early death, though there is a compensatory up-regulation in the expression and accumulation of the cardiac and vascular isoforms (Crawford et al. 2002). In human early fetal skeletal muscle, the predominant actin is α-cardiac

(ACTC) but by 25–27 weeks gestation the α-skeletal (ACTA1) becomes predominant and is the exclusive actin isoform expressed in skeletal muscle for the rest of life (Ilkovski et al. 2005).

Nemaline Rods

Nemaline myopathy was first described in 1963 (Shy et al. 1963; Conen, Murphy and Donohue 1963). It is characterized by the presence of rods (nema = thread, Greek) in the sarcoplasm. It is the most common feature of patients with ACTA1 congenital myopathies. For light microscopy of biopsy samples these rods are made visible using Gomori trichrome staining (Engel and Cunningham 1963; Nienhuis et al. 1967). In electron micrographs, following standard staining procedures, the nemaline rods are visible as electron dense bodies, typically measuring 1–7 μm in length and 0.3–2.0 μm in width with a similar contrast to sarcomeric Z-disks.

Nemaline rods are considered to derive from lateral expansion of the Z-disks. The reasons for this interpretation are that the rods (a) can often be seen to have structural continuity with Z-disks; (b) their ultrastructure resembles Z-disk lattice pattern (Engel and Gomez 1967; Yamaguchi et al. 1982; Morris, Nneji and Squire 1990; Luther and Squire 2002); (c) they are penetrated by thin filaments, which run parallel to their long axis and which behave like actin filaments (Yamaguchi et al. 1978); (d) α-actinin, a major component of Z-disks is also a major component of nemaline rods (Jockusch et al. 1980; Wallgren-Pettersson et al. 1995); (e) myotilin, another component of the Z-disk, is also found in nemaline bodies (Schröder et al. 2003); and (f) desmin, the structural component of muscle-specific intermediate filaments accumulates at the periphery of Z-disks and of nemaline bodies (Jockusch et al. 1980).

An occasional phenotype in nemaline myopathy is the appearance of "zebra-bodies" (Lake and Wilson 1975), linear stacks of Z-disk-like material separated by only 200 nm or so. These have not been reported in missense ACTA1 mutations (Sparrow et al. 2003). However, they do occur in ACTA1 null mutant homozygous patients (Nowak et al. 2007). Again these structures appear to be related to Z-disks, and in *Drosophila* expressing human nemaline mutations in the Act88F gene zebra-bodies are seen in the indirect flight muscles associated with complete Z-disks within the sarcomeres themselves (Sparrow, Peckham and Kumar unpublished observations).

When do nemaline rods arise? Nemaline rods show much the same appearance independent of age or severity (Shimomura and Nonaka 1989) and the phenotype is already apparent at birth in most patients, so biopsies even at these early stages do not show the etiology of the disease. However, it is clear that ultrastructurally the disease is not progressive, except for the uncommon late-onset cases. Nemaline myopathy leads to changes in Z-band structure, possibly during the fetal cardiac to skeletal actin isoform switch. The characteristic nemaline structures almost certainly derive from either defective sarcomeric assembly or incomplete disassembly. The nemaline rods are usually much larger than Z-disks, implying that continued aberrant assembly of Z-disk material occurs.

Intranuclear Actin Rods

A minority of ACTA1 mutations lead to the appearance of intranuclear rods (Sparrow et al. 2003; Ilkovski et al. 2004; Schröder, Durling and Laing 2004; Hutchinson et al. 2006), but they usually occur in patients that also have cytoplasmic nemaline rods. They do not appear to be a feature of mutations causing nemaline myopathy in other thin filament protein genes. They appear across the range of disease severity (Hutchinson et al. 2006). In one family with the ACTA1 mutation V163M (Hutchinson et al. 2006) and two unre-lated patients with the V163L mutation, the myopathy appears almost exclu-sively as intranuclear rods, without other associated ACTA1 nemaline phenotypes (Nowak et al. 1999). This suggests that alteration of actin inter-actions involving actin residue V163 are especially important in the forma-tion of intranuclear rods.

By phalloidin and antibody staining intranuclear rods are enriched in F-actin and α-actinin-2, respectively (Hutchinson et al. 2006), but stain only faintly with a polyclonal antibody specific to alpha-skeletal actin (Domazetovska et al. unpublished data). However, this may reflect antibody access to specific epitopes as both an antibody recognizing all α-actin isoforms and another spe-cific to cardiac and α-skeletal actin failed to stain intranuclear rods (Domazetovska et al. unpublished data). The intranuclear rods also stain with antibodies to the Z-disk proteins γ-filamin and myotillin, neither of which have been found previously in nuclei, but not with antibodies to nebulin, tropomyosin, cofilin, [H2]dystrophin, lamin A/C, desmin, or α-tubulin (Domazetovska et al. unpublished data).

Intranuclear rods vary in size and shape. This can depend on nuclear size and the age of the cell, but they usually remain straight or solitary. Multiple rods in a single nucleus are uncommon. The rods are usually a few microns in length; occasionally they are sufficiently long to distort the shape of the nuclei.

The intranuclear rods have been less well studied than the nemaline rods as they are a rare phenotype. They show considerable homology to intranuclear structures found in vertebrate cell cultures stressed by exposure to DMSO or heat shock (Fukui 1978; Fukui and Katsumaru 1979; Iida, Iida and Yahara 1986; Welch and Suhan, 1985). Stress can induce these structures in cultured muscle cells (Ono et al. 1993). In spores of the slime mould *Dictyostelium discoideum* they occur naturally during spore formation (Fukui and Katsumaru 1979; Sameshima et al. 1994) and have been shown to contain regular arrays of tubular actin filaments (Sameshima et al. 2000). The degree of molecular homology between these structures and intranuclear ACTA1 rods is not clear. By expressing GFP-tagged ACTA1 mutants in fibroblasts,

Ilkovski et al. (2005) have shown that mutant skeletal α-actins can accumulate intranuclearly and assemble into compact rod structures. This suggests that in human muscle the rods form due to the presence of mutant actins in the nucleus, though whether wild-type α- or β-actins are also components of these rods remains unknown. The presence of actin in normal nuclei was debated for many years but is now accepted and actin seems to have diverse nuclear activities (see Bettinger, Gilbert and Amberg 2004). Actin contains two nuclear export signal (NES) sequences, residues 170–181 and 211–222, that are highly conserved in all actins, and by mutational analysis and transfection appear to be functional (Wada et al. 1998). ACTA1 mutations that lead to intranuclear rods in patients, in transfected C2C12 myoblast cells (Ilkovski et al. 2005; Domazetovska et al. submitted) or in transfected NIH 3T3 fibroblasts (Costa et al. 2004) are not in these sequences, though some, e.g., V163M are close (Hutchinson et al. 2006). Why specific ACTA1 mutations cause the mutant actins to accumulate in the nucleus is far from clear, nor is there any evidence that α-skeletal actin, rather than the cytoplasmic actins, normally traffic through the nucleus.

Congenital Fiber-Type Disproportion (CFTD)

Congenital fiber-type disproportion (CFTD) is a rare genetic disease that causes generalized skeletal muscle weakness in which a subset of muscle fibers is significantly reduced in size (Brooke 1973). CFTD, like other congenital myopathies, shows highly variable clinical severity, and is otherwise similar to the other congenital myopathies, including nemaline myopathy phenotypically, but the histopathology does not include nemaline rods or other aberrant structures. The overlap in the diseases is emphasized by the fact that fiber-type disproportion is frequently associated with clear cases of clinical nemaline myopathy (Agrawal et al. 2004) at least in some instances because the mutated gene, such as slow α-tropomyosin (TPM3) is expressed only in slow muscle fibers (Durling et al. 2002; North, 2004).

Skeletal muscle fibers are elongated single, multinucleated cells containing the contractile myofibrils. Using cytochemical techniques it has long been known that human skeletal muscle fibers exist in two broad types, type 1 and type 2, which can be further subdivided into, for instance, type 2a and 2b. This means, and it is well established, that each fiber is expressing a specific subset of fiber-type specific proteins. The genetic basis of CFTD still remains largely unknown and it has been suggested (Clarke and North 2003) that the heterogeneity in the clinical phenotype and inheritance patterns indicates that there is likely more than one cause.

The ACTA1 genes for a cohort of 50 patients exhibiting the clinical and histological features of CFTD showed that three of them were heterozygous for missense mutations in this gene (Laing et al. 2004). These data indicate that ACTA1 mutations are not a common cause of the disease and confirm the proposal (Clarke and North 2003) that mutation of other, probably many more, genes can cause this disease. Again, as with most actin diseases these are dominant mutations. The CFTD mutations were L221P, D292V, and P332S.

The observations raise two major questions:

(1) Since both fiber types express ACTA1 as the major actin in postnatal skeletal muscle why do type 1 fibers show hypotrophy but type 2 fibers do not?
(2) Since these patients did not show the nemaline myopathy (see above) typical of the majority of ACTA1 mutations do the CFTD mutations affect a specific actin function?

At present one can only speculate on why these ACTA1 mutations specifically affect type 1 fiber size. The histochemical differentiation of the two fiber types, usually on the basis of staining for ATPase activity, reflects the different myosin isoforms present in each fiber type. In the monomeric actin atomic structure (Kabsch et al. 1990; Otterbein, Graceffa and Dominguez 2001) the amino acid residues affected by the three mutations (Leu221Pro, Asp292Val and Pro332Ser) form part of the monomeric actin surface and will remain exposed in the F-actin polymer. By fitting the atomic structures of both actin and the myosin S1 fragment into EM reconstructions of rigor-state S1-decorated F-actin (Rayment et al. 1993; Holmes et al. 1990; Milligan, Whittaker and Safer 1990; Lorenz, Popp and Holmes 1993; reviewed by Milligan 1996) the myosin-binding site on actin resolves into primary and secondary sites. The primary site consists of residues 1–4, 24–25, 332–334, and residues 144, 341, 345, and 355. The secondary site is on the actin monomer lying at $n-2$ in the F-actin helix, and consists of residues 40,42 and a loop from residues 95–100. One of the CFTD mutant residues, P332, thus occurs in the strong myosin binding site on actin. This mutant actin might cause type 1 fiber-specific hypotrophy by interactions with a different myosin isoform, but is unlikely to be an explanation for the two other mutant residues.

While replacement of a surface aspartate residue by a valine (D292V) might have subtle effects, the insertion of a proline (L221P) is likely to be much more disruptive. It is likely that a number of different sarcomeric proteins have fiber-type specific isoforms. It may be that in CFTD patients with ACTA1 mutations, thin filament assembly or stability is differentially affected by other thin filament proteins. All three residues affected in these CFTD patients occur on the actin surface over which tropomyosin moves during muscle activation (Laing et al. 2004). Tropomyosin is well known to enhance F-actin polymerization *in vitro* and to increase F-actin stability. There are fiber-type specific tropomyosin isoforms; slow alpha-tropomyosin TPM3 is expressed in slow (type 1) muscle fibers – hence the pathology in the described TPM3 null patient being restricted to type 1 muscle fibers (Tan et al. 1999). Fast α-tropomyosin, TPM1 is supposedly expressed in fast muscle fibers and the β-tropomyosin, TPM2 is expressed in both muscle fiber types.

However, other proteins are differentially expressed in type 1 fibers, including calsarcin-1 that binds to both α-actinin and calcineurin, thus anchoring calcineurin to the sarcomeric Z-disk only in type 1 muscle fibers (Frey, Richardson and Olson 2000). Calcineurin, a serine/threonine phosphatase, is an important player in the NFAT signaling pathway. This pathway is known to mediate the conversion of type 2 to type 1 fibers under certain physiological conditions (Semsarian et al. 1999). There is also evidence that calcineurin can influence fiber size (Serrano et al. 2001). Clearly those ACTA1 mutants causing CFTD do so by interactions with protein isoforms not found in type 2 fibers. Mutations in other, unknown, genes must account for the majority of CFTD occurrences. At last one of the players, ACTA1 has now been identified. Studies of these mutations may lead to an understanding of how the specific hypotrophy of type 1 fiber arises, and the likely genes in which the other CFTD mutations occur. Recently the CFTD phenotype has also been linked to SEPN1, selenoprotein N1 gene mutations (Clarke et al. 2006).

Many patients with nemaline rod myopathy exhibit CFTD of type 1 fibers (Iannaccone et al. 1987; Miike et al. 1986; Ryan et al. 2003). This suggests that all genes that cause nemaline myopathy are candidate genes for CFTD (Laing et al. 2004). Why some mutations are specific to CFTD or nemaline myopathy, and others cause both, remains an issue that might explain the ontogeny of both diseases.

Molecular Effects of ACTA1 Myopathy Mutations

What functions of actin are affected by these congenital ACTA1 myopathy mutations? The fact that mutations within ACTA1 can produce different phenotypes – nemaline rods, actinopathy, intranuclear rods, and CFTD suggests that these might arise because different mutants affect the interaction of actin with different sarcomeric proteins. If so, then the mutations producing specific phenotypes might be expected to cluster within the actin monomer.

The distribution of 67 ACTA1 mutations was examined (Sparrow et al. 2003) for clustering in the atomic structure of the actin monomer (Kabsch et al. 1990; Otterbein, Graceffa and Dominguez 2001). Mutations causing actinopathy seemed to cluster around the "cleft" of actin, specifically close to the nucleotide binding site. The same was largely true of the intranuclear rod mutants, though at least two were well outside this region and seemed more likely to affect the mobility of subdomain-2 within the F-actin structure. It has been claimed (Orlova and Egelman 1993) that movement of subdomain-2 is dependent on the nucleotide bound, indicating a "signaling" response within the actin monomer that might allow all intranuclear rod mutants to affect a single actin function. However, the large majority of the ACTA1 mutants causing nemaline myopathy of varying degrees were dispersed throughout the G-actin molecule (Sparrow et al. 2003) and many occurred in putative actin–actin, actomyosin, actin–tropomyosin, actin–α-actinin, and actin–nebulin binding sites (Sparrow et al. 2003). This suggests either that changes in many different actin binding protein sites can lead to nemaline myopathies, or that any major defect in actin monomer stability, polymerization or thin filament integrity can lead to myofibrillar disruption and the appearance of a small number of phenotypic end states characterized by the presence of nemaline rods, intranuclear rods, zebra bodies, actinopathy, or occasionally CFTD. It is not clear whether mutations specifically causing CFTD cluster, as only three such mutations are known, although two are within actin sub-domain 3 and one in subdomain 4 but close to subdomain 3. These are all surface residues and it has been proposed they could affect interactions with tropomyosin (Laing et al. 2004).

Little work has yet been done on the biochemistry of ACTA1 mutants. This is undoubtedly due to the longstanding problems in expressing actins and purifying actins from heterologous systems. One approach, that produces labeled actins in small quantities, but suitable for actin monomer binding assays, is *in vitro* transcription/translation (Hennessey, Drummond and Sparrow 1991; Drummond, Hennessey and Sparrow 1992; Rommelaere et al. 2003). By *in vitro* transcription/translation of 19 ACTA1 nemaline mutations sufficient ^{35}S-methionine-labeled actins were recovered for studies of native conformation (in native gels), copolymerization with excess unlabeled, wild-type actin, and binding to G-actin binding proteins (thymosin-β4, DNase I, profilin-IIa, and Vitamin-D-binding protein) (Costa et al. 2004). The different mutant actins showed a range of different biochemical properties from four of them (H40Y, M132V, R183C, and D286G) behaving as wild-type actin in all the biochemical assays to two that failed to fold *in vitro*. These two exceptions (L94P, E259V) cause severe recessive nemaline myopathy, and the *in vitro* data suggest that *in vivo* these alleles produce nonfunctional actin (Costa et al. 2004), which is compatible with the recessive inheritance. Other mutants (G15R, N115S, V163L, and G182D) showed increased binding to the CAP protein, a component of the reticulocyte lysate expression system that is necessary for actin folding (Rommelaere et al. 2003). This probably indicates that these actins are less stable than wild type and it is proposed that this may explain the various actinopathy phenotypes these mutants exhibit *in vivo* and the aggregate structures seen *in vitro* when they are expressed in fibro-blasts (Costa et al. 2004). A further group of mutants (I64N, Q263L, G268C, G268R, and N280K) expressed *in vitro* showed reduced ability to copoly-merize with an excess of wild type.

Human muscle actins have been expressed in cultured insect Sf9 cells (Akkari et al. 2003; Joel, Fagnant and Trybus 2004; Bookwalter and Trybus 2006) in sufficient yields for assays of actin polymerization, monomer stability, and studies of actomyosin kinetics, including *in vitro* motility assays (Bookwalter and Trybus 2006). The purified actins include the expressed human actin and the endogenous insect cell β-actin, but using western blots with actin isoform-specific antibodies it is claimed that the endogenous actin is only a small fraction of the total (Bookwalter and Trybus 2006). Expression of ACTA1 actins, wild type and mutant, has been achieved in this system, with and without N-terminal His-tags or eGFP fusions at the carboxy terminus (Akkari et al. 2003). To date there have been no reports showing that the His-tags can be removed or comparing the biochemical properties of the wild type and nemaline mutant ACTA1 actins expressed in this system.

An alternate approach to obtain biochemical insights into ACTA1 nemaline mutant actins is to study the actins and their location within transfected cells by immunohistochemistry and to use cell extracts on native gels to look at actin folding/stability (Ilkovski et al. 2004). Comparisons of the relative contribution of mutant nemaline EGFP-tagged actins (I136M, V163L, R183G, C268G, and I357L) transfected into C2C12 cells compared to a wild-type transfected control showed that only R183G contributed less efficiently to the insoluble actin filaments, indicating a reduction in accumulation and polymerizing into F-actin. V163L actin contributed to the insoluble fraction at high levels consistent with its induction of insoluble aggregates and intranuclear rods in transfected cells, as also seen in patient biopsies. Expression of untagged ACTA1 mutant (I357L and R183G) constructs transfected into 3T3 cells allowed recovery of soluble G-actin. Only I357L actin showed any retardation on the native gels, suggesting a less compact, or varying, conformation (Ilkovski et al. 2004).

In a technical tour de force Marston et al. (2004) isolated actins from small (2.5 mg) human biopsies taken from normal controls and an individual with mild nemaline (M132V) myopathy. Actin was obtained in sufficient quantities to estimate polymerization ability and *in vitro* motility. The mutant actin showed a reduced capacity to polymerize *in vitro*, and increased velocity in the *in vitro* actomyosin motility assay. Interestingly Costa et al. (2004) showed that this mutant was without effect on polymerization, though the studies differ in that in the latter experiments a large excess of wild-type actin is added to achieve the critical concentration for polymerization, whereas in the Marston et al. study the actins were present at an approximate ratio of 1:1. While this biopsy approach provides native actin from heterozygotes, the disease relevant situation for most mutations, it is difficult to repeat experiments and to interpret the data. As these authors point out it is important that one knows the proportions of polymerized wild type and mutant actins in the F-actin assays.

M132V is buried within actin subdomain 1 and is not within the deduced myosin binding site on actin (Holmes et al. 1990; Milligan, Whittaker and Safer 1990; Rayment et al. 1993; Lorenz, Popp and Holmes 1993; reviewed by Milligan 1996). The effect of the M132V mutation on actomyosin interactions in the *in vitro* motility assay is explained as indirect effects due to conformational changes within subdomain 1, on which the larger part of the myosin binding site resides (Marston et al. 2004).

The analyses of nemaline actin mutants transfected into fibroblasts or myoblasts (Costa et al. 2004; Ilkovski et al. 2004) show that aspects of the nemaline phenotype – nuclear rods and cytoplasmic aggregates – can be readily generated for experimental studies. These systems and perhaps increasingly in the muscle of whole animal models (e.g., mouse, *Drosophila*) where the complete phenotype can be produced in muscles will increasingly provide insights into how the phenotypes arise. Already the, as yet, limited biochemical studies have revealed a range of mutant molecular defects in ACTA1 mutant actins. As predicted by the modeling of mutants in the actin atomic structure (Sparrow et al. 2003) almost any defect in actin folding, stability, polymerisation, or interactions with a range of actin-binding proteins seems capable of generating the phenotypes seen in the actinopathy patients.

Gamma Actin (ACTG1) Gene Mutations and Deafness

Mutations in the ACTG1 gene causing a late-onset, progressive sensorineural hearing loss have recently been described (Zhu et al. 2003; van Wijk et al. 2003). These are the first known mutations in a human cytoplasmic actin gene. Although hereditary hearing loss is caused by mutations in many genes, a number of studies showed a dominant autosomal hearing loss trait is located, as DFNA20, to human chromosome region 17q25 (Morell et al. 2000; Yang and Smith 2000; DeWan, Parrado and Leal 2000). The trait mapped to a 6 cM interval containing 62 known genes (DeWan, Parrado and Leal 2000) including ACTG1. A number of likely genes were sequenced. Sequencing of the ACTG1 genes in different kindreds (Zhu et al. 2003; van Wijk et al. 2003) showed consistent nucleotide changes segregated in family members exhibiting the deafness trait that would lead to amino acid changes in the conserved actin sequence. The mutations are T89I, K118M, P332A, P264L (Zhu et al. 2003), and T278I (van Wijk et al. 2003). These changes were not seen in 103 and 300 controls, respectively, from normal hearing individuals. On this basis it seems likely that ACTG1 mutations cause this deafness.

Since β- and γ-actins are ubiquitously expressed in human cells a specific effect of an ACTG1 mutation on sensorineural hearing loss, apparently without any other effects, might seem rather odd. The specific sensorineural hearing loss in these individuals suggests an association with cochlear structures (Elfenbein et al. 2001). In particular the stereocilia of the hair cells have been suggested as a likely site of progressive dysfunction (Zhu et al. 2003; van Wijk et al. 2003). The hair cells function to detect movements of cochlear fluid and generate signals that are transmitted to the nervous system. Cochlear fluid movements are generated by sound waves reaching the eardrum and then being transmitted through the three small bones (malleus, incus, and stapes) to the cochlear window. It is the stereocilia at the apex of the hair cells that sense fluid movement. The stereocilia contain a rigid core of actin filaments and regular clusters of these stereocilia are anchored into a cuticular plate containing a rich F-actin cytoskeletal mesh work, surrounded by a ring of parallel F-actin filaments connected to the zona adherens of the cell (Slepecky 1996). Movements of the stereocilia lead to changes in membrane potential and release of calcium. Resting tensions between the stereocilia are maintained by "tip" links, a protein connection between the tips of neighboring stereocilia so that small relative movements of neighboring stereocilia generate membrane depolarizations as at one end of each tip link is a strain-sensitive channel (for review see Gillespie, Dumont and Kachar 2005). The resting tensions are generated by the movements of a myosin 1 on F-actin filaments within the hair cell. Studies on the distribution of the cytoskeletal actins (Erba et al. 1988) have shown that β-actin is the major isoform in most human tissues. In chicken auditory cells γ-actin is the major isoform and is found in all three F-actin rich cytoskeletal structures of the hair cell (Höfer, Ness and Drenckhahn 1997). If this is also true in humans, then the prevalence of γ-actin in the hair cells might explain why γ-actin mutations can cause dominant hearing loss without other phenotypic effects. Hearing loss at high frequencies is most evident early in disease progression and it can be argued that this might be because of the higher stereociliary stiffness required to detect these frequencies (van Wijk et al. 2003).

From our current knowledge of actin biochemistry (see summary in Sheterline, Clayton and Sparrow (1998)) or atomic structures and modeling (Holmes et al. 1990; Milligan, Whittaker and Safer 1990; Rayment et al. 1993; Lorenz, Popp and Holmes 1993; Milligan 1996) there appears to be no single function of actin that is encompassed by the ACTG1 mutants that effect sensorineural loss. Residue T89 is in subdomain 2, within an α-helix that lies on the outer edge of the F-actin helix. This α-helix extends to form a loop, residues 93, 95, 99, and 100, that is part of the secondary myosin binding site (Rayment et al. 1993; Lorenz, Popp and Holmes 1993; Milligan 1996). Through local conformational changes the T89I mutation could affect actomyosin interactions. The E93K mutation in the *Drosophila* Act88F gene not only causes effects on actomyosin binding (Sparrow et al. 1991; Razzaq et al. 1999) but also affects tropomyosin movement (Bing et al. 1998).

Residue K118 lies on the opposite surface of subdomain 1 from that of the myosin binding site, so the K118M mutation is unlikely to affect actomyosin binding. It lies close to one of the α-actinin binding sites. Residues 83–117 have been implicated in α-actinin binding in a number of studies (Mimura and Asano 1987; Lebart et al. 1990). This is consistent with EM reconstructions of α-actinin bound to F-actin (McGough, Way and DeRosier 1994). α-actinin is a major F-actin crosslinking protein and is likely to be an important component of the actin cytoskeleton within the stereocilia of the hair cells. The P332A will affect actomyosin interactions as it occurs within the P–P–E loop that is part of the primary myosin binding site and is important in "strong" myosin binding states (Rayment et al. 1993; Lorenz, Popp and Holmes 1993; Milligan 1996). It also forms part of the actin surface over which tropomyosin moves during thin filament activation in striated muscle, but this is an unlikely explanation for hearing loss due to changes in cytoskeletal structures.

Residue P264 lies within a loop that in crystal structures of actin (Kabsch et al. 1990; Otterbein, Graceffa and Dominguez 2001) lies close to the actin surface. However, modeling studies (Holmes et al. 1990) suggested that this loop forms a four residue (266–269) hydrophobic "plug" that can insert into a hydrophobic pocket formed by two adjacent monomers on the opposing strand thereby stabilizing the F-actin helix. Direct supporting evidence for the proposed loop function in F-actin comes from mutagenesis studies of yeast actin (Chen, Cook and Rubenstein 1993; Kuang and Rubenstein 1997) where mutations reducing plug hydrophobicity led to cold-sensitivity of *in vivo* growth and *in vitro* effects on actin polymerization kinetics. So the P264L mutation may affect F-actin stability or compliance producing effects on the function of the stereocilia.

Threonine-278 is conserved in almost all vertebrate actins, but substitution by valine, cysteine, alanine, or serine is common (Sheterline, Clayton and Sparrow 1998). A V278T mutation was described in the *Drosophila* Act88F muscle actin gene (Reedy, Beall and Fyrberg 1991) but was apparently without effect on muscle structure. In the ACTA1 gene nearby mutations Y279H and N280K cause nemaline myopathy (Nowak et al. 1999; Sparrow et al. 2003). Residue 278 is in actin subdomain 3 and helix 9. Modeling by van Wijk et al. (2003) suggests that the T278I substitution may affect interactions with helix 11 and possibly actin polymerization. As they point out, helix 11 contains a number of residues in which mutations in other actins have severe effects. The R312H mutation in ACTC leads to DCM (Olson et al. 1998) and the E316K mutation in the *Drosophila* Act88F gene causes a dominant flightless phenotype associated with some disruption of muscle ultrastructure, altered actomyosin force kinetics and reduced *in vitro* protein stability (Drummond et al. 1990; Drummond, Hennessey and Sparrow 1992).

Actin Ubiquitination: Limb Girdle Disease

Actin undergoes a number of posttranslational modifications (for review see Sheterline, Clayton and Sparrow 1998). Mature vertebrate actin has undergone posttranslational deletion of the N-terminal two amino acids, invariably Met and Cys followed by N-terminal acetylation. Residue His-73 is also methylated. The function and importance of these changes remain undetermined, though some evidence indicates that inhibition of N-terminal processing affects the initiation of actin polymerization (Hennessey, Drummond and Sparrow 1991). No human mutations affecting actin modification have yet been found.

Some nonvertebrate actins are stably monoubiquitinated, notably in some insect flight muscles (Ball et al. 1987), but always at residue K118 (Schmitz et al. 2003); these modified actins are regularly spaced, occurring every seventh actin within the thin filament (Burgess et al. 2004), the same spacing as occurs with the troponin complex along these filaments.

A mutation (D489N) in a human protein, Trim32, is associated with a limb girdle muscular dystrophy and sarcotubular myopathy (Kudryoshova et al. 2005). Trim32 is a member of the TRIM (tripartite motif) protein family, having three structural domains – a RING finger, a B-box, and a coiled coil motif (Reymond et al. 2001). Expression of this gene appears to be muscle-specific and the protein, a ubiquitin E3 ligase capable of ubiquitinating actin though not myosin *in vitro*, is associated with sarcomeric thick filaments (Kudryashova et al. 2005). Expression of the wild type and mutant (D489N) gene in transfected cultured cells not normally expressing Trim32, produced equally elevated multiple ubiquitinated proteins. Induced pathological atrophy is associated with increased Trim32 expression, which correlates with increased ubiquitination and targeting for proteasomal degradation of many muscle proteins under these conditions (Kudryashova et al. 2005). Since the mutant Trim32 retains ligase activity with an actin substrate it is unlikely that this form of limb girdle myopathy is specific to actin.

Conclusions

An increasing number of mutations in three human actin genes, ACTC, ACTA1, and ACTG1 have been described recently. They each cause specific disease phenotypes, related to the tissues in which they are expressed. However, in none of these diseases do the mutant residues cluster in the actin monomer structure. This argues against any proposal that the diseases are caused by effects on specific actin binding sites. In the case of the ACTA1 gene, where a very large number of mutant alleles are known, this seems particularly to be the case. The major biological functions of actin are as F-actin. Mutations that affect polymerization are likely to affect both assembly/disassembly of thin filaments or the cytoskeleton and the structural properties of both. A major function of actin apart from its direct involvement in actomyosin molecular motor function is to transmit force. It is therefore interesting to consider for these actin diseases that the transmission of large forces (skeletal and cardiac muscle) or a requirement for low compliance to stiffen the stereocilia of hair cells is important.

The biochemical and biophysical properties of actin disease mutants have been difficult to study in the past because of problems of recovering human actins in sufficient quantities. Model systems such as yeast and *Drosophila* flight muscles have provided some insights. The demonstration that mutant human actins can be expressed in Sf9 cells should now accelerate investigations of the effects of these mutations and give further insights into actin structure/function and whether specific actin dysfunction can be correlated with disease states.

References

Agrawal, P. B., Strickland, C. D., Midgett, C., Morales, A., Newburger, D. E., Poulos, M. A., Tomczak, K. K., Ryan, M. M., Iannaccone, S. T., Crawford, T. O., Laing, N. G. and Beggs, A. H. 2004. Heterogeneity of nemaline myopathy cases with skeletal muscle alpha-actin gene mutations. Ann. Neurol. 56, 86–96.

Agrawal, P. B., Greenleaf, R. S., Tomczak, K. K., Lehtokari, V. L., Wallgren-Pettersson, C., Wallefeld, W., Laing, N. G., Darras, B. T., Maciver, S. K., Dormitzer, P. R., Beggs, A. H. 2007. Nemaline myopathy with minicores caused by mutation of the CFL2 gene encoding the skeletal muscle actin-binding protein, cofilin-2. Am. J. Hum. Genet. 80, 162–167.

Akkari, P. A., Nowak, K. J., Beckman, K., Walker, K. R., Schachat, F. and Laing, N. G. 2003. Production of human skeletal α-actin proteins by the baculovirus expression system. Biochem. Biophys. Res. Commun. 307, 74–79.

An, H. and Mogami, K. 1996. Isolation of 88F actin mutants of *Drosophila melanogaster* and possible alterations in the mutant structures. J. Mol. Biol. 260, 492–505.

Ball. E., Karlik. C. C., Beall, C. J., Saville, D. L., Sparrow, J. C., Bullard, B. and Fyrberg, E. A. 1987. Arthrin, a myofibrillar protein of insect flight muscle, is an actin-ubiquitin conjugate. Cell 51, 221–228.

Bettinger, B. T., Gilbert, D. M. and Amberg, D. C. 2004. Actin in the nucleus. Nat. Rev. Mol. Cell Biol. 5, 410–415.

Bing, W., Razzaq, R., Sparrow, J. C. and Marston, S. 1998. Tropomyosin and troponin regulation of wild-type and E93K mutant actin filaments from *Drosophila* flight muscle: Charge reversal on actin changes actin-tropomyosin from ON to OFF state. J. Biol. Chem. 273, 15016–15021.

Bookwalter, C. S. and Trybus, K. M. 2006. Functional consequences of a mutation in an expressed human α-cardiac actin at a site implicated in familial hypertrophic cardiomyopathy. J. Biol. Chem. 281, 16777–16784.

Brooke, M. H. 1973. Congenital fiber type disproportion. In Clinical Studies in Myology. International Congress Series No. 295. B. A. Kakulas (Editor). Excerpta Medica, Amsterdam. pp. 147–159.

Burgess, S., Walker, M., Knight, P. J., Sparrow, J. C., Schmitz, S., Offer, G., Bullard, B., Leonard, K., Holt, J. and Trinick, J. 2004. Structural studies of arthrin: Monoubiquitinated actin. J. Mol. Biol. 341, 1161–1173.

Chen, X., Cook, R. K. and Rubenstein, P. A. 1993. Yeast actin with a mutation in the hydrophobic plug between subdomain-3 and subdomain-4 (L(266)D) displays a cold-sensitive polymerization defect. J. Cell Biol. 123, 1185–1195.

Clarke, N. F., Kidson, W., Quijano-Roy, S., Estournet, B., Ferreiro, A., Guicheney, P., Manson, J. I., Kornberg, A. J., Shield, L. K. and North, K. N. (2006). SEPN1: Associated with congenital fiber-type disproportion and insulin resistance. Ann. Neurol. 59, 546–552.

Clarke, N. F. and North, K. N. 2003. Congenital fiber type disproportion – 30 years on. J. Neuropathol. Exp. Neurol. 62, 977–989.

Conen, P. E., Murphy, E. G. and Donohue, W. L. 1963. Light and electron microscopic studies of 'myogranules' in a child with hypotonia and muscle weakness. Can. Med. Assoc. J. 89, 893–896.

Costa, C. F., Rommelaere, H., Waterschoot, D., Sethi, K. K., Nowak, K. J., Laing, N. G., Ampe, C. and Machesky, L. M. 2004. Myopathy mutations in α-skeletal-muscle actin cause a range of molecular effects. J. Cell Sci. 117, 3367–3377.

Crawford, K., Flick, R., Cloe, L., Shelly, D., Paul, R., Bove, K., Kumar, A. and Lessard, J. 2002. Mice lacking a skeletal muscle actin show reduced muscle strength and growth deficits and die during the neonatal period. Mol. Cell. Biol. 22, 5887–5896.

Cripps, R. M., Ball E., Stark, M. and Sparrow, J. C. 1994. Dominant flightless mutants of *Drosophila melanogaster* and identification of a new gene required for normal muscle structure and function. Genetics 137, 151–164.

DeWan, A. T., Parrado, A. R. and Leal, S. M. 2000. A second kindred linked to DFNA20 (17q25.3) reduces the genetic interval. Clin. Genet. 63, 39–45.

Drummond, D. R., Hennessey, E. S. and Sparrow, J. C. 1992. The binding of mutant actins to profilin, ATP and DNase I. Eur. J. Biochem. 209, 171–179.

Drummond, D. R., Peckham, M., Sparrow, J. C. and White, D. C. S. 1990. Actin mutants causing changed muscle kinetics. Nature 348, 440–442.

Durling, H. J., Reilich, P., Muller-Hocker, J., Mendel, B., Pongratz, D., Wallgren-Pettersson, C., Gunning, P., Lochmuller, H. and Laing, N. G. 2002. De novo missense mutation in a constitutively expressed exon of the slow alpha-tropomyosin gene TPM3 associated with an atypical, sporadic case of nemaline myopathy. Neuromuscul. Dis. 12, 947–951.

Elfenbein, J. L., Fisher, R. A., Wei, S., Morrell, R. J., Stewart, C., Friedman, T. B. and Friderici, K. 2001. Audiological aspects of the search for DFNA20: A gene causing late-onset, progressive sensoneural hearing loss. Ear Hear. 22, 279–288.

Engel, W. K. and Cunningham, G. G. 1963. Rapid examination of muscle tissue. An improved trichrome method for fresh frozen biopsy sections. Neurology 13, 919–923.

Engel, A. G. and Gomez, M. R. 1967. Nemaline (Z disk) myopathy: Observations on the origin, structure, and solubility properties of the nemaline structures. J. Neuropathol. Exp. Neurol. 2, 601–619.

Erba, H. P., Eddy, R., Shows, T., Kedes, L. and Gunning, P. 1988. Structure, chromosome location and expression of the human gamma-actin gene: Differential evolution, location and expression of the cytoskeletal beta- and gamma-actin genes. Mol. Cell. Biol. 8, 1775–1789.

Frey, N., Richardson, J. A. and Olson, E. N. 2000. Calsarcins, a novel family of sarcomeric calcineurin-binding proteins. Proc. Natl Acad. Sci. USA 9, 14632–14637.

Fukui, Y. 1978. Intranuclear actin bundles induced by dimethyl sulfoxide in interphase nucleus of *Dictyostelium*. J. Cell Biol. 76, 146–157.

Fukui, Y. and Katsumaru, H. 1979. Nuclear actin bundles in *Amoeba*, *Dictyostelium* and human HeLa cells induced by dimethyl sulfoxide. Exp. Cell Res. 120, 451–455.

Gillespie, P. G., Dumont, R. A. and Kachar, B. 2005. Have we found the tip link, a trasnsduction channel and gating spring of the hair cell? Curr. Opin. Neurobiol. 15, 389–396.

Hennessey, E. S., Drummond, D. R. and Sparrow, J. C. 1991. Post-translational processing of the amino terminus affects actin function. Eur. J. Biochem. 197, 345–352.

Hennessey, E. S., Harrison, A., Drummond, D. R. and Sparrow, J. C. 1992. Mutant actin: A dead end? J. Muscle Res. Cell Motil. 13, 127–131.

Hiromi, Y. and Hotta, Y. 1985. Actin gene mutations in *Drosophila*: Heat shock activation in the indirect flight muscles. EMBO J. 4, 1681–1687.

Höfer, D., Ness, W. and Drenckhahn, D. 1997. Sorting of actin isoforms in chicken auditory hair cells. J. Cell Sci. 110, 765–770.

Holmes, K. C., Popp, D., Gebhard, W. and Kabsch, W. 1990. Atomic model of the actin filament. Nature 288, 44–49.

Hutchinson, D. O., Charlton, A., Laing, N. G., Ilkovski, B. and North, K. N. 2006. Autosomal dominant nemaline myopathy with intranuclear rods due to mutation of the skeletal muscle ACTA1 gene: Clinical and pathological variability within a kindred. Neuromuscul. Disord. 16, 113–121.

Iannaccone, S. T., Bove, K. E., Vogler, C. A. and Buchino, J. J. 1987. Type 1 fiber size disproportion: Morphometric data from 37 children with myopathic, neuropathic, or idiopathic hypotonia. Pediatr. Pathol. 7, 395–419.

Iida, K., Iida, H. and Yahara, I. 1986. Heat shock induction of intranuclear actin rods in cultured mammalian cells. Exp. Cell Res. 165, 207–215.

Ilkovski, B., Clement, S., Sewry, C., North, K. N. and Cooper, S. T. 2005. Defining alpha-skeletal and alpha-cardiac actin expression in human heart and skeletal muscle explains the absence of cardiac involvement in ACTA1 nemaline myopathy. Neuromuscul. Disord. 15, 829–835.

Ilkovski, B., Nowak, K. J., Domazetovska, A., Maxwell, A. L., Clement, S., Davies, K. E., Laing, N. G., North, K. N. and Cooper, S. T. 2004. Evidence for dominant-negative effects in ACTA1 nemaline myopathy by abnormal folding, aggregation and latered polymerization of mutant actin isoforms. Hum. Mol. Genet. 13, 1727–1743.

Jockusch, B. M., Veldman, H., Griffiths, G., van Oost, B. A. and Jennekens, F. G. I. 1980. Immnuofluorescence microscopy of a myopathy. α-actinin is a major constituent of nemaline rods. Exp. Cell Res. 127, 409–420.

Joel, P. B., Fagnant, P. M. and Trybus, K. M. 2004. Expression of a nonpolymerizable actin mutant in Sf9 cells. Biochemistry 43, 11554–11559.

Kabsch, W., Mannherz, H. G., Suck, D., Pai, E. F. and Holmes, K. C. 1990. Atomic structure of the actin:DNase I complex. Nature 347, 37–44.

Kaindl, A. M., Ruschendorf, F., Krause, S., Goebel, H. H., Koehler, K., Becker, C., Pongratz, D., Muller-Hocker, J., Nurnberg, P., Stoltenburg-Didinger, G., Lochmuller, H. and Huebner, A. 2004. Missense mutations of ACTA1 cause dominant congenital myopathy with cores. J. Med. Genet. 41, 842–848.

Kuang, B. and Rubenstein. P. A. 1997. Beryllium fluoride and phalloidin restore polymerizability of a mutant yeast actin (V266G,L267G) with severely decreased hydrophobicity in a subdomain 3/4 loop. J. Biol. Chem. 272, 1237–1247.

Kudryashova, E., Kudryashov, D., Kramerova, I. and Spencer, M. J. 2005. Trim32 is a ubiquitin ligase mutated in limb girdle muscular dystrophy type 2H that binds to skeletal myosin and ubiquitinates actin. J. Mol. Biol. 354, 413–424.

Laing, N. G., Clarke, N. F., Dye, D. E., Liyanage, K., Walker, K. R., Kobayashi, Y., Shimakawa, S., Hagiwara, T., Ouvrier, R., Sparrow, J. C., Nishino, I., North, K. N. and Nonaka, I. 2004. Actin mutations are one cause of congenital fiber type disproportion. Ann. Neurol. 56, 689–694.

Laing, N. G. and Nowak, K. J. 2005. When contractile proteins go bad: The sarcomere and skeletal muscle disease. BioEssays 27, 809–822.

Lake, B. D. and Wilson, J. 1975. Zebra body myopathy. Clinical, histochemical and ultrastructural studies. J. Neurol. Sci. 24, 437–446.

Lebart, M. C., Méjean, C., Boyer, M., Roustan, C. and Benyamin, Y. 1990. Localization of a new α-actinin binding site in the COOH-terminal part of actin sequence. Biochem. Biophys. Res. Commun. 173, 120–126.

Lorenz, M., Popp, D. and Holmes, K. C. 1993. Refinement of the F-actin model against X-ray fiber diffraction data by the use of a directed mutation algorithm. J. Mol. Biol. 234, 826–836.

Luther, P. and Squire, J. M. 2002. Muscle Z-band ultrastructure: Titin Z-repeats and Z-band periodicities do not match. J. Mol. Biol. 319, 1157–1164.

Maron, B. J. 1997. Hypertrophic cardiomyopathy. Lancet, 350, 127–133.

Marston, S., Mirza, M., Abdulrazzak, H. and Sewry, C. 2004. Functional characterisation of a mutant actin (Met132Val) from a patient with nemaline myopathy. Neuromuscul. Disord. 14, 167–174.

McGough, A., Way, M. and DeRosier, D. 1994. Determination of the α-actinin-binding site on actin filaments by cryoelectron microscopy and image analysis. J. Cell Biol. 126, 1231–1240.

Miike, T., Ohtani, Y., Tamari, H., Ishitsu, T. and Une, Y. 1986. Muscle fiber type transformation in nemaline myopathy and congenital fiber type disproportion. Brain Dev. 8, 526–532.

Milligan, R. A. 1996. Protein–protein interactions in the rigor actomyosin complex. Proc. Natl Acad. Sci. USA 93, 21–26.

Milligan, R. A., Whittaker, M. and Safer, D. 1990. Molecular structure of F-actin and location of surface binding sites. Nature 348, 217–221.

Mimura, N. and Asano, A. 1987. Further characterization of a conserved actin-binding 27-kDa fragment of actinogelin and α-actinins and mapping their binding sites on the actin molecule by chemical cross-linking. J. Biol. Chem. 262, 4717–4723.

Mogensen, J., Clausen, I. C., Pedersen, A. K., Egeblad, H., Bross, P., Kruse, T. A., Gregersen, N., Hansen, P. S., Baandrup, U. and Børglum, A. D. 1999. α-cardiac actin is a novel disease gene in familial hypertrophic cardiomyopathy. J. Clin. Invest. 103, R39–R43.

Morell, R. J., Friderici, K. H., Wei, S., Friedman, T. B. and Fisher, R. A. 2000. A new locus for late-onset, progressive, hereditary hearing loss DFNA20 maps to 17q25. Genomics 63, 1–6.

Morris, E. P., Nneji, G. and Squire, J. M. 1990. The 3-dimensional structure of the nemaline rod Z-band. J. Cell Biol. 111, 2961–2978.

Nienhuis, A. W., Coleman, R. F., Brown, W. J., Munsat, T. L. and Pearson, C. M. 1967. Nemaline myopathy. A histopathologic and histochemical study. Am. J. Clin. Pathol. 48, 1–13.

Nongthomba, U., Pasalodos-Sanchez, S., Clark, S., Clayton, J. D. and Sparrow, J. C. 2001. Expression and function of the Drosophila ACT88F actin isoform is not restricted to the indirect flight muscles. J. Muscle Res. Cell Motil. 22, 111–119.

North, K. 2004. Congenital myopathies. In Myology. A. G. Engel and L. Franzini-Armstrong (Editors), vol. 2. McGraw-Hill, New York. pp. 1473–1533.

North, K. N., Laing, N. G., Wallgren-Pettersson, C. and the ENMC International Consortium on Nemaline myopathy (1997). J. Mol. Genet. 34, 705–713.

Nowak, K. J., Sewry, C. A., Navarro, C., Squier, W., Reina, C., Ricoy, J. C., Jayawant S., Childs, A.-M., Dobbie, J. A., Appleton, R. E., Mountford, R. C., Walker K. R., Clement, S., Barois, A., Muntoni, F. and Laing, N. G. 2007. Nemaline myopathy caused by absence of alpha-skeletal muscle actin. Ann. Neurol. 61, 175–184.

Nowak, K. J., Wattanasirichaigoon, D., Goebel, H. H., Wilce, M., Pelin, K., Donner, K., Jacob, R. L., Hubner, C., Oexle, K., Anderson, J. R., Verity, C. M., North, K. N., Iannaccone, S. T., Muller, C. R., Nurnberg, P., Muntoni, F., Sewry, C., Hughes, I., Sutphen, R., Lacson, A. G., Swoboda, K. J., Vigneron, J., Wallgren-Pettersson, C., Beggs, A. H. and Laing, N. G. 1999. Mutations in the skeletal muscle α-actin gene in patients with actin myopathy and nemaline myopathy. Nat. Genet. 23, 208–212.

Olson, T. M., Kishimoto, N. Y., Whitby, F. G. and Michels, V. V. 2001. Mutations that alter the surface charge of alpha-tropomyosin are associated with dilated cardiomyopathy. J. Mol. Cell. Cardiol. 33, 723–732.

Olson, T. M., Michels, V. V., Thibodeau, S. N., Tai, Y.-S. and Keating, M. T. 1998. Actin mutations in dilated cardiomyopathy, a heritable form of heart failure. Science 280, 750–752.

Ono, S., Abe, H., Nagaoka, R. and Obinata, T. 1993. Co-localization of ADF and cofilin in intranuclear actin rods of cultured muscle cells. J. Muscle Res. Cell Motil. 14, 195–204.

Orlova, A. and Egelman, E. H. 1993. A conformational change in the actin subunit can change the flexibility of the actin filament. J. Mol. Biol. 232, 334–341.

Otterbein, L. R., Graceffa, P. and Dominguez, R. 2001. The crystal structure of uncomplexed actin in the ADP state. Science 293, 708–711.

Rayment, I., Holden, H. M., Whittaker, C. B., Yohn, M., Lorenz, K., Holmes, K. C. and Milligan, R. A. 1993. Structure of the actin–myosin complex and its implications for muscle contraction. Science 262, 58–65.

Razzaq, A., Schmitz, S., Veigel, C., Molloy. J. E., Geeves, M. A. and Sparrow, J. C. 1999. Actin residue E93 is identified as an amino acid affecting myosin binding. J. Biol. Chem. 274, 28321–28328.

Reedy, M. C., Beall, C. and Fyrberg E. A. 1991. Mutations at the N-terminus of *Drosophila* actin. Biophys. J. 59, 187a.

Reymond, A., Meroni, G., Fantozzi, A., Merla, G., Cairo, S., Luzi, L., Riganelli, D., Zanaria, E., Messali, S., Caincara, S., Guffanti, A., Minucci, S., Pelicci, C. G. and Ballabio, A. 2001. The tripartite motif family identifies cell compartments. EMBO J. 20, 2140–2151.

Rommelaere, H., Waterschoot, D., Neirynck, K., Vanderkerckhove, J. and Ampe, C. 2003. Structural plasticity of functional actin: Pictures of actin binding protein and polymer interfaces. Structure 11, 1279–1289.

Ryan, M. M., Ilkovski, B., Strickland, C. D., Schnell, C., Sanoudou, D., Midgett, C., Houston, R., Muirhead, D., Dennett, X., Shield, L. K., De Girolami, U., Iannaccone, S. T., Laing, N. G., North, K. N. and Beggs, A. H. 2003. Clinical course correlates poorly with muscle pathology in nemaline myopathy. Neurology 60, 665–673.

Sameshima, M., Chijiiwa, Y., Kishi, Y. and Hashimoto, Y. 1994. Novel actin rods appeared in the spores of *Dictyostelium discoideum*. Cell Struct. Funct. 19, 189–194.

Sameshima, M., Kishi, Y., Osumi, M., Mahadeo, D. and Cotter, D. A. 2000. Novel actin cytoskeleton: Actin tubules. Cell Struct. Funct. 25, 291–295.

Schmitz, S., Schankin, C. J., Prinz, H., Curwen, R. S., Ashton, P., Caves, L. S. D., Fink, R. H. A., Sparrow, J. C., Mayhew, P. J. and Veigel, C. 2003. Molecular evolutionary convergence of the flight muscle protein arthrin in Diptera and Hemiptera. Mol. Biol. Evol. 20, 2019–2033.

Schröder, J. M., Durling, H. and Laing, N. G. 2004. Actin myopathy with nemaline bodies, intranuclear rods and a heterozygous mutations in ACTA1 (Asp154Asn). Acta Neuropathol. 108, 250–256.

Schröder, R., Reimann, J., Salmikangas, P., Clemen, C. S., Hayashi, Y. K., Nonaka, I., Arahata, K. and Carpen, O. 2003. Beyond LGMD1A: Myotilin is a component of central core lesions and nemaline rods. Neuromuscul. Disord. 13, 451–455.

Semsarian, C., Wu, M. J., Ju, Y. K., Maciniec, T., Yeoh, T., Allen, D. G., Harvey, R. P. and Graham, R. M. 1999. Skeletal muscle hypertrophy is mediated by a Ca^{2+}-dependent calcineurin signalling pathway. Nature 400, 576–581.

Serrano, A. L., Murgia, M., Pallafacchina, G., Calabria, E., Coniglio, P., Lomo, T. and Schiaffino, S. 2001. Calcineurin controls nerve activity-dependent specification of slow skeletal muscle fibers but not muscle growth. Proc. Natl Acad. Sci. USA 98, 13108–13113.

Sheterline, P., Clayton J. and Sparrow J. 1998. Actin. Oxford University Press, Oxford.

Shimomura, C. and Nonaka, I. 1989. Nemaline myopathy: Comparative muscle histochemistry in the severe neonatal, moderate congenital, and adult-onset forms. Pediatr. Neurol. 1, 25–31.

Shy, G. M., Engel, W. K., Somers, J. E. and Wanko, T. 1963. A new congenital myopathy. Brain 86, 793–810.

Slepecky, N. B. 1996. Structure of the mammalian cochlea. In The Cochlea. P. Dallas, A. N. Popper and R. R. Fay (Editors). Springer-Verlag, New York.

Sparrow, J. C., Drummond, D. R., Hennessey, E. S., Clayton, J. D. and Lindegaard, F. B. 1992. *Drosophila* actin mutants and the study of myofibrillar assembly and function. Soc. Exp. Biol. Symp. 46, 111–129.

Sparrow, J. C., Nowak, K., Durling, H. J., Beggs, A., Wallgren-Pettersson, C., Romero, N., Nonaka, I. and Laing, N. G. 2003. Muscle disease caused by mutations in the skeletal muscle alpha-actin gene, ACTA1. Neuromuscul. Disord. 13, 519–531.

Sparrow, J., Reedy, M., Ball, E., Kyrtatas, V., Molloy, J., Durston, J., Hennessey E. and White, D. 1991. Functional and ultrastructural effects of a missense mutation in the indirect flight muscle-specific actin gene of *Drosophila melanogaster*. J. Mol. Biol. 222, 963–982.

Tan, P., Briner, J., Boltshauser, E., Davis, M. R., Wilton, S. D., North, K., Wallgren-Pettersson, C. and Laing, N. G. 1999. Homozygosity for a nonsense mutation in the alpha-tropomyosin gene TPM3 in a patient with severe infantile nemaline myopathy. Neuromuscul. Disord. 9, 573–579.

Wada, A., Fukuda, M., Mishima, M. and Nishida, E. 1998. Nuclear export of actin: A novel mechanism regulating the subcellular localization of a major cytoskeletal protein. EMBO J. 17, 1635–1641.

Wallgren-Pettersson, C., Jasani, B., Newman, G. R., Morris, G. E., Jones, S., Singrao, S., Clarke, A., Virtanen, I., Holmberg, C. and Rapola, J. 1995. α-actinin in nemaline rods in congenital nemaline myopathy: Immunological confirmation by light and electron microscopy. Neuromuscul. Disord. 5, 93–104.

Wallgren-Pettersson, C. and Laing, N. G. 2006. 138th ENMC Workshop: Nemaline Myopathy, 20–22 May 2005, Naarden, The Netherlands. Neuromuscul. Disord. 16, 54–60.

Waterston, R. H., Hirsh, D. and Lane, T. R. 1984. Dominant mutations affecting muscle structure in *Caenorhabditis elegans* that map near the actin gene cluster. J. Mol. Biol. 180, 473–496.

Welch, W. J. and Suhan, J. P. 1985. Morphological study of the mammalian stress response: Characterization of changes in cytoplasmic organelles, cytoskeleton, and nucleoli, and appearance of intranuclear actin filaments in rat fibroblasts after heat-shock treatment, J. Cell Biol. 101, 1198–1211.

Wertman, K. F., Drubin, D. G. and Botstein, D. 1992. Systematic mutational analysis of the yeast ACT1 gene. Proc. Natl Acad. Sci. USA 93, 91–95.

van Wijk, E., Kreiger, E., Kemperman, M. H., De Leenheer, E. M. R., Huygen, P. L. M., Cremers, C. W. R. J., Cremers, F. P. M. and Kremer, H. (2003). A mutation in the gamma actin gene 1 (ACTG1) gene causes autosomal dominant hearing loss (DFND20/26). J. Med. Genet. 40, 879–884.

Yamaguchi, M., Robson, R., Stromer, M. H. and Dahl, D. S. 1978. Actin filaments form the backbone of nemaline myopathy rods. Nature 271, 265–267.

Yamaguchi, M., Robson, R., Stromer, M. H., Dahl, D. S. and Oda, T. 1982. Nemaline myopathy rod bodies. Structure and composition. J. Neurol. Sci. 56, 35–36.

Yang, T. and Smith, R. 2000. A novel locus of DFNA 26 maps to chromosome 17q25in two unrelated families with progressive autosomal dominant hearing loss. Am. J. Hum. Genet. 67(Suppl. 2), 300.

Zhu, M., Yang, T., Wei, S., DeWan, A. T., Morrell, R. J., Elfenbein, J. L., Fisher, R. A., Leal, S. M., Smith, R. J. H. and Friderici, K. H. 2003. Mutations in the γ-actin gene (ACTG1) are associated with dominant progressive deafness (DFNA20/26). Am. J. Hum. Genet. 73, 1082–1091.

3
Actin-Binding Drugs: An Elegant Tool to Dissect Subcellular Processes in Endothelial and Cancer Cells

Filip Braet, Lilian Soon, Katrien Vekemans, Pall Thordarson, and Ilan Spector

Introduction

Until a decade or so ago there were only a few agents available that interfered with cellular activities by binding to actin. In fact, most of our initial knowledge concerning the involvement of actin in basic cellular processes was based on the extensive use of the mold metabolites cytochalasins. However, the actin-binding activities and cellular effects of cytochalasins are complex and difficult to interpret (Cooper 1987; Sampath and Pollard 1991), so that the functions and dynamics of the actin cytoskeleton in various organisms remained elusive. Another widely used class of actin-binding drugs, the mushroom-derived phallotoxins, stabilizes actin filaments and promotes actin polymerization (Cooper 1987; Sampath and Pollard 1991), but they do not enter most cell types and are predominantly used as fluorescent derivatives to visualize actin filaments in fixed cells.

With the growing awareness that the actin cytoskeleton is involved in practically all aspects of cell behavior, as well as in cancer and other human diseases, it has become increasingly important to identify new agents with well-defined actin-binding properties that preferentially affect certain aspects of actin filament organization and dynamics in a reversible manner (Table 1). Such agents are not only necessary as research tools to better understand how the different actin structures are assembled, organized, and function in cells, but may also be potentially useful as therapeutic agents in the treatment of diseases (Spector et al. 1999; Yeung and Paterson 2002; Fenteany and Zhu 2003; Giganti and Friederich 2003).

So far, however, there is no actin-binding drug in the clinic, and there are only limited attempts to explore the pharmaceutical usefulness of antiactin drugs for the treatment of diseases (Table 2). Inhibition of cell growth and motility, control of viral budding, eye outflow, transendothelial transport, and neointimal hyperplasia are a few examples in which actin-binding drugs were found to have some effects. Of these, the topical application of microfilament-disrupting drugs (mainly the family of latrunculins) to the eye seems to reduce intraocular pressure in such an effective way that this might

TABLE 1. Actin-binding drugs and mode of action.

Mode of action/actin-binding protein (reference)
Jasplakinolides
Polymerization inducer and filament stabilizer/Arp2/3 and tropomysosin (Bubb et al. 1994)
Latrunculins
Monomer binding/profilin and thymosin β (Coue et al. 1987)
Swinholide
Severing and dimer binding/gelsolin (Bubb et al. 1995)
Misakinolide
Capping and dimer binding/capping proteins (Terry et al. 1997)
Halichondramide
Monomer binding, capping and severing/gelsolin, cofilin/ADF (Spector et al. 1999)
Dihydrohalichondramide
Capping/capping proteins (Spector et al. 1999)

TABLE 2. Actin-binding drugs in diseases.

Species/relevant model/disease (reference)
Jasplakinolide
Human/prostate carcinoma cells/metastasis (Senderowicz et al. 1995)
Mouse/lung carcinoma cells/metastasis (Takeuchi et al. 1998)
Human/lung carcinoma cells/metastasis (Hayot et al. 2006)
Mouse/breast carcinoma cells/metastasis (Hayot et al. 2006)
Monkey/kidney cells/viral infection (Eash and Atwood 2005)
Rat/colon carcinoma cells/metastasis (Figs. 3 and 4)
Latrunculins
Monkey/eye/mydriasis and cycloplegia (Peterson et al. 1999)
Monkey/eye/intraocular pressure and outflow (Okka, Tian and Kaufman 2004)
Rat/hippocampus/epileptogenesis (Vázquez-López, Sierra-Paredes and Sierra-Marcuno 2005)
Cytochalasins
Human/eye/intraocular pressure and outflow (Johnson 1997)
Pig/coronary model of stent-coating/neointimal hyperplasia (Salu et al. 2003)
Rat/liver endothelial cells/transendothelial transport (Braet et al. 2004)
Swinholide
Monkey/eye/intraocular pressure and outflow (Tian, Kiland and Kaufman 2001)

improve the quality of life for glaucoma patients (Table 2). On the other hand, actin-binding drugs which are effective in reducing migratory behavior of cancer cells (Hayot et al. 2006) may be used as additional treatment in cancer metastasis. Especially, jasplakinolide seems to possess strong anti-migratory effects on a variety of cancer cells *in vitro* (see chapter by Bearer).

This chapter does not intend to provide a comprehensive overview on all actin-binding drugs identified so far, nor does it describe in detail their different biochemical properties and cellular effects. Instead we refer to the reviews by Spector et al. (1999), Yeung and Paterson (2002), Fenteany and Zhu (2003), and Giganti and Friederich (2003), which extensively discuss the current state of knowledge concerning their mechanism of action and chemical structure. In the following sections we will discuss an example of our own studies in which only the use of different actin-binding drugs, such as cytochalasin B, latrunculin A, jasplakinolide A, swinholide A, misaki-nolide A, halichondramide, and dihydrohalichondramide (Fig. 1), played a key role in revealing a particular cellular process – the new formation of hepatic endothelial fenestrae. We also present some data on the effects of cytochalasin B, latrunculin A, and jasplakinolide A on cultured rat colorectal cancer cells.

Actin-Binding Drugs and Transendothelial Transport

The effects of a number of actin-binding drugs on actin organization in hepatic endothelial cells showed that each class of drugs alters the distribution patterns of actin in a unique way, and that even within a chemical class structurally similar compounds can have qualitatively different effects on actin organization, indicating different actin-binding properties (Braet et al. 1998, 2002; Spector et al. 1999). The study of the numerical dynamics of fenestrae in hepatic endothelial cells is an example of an actin-mediated cellular process, in which only the use of several compounds was necessary to identify a family of functional-related antiactin drugs that could selectively reveal the different steps of fenestrae formation (reviewed in Braet 2004). Hepatic endothelial fenestrae are dynamic membrane-bound open pores that act as a sieving barrier to control the extensive exchange of material between the blood and the liver parenchyma, and their biological relevance in various diseases has been widely recognized (for reviews see Fraser, Dobbs and Rogers 1995; Braet and Wisse 2002).

The involvement of the actin cytoskeleton in regulating fenestrae dynamics was initially shown by treating cultured hepatic endothelial cells with the actin inhibitor cytochalasin B (Steffan, Gendrault and Kirn 1987) that resulted in an increase in the number of fenestrae indicating that these membrane-bound holes are inducible structures. Later on the microfilament-disrupting drug latrunculin A was applied to elaborate and confirm the earlier observations made by Steffan, Gendrault and Kirn (1987). From the latrunculin A study it became clear that if the increase in the number of fenestrae was a side effect of cytochalasin B, then it was most unlikely that latrunculin A would have the same side effect (Braet et al. 1996). At that

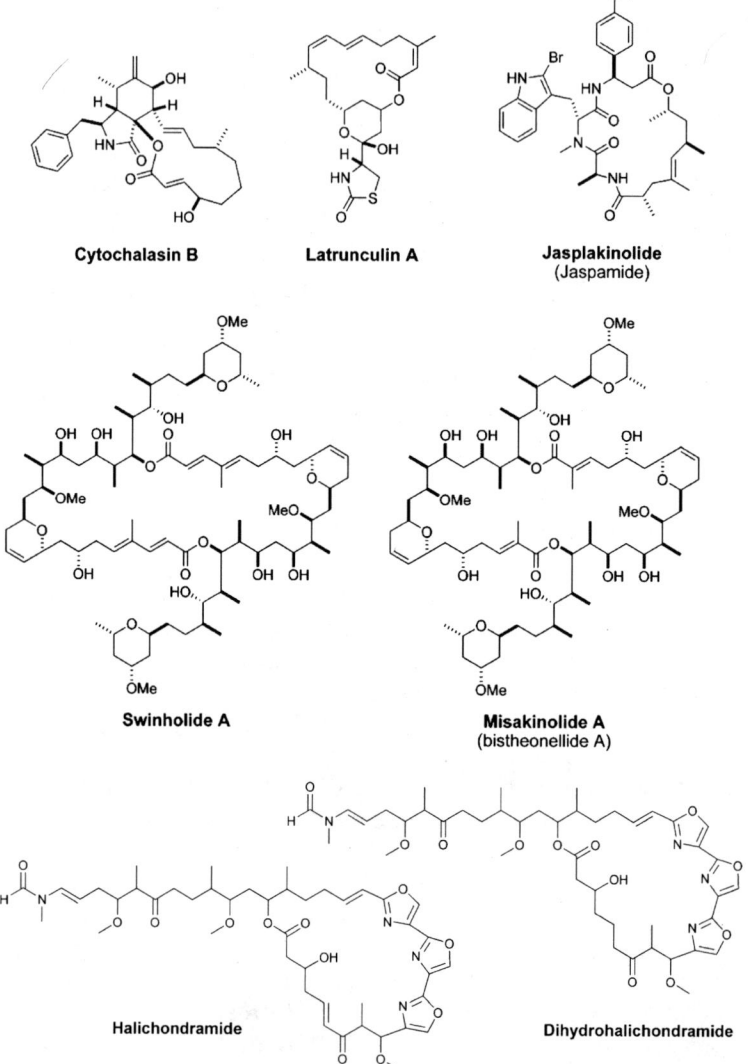

Cytochalasin B

Latrunculin A

Jasplakinolide
(Jaspamide)

Swinholide A

Misakinolide A
(bistheonellide A)

Halichondramide

Dihydrohalichondramide

Fɪɢ. 1. Structures of actin-binding drugs. Note the similar structures of swinholide A and misakinolide A, the later differs from swinholide A only by a subtraction of two double-bonded carbon in each repeated lactone moiety. Noteworthy is that these structurally comparable dimeric macrolides both bind to two actin monomers, but at the same time having different binding properties to actin filaments; i.e., swinholide A severs actin filaments, while misakinolide A caps the barbed ends. A similar observation can be made when comparing the structures of halichondramide and dihydrohalichondramide; i.e., dihydrohalichondramide differs from halichondramide only in having a single bond in the macrolide at positions 4–5 instead of a double bond. Interestingly is that halichondramide and dihydrohalichondramide possess biochemical similarities to the actin-binding drugs swinholide A and misakinolide A which have severing and barbed-end capping activities, respectively

time, the obtained data permitted concluding that an increase in the number of fenestrae occurs when actin filaments are depolymerized by cytochalasin B or by latrunculin A.

Considering the complexity of the actin cytoskeleton and the difficulties in understanding how different actin filament populations were involved in the numerical dynamics of fenestrae as demonstrated by the generic actin disrupters cytochalasin B and latrunculin A, it was deemed necessary to assess the effects of other new generation actin-targeted drugs with different, more specific, mechanisms of action on the structure and dynamics of fenestrae (Spector et al. 1999). Initially, it was concluded that the disassembly of actin filaments in hepatic endothelial cells by all compounds tested (jasplakinolide, swinholide A, misakinolide A, halichondramide, and dihydrohalichondramide), together with the increase in the number of fenestrae, suggested a common mechanism of fenestrae formation for all actin-binding agents. Unexpectedly however, by treating hepatic endothelial cells with the actin inhibitors misakinolide A (Braet et al. 1998) or dihydrohalichondramide (Braet et al. 2002), a structure indicative of fenestrae formation, designated "fenestrae-forming center" (FFC), could be captured. From these FFCs rows of newly formed fenestrae are connected fanning out into the surrounding fenestrated cytoplasm as a whirlwind (Fig. 2). Therefore, the unmasking of nascent fenestrae emerging from the FFCs by only two of the seven agents tested indicate that specific alterations in actin organization at particular locations and times are required to bring to light the process of fenestrae formation. Furthermore, based on the unique biochemical properties of all antiactin-binding tested, it could be concluded that the only biochemical activity that misakinolide A and dihydrohalichondramide have in common is their barbed-end capping activity. This mode of action seems to slow down the process of fenestrae formation to such an extent that it becomes possible to resolve FFCs. Recently, supplementary insights at the molecular and structural level about the process of hepatic endothelial fenestrae formation could been gathered by applying the methodology of correlative microscopy (Braet et al. 2007), unambiguously illustrating that microfilament-disruption induces translocation of preexisting three-dimensional organized FFCs from the perinuclear area toward the peripheral cytoplasm.

The insights gathered with the antiactin drugs may open up an attractive possibility of treating a "fenestrae-related disorder" in which actin filaments may serve as a possible molecular target for the treatment of hyperlipoproteinemia. For example, the rise of the liver sieve's porosity by increasing the number of fenestrae with these drugs may improve the extraction of atherogenic lipoproteins from the circulation (Fraser, Dobbs and Rogers 1995). Recently, an attempt to modulate the liver sieve *in vivo* using liposome-encapsulated cytochalasin B was briefly described (Braet et al. 2004). On the other hand, an increase in the porosity of the hepatic endothelium can be of

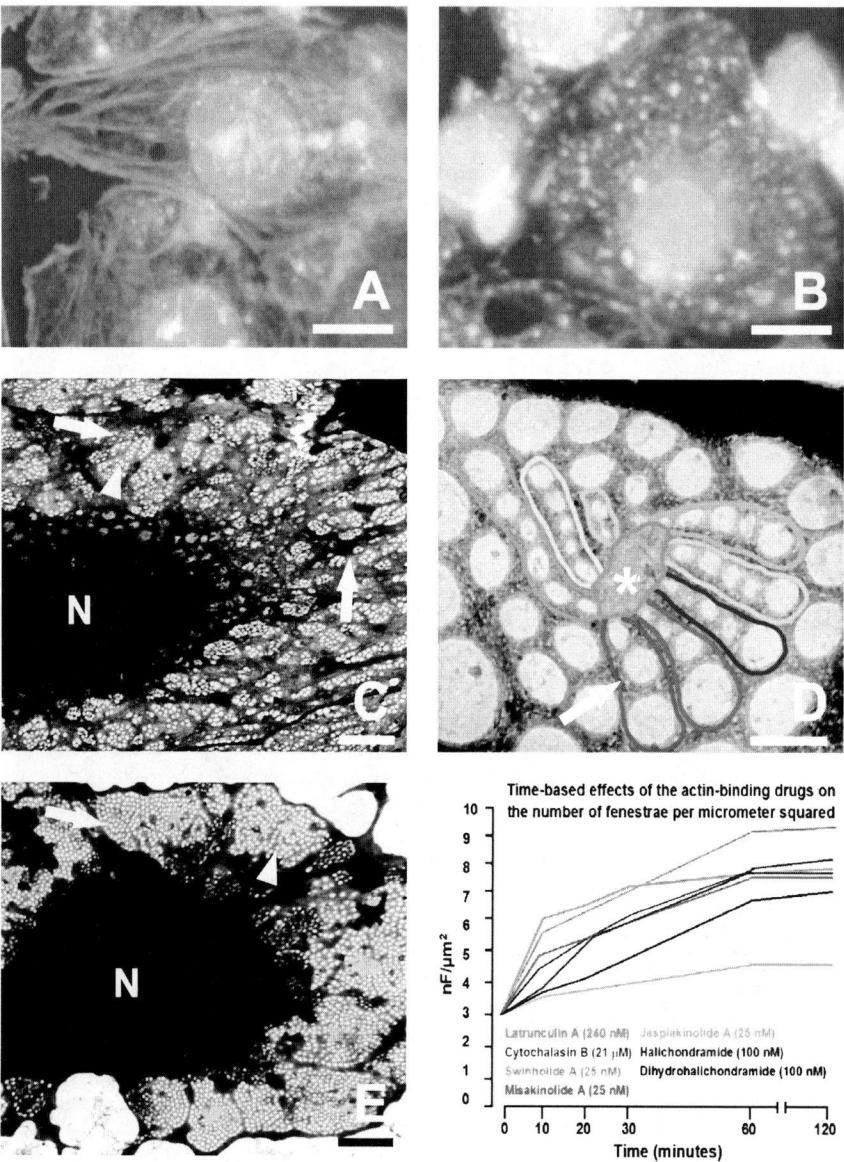

FIG. 2. Actin-mediated changes and numerical fenestrae dynamics in hepatic endothelial cells. (**a–b**) Fluorescence micrographs showing the effect of dihydrohalichondramide on actin organization in hepatic endothelial cells when compared to control cells, monitored with rhodamine-phalloidin (filamentous-actin/red) and fluorescein-DNase I staining (globular actin/green). *Blue color* represents the nuclei stained with DAPI. (**a**) Filamentous-actin distribution in control cells shows the presence of cytoplasmic stress fibers and peripheral bands of actin bundles that line the cell margin.

great benefit for enhancing gene delivery to liver parenchymal cells, and can consequently be exploited as a new therapeutic approach for treating various liver diseases (Lievens et al. 2004).

Actin-Binding Drugs and Cancer

Mounting evidence discloses that alterations of actin polymerization, or actin remodeling, play a key role in regulating the metastatic behavior of a malignant cell (for reviews, Lambrechts, Van Troys and Ampe 2004; Rao and Li 2004). This evidence is discussed in considerable detail in the chapter by Van Trays et al. in this volume. It has long been recognized that control of the actin cytoskeleton must be coordinated with control of cell cycle events and that actin-related events, such as adhesion and cell motility, can affect mitogenic signals (Carragher and Frame 2004); and consequently, contribute to cancer metastasis. Mitogenic signal transduction pathways and cell death mechanisms are closely intertwined and are responsible for maintaining tissue homeostasis. These processes that require coordinated modification in cellular architecture and the actin cytoskeleton have been shown to play a key role in apoptosis (for a review see Gourlay and Ayscough 2005). Interestingly, the vast majority of chemotherapeutic drugs exhibit their cytotoxic effect by triggering the apoptotic cascade in cancer cells (Kim 2005). Taking these findings into

---

FIG. 2. Continued Globular actin is mainly localized in the perinuclear region. (b) Hepatic endothelial cells treated with 100 nM of dihydrohalichondramide for 10 min show loss of filamentous-actin bundles and appearance of brightly stained filamentous-actin patches. Furthermore, globular actin is diffuse and faintly stained. Scale bars, 5 μm. (c–e). Transmission electron micrographs of whole mount formaldehyde prefixed and detergent-extracted hepatic endothelial cells. (c) Low magnification showing the nuclear area (*N*) and surrounding extracted cytoplasm. Note that the sieve plates are well defined by a darker border (*arrow*) and that inside the sieve plates fenestrae (*arrowhead*) can be observed. Scale bar, 5 μm. (d) High magnification micrograph of a hepatic endothelial cell treated with 100 nM dihydrohalichondramide for 30 min, showing a fenestrae-forming center (*asterisk*) (FFC) to which rows (*arrow*) of fenestrae (as indicated by *artificial colors*) with increasing diameter are connected. Scale bar, 200 nm. (e) Low magnification image illustrating a hepatic endothelial cell treated with 100 nM dihydrohalichondramide for 120 min, showing huge fenestrated areas (*arrowhead*), sieve plates (*arrow*). Compare with (c) for the difference. Scale bar, 5 μm. (f) Effect of the different antiactin-binding drugs tested on the number of fenestrae per micrometer squared (nF μm^{-2}) in time. From this graph, we can conclude that all agents increase the number of fenestrae, although at a different rate and different maximum (*See Color Plates*)

consideration it has become increasingly important to identify new agents that preferentially affect certain aspects of actin filament organization and cellular functions. These agents can be added to the battery of anti-cancer drugs as a potential new therapeutic approach in the treatment of cancer.

Earlier studies have demonstrated a clear relationship between actin-mediated fine structural changes and the inhibition of prostate cancer cell growth *in vitro* by exposing cancer cells to jasplakinolide A for 48 h at nanomolar concentrations (Senderowicz et al. 1995). Interestingly to note is that, unlike the other known actin-stabilizer phalloidin, jasplakinolide appears to be cell-permeant (Stingl, Andersen and Emerman 1992; Bubb et al. 1994). Recently, Hayot et al. (2006) studied the effects of cytocha-lasin D and jasplakinolide on cell growth and cell motility of two cancer cell lines and found differences in the effects of the two drugs on the motility of two cell lines. The above findings are in line with our recent studies (Figs. 3 and 4). Furthermore, we found that prolonged exposure of colorectal cancer cells to the actin-binding drug jasplakinolide resulted in an apoptotic rate of about 20% at concentrations of 400 nM (Fig. 4). Similar findings about the apoptosis-inducing effects of jasplakinolide were reported initially by Posey and Bierer (1999). The later is an impor-tant observation as there is mounting evidence showing a clear link between actin remodeling and apoptosis (Gourlay and Ayscough 2005). The actin cytoskeleton should therefore be considered as an appealing molecular target not only for inhibiting cell migration, but also for pro-voking cell death, and consequently eliminating malignant cells. As a result, this opens up an entirely new avenue for drug designers in which the existing actin-binding drugs or their analogues should be considered as the next generation of anticancer drugs in conjugation with existing pharmaceutical compounds.

Interestingly, we found that latrunculin A showed an overall low cyto-toxic action and moderate apoptosis-inducing effect at similar concentra-tions and exposure times (Fig. 4). This observation is in line with earlier *in vitro* and *in vivo* findings, although by using other functional measures (Spector et al. 1983, 1989). Furthermore, it has been reported that the effects of latrunculin A on the actin cytoskeleton are reversible, and do not affect the viability of the cells (for a review see Ayscough 1998). These advantages make latrunculin A an ideal tool to study the structure and function of actin *in vitro* and *in vivo*.

The above data indicate important differences in the effects of actin-binding drugs on cancer cells, so that certain actin-binding agents can induce apoptosis in cancer cells, whereas others cannot. Furthermore, in the case of colorectal liver metastasis, malignant cells from the vascular bed could be eliminated with the aid of liposome-encapsulated microfilament-disrupting drugs (Braet et al. 2004). This possible therapeutic implication is supported by the fact that polyanionized proteoliposomes can be targeted with high

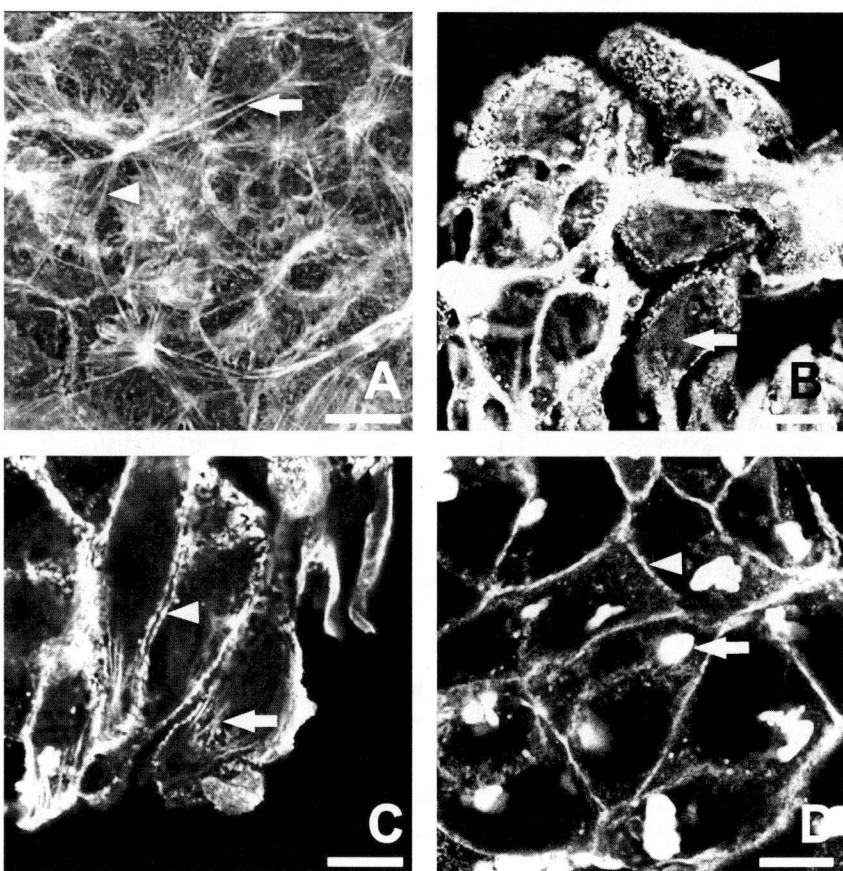

FIG. 3. Actin filament-mediated changes in rat CC531s colon carcinoma cells *in vitro* as observed by rhodamine-phalloidin staining and confocal scanning laser microscopy. (**a**) Control cells show the presence of an intricate network of cytoplasmic stress fibers (*arrow*) and peripheral bands of actin bundles that line the cell margins (*arrowhead*). (**b**) CC531s cells treated with 400 nM cytochalasin B for 10 min show a loss of cytoplasmic filamentous-actin bundles (*arrow*) and instead a diffuse punctuate fluorescence within the cytoplasm can be observed. Note that peripheral actin filaments accumulate at the cell margins (*arrowhead*). (**c**) CC531s cells exposed to 400 nM latrunculin A for 10 min show overall a loss of stress fibers (*arrow*) and an accumulation of peripheral laying actin filaments (*arrowhead*). Compare with the control situation for the difference. However the buildup of actin at the cell margins is less pronounced when compared to the cytochalasin B-treated cells. (**d**) Filamentous-actin distribution in CC531s colorectal cancer cells after 10 min treatment with 400 nM jasplakinolide A. The cytoplasm lacks filamentous organized actin, and instead large lumps of actin dots could be detected (*arrow*). Actin filaments at the cell margin are less dense (*arrowhead*) (courtesy of S. Hofmans). Scale bars 20 μm

Cytochalasin B

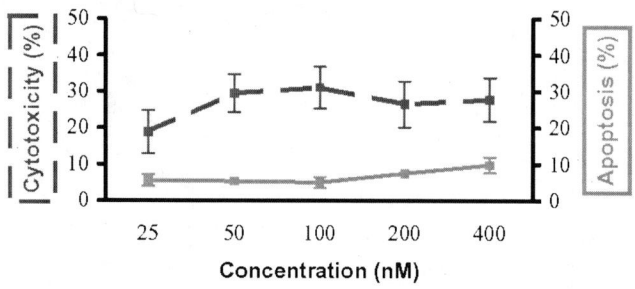

Latrunculin A

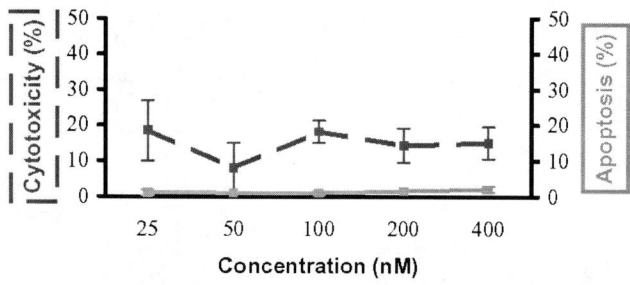

Jasplakinolide A

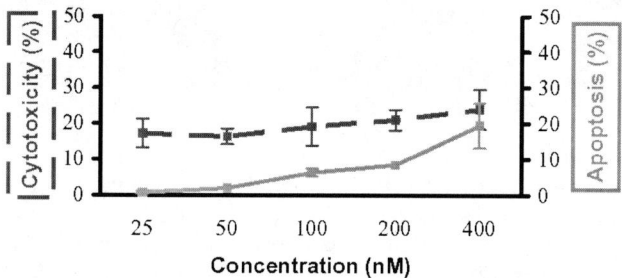

Fɪɢ. 4. Concentration-dependent cytotoxic (*dashed lines*) and apoptosis-inducing (*solid lines*) effects of cytochalasin B, latrunculin A, and jasplakinolide A on CC531s colorectal cancer cells *in vitro* as ascertained by crystal violet and propidium iodide/hoechst staining, respectively. Briefly, CC531s cells cultured in 96-multiwell plates were exposed to the actin-binding agents at different concentrations for 24 h and next incubated with viability stains according to standard protocols. Subsequently, data were recorded and analyzed using a microplate reader and processed by the accompanying data analysis software (Wallac, Victor³ V). From these graphs it becomes clear that of the three actin-binding agents studied cytochalasin B exerts the most cytotoxic effect at all concentrations tested; whereas jasplakinolide A reveals a dose-dependent cytotoxic effect and simultaneously is a potent apoptosis-inducing drug. Noteworthy is the overall low cytotoxic action and moderate apoptosis-inducing effect of latrunculin A, making this drug an ideal tool to dissect actin function and to study actin organization *in vitro* and *in vivo*

efficiency to metastatic colorectal cells in the liver (Richardso, Kren and Steer 2002; Koning et al. 2003). Intravenous infusion of these lipid-encapsulated drugs would definitely be a huge benefit during the surgical procedure, as a significant number of secondary colorectal metastases (i.e., liver) are caused during resection of the affected colon.

Conclusion

The availability of novel actin-binding agents with a wide range of cellular effects provides a unique opportunity to gain new insights into specific aspects of actin-mediated cellular and subcellular processes that so far remained elusive (Spector et al. 1999). New compounds have currently been isolated, identified, and synthesized (Yeung and Paterson 2002), and we foresee that they will become increasingly important in further advancing our understanding of the actin cytoskeleton as well as potential new therapeutic compounds in diseases were actin plays an important role.

Acknowledgments. The authors acknowledge the facilities as well as technical and administrative assistance from staff in the NANO Major National Research Facility at the "Australian Key Centre for Microscopy and Microanalysis" of The University of Sydney.

This work was supported by the "Cancer Research Fund of The University of Sydney" grant R5481 (F. Braet and L. Soon) and partially by the "NHMRC" grant 402510 (L. Soon and F. Braet). K. Vekemans is a Postdoctoral Fellow of the "FSR-Flanders" and P. Thordarson is an "Australian Research Fellow – ARC Australia."

References

Ayscough, K. 1998. Use of latrunculin-A, an actin monomer-binding drug. Meth. Enzymol., 298, 18–25.

Braet, F. 2004. How molecular microscopy revealed new insights in the dynamics of hepatic endothelial fenestrae in the past decade. Liver Int. 24, 532–539.

Braet, F., De Zanger, R., Jans, D., Spector, I. and Wisse, E. 1996. Microfilament disrupting agent latrunculin A induces an increased number of fenestrae in rat liver sinusoidal endothelial cells: Comparison with cytochalasin B. Hepatology 24, 627–635.

Braet, F., Spector, I., De Zanger, R. and Wisse, E. 1998. A novel structure involved in the formation of liver endothelial cell fenestrae revealed using the actin inhibitor misakinolide. Proc. Natl Acad. Sci. USA 95, 13635–13640.

Braet, F., Spector, I., Shochet, N. R., Crews, P., Higa, T., Menu, E., De Zanger, R. and Wisse, E. 2002. The new anti-actin agent dihydrohalichondramide reveals fenestrae-forming centers in hepatic endothelial cells. BMC Cell Biol. 3, 7.

Braet, F., Vekemans, K., Morselt, H., De Zanger, R., Wisse, E., Scherphof, G. and Kamps, J. 2004. The effect of cytochalasin B-loaded liposomes on the ultrastructure of the liver sieve. Comp. Hepatol. 3, S27.

Braet, F. and Wisse, E. 2002. Structural and functional aspects of liver sinusoidal endothelial cell fenestrae: A review. Comp. Hepatol. 1, 1.

Braet, F., Wisse, E., Bomans, P., Frederik, P., Geerts, W., Koster, A., Soon, L. and Ringer, S. 2007. Contribution of high-resolution correlative imaging techniques in the study of the hepatic sieve in three-dimensions. Microsc. Res. Tech. 70, 230–242.

Bubb, M. R., Senderowicz, A. M., Sausville, E. A., Duncan, K. L. and Korn, E. D. 1994. Jasplakinolide, a cytotoxic natural product, induces actin polymerization and competitively inhibits the binding of phalloidin to F-actin. J. Biol. Chem. 269, 14869–14871.

Bubb, M. R., Spector, I., Bershadsky, A. D. and Korn, E. D. 1995. Swinholide A is a microfilament disrupting marine toxin that stabilizes actin dimers and severs actin filaments. J. Biol. Chem. 270, 3463–3466.

Carragher, N. O. and Frame, M. C. 2004. Focal adhesion and actin dynamics: A place where kinases and proteases meet to promote invasion. Trends Cell Biol. 14, 241–249.

Cooper, J. A. 1987. Effects of cytochalasin and phalloidin on actin. J. Cell Biol. 105, 1473–1478.

Coue, M., Brenner, S. L., Spector, I. and Korn, E. D. 1987. Inhibition of actin polymerization by latrunculin A. FEBS Lett. 213, 316–318.

Eash, S., and Atwood, W. J. 2005. Involvement of cytoskeletal components in BK virus infectious entry. J. Virol. 79, 11734–11741.

Fenteany, G. and Zhu, S. 2003. Small-molecule inhibitors of actin dynamics and cell motility. Curr. Top. Med. Chem. 3, 593–616.

Fraser, R., Dobbs, B. R. and Rogers, G. W. T. 1995. Lipoproteins and the liver sieve: The role of the fenestrated sinusoidal endothelium in lipoprotein metabolism, atherosclerosis, and cirrhosis. Hepatology 21, 863–874.

Giganti, A. and Friederich, E. 2003. The actin cytoskeleton as a therapeutic target: State of the art and future directions. Prog. Cell Cycle Res. 5, 511–525.

Gourlay, C. W. and Ayscough, K. R. 2005. The actin cytoskeleton: A key regulator of apoptosis and ageing? Nat. Rev. Mol. Cell Biol. 6, 583–589.

Hayot, C., Debeir, O., Van Ham, P., Van Damme, M., Kiss, R. and Decaestecker, C. 2006. Characterization of the activities of actin-affecting drugs on tumor cell migration. Toxicol. Appl. Pharmacol. 211, 30–40.

Johnson, D. H. 1997. The effect of cytochalasin D on outflow facility and the trabecular meshwork of the human eye in perfusion organ culture. Invest. Ophthalmol. Vis. Sci. 38, 2790–2799.

Kim, R. 2005. Recent advances in understanding the cell death pathways activated by anticancer therapy. Cancer 103, 1551–1560.

Koning, G. A., Morselt, H. W. M., Gorter, A., Allen, T. M., Zalipsky, S., Scherphof, G. L. and Kamps, J. A. A. M. 2003. Interaction of differently designed immunoliposomes with colon cancer cells and Kupffer cells: An in vitro comparison. Pharm. Res. 20, 1249–1257.

Lambrechts, A., Van Troys, M. and Ampe, C. 2004. The actin cytoskeleton in normal and pathological cell motility. Int. J. Biochem. Cell Biol. 36, 1890–1909.

Lievens, J., Snoeys, J., Vekemans, K., Van Linthout, S., De Zanger, R., Collen, D., Wisse, E. and De Geest, B. 2004. The size of sinusoidal fenestrae is a critical determinant of hepatocyte transduction after adenoviral gene transfer. Gene Ther. 11, 1523–1531.

Okka, M., Tian, B. and Kaufman, P. L. 2004. Effect of low-dose latrunculin B on anterior segment physiologic features in the monkey eye. Arch. Ophthalmol. 122, 482–488.

Peterson, J. A., Tian, B., Geiger, B. and Kaufman, P. L. 1999. Latrunculin-A causes mydriasis and cycloplegia in the cynomolgus monkey. Invest. Ophthalmol. Vis. Sci. 40, 631–938.

Posey, S. C., Bierer, B. E. 1999. Actin stabilization by jasplakinolide enhances apoptosis induced by cytokine deprivation. J. Biol. Chem. 274, 4259–4265.

Rao, J. and Li, N. 2004. Microfilament actin remodeling as a potential target for cancer drug development. Curr. Cancer Drug Targets 44, 345–354.

Richardso, P., Kren, B. T. and Steer, C. J. 2002. In vivo application of non-viral vectors to the liver. J. Drug Target, 10, 123–131.

Salu, K. J., Huang, Y., Bosmans, J. M., Liu, X., Li, S., Wang, L., Verbeken, E., Bult, H., Vrints, C. J. and De Scheerder, I. K. 2003. Addition of cytochalasin D to a biocompatible oil stent coating inhibits intimal hyperplasia in a porcine coronary model. Coron. Artery Dis. 14, 545–555.

Sampath, P. and Pollard, T. D. 1991. Effects of cytochalasin, phalloidin, and pH on the elongation of actin filaments. Biochemistry 30, 1973–1980.

Senderowicz, A. M., Kaur, G., Sainz, E., Laing, C., Inman, W. D., Rodriguez, J., Crews, P., Malspeis, L., Grever, M. R., Sausville, E. A. and Duncan, L. K. 1995. Jasplakinolide's inhibition of the growth of prostate carcinoma cells in vitro with disruption of the actin cytoskeleton. J. Natl Cancer Inst. 87, 46–51.

Spector, I., Braet, F., Shochet, N. R. and Bubb, M. R. 1999. New anti-actin drugs in the study of the organization and function of the actin cytoskeleton. Microsc. Res. Tech. 47, 18–37.

Spector, I., Shochet, N. R., Blasberger, D. and Kashman, Y. 1989. Latrunculins: Novel marine macrolides that disrupt microfilament organization and affect cell growth: I. Comparison with cytochalasin D. Cell Motil. Cytoskel. 13, 127–144.

Spector, I., Shochet, N. R., Kashman, Y. and Groweiss, A. 1983. Latrunculins: Novel marine toxins that disrupt microfilament organization in cultured cells. Science 219, 493–495.

Steffan, A. M., Gendrault, J. L., Kirn, A. 1987. Increase in the number of fenestrae in mouse endothelial liver cells by altering the cytoskeleton with cytochalasin B. Hepatology 7, 1230–1238.

Stingl, J., Andersen, R. J. and Emerman, J. T. 1992. In vitro screening of crude extracts and pure metabolites obtained from marine invertebrates for the treatment of breast cancer. Cancer Chemother. Pharmacol. 30, 401–406.

Takeuchi, H., Ara, G., Sausville, E. A. and Teicher, B. 1998. Jasplakinolide: Interaction with radiation and hyperthermia in human prostate carcinoma and Lewis lung carcinoma. Cancer Chemother. Pharmacol. 42, 491–496.

Terry, D. R., Spector, I., Higa, T. and Bubb, M. R. 1997. Misakinolide A is a marine macrolide that caps but does not sever filamentous actin. J. Biol. Chem. 272, 7841–7845.

Tian, B., Kiland, J. A. and Kaufman, P. L. 2001. Effects of the marine macrolides swinholide A and jasplakinolide on outflow facility in monkeys. Invest. Ophthalmol. Vis. Sci. 42, 3187–3192.

Vázquez-López, A., Sierra-Paredes, G. and Sierra-Marcuno, G. 2005. Seizures induced by microperfusion of glutamate and glycine in the hippocampus of rats pretreated with latrunculin A. Neurosci. Lett. 388, 81–85.

Yeung, K. S. and Paterson, I. 2002. Actin-binding marine macrolides: Total synthesis and biological importance. Angew. Chem. Int. Ed. Engl. 41, 4632–4653.

4
Autoantigenicity of Actin

Ian R. Mackay, Roberto Martinez-Neira, Senga Whittingham,
Dan Nicolau, and Ban-Hock Toh

Introduction

Actin is one of the most abundant of the constituents of the proteome. In its polymeric filamentous form, actin is a critical component of the cytoskeleton, so influencing cell shape and motility. It is viewed by the research community from several different perspectives. For example, cell biologists view actin as the major element in a complex suite of molecules that maintain cell form and mediate cellular motility responses to various stimuli, e.g., chemotaxis, covered in detail in other chapters of this book. Analytical biochemists take advantage of the cellular abundance and high conservation of structure of actin to use it as a "housekeeping" protein control for particular functional studies on other elements of the proteome. Immunopathologists and diagnostic serologists are interested in the propensity of actin to behave as an autoantigenic molecule, which it does in virtually just one disease, autoimmune hepatitis (AIH). This chapter is written from this third perspective. The chapter is introduced by a brief survey of immunity and autoimmunity, and the nature of the disease AIH in which F-actin functions as an autoantigen. The chapter examines anti-F-actin in terms of its origins and significance, and possible role in immune-mediated destruction of hepatocytes. Functional effects *in vitro* of anti-F-actin (delivered as Fabs of IgG) on actin motility are demonstrable using an actin–myosin system in a motility chamber.

Immunity and Autoimmunity

Background

Immunology developed around year 1900 in accompaniment with bacteriology, and thus is relatively new among the biosciences. The subsection of immunopathology including autoimmunity developed even later, in the 1950s, based on clinical insights and technical developments in applied (clinical) immunology laboratories. As of today, there are some 80 diseases in which autoimmunity is causal or deeply implicated, yet the origins and nature of autoimmune responsiveness still remain unclear (Rose and Mackay 2006). In particular, there is no convincing explanation why only a relatively limited number of molecules among the myriads of "self"-proteins become selected as autoantigens. Actin is one of this limited number of autoantigenic protein molecules.

Innate and Adaptive Immunity

The innate immune system includes various phagocytic cells, macrophages, and dendritic cells that repel microbial invaders, dispose of tissue debris, and present antigen to cells of the adaptive immune system. Additionally, by virtue of encounter of microbial products with specific receptors called "Toll-like" (TLR), such antigen-presenting cells become "activated" and capable of presenting antigen in immunogenic mode, due in part to their production of proimmunogenic cytokines. This is the first step in an immune response. Self-products would not be expected to engage the TLRs of antigen-presenting cells (APCs), although this is in fact reported to occur in certain experimental settings (Lang et al. 2005), and so could occur naturally.

The adaptive immune system provides for precision of immune responsiveness. The components comprise B lymphocytes that develop in *B*one marrow and later in spleen, and subserve antibody production, and T lymphocytes that develop in *T*hymus and subserve cell-based immunity. As specified by Clonal Selection Theory, each B and T lymphocyte expresses, before any antigenic contact, just one of the myriad of potential receptors for extrinsic antigens, such that the "antigenic universe" would be covered by the entire T and B lymphocyte populations. Antigen acts by becoming "selected" by the lymphocyte bearing the receptor of best fit, so promoting clonal proliferation of that cell, and a progeny that serves both to eliminate the antigen and establish a residual population of memory cells geared up for the next antigen encounter.

T lymphocytes fall into several functional subsets. These include, primarily, "helper" CD4 positive (Th) and cytotoxic CD8 positive (Tc) cells; the Th (and likely also the Tc) subsets are further divided into types 1 and 2 according to functionality, itself dependent on cytokines secreted. Th1 cells secrete cytokines that drive inflammatory/destructive responses and Th2 cells secrete cytokines that are down regulatory for Th1, and drive B cell and proallergic responses. A third subset that secretes the proinflammatory cytokine IL17 and is designated as Th17 has recently been identified and is currently the subject of active investigation (Harrington, Mangan and Weaver 2006). Additionally there is a T regulatory class, described below.

Finally, the immune response has two distinct phases: an inductor phase during which there is recruitment of antigen-specific cells in a regional lymph node or spleen, and an effector phase during which an expanded cell population reacts with the inciting antigen (or autoantigen) in order to eliminate it.

Tolerance and Autoimmunity

Complex mechanisms have evolved to ensure that, in the developmental processes of adaptive immunity, the integrity of the individual's own tissues is not compromised: this constitutes natural immune tolerance. Autoimmunity reflects a failure of one or more of these complex mechanisms that collectively ensure a state of nonresponsiveness to the "self" proteins of the organism, in contrast to the necessary immune responsiveness to nonself or foreign proteins, notably proteins associated with microbial pathogens. There are two main processes involved in immune tolerance. One process, *central tolerance* or *recessive tolerance*, is mediated by deletion, also called *negative selection*, of self-reactive lymphocytes in the primary central organs of lymphopoiesis, the thymus and bone marrow, wherein there occurs maturation and education of T lymphocytes (Kyewski and Klein 2006) and B lymphocytes, respectively. Concurrently there is proliferation of lymphocytes with "antiforeign" reactivity, called *positive selection*. The other tolerogenic process depends on activities of down-regulatory subsets of T cells (T_{regs}) that develop in the thymus but operate in the periphery, and hence is called *peripheral tolerance* or *dominant tolerance* (Kronenberg and Rodensky 2005). The respective roles in self-tolerance and autoimmunity of these two types of tolerance likely vary from tissue to tissue, and await further investigation.

Lymphocytes that escape, by whatever means, from these tolerance mechanisms can expand clonally by reason of ongoing reactivity with a cognate autoantigen, in the course of which they acquire characteristics of "memory" T or B lymphocytes. A particular characteristic of memory cells is "persistibility" conferred by enhanced resistance to down-regulation or elimination, whether by natural immunoregulation or by therapeutic immunosuppressive drugs. Self-reactive lymphocyte populations were formerly called *forbidden clones* but this term fell into disuse after findings of low expression of self-reactive T and B lymphocytes in healthy individuals. This presumably reflects "leakiness" of central tolerance and is compensated for by competent activity of T_{reg} cells that provide integrity of peripheral tolerance. Although it is clearly desirable to investigate concurrently the nature of failed self-tolerance in both T and B lymphocyte populations, in practice B lymphocytes receive more attention simply because these and their autoantibodies are more amenable to study.

Autoimmune Hepatitis

Autoimmune hepatitis was one of the earliest recognized autoimmune diseases. This developed from descriptions during 1948–1956 of a disease affecting mostly young women and presenting as a nonresolving "active" hepatitis with greatly increased levels of serum gamma globulin (Mackay et al. 2006; Manns and Vogel 2006), a feature still unexplained. Activity in the context of this disease denotes progressive damage to hepatocytes, reflected biochemically by increased levels in serum of aminotransferase enzymes released from leaky liver cells, histologically by lymphoid infiltration and hepatocellular damage at the interface of portal tracts and hepatic lobules, and serologically by extreme hypergammaglobulinaemia and various autoantibodies in blood. This new disease first became known in the 1950s as "chronic active hepatitis" (CAH). The autoantibody initially recognized was reactive with nuclear antigens, as seen in the prototypic autoimmune disease systemic lupus erythematosus (SLE), and hence leading to the appellation of "lupoid hepatitis," later revised by Mackay, Weiden and Hasker (1965) to AIH.

Other autoantigenic reactants were soon identified including in 1965 gastric smooth muscle antigen, in 1973 an antigen enriched in liver and kidney microsomes (LKM) (Rizetto, Swana and Doniach 1973), and subsequently various others (Mackay et al. 2006). Initially AIH was a serious affliction because of progressive deterioration of liver function and development of macronodular cirrhosis over 3–4 years: survival was very limited. With the introduction of long-term immunosuppressive regimens in the 1960s, there was dramatic clinical, histological, and serological improvement, and survival among some populations studied approached that of an age-matched normal population (Mackay et al. 2006).

A study in the 1980s indicated that there were two mutually exclusive serological subtypes of AIH: type 1 (AIH-1) was distinguished by antinuclear antibody (ANA) and smooth muscle antibody (SMA) (see below), and type 2 (AIH-2) by antibody to LKM (Homberg et al. 1987). This serological segregation is curious because otherwise the two types are similar, clinically and histopathologically.

AIH-2 is the less frequent of the two types, the incidence ratio being about 1:10 but, paradoxically, research advances have predominantly occurred in AIH-2. These have included molecular characterization and B-cell epitope analysis of LKM and other microsomal autoantigens as isoforms of the large family of cytochrome P450 oxidase enzymes (Manns and Vogel 2006), analysis of T-cell epitopes on the LKM antigen (Ma et al. 2006), and the development of a valid animal model of AIH-2 (Lapierre et al. 2004). Equivalent progress has not been made with AIH-1.

Smooth Muscle Antibody

A major impetus in the 1960s for the concept of autoimmunity as a cause of disease came from the use of indirect immunofluorescence (IIF) for detection of autoantibodies in serum, using frozen sections of blocks of multiple tissues, of which one was rodent stomach. Sera from cases of AIH-1, already known to react by IIF with cell nuclei, were found by Johnson, Holborow and Glynn (1965), to react also with smooth muscle constituents of rodent gastric mucosa, the muscularis and muscularis mucosae. The serum reactant was further defined, and designated as *smooth muscle antibody* (SMA) by Whittingham, Mackay and Irwin (1966). This same IIF procedure is retained today for discriminating AIH-1 from other diseases (Vergani et al. 2004).

SMA-positive sera were then found to react with tissues other than gastric mucosa, notably renal glomerular mesangium and the smooth muscle of blood vessels (Whittingham, Mackay and Irwin 1966), as shown in Fig. 1. Notably, some SMA-positive sera reacted only with the smooth muscle layer of the arterial wall (SMAv), and others additionally with antigens in renal glomeruli (SMAg), liver canaliculi in a "polygonal" pattern, and brush border of renal

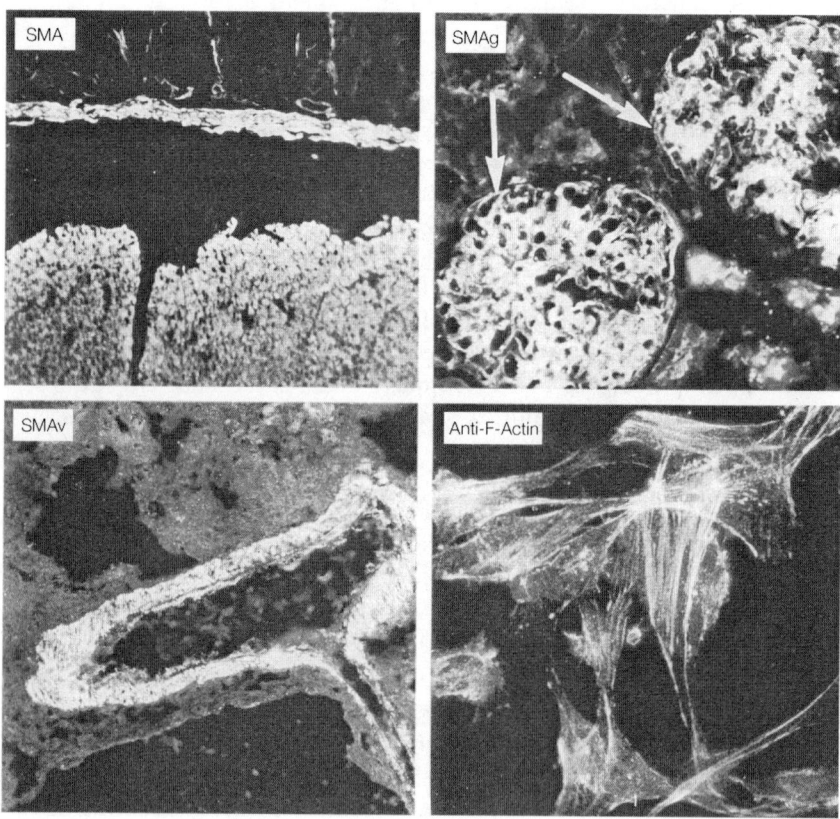

FIG. 1. Reactivity by IIF of SMA of antiactin specificity. *Top left*: reaction with smooth muscle of mouse gastric mucosa; *top right*: reactivity with actin in glomerular mesangium (SMAg), considered indicative of antibody to filamentous actin; *bottom left*: reaction with media of blood vessel, indicative of antibody to F-actin and/or other cellular filaments; *bottom right*: reactivity with "actin cables," indicative of antibody to F-actin (from Whittingham and Mackay (1996). Smooth muscle autoantibodies. In Autoantibodies. J. B. Peter and Y. Shoenfeld (Editors). pp. 767–773 with permission from Elsevier)

tubular cells (SMAt), indicating that SMA could be directed to multiple antigenic structures (Bottazzo et al. 1976) or to an antigen shared in common in
the reactive tissues. Moreover, SMAg/t antibodies with a broad tissue reactivity aligned closely with AIH, particularly in its active stages, whereas SMAv
with reactivity restricted to vascular smooth muscle occurred in AIH and also
in various infectious/inflammatory disorders. Even decades later, this difference
is claimed to be insufficiently appreciated (Silvestrini and Benson 2001).

Actin as a Major Reactant for SMA

In 1973, actin was proposed to be a major reactant for SMA (Gabbiani, Ryan
and Lamelin 1973), substantiated by various procedures including neutral-
ization of IIF reactivity by serum absorption with actin. The use of vinblas-
tine-arrested cultured fibroblasts for IIF tests with SMAt-positive sera from
cases of AIH-1 showed decoration of the parallel actin-containing microfilaments (stress fibers or "actin cables") spanning the long axis of the cell
(Fig. 1) whereas, conversely, SMAv-positive sera from individuals with diseases other than AIH-1 reacted with nonactin cellular filaments, particularly
intermediate filaments such as vimentin and desmin (Toh 1979; Kurki et al.
1980; Kurki and Virtanen 1984; Kurki 1994). However, as microfilaments
also contain contractile proteins other than actin such as myosin, troponin,
and tropomyosin as well as various actin-binding proteins, the decoration of
these "actin cables" by IIF cannot be considered diagnostic of the presence
of anti-F-actin antibody. It is also possible that SMAg/t-positive sera may
contain populations of antibodies that react not only with F-actin but also
with these various other microfilament-associated proteins.

The dosing of rats with phalloidin, that stabilizes polymeric actin,
enhanced the utility of rodent liver as a source of sections applicable to diagnostic immunofluorescence (Fusconi et al. 1990). These studies further indicated that antigenicity likely resided in polymeric filamentous (F)-actin
rather than in monomeric globular (G)-actin. Further evidence for this was
provided by a study of sera from 94 cases of alcoholic liver disease, tested by
ELISA on monomeric G-actin (Cunningham et al. 1985). The positive
responses obtained, mostly in alcoholic cirrhosis, were limited to G-actin; in
fact, only one of the 94 cases reacted (weakly) with F-actin as judged by IIF
patterns. On the other hand, cases of AIH-1 reacted with F- and G-actin.

Technical refinements and new assay systems for anti-F-actin have been
introduced over the past two to three decades. The presence of a heat-labile
actin-depolymerizing agent in human sera, first detected in the 1970s, was
found to interfere with readouts based on IIF, and could be overcome by
heating sera to 60°C before the assay (Cancado et al. 2001). However, others
claim that this does not improve assay results (Silvestrini and Benson 2001).
Given that IIF showing SMAg/t reactivity in tissue sections coupled with decoration of "actin cables" in cultured cells provides only presumptive but not
definitive evidence of the presence of antibody to F-actin, there is a pressing

need for a confirmatory assay such as a Western blot, line blot, or ELISA for the presence of anti-F-actin antibody.

Diagnostic assay kits based on an ELISA format with highly purified F-actin have been introduced and in their modern format are performing as well as testing by IIF (Granito et al 2006). Data for the actual frequency of SMA, and anti-F-actin, in AIH-1 and other disease sera are influenced by variability in both case selection and assay formats used in different laboratories. For example, commercially available sections for IIF are treated by mild fixation, to improve shelf life, but such fixation is prejudicial to optimal readouts from fluorescence microscopy.

Whilst data from published studies vary on sensitivity/specificity of anti-F-actin for the diagnosis of AIH-1, the earliest report in the 1960s (Whittingham et al. 1966), and the latest (Granito et al. 2006), both indicate a sensitivity ~80% and a specificity of >90%. However, there is seen to be a need for a rigorous multicenter study based on adequate numbers of sera from well-diagnosed cases of AIH-1 in relapse, a quantitative assay readout in units based on a potent well-calibrated antiserum, and programs for inter-laboratory serum exchange workshops to maintain quality control (Vergani et al. 2004). Meanwhile detection of presumptive anti-F-actin allows some important distinctions between AIH-1 and other diseases to be made, including AIH-2, SLE, and liver diseases in which a positive IIF test for SMA may depend on reactivity with G-actin or filaments other than F-actin. However this area is still controversial.

To sum up the serologic data, a presumptive anti-F-actin type of SMA is characteristic of AIH, but is lacking in some 20% of cases (Czaja et al. 1996). Possible reasons for this include technical factors, or that free antibody is adsorbed *in vivo* by actin released from degraded cells; however, studies are lacking on whether F-actin is demonstrable in serum in cases of AIH-1, and whether anti-F-actin can be eluted from hepatocytes from liver biopsies. Also, presumptive anti-F-actin is reported, albeit at low-titer and frequency, in some diseases other than AIH-1, notably chronic hepatitis C (Chazouillères et al. 1996) and multisystem autoimmune diseases (Senécal et al. 1985; Chretien-Leprince et al. 2005). SMA-positive sera are likely to contain populations of antibodies that can react with actin and perhaps various actin-binding proteins, or with cytoskeletal filaments other than actin. There is reported a higher frequency of presumptive anti-F-actin in patients bearing HLA alleles B8, DR3 (Czaja et al. 1996), suggesting that epitopes of actin are more readily presentable to T cells by these "severe disease" accompaniments of AIH-1.

The specifically anti-F-actin population of autoantibodies reacts with polymeric F-actin derived from any of the four actin genes, i.e., with muscle actins (α-cardiac muscle actin, α-striated muscle actin, α-smooth muscle and vascular actin, and γ-smooth muscle enteric actin), and also with the products of two genes for the two nonmuscle actins, β- and γ-cytoplasmic actin. This is consistent with each of the actin isoforms having close (>90%) homology, and therefore expressing an epitope in common for anti-F-actin autoantibody, and likely accounts for the broad IIF tissue reactivity of SMAg/t antibody. Each of the actin isoforms is in a state of continual assembly–disassembly from monomer to polymer under the influence of numerous actin-binding proteins (dos Remedios et al. 2003) (and see other chapters in this book). The epitope of interest for an autoantibody to actin is expressed by assembled polymeric F-actin rather than nonassembled monomeric G-actin, as discussed above.

Immunogenic Stimulus for Anticytoskeletal Antibodies

An immunogenic stimulus for production of anticytoskeletal antibodies is not identifiable, which is often the case for autoantibodies in general. We preface this section by mentioning that normal human sera contain "natural" autoantibodies to actin detectable at low-titer on smooth muscle substrates, and likely reactive with G-actin. Coming to disease-associated anti-F-actin, there are two possibilities to consider, and each requires the existence of a degree of genetic tilting toward autoimmunity.

First, antiself-reactivity and autoimmunity are due to exposure of cells of the immune system to autoantigen derived from "spillage": this can occur in the course of any type of tissue damage, whether expressed as apoptosis or necrosis, and even in the setting of apoptosis that occurs during tissue remodeling (O'Brien et al. 2002). In this context, there is an abundance of submembranous actin in hepatocytes as shown in Fig. 2, so providing for ready exposure of spilled antigen (actin) to the various types of phagocytic APCs that are present in liver sinusoids. And, as mentioned above, certain autoantigens may be capable of activating APCs by engaging TLRs, thereby

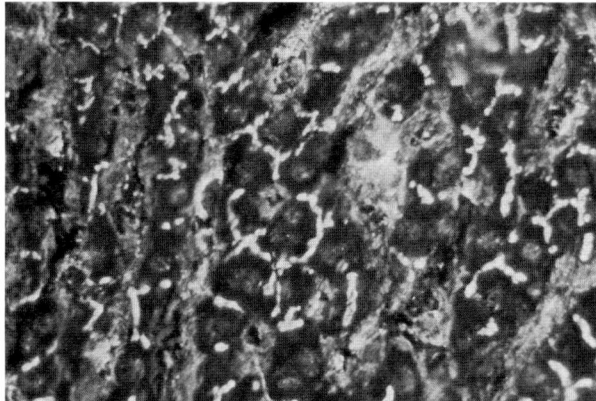

FIG. 2. Immunofluorescence photomicrograph of antiactin serum on frozen section of mouse liver, showing the typical "polygonal" pattern given by reactivity of serum with submembranous actin in hepatocytes (*See Color Plates*)

potentiating the autoimmunogenic stimulus. But there still remains the possibility that the anticytoskeletal (including antiactin) response is simply a "housekeeping" activity directed to disposal as a "transporteur" of cellular debris, as proposed by Grabar (1975). Also, of note, levels of antiactin in AIH-1 fluctuate with phases of activity of disease, as judged biochemically and histologically, such that antiactin is seen by some authors as a consequence of this.

The second stimulus for autoimmunity, conjectured in many settings but proven in few, is that an environmental antigen, usually microbial, is capable of presenting an epitope that sufficiently resembles a host protein for a crossreactive response to be initiated. Whilst this idea of epitope mimicry is attractive to many, three conditions should be met to validate it. A credible epitope mimic should closely but not completely resemble the culprit autoantigen molecule; reliable evidence should exist for natural environmental exposure of individuals to this mimicking molecule; and there should be experimental evidence that animals immunologically exposed to the epitope mimic develop autoreactivity involving T and B cells and disease. This latter requirement is seldom if ever fulfilled. A microbial source of a mimicking antigen for actin would seem unlikely, since actin is represented only in eukaryotic cells. However, bacteria express molecules with close structural homologies, albeit not closely homologous sequences, with eukaryotic actin, MreB (Doolittle and York 2002), and parM (van den Ent et al. 2002).

Epitopes of Actin

Clinical immunologists are interested in the identification and localization (mapping) of epitopes for antibodies on autoantigenic molecules, at least for providing some insight into the mimicry hypothesis (see above). This has been approached in several ways. These include (a) expression of protein from truncated cDNAs that encode segments of recombinant autoantigenic molecules against which can be tested reactive T cells or autoantibodies, (b) synthesis of sequential peptides along the length of the autoantigenic molecule, and (c) biopanning of random phage-displayed peptide libraries with autoantigenic sera, which can inform on linear epitopes or mimotopes of conformational epitopes (Rowley, O'Connor and Wijewickrama, 2004).

In one study, Western blotting was used for testing actin autoantibodies against monomeric G-actin and proteolytic fragments thereof, and against actin-binding proteins, including myosin, tropomyosin, troponin, α-actinin, and desmin (Zamanou et al. 2003). Results therein cited the presence of "natural" and "disease-associated" antibodies to actin, the reactivity of human antisera with a C-terminal fragment of G-actin, and the absence of reactivity of antisera to the various actin-binding proteins tested. Although a

C-terminal antibody epitope of G-actin was specified, this study did not test for reactivity against polymeric F-actin which is the reactant demonstrable by IIF and likely to be engaged by antisera *in vivo*. This F-actin epitope is presumably conformational, and generated by the quaternary assembly of putative trimeric actin polymers and, consistent with this, reactivity of autoimmune antisera with polymeric F-actin is not readily demonstrable by Western blot.

Characteristics and Functional Effects of F-actin Autoantibodies

It was ascertained that the autoantibody to actin (monomeric G-actin was studied) is an IgG, of the IgG1 subclass (Zamanou et al. 2003). Numerous antisera to actin have been raised, including those developed commercially by immunization of animals with actin-containing tissues, rabbit skeletal muscle, human platelets, and others, and the ensuing antibodies are specific for G- and F-actin; whether the spontaneous human antiactin antisera and deliberately raised antisera in animals are crossinhibitory by IIF on the conventional substrates seems worthy of further investigation. Also, an obvious area of interest would be whether T cells from individuals with AIH-1 and anti-F-actin antibodies respond to F-actin. We are not aware of such studies, using either intact G- or F-actin, or fragments thereof, so that data on T-cell epitope sites are lacking. Possible epitope sites for B cells are mentioned below. The location of actin-reactive T and B lymphocytes in AIH-1 is likely to be predominantly in the liver, although this seems not to have been ascertained, e.g., by histochemistry on frozen sections of liver derived from a patient with AIH-1.

Autoantibodies to F-actin are considered nonpathogenic for several reasons. Some 20–30% of individuals with AIH-1 lack detectable anti-F-actin antibody (see above), and there are a few individuals ascertained in population or clinical surveys that have high levels in serum of antiactin yet are apparently healthy. Also, levels of anti-F-actin fall during disease remission yet hepatocyte damage continues, although at a much retarded rate.

There are no clinical reports of pregnant women with anti-F-actin transmitting liver disease or any other disease to the fetus, in accord with our own experience. On the other hand, actin is abundant submembranously in hepatocytes, such that any loss of cell-membrane integrity for whatever reason could allow ready access of serum antibody to intracellular actin, with adverse effects on hepatocytes.

Functional Effects of Autoantibodies to F-actin

There are many autoantibodies that have demonstrable functional effects on the autoantigen with which they react *in vitro*. This is most convincingly illustrated by effects *in vivo* of autoantibodies that cross the placenta of pregnant women and adversely affect the fetus. Examples include stimulating antibodies to the thyrotropin receptor (thyrotoxicosis), and blocking antibodies to the acetyl choline receptor (myasthenia gravis), and to cardiac conducting tissue (congenital heart block). There are other autoantibodies, mostly to enzyme autoantigens, that have a potent capacity *in vitro* to inhibit the catalytic effects of their enzyme reactant, including antibodies to pyruvate dehydrogenase complex E2 subunit, cytochrome P450 2D6, thyroid peroxidase, and gastric H+/K+-ATPase, although adverse effects on the fetus upon placental transfer are not observed.

Antiactin antibodies also fall into the category of functionally active antibodies, as recently described by Martinez-Neira (manuscript submitted, 2007). In these studies the effects were measured of serum immunoglobulin G and Fab fragments of antiactin-positive sera on chemical and mechanical activities using an actin–myosin motor system. Sera of two patients identified by IIF as SMA-positive with an antiactin microfilament pattern of tissue staining, and ten healthy subjects, were investigated. The two assays used were (a) inhibitory effects of antisera on actin-activated myosin ATPase activity, and (b) motility measurements of labeled actin filaments moving on a myosin-coated surface that is equivalent to muscle contraction *in vitro*.

It was found first that Fab fragments derived from the two SMA and anti-F-actin-positive sera strongly inhibited the motor activity of myosin as judged by a dose-dependent decrease of myosin ATPase release, and second that Fab fragments of the test sera totally abrogated the myosin-dependent motility of F-actin filaments in an actin–myosin motility chamber. These effects were due to binding and clumping of actin filaments, as well as possible binding to and inactivation of myosin. The set-up and results of the actin-activated myosin ATPase activity are shown in Fig. 3 and the legend thereto.

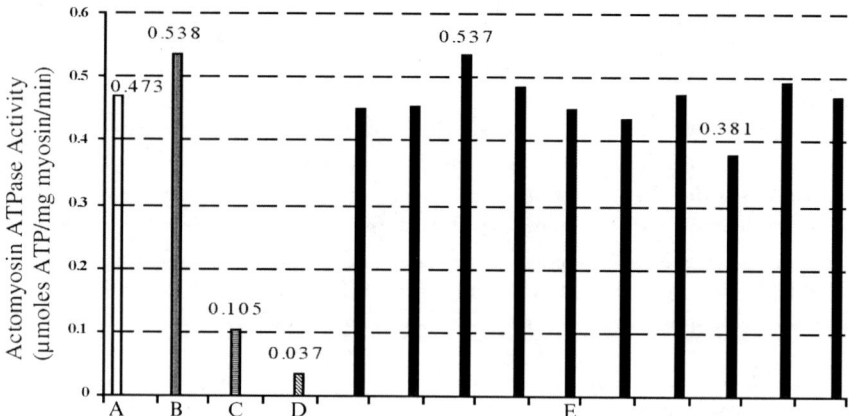

FIG. 3. Mean actomyosin ATPase activity after exposure to Fab from various plasma samples. All assays were run in triplicates at 1:1 actin to Fab molar rations. (A) blank, (B) BSA, (C) and (D) Fabs from plasma of two patients with antimicrofilament activity, and (E) Fabs from plasma of 10 normal subjects (Martinez-Neira et al. unpublished data)

A motility assay was performed using myosin immobilized on a nitrocellu-lose-coated coverslip in a motility chamber assembly as described by Sellers et al. (1993) and Trybus (2000), and procedures for inhibition by antisera of actin motility were as described by Martinez-Neira et al. (manuscript submitted, 2007). Notably, exposure of F-actin/myosin to Fab before motility was initiated did not affect actin motility, whereas exposure to Fabs after motility had been started caused clumping of actin fibrils (Fig. 4). There was a slowing of filament velocity in proportion to the concentrations of Fabs in the chamber, until motility was completely halted (Fig. 5). Thus, when actomyosin

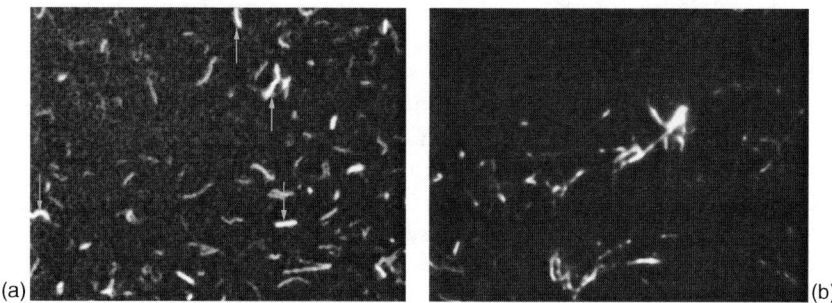

(a)   (b)

FIG. 4. *In vitro* motility images. (**a**) Control image showing motile actin filaments (indicated by *arrows*); (**b**) after washing with IgG in M buffer, actin filaments are immobilized and clumped (Martinez-Neira et al. unpublished data) (*See Color Plates*)

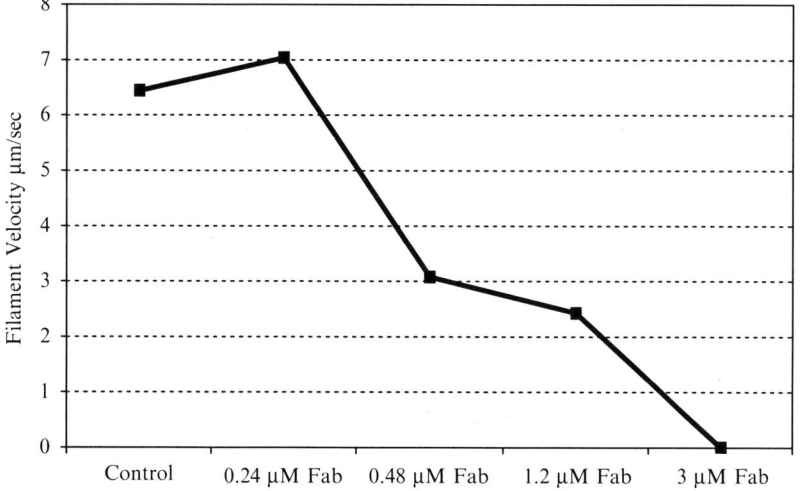

FIG. 5. Decreased actin filament velocity upon exposure to increasing concentrations of Fab. After motility was initiated, the motility chamber was washed, with increasing concentrations of Fab from plasma with antimicrofilament activity. There was a progressive slowing of filament velocity to a complete halt at a concentration of Fab of 3 μM (Martinez-Neira et al. unpublished data)

is exposed to Fab fragments of IgG from patients seropositive to F-actin, there are clear dose-dependent functional effects that include inhibition of myosin ATPase release, and clumping of actin filaments with retarded motility. Further data are needed, however, to ascertain whether these functional assays will discriminate between SMA reactivity that is actin-specific or actin nonspecific, and whether the functional effects are mediated by reactivity with F-actin, myosin, or both.

Conclusions

The cytoskeletal proteins are a major target for autoimmune reactions that are expressed clinically as autoantibodies to actin and/or intermediate filaments. Production of autoantibodies to F-actin in its polymeric, i.e., filamentous (F) assembled state characterizes AIH, whereas autoantibodies to actin in its monomeric, i.e., globular state are rather nonspecific in terms of disease associations. Diagnostic laboratories currently rely on IIF for detection of antibodies to presumptive F-actin, but various technical obstacles have precluded collection of good data sets. Also, there is a need for a definitive and reliable confirmatory assay for presumptive F-actin demonstrated by IIF. Toward this end, we describe preliminary data indicating that human autoimmune antisera (Fab fragments) to F-actin have functional effects *in vitro* on actin motility in an actin–myosin motility assay. Further studies are needed to assess whether this system will cleanly select out specific anti-F-actin reactivity. Other pressing questions include the initial stimulus for provocation of anti-F-actin antibody production, the nature of epitopes engaged by T and B lymphocytes including overlap of these with other functional sites on the molecule, and circumstances in which human anti-F-actin may have pathogenic effects *in vivo*.

References

Bottazzo, G. F., Florin-Christensen, A., Fairfax, A., Swana, G., Doniach, D. and Groeschel-Stewart, U. 1976. Classification of smooth muscle autoantibodies detected by immunofluorescence. J. Clin. Pathol. 29, 403–410.

Cancado, E. L., Abrantes-Lemos, C. P., Vilas-Boas, L. S., Novo, N. F., Carrilho, F. J. and Laudanna, A. A. 2001. Thermolabile and calcium-dependent serum factor interferes with polymerized actin, and impairs anti-actin antibody detection. J. Autoimmun. 17, 223–228.

Chazouillères, O., Johanet, C., Sefaty, L., Carbonnel, F., Smadja, M., Naudin, G., Dubel, L. and Poupon, R. 1996. Anti-actin autoantibodies in patients with chronic hepatitis C. J. Hepatol. 24, 513.

Chretien-Leprince, P., Ballot, E., Andre, C., Olsson, N. O., Fabien, N., Escande, A., Oksman, F., Dubuquoi, S., Jego, S., Goetz, J., Chevaller, A., Sanmarco, M., Humbel, R. L. and Johanet, C. 2005. Diagnostic value of anti-F-actin antibodies in a French multi-center study. Ann. NY Acad. Sci. 1050, 266–273.

Cunningham, A. L., Mackay, I. R., Frazer, I. H., Brown, C., Pederson, J. S., Toh, B. -H., Tait, B. D. and Clarke, F. M. 1985. Antibody to G-actin in different categories of alcoholic liver disease: Quantification by an ELISA and significance for alcoholic cirrhosis. Clin. Immunol. Immunopathol. 34, 158–164.

Czaja, A. J., Cassani, F., Cataleta, M., Valentini, P. and Bianchi, F. B. 1996. Frequency and significance of antibodies to actin in type 1 autoimmune hepatitis. Hepatology 24, 1068–1073.

Doolittle, R. F. and York, A. L. 2002. Bacterial actins? An evolutionary perspective. BioEssays 24, 293–296.

van den Ent, F., Moller-Jensen, J., Amos, L., Gerdes, K. and Lowe, J. 2002. F-actin-like filaments formed by plasmid segregation protein ParM. EMBO J. 21, 6935–6943.

Frenzel, C., Herkel, J., Lüth, S., Galle, P. R., Schramm, C. and Lohse, A. W. 2006. Evaluation of F-Actin ELISA for the diagonsis of autoimmune hepatitis. Am. J. Gastroenterol. 101, 2731–2736.

Fusconi, M., Cassani, D., Zauli, D., Lenzi, M., Ballarchini, G., Volta, U. and Bianchi, F. B. 1990. Antiactin antibodies: A new test for an old problem. J. Immunol. Meth. 130, 1–8.

Gabbiani, G., Ryan, G. B. and Lamelin, J. P. 1973. Human smooth muscle antibody, its identification as anti-actin antibody, and a study of its binding to non-muscular cells. Am. J. Pathol. 72, 473–484.

Grabar, P. 1975. Hypothesis. Autoantibodies and immunological theories: An analytic review. Clin. Immunol. Immunopathol. 4, 453–466.

Granito, A., Moratori, L., Pappas, G., Guidi, M., Cassani, F., Volta, G., Ferri, A., Lenzi, M. and Bianchi, F. B. 2006. Antibodies to filamentous actin (F-actin) in type 1 autoimmune hepatitis. J. Clin. Pathol. 59, 280–284.

Harrington, L. E., Mangan, P. R. and Weaver, C. T. 2006. Expanding the effector CD4 T-cell repertoire: The Th17 lineage. Curr. Opin. Immunol. 18, 349–356.

Homberg, J.-C., Abuaf, N., Bernard, O., Alvarez, F., Khalil, S. H., Poupon, R., Darnis, F., Levy, F.-G., Grippon, P., Opolon, P., Bernuau, J., Benhamou, J.-P. and Alagille, D. 1987. Chronic active hepatitis with anti-liver/kidney microsome antibody Type 1: A second type of autoimmune hepatitis. Hepatology 7, 1333–1339.

Johnson, G. D., Holborow, E. J. and Glynn, L. E. 1965. Antibody to smooth muscle in patients with liver disease. Lancet 2, 878–879.

Kronenberg, M. and Rodensky, A. 2005. Regulation of immunity by self-reactive T cells. Nature 435, 598–604.

Kurki, P. 1994. Cytoskeleton antibodies. In Autoimmune Hepatitis. M. Nishioka, G. Toda and M. Zeniya (Editors). Elsevier Science, Netherlands. pp. 185–198.

Kurki, P., Mietinnen, A., Linder, E., Pikkaramen, P., Vuoristo, M. and Salaspuro, M. 1980. Different types of smooth muscle antibodies in chronic active hepatitis and primary biliary cirrhosis: Their diagnostic and prognostic significance. Gut 21, 878–884.

Kurki, P. and Virtanen, C. 1984. The detection of human antibodies against cytoskeletal components. J. Immunol. Meth. 67, 209–223.

Kyewski, B. and Klein, L. 2006. A central role for central tolerance. Annu. Rev. Immunol. 24, 571–606.

Lang, K. S., Rechner, M., Junt, T., Navarini, A. A., Harris, N. L., Freigang, S., Odermatt, B., Conrad, C., Ittner, L. M., Bauer, S., Luther, S. A., Uematsu, S., Akira, S., Hengartner, H. and Zinkernagel, R. M. 2005. Toll-like receptor engagement converts T-cell autoreactivity into overt autoimmune disease. Nat. Med. 11, 138–145.

Lapierre, P., Djilali-saiah, I., Votozzi, S. and Alvarez, F. 2004. A murine model of type 2 autoimmune hepatitis: Xenoimmunization with human antigens. Hepatology 39, 1066–1074.

Ma, Y., Bogdanos, P., Hussein, M. J., Underhill, J., Bansal, S., Longhi, M. S., Cheeseman, P., Miele-Vergani, G. and Vergani, D. 2006. Polyclonal T-cell responses to cytochrome P450 IID2 are associated with disease activity in autoimmune hepatitis type 2. Gastroenterology 130, 868–882.

Mackay, I. R., Czaja, A. J., McFarlane, I. G. and Manns, M. P. 2006. Chronic hepatitis. In The Autoimmune Diseases. N. R. Rose and I. R. Mackay (Editors). Elsevier, London. pp 729–747.

Mackay, I. R., Weiden, S. and Hasker, J. 1965. Autoimmune hepatitis. Ann. NY Acad. Sci. 24, 767–780.

Manns, M. P. and Vogel, A. 2006. Autoimmune hepatitis, from mechanisms to therapy. Hepatology 43 (Supplement), S132–S144.

O'Brien, B. A., Huang, Y., Geng, X., Dutz, J. P. and Finegood, D. T. 2002. Phagocytosis of apoptotic cells by macrophages from NOD mice is reduced. Diabetes 51, 2481–2488.

dos Remedios, C. G., Chhabra, D., Kekic, M., Dedova, M. V., Tsubakihara, M., Berry, D. A. and Nosworthy, N. J. 2003. Actin binding proteins: Regulation of cytoskeletal microfilaments. Physiol. Rev. 83, 433–472.

Rizetto, M., Swana, G. and Doniach, D. 1973. Microsomal antibodies in active chronic hepatitis and other disorders. Clin. Exp. Immunol. 15, 331–344.

Rose, N. R. and Mackay, I. R. 2006. Prospectus: The road to autoimmune disease. In The Autoimmune Diseases. N. R. Rose and I. R. Mackay (Editors). Elsevier, London. pp. xix–xxv.

Rowley, M. J., O'Connor, K. and Wijewickrama, L. 2004. Phage display for eptiope determination: A paradigm for identifying receptor–ligand interactions. Biotechnol. Annu. Rev. 10, 151–188.

Sellers, J. R., Cuda, G., Wang, F. and Homsher, E. 1993. Myosin-specific adaptations of the motility assay. Meth. Cell Biol. 39, 23–49.

Senécal, J. -L., Oliver, J. M. and Rothfield, N. 1985. Anti-cytoskeletal antibodies in the connective tissue diseases. Arthritis Rheum. 28, 889–898.

Silvestrini, R. A. and Benson, E. M. 2001. Whither smooth muscle antibodies in the new millennium. J. Clin. Pathol. 54, 677–678.

Toh, B.-H. 1979. Smooth muscle autoantibodies and autoantigens. Clin. Exp. Immunol. 38, 621–628.

Trybus, K. M. 2000. Biochemical studies of myosin. Methods 22, 327–335.

Vergani, D., Alvarez, F., Bianchi, F. B., Cancado, E. L. R., Mackay, I. R., Manns, M. P., Nishioka, M. and Penner, E. 2004. Autoimmune serology: A consensus statement from the committee for autoimmune serology of the International Autoimmune Hepatitis Group. J. Hepatol. 41, 677–683.

Whittingham, S., Irwin, J., Mackay, I. R. and Smalley, M. 1966. Smooth muscle antibody in "autoimmune" hepatitis. Gastroenterology 51, 499–505.

Whittingham, S. and Mackay, I. R. 1996. Smooth muscle autoantibodies. In Autoantibodies. J. B. Peter and Y. Shoenfeld (Editors). Elsevier, Amsterdam. pp. 767–773.

Whittingham, S., Mackay, I. R. and Irwin, J. 1966. Autoimmune hepatitis. Immuno fluorescence reactions with cytoplasm of smooth muscle and renal glomerular cells. Lancet 1, 1333–1335.

Zamanou, A., Samiotaki, M., Panayotou, G., Margaritis, L. and Lymberi, P. 2003. Fine specificity and subclasses of IgG anti-actin autoantibodies differ in health and disease. J. Autoimmun. 20, 333–344.

5
Overview: Actin-Binding Protein Function and Its Relation to Disease Pathology

Mira Krendel and Enrique M. De La Cruz

Introduction

The actin cytoskeleton generates force and movement responsible for many critical and fundamental cellular processes (see Chap. 1). Force generation and motility are produced by two distinct mechanisms (1) the self-assembly of actin monomers into filaments, which can exert forces against boundaries and particles such as cell membranes, vesicles, organelles, or pathogenic bacteria, and generate movement of these boundaries and (2) through the activity of contractile motor proteins of the myosin family, which generate force and motility along actin filaments. Both mechanisms utilize chemical energy in the form of ATP although hydrolysis of ATP by actin does not contribute to the force generated by actomyosin. Each monomer incorporated into a filament and each myosin mechanical "step" consumes one ATP molecule, generating ADP and P_i as the hydrolysis products.

The actin cytoskeleton also helps form stable structures, often linked to membranes and the extracellular matrix (ECM) through membrane-associated proteins, which give the cell mechanical strength and polarity. Although these structures are dynamic and constantly undergoing remodeling through the addition and exchange of filament subunits, the principal functions of these structures are attributed to the large-scale, mechanical properties of actin filaments and filament networks.

The assembly and three-dimensional organization of actin filaments in cells is controlled through interactions with a large number of distinct, regulatory actin-binding proteins (ABPs) that are classified as monomer- or filament-binding based on which assembly state of actin they bind more strongly. Well-characterized disease phenotypes in humans and higher vertebrates that arise from disturbed ABP function generally interfere with one or more of the following cellular functions of the actin cytoskeleton (1) polymerization and assembly dynamics; (2) formation of filament assemblies and networks; (3) interaction with membranes and the ECM; and/or (4) actomyosin-based contractility.

In this overview, we turn our attention to these processes and general functions of the actin cytoskeleton, and how they contribute to disease development and pathology. We focus on specific, well-characterized examples, understanding that proteins with related functions in different organisms will likely contribute to analogous phenotypes, depending on the specific function of the actin cytoskeleton in those cells.

Modulation of Actin Assembly and Polymerization Dynamics

Because actin filaments elongate only from their ends, any process that changes the number of accessible filament ends will influence actin assembly driven processes. The Arp2/3 complex and members of the formin protein family nucleate actin filaments. Isolated Arp2/3 complex is a weak nucleator, but is activated by regulatory, WASp/SCAR family proteins that bind the complex and promote formation of the active, nucleating conformation. Rho family GTPase proteins activate WASp/SCAR family proteins in response to extracellular cues and signals. Formin activity is also regulated through Rho GTPases.

Capping proteins bind terminal subunits at filament ends and reduce the rates of filament polymerization and depolymerization. The affinity of capping proteins for filament ends depends on calcium, phospholipid binding (see chapter by dos Remedios), and various other signals.

Cofilin (see chapter by Maloney et al.) and members of the villin/gelsolin family (see chapter by Burtnick and Robinson) sever filaments and increase the number of filament ends. The actin monomer concentration determines whether these ends grow or shrink. Cofilin is regulated through phosphorylation by LIM kinase, which in turn, is regulated by Rho GTPases through PAK kinase activity. Actin filament binding and severing activity of villin/gelsolin family proteins is regulated by calcium and phospholipids.

Thus, both actin polymerization and depolymerization are regulated via complex signaling pathways (Fig. 1), which create numerous possibilities for finely tuned modulation of cytoskeletal organization and cell motility. Inactivation or upregulation of any of the components of these pathways can result in disruption of actin organization and cell motility. Such modulation of actin assembly is associated with a number of diverse disease phenotypes and pathologies.

Pathogenic Bacteria and Viruses

Many cell pathogens (see chapter by Bearer) do not disrupt ABP function, but rather utilize normal functions of proteins involved in regulating actin polymerization to promote infectivity. *Listeria*, *Shigella*, *Rickettsia*, *Burkholderia*, and *Mycobacterium* bacterial pathogens enter human cells and use growing actin filament "tails" to propel themselves through the

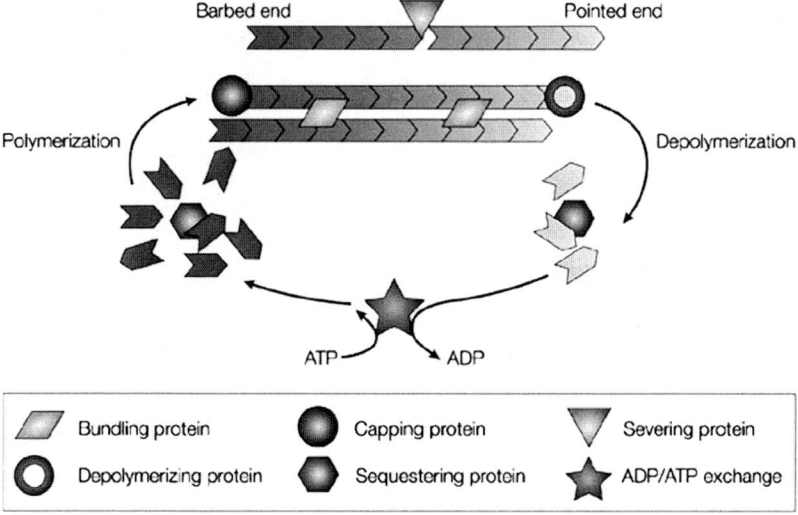

Barbed end   Pointed end

Polymerization   Depolymerization

ATP ———→ ADP

| Bundling protein | Capping protein | Severing protein |
| Depolymerizing protein | Sequestering protein | ADP/ATP exchange |

FIG. 1. Regulation of actin dynamics and organization by actin-binding proteins. Assembly of actin monomers into a filament is dependent on actin monomer concentration (the fast-growing, or barbed, end of the filament assembles at a lower critical concentration than the slow-growing, or pointed, end) and on ATP exchange and hydrolysis (ATP-bound monomers tend to associate with the filament and ADP-bound monomers readily dissociate from the filament). These properties of actin result in the process known as treadmilling, in which actin monomers are added to the barbed end of the filament and removed from the pointing end, allowing the barbed end of the filament to exert pushing force against the plasma membrane or other structures. Actin assembly and disassembly are regulated by actin-binding proteins that modulate effective concentration of actin monomers (monomer sequestering proteins) or exchange of monomer-bound nucleotides, as well as by proteins that prevent monomer addition to the barbed end (end capping proteins) and promote filament depolymerization (filament severing and depolymerizing proteins). Assembled filaments are organized into rigid structures by filament bundling and crosslinking proteins. Reproduced with permission from Macmillan Publishers Ltd (Revenu et al. 2004)

cytoplasm (Fig. 2) and to induce protrusions at the plasma membrane that allow invasion of adjacent cells (Stevens, Galyov and Stevens 2006). Enteropathogenic *E. coli* and enterohemorrhagic *E. coli* do not enter host cells, but instead induce formation of large membrane protrusions filled with actin filaments, called "pedestals." Pedestals move along the cell surface and may promote spreading of bacteria from one enterocyte to another. A similar strategy for movement along the cell surface using submembrane actin assemblies is employed by *Vaccinia* virus to move from one cell to another, whereas for intracellular movement this virus utilizes microtubule-dependent transport (Smith, Murphy and Law 2003).

Some bacterial pathogens, such as *Listeria* and *Rickettsia*, express on their surface proteins that recruit Arp2/3 complex and serve as nucleation-promoting

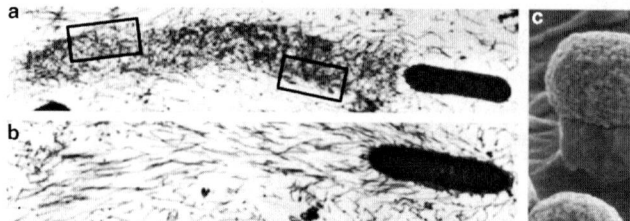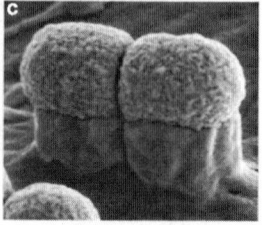

FIG. 2. Bacterial pathogens hijack cellular actin and actin-binding proteins. Upon cell entry, pathogenic bacteria *Listeria monocytogenes* (**a**) and *Rickettsia conorii* (**b**) induce formation of actin tails, shown here in the transmission electron micrographs. Actin tails induced by these intracellular pathogens represent either branched and crosslinked meshworks of short actin filaments (*Listeria*) or parallel bundles of long actin filaments (*Rickettsia*). Actin filaments are more densely packed near the surface of the bacterium than in the rest of the tail (*boxes*). (**c**) Attachment of enteropathogenic *Escherichia coli* to the surface of enterocytes promotes formation of actin-rich pedestals that can be observed by scanning electron microscopy. Formation of actin tails and pedestals relies on the functions of intracellular actin-binding proteins that regulate actin assembly and cross-linking. Reproduced with permission from Macmillan Publishers Ltd (Stevens, Galyov and Stevens 2004), Company of Biologists (Gouin et al. 1999) (**a, b**), and Wiley Interscience (Rottner et al. 2004) (**c**)

factors. Others, including *Shigella*, *Mycobacterium marinum*, *Vaccinia*, and *E. coli*, recruit endogenous, host cellular nucleation-promoting factors, primarily N-WASp. Both RickA protein of *Rickettsia* and ActA protein of *Listeria* contain protein domains similar to those in the WASp family proteins: proline-rich motifs, WH2 domains that bind actin monomers, and a CA domain that interacts with the Arp2/3 complex.

Actin tails of *Shigella*, *Listeria*, and *Mycobacterium* consist of short, orthogonally crosslinked actin filament networks, whereas actin tails of *Rickettsia* contain long parallel bundles of actin filaments, similar to the actin bundles of microvilli or filopodia. It is unclear what factors are responsible for this difference in actin tail organization. One of the explanations that has been proposed for the unusual morphology of *Rickettsia* actin tails is that VASP protein, which is known to promote formation of long actin filaments and decrease the density of actin branches, is located along the entire length of *Rickettsia* actin tail, and not just on the surface of the bacteria, as is the case for *Listeria*.

Defects in Cell Motility Induced by Changes in Expression or Activity of ABPs

Mutations in the gene encoding WASp that abrogate expression cause Wiskott–Aldrich syndrome, an X-linked syndrome characterized by immunodeficiency and reduced platelet number (thrombocytopenia) and size, which lead to frequent infections and bleeding (Ochs and Thrasher 2006). Immune system disorder in Wiskott–Aldrich syndrome is attributed to defects in actin polymerization, since macrophages, dendritic cells, and lymphocytes in WAS patients exhibit defects in cell adhesion and cell motility.

The precise pathogenesis of thrombocytopenia is unknown, but proposed mechanisms include defects in platelet formation and increased platelet fragility. Most missense mutations in WASp result in X-linked thrombocytopenia, a milder disorder characterized by decreased platelet number and size but with less pronounced immunodeficiency. Missense mutations in the Cdc42-binding region of WASp, which is necessary for regulation of WASp activity by the Rho family member Cdc42, result in X-linked neutropenia, which leads to frequent infections and may be caused by a defect in neutrophil formation (see chapter by Nunoi).

Enhanced cell motility is also frequently observed in tumor cells (see chapter by Van Troys et al.). This change is implicated in promoting invasion of surrounding tissues, intravasation, and tumor metastasis. Certain ABPs are upregulated in invasive tumor cells (Condeelis, Singer and Segall 2005), including cofilin and its upstream regulators (LIM kinase and PAK), capping protein, Arp2/3 subunits, and N-WASp. On the other hand, decreased expression of actin-severing protein gelsolin is observed in various cancers (Dosaka-Akita et al. 1998; Lee et al. 1999; Tanaka et al. 1995), and it has been suggested that gelsolin may act as a tumor suppressor.

Neurological Defects

Actin dynamics play an important role in neuronal migration, growth cone guidance, and the establishment of new connections in the developing brain and nervous system. Disruption of some actin-regulatory proteins leads to severe nervous system defects. Haploinsufficiency of WRP, a WAVE-binding protein that also acts as a GAP for Rac, and mutations in PAK3, which may regulate cofilin activity downstream of Rho GTPases, cause X-linked mental retardation (Allen et al. 1998; Endris et al. 2002). Deletion of LIMK1, a regulator of cofilin, has been implicated in Williams syndrome, a developmental disorder characterized by cognitive defects (Frangiskakis et al. 1996).

Disruption of Actin Filament Bundles and Networks

Crosslinking proteins organize actin filaments into complex, higher ordered assemblies and networks that provide mechanical strength and introduce a well-defined polarity to the cell. Most, but not all, actin cross-linking proteins contain characteristic actin-binding domains (ABDs) consisting of two calponin homology (CH) domains arranged in tandem (Gimona et al. 2002)

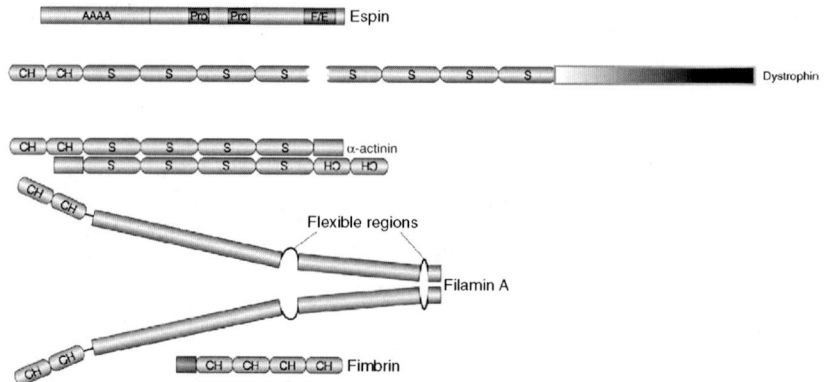

FIG. 3. Domain organization of actin filament cross-linkers. Most actin-bundling proteins contain actin-binding domains (ABDs) consisting of two calponin homology (CH) domains. The number of ABDs and the length of the linkers between them determine the type of actin filament assembly that can be formed by a particular protein. Fimbrin, which contains two adjacent ABDs within the same polypeptide chain, forms tightly packed actin bundles. α-actinin, in which ABDs are separated by a rod domain consisting of several spectrin repeats (S), and filamin, which also contains a long spacer between the ABDs, form loose actin networks. Espins and closely related forked protein contain unusual actin filament-binding modules (espin/forked homology domains, E/F) in their C-termini, which are necessary for their actin-bundling activity. The N-terminal regions of various espin isoforms are variable and may contain multiple ankyrin-like repeats (AAAA) and Pro-rich regions. Dystrophin and related muscle proteins that link actin filaments to the ECM include multiple spectrin repeats, an N-terminal ABD, and various protein interaction motifs in the C-terminus, which are responsible for binding to transmembrane protein complexes in the muscle sarcolemma. Modified with permission from Macmillan Publishers Ltd (Revenu et al. 2004)

(Fig. 3). The presence of two or more ABDs is necessary for them to function as an actin filament crosslinking protein. This multiplication of ABDs is typically achieved through oligomerization.

The length of the spacer separating the ABDs determines the type of actin filament structure formed by a particular crosslinking protein. Crosslinking proteins with short spacers, such as fimbrin, produce tight, parallel bundles of actin filaments. Other crosslinkers such as α-actinin and related proteins have long spacers separating ABDs and tend to form more loosely organized actin assemblies and meshworks (Puius, Mahoney and Almo 1998).

The type of actin assemblies formed by the crosslinking proteins also depends on their absolute concentration, and the relative concentrations of ABP and actin. At low [ABP]:[actin] ratios, orthogonal networks are favored. Parallel bundles form preferentially at high [ABP]:[actin] ratios (Wachsstock, Schwartz and Pollard 1993).

Mutations in human actin filament crosslinking proteins have been linked to various hereditary disorders (see below). In only one of these diseases, hereditary deafness is the pathology directly traced to malformation of actin filament arrays due to loss of the filament crosslinking protein. In other cases, the disease phenotype is thought to arise from (1) the loss of actin-membrane linkage through a multifunctional crosslinker (muscular dystrophy); (2) disruption of cell migration, which may be caused by the loss of actin meshworks and lack of coordination of various signaling inputs (filamin-related disorders); and (3) cell degeneration due to protein degradation (focal glomerulosclerosis).

Deafness

Specialized actin-rich, microvilli-like projections, called stereocilia, play a prominent role in sound and balance perception in the inner ear (Fig. 4). Stereocilia contain parallel bundles of actin filaments, which are connected to ion channels, cell adhesion receptors, and signaling complexes in the plasma membrane. Deflection of the bundle of stereocilia on the surface of a hair cell by the sound waves initiates opening of mechanically gated ion channels in the hair cell membrane, resulting in the generation of a nerve

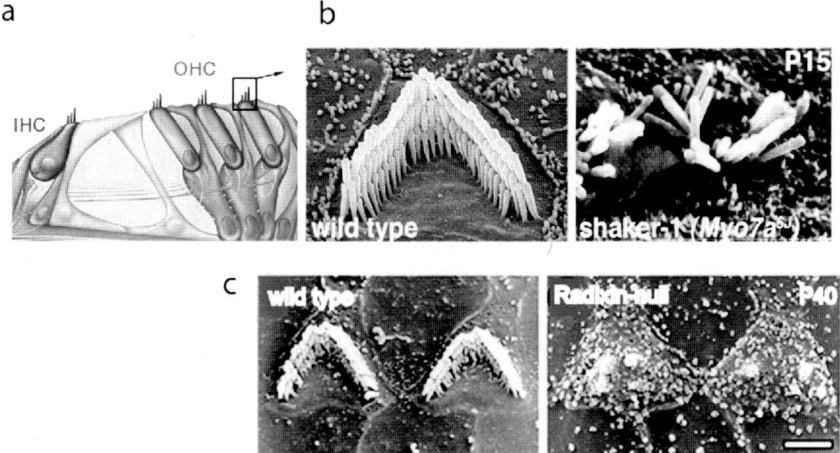

FIG. 4. Actin-binding proteins and structural organization of the sensory stereocilia. (a) Mammalian organ of Corti, the sensory element of the inner ear, contains inner and outer hair cells (IHC and OHC), which form numerous mechanosensory stereocilia on their surface. In OHCs of wild-type mice stereocilia form well-organized parallel arrays (b and c, visualized using scanning electron microscopy). In myosin7-mutant mice (Shaker-1, b) or mice lacking actin-bundling protein radixin (c) stereocilia are disorganized and reduced in number. Parts a and b are reproduced with permission from the Company of Biologists (El-Amraoui and Petit 2005), and part c with permission from the Rockefeller University Press (Kitajiri et al. 2004) (*See Color Plates*)

impulse. Mice lacking the bundling proteins radixin and espin exhibit hearing loss and vestibular dysfunction due to degeneration of stereocilia (Kitajiri et al. 2004; Zheng et al. 2000). Mutations in the actin-bundling proteins, espin, lead to inherited deafness in humans (Donaudy et al. 2006; Naz et al. 2004). Hereditary deafness has also been linked to mutations in various myosins expressed in the inner ear (discussed below).

Muscular Dystrophy

Dystrophin is a member of the α-actinin protein family whose expression is restricted to muscle and neuronal cells. In muscle cells dystrophin links actin filaments to the transmembrane cell adhesion complex, which interacts with laminin, a component of the ECM. Mutations in dystrophin result in increased susceptibility of the muscle membrane to tearing and abnormalities in intracellular signaling, which eventually lead to muscle degeneration. Complete loss of dystrophin expression causes Duchenne muscular dystrophy (Blake et al. 2002) and patients die from cardiac and respiratory muscle dysfunction. Truncated or reduced dystrophin expression levels lead to Becker's muscular dystrophy, a milder form of muscular dystrophy (Blake et al. 2002). Both forms are highly prevalent and affect mostly males since the dystrophin gene resides on the X-chromosome.

Utrophin, an ABP closely related to dystrophin, is also expressed in muscle, although unlike dystrophin it is found mostly at the neuromuscular and myotendonious junctions rather than the entire sarcolemma (Blake et al. 2002). One potential therapeutic intervention for muscular dystrophy currently being investigated is the artificial upregulation of utrophin expression to perform dystrophin's normal functions (Miura and Jasmin 2006).

Filamin-Related Diseases

FLNA, a gene encoding filamin A in man, is located on the X-chromosome. Mutations in FLNA result in X-linked disorders, whose features are quite diverse and may include skeletal dysplasia as well as malformations of the heart, brain, and genitourinary tract (Robertson 2005). Males that have no functional filamin A protein typically die before birth, while females heterozygous for functionally null mutations exhibit bilateral periventricular nodular heterotopia, a disorder or neuronal migration that leads to seizures. The precise pathogenesis of the FLNA-linked disorders is unclear, but is probably related to abnormal cell migration, particularly neuronal migration, during development, consistent with filamin playing an important role in cell motility.

Renal Disorders

Mutations in the gene encoding α-actinin 4 cause familial focal segmental glomerulosclerosis, a disease characterized by defects in renal glomeruli, the elements responsible for urinary filtration (Kaplan et al. 2000). The main effect of α-actinin-4 mutations associated with focal segmental glomerulosclerosis is destabilization of the protein leading to formation of protein aggregates and protein degradation, which, in turn, is thought to lead to degeneration of the podocytes, specialized epithelial cells of the glomerulus (Yao et al. 2004).

Disruption of Actin-Membrane and Actin-ECM Crosslinking

In addition to actin–actin crosslinkers, an important structural role is performed by crosslinking ABPs that link the actin cytoskeleton with the plasma membrane. Many of these proteins belong to the protein 4.1 superfamily and contain a FERM (4.1 protein, ezrin, radixin, moesin) domain as well as a spectrin/actin-binding domain. Band 4.1 is a component of the red blood cell membrane skeleton, and its name is based on its position on the 2D gel of erythrocyte proteins. Band 4.1 proteins bind integral membrane proteins, spectrin, and actin, and link the actin cytoskeleton to proteins of the plasma membrane. Mutations causing instability or truncations of band 4.1 family members or their binding partners compromise the shape and mechanical properties of red blood cells, resulting in anemia and spleen enlargement (Anstee et al. 1984; Daniels et al. 1986; Tchernia, Mohandas and Shohet 1981).

Kindlin-1, a focal adhesion protein, contains an unusual bipartite FERM domain, separated by the insertion of a PH domain. It links actin filaments to integrins and plays an important role in keratinocyte adhesion to the ECM (Herz et al. 2006). Mutations in kindlin are associated with Kindler syndrome, the first skin blistering disease to be linked to defects in ABPs (Jobard et al. 2003; Siegel et al. 2003) (Fig. 5).

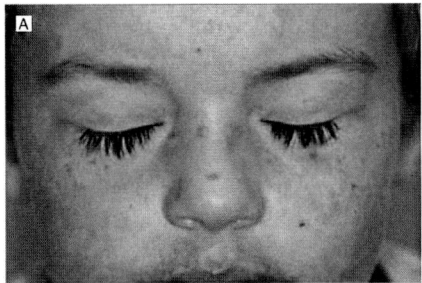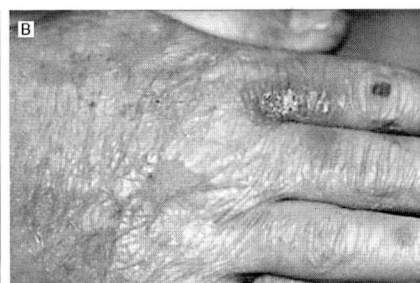

FIG. 5. Mutations in actin-ECM crosslinker kindlin-1 result in Kindler syndrome, a skin-blistering disease. A Kindler syndrome patient exhibits healing blisters on the nose and facial telangiectasia (red blotches created by dilated capillaries) (**a**) and fine wrinkling of the skin and thickened brownish patches of the outer layer of skin (keratoses) on the forearms and hands (**b**). Reproduced with permission from Burch et al. (2006). Copyright AMA (*See Color Plates*)

The ankyrin family of membrane-actin crosslinkers connect transmembrane ion channels, transporters, and cell adhesion receptors to spectrin (Bennett and Chen 2001). Ankyrin family members are characterized by numerous amino acid repeats, referred to as ankyrin repeats, which bind to membrane proteins. Mutations in human ankyrins lead to red blood cell abnormalities and anemia, as well as to cardiac arrythmia that result from mislocalization of ion channels (Eber et al. 1996; Mohler et al. 2003).

Disruption of Actomyosin-Based Contractility

A principal function of the actin cytoskeleton is to serve as tracks for intra-cellular movement of myosins – molecular motors that generate force and movement along actin filaments using the energy of ATP hydrolysis. Myosins are responsible for muscle contraction, regulation of cell motility, cell adhe-sion, and actin organization, as well as the transport of a variety of intracel-lular components.

All human myosins share a common molecular organization. Myosin heavy chains encode an N-terminal motor domain, responsible for ATP hydrolysis and motor activity, a neck domain, which binds myosin light chains, and a C-terminal tail domain, which may contain regions responsible for heavy chain dimerization, cargo binding, and intracellular localization. Calmodulin-like light chains bind to the neck domains via conserved IQ motifs and contribute to maximum force generation and regulation of myosin enzymatic activity. The number of IQ motifs in the neck region of a particular myosin heavy chain determines the number of bound light chains and influences the working stroke length and step size of a particular myosin. Variations in the enzymatic properties of the motor domain, the number of the IQ motifs, and structural motifs present in the tail domain result in a remarkable functional diversity of various myosins.

Human myosins are grouped into 12 classes based on the comparison of the structures of their head and tail domains (Berg, Powell and Cheney 2001). Mutations in genes encoding myosins and their regulators are asso-ciated with disruption of muscle contractility, cell migration, organelle transport, and actin organization in cells where these myosins are normally expressed.

Diseases Associated with Muscle Myosin and Regulatory Proteins

Muscle myosins belong to class II of the myosin family. Class II myosins con-tain coiled-coil motifs in their tails, which allows them to dimerize and form two-headed motors. Muscle myosin II self-associates further into thick fila-ments, such as those of the skeletal muscle sarcomere that are responsible for muscle contraction.

In humans, the various muscle fiber types contain different myosin II isoforms. Skeletal muscles express three myosin heavy chain isoforms, encoded by MYH1, MYH2, and MYH7 genes. MYH7 is also highly expressed in the heart and is known as cardiac MHC β. Smooth muscles express several splice variants of MHC, encoded by a single gene, MYH11. Two myosin heavy chain isoforms (cardiac myosin heavy chains α and β) are expressed in the heart. In humans, ventricles express mostly β-MHC (MYH7), while α-MHC, encoded by MYH6 gene, is highly expressed in the atria (Kurabayashi et al. 1988).

Mutations in genes encoding muscle MHCs and associated light chains cause myopathies, diseases characterized by disruption of structure and functions of muscle fibers, which leads to muscle weakness and progressive degeneration in the case of skeletal muscle myosins and to cardiac dysfunction (decreased contractility, arrhythmias, etc.) accompanied by ventricular dilation or enlargement of ventricular muscle mass (dilated or hypertrophic cardiomyopathy) in case of cardiac myosins (Chang and Potter 2005) (see chapter by Stefani et al.).

Many of the known MYH7 mutations (see chapter by Stefani et al.) are located in the actin- and ATP-binding regions of the head domain and the LC-binding region in the neck (Rayment et al. 1995). Some of these mutations correlate with a reduction in the rate of myosin translocation along actin filaments *in vitro* (Cuda et al. 1997). Other myosin mutations may lead to decreased stability of the protein. In addition to myosin mutations that cause myopathies, a mutation in the LC-binding neck region of the cardiac α-MHC, expressed in the atria, has also been linked to an inherited developmental disorder in which the interatrial septum separating the left and right atria fails to form properly (Ching et al. 2005). Mutations in a nonmuscle myosin VI have also been implicated in hypertrophic cardiomyopathy in humans; the role of myosin VI in the heart remains to be determined, and the mechanism leading to cardiomyopathy in patients with myosin VI mutations is unknown (Mohiddin et al. 2004).

In addition to mutations in myosins, myopathies and cardiomyopathies are also associated with mutations in proteins that regulate actomyosin interactions and muscle contractility. These include cardiac myosin binding protein-C, tropomyosin, and troponin (Bonne et al. 1998). While the precise functions of myosin binding protein-C in regulation of muscle contraction have not been identified, the tropomyosin–troponin complex regulates myosin binding to actin filaments in a calcium-dependent manner by influencing the accessibility of binding sites on actin.

Diseases Associated with Mutations in Nonmuscle Myosins

In addition to muscle myosin II, the myosin superfamily includes a variety of nonmuscle myosins. Human diseases caused by mutations in nonmuscle myosins fall into the following broad categories: hearing loss, typically associated with disruption of structure and functions of the sensory elements of the inner ear, called hair cells; vision disorders; neurological disorders; cancer; cardiomyopathies; and some broad syndromic disorders with multiple symptoms such as those caused by mutations in myosin II.

Hearing Loss

Mutations in at least six different myosins and in several other ABPs have been implicated in syndromic and nonsyndromic deafness in humans. This strong genetic link between myosins and deafness is due to the fact that the presence of a variety of myosins is necessary to maintain normal structural and functional properties of stereocilia in the inner ear. Mutations in myosins VI, VIIa, and XV in humans lead to inherited deafness (Melchionda et al. 2001; Wang et al. 1998; Weil et al. 1995). The roles of these myosins in the development of hearing impairment have been investigated using corresponding mouse mutants, which exhibit defects in stereocilia formation (fused stereocilia in myosin VI mutant Snell's waltzer (Self et al. 1999)), disorganized stereocilia in myosin VIIa mutant shaker-1 (Self et al. 1998), and abnormally short stereocilia in the myosin XV mutant shaker-2 (Probst et al. 1998), suggesting that these myosins play important roles in maintaining spatial organization of stereocilia (Fig. 4).

Myosins may also play a role in delivering or retaining other stereocilia components to the tips of stereocilia. For example, Myo15 is necessary for localization of whirlin, a PDZ-domain-containing scaffolding protein, to the tips of stereocilia (Belyantseva et al. 2005), while Myo7a is required for localization of harmonin, another scaffolding protein containing PDZ domains (Boeda et al. 2002). Mutations in Myo1A (Donaudy et al. 2003) and non-muscle Myo2C (MYH14) have also been linked to human deafness (Donaudy et al. 2004). While these myosins are expressed in the inner ear, their function in hearing is unknown. Additionally, some of the syndromic diseases related to mutations in Myo2a (MYH9) are characterized by deafness in addition to other abnormalities (see below).

Blindness

Myosin VIIa is found in both photoreceptor cells and retinal pigment epithelial cells of the retina. Mutations in myosin VIIa cause the deafness and blindness disorder known as Usher syndrome type I. Loss of vision in Usher syndrome results from retinal photoreceptor cell death (Reiners et al. 2006). Mutations in myosin VIIa have also been linked to abnormal melanosome distribution in retinal pigment epithelial cells (Liu, Ondek and Williams 1998) and to defects in endocytosis of photoreceptor outer rod segments by RPE cells (Gibbs, Kitamoto and Williams 2003). These abnormalities in RPE cell functions may make mutant retina more susceptible to light damage and/or disrupt rod outer segment membrane renewal, which is dependent on phagocytosis by RPE cells. In addition, myosin VIIa has been implicated in transport of opsin through the cilium that connects inner and outer rod segments of photoreceptors, which may decrease the rate of protein renewal in the outer rod segment in the absence of functional myosin VIIa and make the photoreceptors vulnerable to degradation (Liu et al. 1999).

Neurological Dysfunction

Mutations in myosin Va cause neurological problems (mental retardation and delay in motor development) and decreased pigmentation resulting in characteristic silvery hair color (Pastural et al. 1997, 2000). Myosin Va is implicated in trafficking of synaptic vesicles in neurons and pigment granules in melanocytes (Bridgman 1999; Libby et al. 2004; Wu and Hammer 2000). Both neurological symptoms and hypopigmentation are likely due to the disruption of vesicle transport in the absence of functional myosin Va.

Cancer

Myosin XVIIIb is hypothesized to function as a tumor suppressor. Myosin XIIIb is deleted, mutated, or silenced by methylation in lung, ovarian, and colorectal tumors (Nakano et al. 2005; Nishioka et al. 2002; Yanaihara et al. 2004). Restoration of myosin XVIIIb expression in colorectal and lung cancer cell lines suppresses anchorage-independent tumor cell growth, consistent with myosin XVIIIb being involved in anchorage-dependent growth regulation (Nakano et al. 2005; Nishioka et al. 2002). Myosin VI is upregulated in aggressive ovarian tumors and may contribute to enhanced tumor cell migration (Yoshida et al. 2004).

MYH9-Related Disease

The Myo2 subclass includes three Myo2 isoforms expressed in nonmuscle cells: Myo2A, Myo2B, and Myo2C. Myo2A, encoded by MYH9 gene, is expressed in a wide variety of cell types where it is involved in cell migration and cell adhesion. Mutations in Myo2A are associated with a plethora of defects, including platelet and leukocyte dysfunction, renal disease, formation of cataracts, and hearing loss (Seri et al. 2000).

In the following chapters, the roles of ABPs are discussed in more detail. The above brief summary intends to lead the reader to the view that ABPs are rather more complex and certainly more closely associated with diseases that was previously thought.

Acknowledgments. MK is grateful to the American Heart Association (Scientist Development Grant #0335441T); EDLC acknowledges support from the NSF (MCB-0546353), NIH (1 R01 GM071688), and the American Heart Association (0655849T).

References

Allen, K. M., Gleeson, J. G., Bagrodia, S., Partington, M. W., MacMillan, J. C., Cerione, R. A., Mulley, J. C. and Walsh, C. A. 1998. PAK3 mutation in nonsyndromic X-linked mental retardation. Nat. Genet. 20, 25–30.

Anstee, D. J., Ridgwell, K., Tanner, M. J., Daniels, G. L. and Parsons, S. F. 1984. Individuals lacking the Gerbich blood-group antigen have alterations in the human erythrocyte membrane sialoglycoproteins beta and gamma. Biochem. J. 221, 97–104.

Belyantseva, I. A., Boger, E. T., Naz, S., Frolenkov, G. I., Sellers, J. R., Ahmed, Z. M., Griffith, A. J. and Friedman T. B. 2005. Myosin-XVa is required for tip localization of whirlin and differential elongation of hair-cell stereocilia. Nat. Cell Biol. 7, 148–156.

Bennett, V. and Chen, L. 2001. Ankyrins and cellular targeting of diverse membrane proteins to physiological sites. Curr. Opin. Cell Biol. 13, 61–67.

Berg, J. S., Powell, B. C. and Cheney, R. E. 2001. A millennial myosin census. Mol. Biol. Cell 12, 780–794.

Blake, D. J., Weir, A., Newey, S. E. and Davies, K. E. 2002. Function and genetics of dystrophin and dystrophin-related proteins in muscle. Physiol. Rev. 82, 291–329.

Boeda, B., El-Amraoui, A., Bahloul, A., Goodyear, R., Daviet, L., Blanchard, S., Perfettini, I., Fath, K. R., Shorte, S., Reiners, J., Houdusse, A., Legrain, P., Wolfrum, U., Richardson, G. and Petit, C. 2002. Myosin VIIa, harmonin and cadherin 23, three Usher I gene products that cooperate to shape the sensory hair cell bundle. EMBO J. 21, 6689–6699.

Bonne, G., Carrier, L., Richard, P., Hainque, B. and Schwartz, K. 1998. Familial hypertrophic cardiomyopathy: From mutations to functional defects. Circ. Res. 83, 580–593.

Bridgman, P. C. 1999. Myosin Va movements in normal and dilute-lethal axons provide support for a dual filament motor complex. J. Cell Biol. 146, 1045–1060.

Burch, J. M., Fassihi, H., Jones, C. A., Mengshol, S. C., Fitzpatrick, J. E., and McGrath, J. A. 2006. Kindler syndrome: A new mutation and new diagnostic possibilities. Arch. Dermatol. 142, 620–624.

Chang, A. N. and Potter, J. D. 2005. Sarcomeric protein mutations in dilated cardiomyopathy. Heart Fail. Rev. 10, 225–235.

Ching, Y. H., Ghosh, T. K., Cross, S. J., Packham, E. A., Honeyman, L., Loughna, S., Robinson, T. E., Dearlove, A. M., Ribas, G., Bonser, A. J., Thomas, N. R., Scotter, A. J., Caves, L. S., Tyrrell, G. P., Newbury-Ecob, R. A., Munnich, A., Bonnet, D. and Brook, J. D. 2005. Mutation in myosin heavy chain 6 causes atrial septal defect. Nat. Genet. 37, 423–428.

Condeelis, J., Singer, R. H. and Segall, J. E. 2005. The great escape: When cancer cells hijack the genes for chemotaxis and motility. Annu. Rev. Cell Dev. Biol. 21, 695–718.

Cuda, G., Fananapazir, L., Epstein, N. D. and Sellers, J. R. 1997. The in vitro motility activity of beta-cardiac myosin depends on the nature of the beta-myosin heavy chain gene mutation in hypertrophic cardiomyopathy. J. Muscle Res. Cell Motil. 18, 275–283.

Daniels, G. L., Shaw, M. A., Judson, P. A., Reid, M. E., Anstee, D. J., Colpitts, P., Cornwall, S., Moore, B. P. and Lee, S. 1986. A family demonstrating inheritance of the Leach phenotype: A Gerbich-negative phenotype associated with elliptocytosis. Vox Sang. 50, 117–121.

Donaudy, F., Ferrara, A., Esposito, L., Hertzano, R., Ben-David, O., Bell, R. E., Melchionda, S., Zelante, L., Avraham, K. B. and Gasparini, P. 2003. Multiple mutations of MYO1A, a cochlear-expressed gene, in sensorineural hearing loss. Am. J. Hum. Genet. 72, 1571–1577.

Donaudy, F., Snoeckx, R., Pfister, M., Zenner, H. P., Blin, N., Di Stazio, M., Ferrara, A., Lanzara, C., Ficarella, R., Declau, F., Pusch, C. M., Nurnberg, P., Melchionda, S., Zelante, L., Ballana, E., Estivill, X., Van Camp, G., Gasparini, P. and Savoia, A. 2004. Nonmuscle myosin heavy-chain gene MYH14 is expressed in cochlea and mutated in patients affected by autosomal dominant hearing impairment (DFNA4). Am. J. Hum. Genet. 74, 770–776.

Donaudy, F., Zheng, L., Ficarella, R., Ballana, E., Carella, M., Melchionda, S., Estivill, X., Bartles, J. R. and Gasparini, P. 2006. Espin gene (ESPN) mutations associated with autosomal dominant hearing loss cause defects in microvillar elongation or organisation. J. Med. Genet. 43, 157–161.

Dosaka-Akita, H., Hommura, F., Fujita, H., Kinoshita, I., Nishi, M., Morikawa, T., Katoh, H., Kawakami, Y. and Kuzumaki, N. 1998. Frequent loss of gelsolin expression in non-small cell lung cancers of heavy smokers. Cancer Res. 58, 322–327.

Eber, S. W., Gonzalez, J. M., Lux, M. L., Scarpa, A. L., Tse, W. T., Dornwell, M., Herbers, J., Kugler, W., Ozcan, R., Pekrun, A., Gallagher, P. G., Schroter, W., Forget, B. G. and Lux, S. E. 1996. Ankyrin-1 mutations are a major cause of dominant and recessive hereditary spherocytosis. Nat. Genet. 13, 214–218.

El-Amraoui, A. and Petit, C. 2005. Usher I syndrome: Unravelling the mechanisms that underlie the cohesion of the growing hair bundle in inner ear sensory cells. J. Cell Sci. 118, 4593–4603.

Endris, V., Wogatzky, B., Leimer, U., Bartsch, D., Zatyka, M., Latif, F., Maher, E. R., Tariverdian, G., Kirsch, S., Karch, D. and Rappold, G. A. 2002. The novel Rho-GTPase activating gene MEGAP/srGAP3 has a putative role in severe mental retardation. Proc. Natl Acad. Sci. USA 99, 11754–11759.

Frangiskakis, J. M., Ewart, A. K., Morris, C. A., Mervis, C. B., Bertrand, J., Robinson, B. F., Klein, B. P., Ensing, G. J., Everett, L. A., Green, E. D., Proschel, C., Gutowski, N. J., Noble, M., Atkinson, D. L., Odelberg, S. J. and Keating, M. T. 1996. LIM-kinase1 hemizygosity implicated in impaired visuospatial constructive cognition. Cell 86, 59–69.

Gibbs, D., Kitamoto, J. and Williams, D. S. 2003. Abnormal phagocytosis by retinal pigmented epithelium that lacks myosin VIIa, the Usher syndrome 1B protein. Proc. Natl Acad. Sci. USA 100, 6481–6486.

Gimona, M., Djinovic-Carugo, K., Kranewitter, W. J. and Winder, S. J. 2002. Functional plasticity of CH domains. FEBS Lett. 513, 98–106.

Gouin, E., Gantelet, H., Egile, C., Lasa, I., Ohayon, H., Villiers, V., Gounon, P., Sansonetti, P. J. and Cossart, P. 1999. A comparative study of the actin-based motilities of the pathogenic bacteria Listeria monocytogenes, Shigella flexneri and Rickettsia conorii. J. Cell Sci. 112(Pt 11), 1697–1708.

Herz, C., Aumailley, M., Schulte, C., Schlotzer-Schrehardt, U., Bruckner-Tuderman, L. and Has, C. 2006. Kindlin-1 is a phosphoprotein involved in regulation of polarity, proliferation and motility of epidermal keratinocytes. J. Biol. Chem. 281, 36082–36090.

Jobard, F., Bouadjar, B., Caux, F., Hadj-Rabia, S., Has, C., Matsuda, F., Weissenbach, J., Lathrop, M., Prud'homme, J. F. and Fischer, J. 2003. Identification of mutations in a new gene encoding a FERM family protein with a pleckstrin homology domain in Kindler syndrome. Hum. Mol. Genet. 12, 925–935.

Kaplan, J. M., Kim, S. H., North, K. N., Rennke, H., Correia, L. A., Tong, H. Q., Mathis, B. J., Rodriguez-Perez, J. C., Allen, P. G., Beggs, A. H. and Pollak, M. R. 2000. Mutations in ACTN4, encoding alpha-actinin-4, cause familial focal segmental glomerulosclerosis. Nat. Genet. 24, 251–256.

Kitajiri, S., Fukumoto, K., Hata, M., Sasaki, H., Katsuno, T., Nakagawa, T., Ito, J. and Tsukita, S. 2004. Radixin deficiency causes deafness associated with progressive degeneration of cochlear stereocilia. J. Cell Biol. 166, 559–570.

Kurabayashi, M., Tsuchimochi, H., Komuro, I., Takaku, F. and Yazaki, Y. 1988. Molecular cloning and characterization of human cardiac alpha- and beta-form myosin heavy chain complementary DNA clones. Regulation of expression during development and pressure overload in human atrium. J. Clin. Invest. 82, 524–531.

Lee, H. K., Driscoll, D., Asch, H., Asch, B. and Zhang, P. J. 1999. Downregulated gelsolin expression in hyperplastic and neoplastic lesions of the prostate. Prostate 40, 14–19.

Libby, R. T., Lillo, C., Kitamoto, J., Williams, D. S. and Steel K. P. 2004. Myosin Va is required for normal photoreceptor synaptic activity. J. Cell Sci. 117, 4509–4515.

Liu, X., Ondek, B. and Williams, D. S. 1998. Mutant myosin VIIa causes defective melanosome distribution in the RPE of shaker-1 mice. Nat. Genet. 19, 117–118.

Liu, X., Udovichenko, I. P., Brown, S. D., Steel, K. P. and Williams, D. S. 1999. Myosin VIIa participates in opsin transport through the photoreceptor cilium. J. Neurosci. 19, 6267–6274.

Melchionda, S., Ahituv, N., Bisceglia, L., Sobe, T., Glaser, F., Rabionet, R., Arbones, M. L., Notarangelo, A., Di Iorio, E., Carella, M., Zelante, L., Estivill, X., Avraham, K. B. and Gasparini, P. 2001. MYO6, the human homologue of the gene responsible for deafness in Snell's waltzer mice, is mutated in autosomal dominant nonsyndromic hearing loss. Am. J. Hum. Genet. 69, 635–640.

Miura, P. and Jasmin, B. J. 2006. Utrophin upregulation for treating Duchenne or Becker muscular dystrophy: How close are we? Trends Mol. Med. 12, 122–129.

Mohiddin, S. A., Ahmed, Z. M., Griffith, A. J., Tripodi, D., Friedman, T. B., Fananapazir, L. and Morell, R. J. 2004. Novel association of hypertrophic cardiomyopathy, sensorineural deafness, and a mutation in unconventional myosin VI (MYO6). J. Med. Genet. 41, 309–314.

Mohler, P. J., Schott, J. J., Gramolini, A. O., Dilly, K. W., Guatimosim, S., duBell, W. H., Song, L. S., Haurogne, K., Kyndt, F., Ali, M. E., Rogers, T. B., Lederer, W. J., Escande, D., Le Marec, H. and Bennett. V. 2003. Ankyrin-B mutation causes type 4 long-QT cardiac arrhythmia and sudden cardiac death. Nature 421, 634–639.

Nakano, T., Tani, M., Nishioka, M., Kohno, T., Otsuka, A., Ohwada, S. and Yokota, J. 2005. Genetic and epigenetic alterations of the candidate tumor-suppressor gene MYO18B, on chromosome arm 22q, in colorectal cancer. Genes Chromosomes Cancer 43, 162–171.

Naz, S., Griffith, A. J., Riazuddin, S., Hampton, L. L., Battey, J. F. Jr., Khan, S. N., Wilcox, E. R. and Friedman, T. B. 2004. Mutations of ESPN cause autosomal recessive deafness and vestibular dysfunction. J. Med. Genet. 41, 591–595.

Nishioka, M., Kohno, T., Tani, M., Yanaihara, N., Tomizawa, Y., Otsuka, A., Sasaki, S., Kobayashi, K., Niki, T., Maeshima, A., Sekido, Y., Minna, J. D., Sone, S. and Yokota, J. 2002. MYO18B, a candidate tumor suppressor gene at chromosome 22q12.1, deleted, mutated, and methylated in human lung cancer. Proc. Natl Acad. Sci. USA 99, 12269–12274.

Ochs, H. D. and Thrasher, A. J. 2006. The Wiskott–Aldrich syndrome. J. Allergy Clin. Immunol. 117, 725–738.

Pastural, E., Barrat, F. J., Dufourcq-Lagelouse, R., Certain, S., Sanal, O., Jabado, N., Seger, R., Griscelli, C., Fischer, A. and de Saint Basile, G. 1997. Griscelli disease maps to chromosome 15q21 and is associated with mutations in the myosin-Va gene. Nat. Genet. 16, 289–292.

Pastural, E., Ersoy, F., Yalman, N., Wulffraat, N., Grillo, E., Ozkinay, F., Tezcan, I., Gedikoglu, G., Philippe, N., Fischer, A. and de Saint Basile, G. 2000. Two genes are responsible for Griscelli syndrome at the same 15q21 locus. Genomics 63, 299–306.

Probst, F. J., Fridell, R. A., Raphael, Y., Saunders, T. L., Wang, A., Liang, Y., Morell, R. J., Touchman, J. W., Lyons, R. H., Noben-Trauth, K., Friedman, T. B. and Camper, S. A. 1998. Correction of deafness in shaker-2 mice by an unconventional myosin in a BAC transgene. Science 280, 1444–1447.

Puius, Y. A., Mahoney, N. M. and Almo, S. C. 1998. The modular structure of actin-regulatory proteins. Curr. Opin. Cell Biol. 10, 23–34.

Rayment, I., Holden, H. M., Sellers, J. R., Fananapazir, L. and Epstein, N. D. 1995. Structural interpretation of the mutations in the beta-cardiac myosin that have been implicated in familial hypertrophic cardiomyopathy. Proc. Natl Acad. Sci. USA 92, 3864–3868.

Reiners, J., Nagel-Wolfrum, K., Jurgens, K., Marker, T. and Wolfrum, U. 2006. Molecular basis of human Usher syndrome: Deciphering the meshes of the Usher protein network provides insights into the pathomechanisms of the Usher disease. Exp. Eye Res. 83, 97–119.

Revenu, C., Athman, R., Robine, S. and Louvard, D. 2004. The co-workers of actin filaments: From cell structures to signals. Nat. Rev. Mol. Cell Biol. 5, 635–646.

Robertson, S. P. 2005. Filamin A: Phenotypic diversity. Curr. Opin. Genet. Dev. 15, 301–307.

Rottner, K., Lommel, S., Wehland, J. and Stradal, T. E. 2004. Pathogen-induced actin filament rearrangement in infectious diseases. J. Pathol. 204, 396–406.

Self, T., Mahony, M., Fleming, J., Walsh, J., Brown, S. D. and Steel, K. P. 1998. Shaker-1 mutations reveal roles for myosin VIIA in both development and function of cochlear hair cells. Development 125, 557–566.

Self, T., Sobe, T., Copeland, N. G., Jenkins, N. A., Avraham, K. B. and Steel, K. P. 1999. Role of myosin VI in the differentiation of cochlear hair cells. Dev. Biol. 214, 331–341.

Seri, M., Cusano, R., Gangarossa, S., Caridi, G., Bordo, D., Lo Nigro, C., Ghiggeri, G. M., Ravazzolo, R., Savino, M., Del Vecchio, M., d'Apolito, M., Iolascon, A., Zelante, L. L., Savoia, A., Balduini, C. L., Noris, P., Magrini, U., Belletti, S., Heath, K. E., Babcock, M., Glucksman, M. J., Aliprandis, E., Bizzaro, N., Desnick, R. J. and Martignetti, J. A. 2000. Mutations in MYH9 result in the May-Hegglin anomaly, and Fechtner and Sebastian syndromes. The May-Heggllin/Fechtner Syndrome Consortium. Nat. Genet. 26, 103–105.

Siegel, D. H., Ashton, G. H., Penagos, H. G., Lee, J. V., Feiler, H. S., Wilhelmsen, K. C., South, A. P., Smith, F. J., Prescott, A. R., Wessagowit, V., Oyama, N., Akiyama, M., Al Aboud, D., Al Aboud, K., Al Githami, A., Al Hawsawi, K., Al Ismaily, A., Al-Suwaid, R., Atherton, D. J., Caputo, R., Fine, J. D., Frieden, I. J., Fuchs, E., Haber, R. M., Harada, T., Kitajima, Y., Mallory, S. B., Ogawa, H., Sahin, S., Shimizu, H., Suga, Y., Tadini, G., Tsuchiya, K., Wiebe, C. B., Wojnarowska, F., Zaghloul, A. B., Hamada, T., Mallipeddi, R., Eady, R. A., McLean, W. H., McGrath, J. A. and Epstein E. H. 2003. Loss of kindlin-1, a human homolog of the Caenorhabditis elegans actin-extracellular-matrix linker protein UNC-112, causes Kindler syndrome. Am. J. Hum. Genet. 73, 174–187.

Smith, G. L., Murphy, B. J. and Law, M. 2003. Vaccinia virus motility. Annu. Rev. Microbiol. 57, 323–342.

Stevens, J. M., Galyov, E. E. and Stevens, M. P. 2006. Actin-dependent movement of bacterial pathogens. Nat. Rev. Microbiol. 4, 91–101.

Tanaka, M., Mullauer, L., Ogiso, Y., Fujita, H., Moriya, S., Furuuchi, K., Harabayashi, T., Shinohara, N., Koyanagi, T. and Kuzumaki, N. 1995. Gelsolin: A candidate for suppressor of human bladder cancer. Cancer Res. 55, 3228–3232.

Tchernia, G., Mohandas, N. and Shohet, S. B. 1981. Deficiency of skeletal membrane protein band 4.1 in homozygous hereditary elliptocytosis. Implications for erythrocyte membrane stability. J. Clin. Invest. 68, 454–460.

Wachsstock, D. H., Schwartz, W. H. and Pollard, T. D. 1993. Affinity of alpha-actinin for actin determines the structure and mechanical properties of actin filament gels. Biophys. J. 65, 205–214.

Wang, A., Liang, Y., Fridell, R. A., Probst, F. J., Wilcox, E. R., Touchman, J. W., Morton, C. C., Morell, R. J., Noben-Trauth, K., Camper, S. A. and Friedman, T. B. 1998. Association of unconventional myosin MYO15 mutations with human non-syndromic deafness DFNB3. Science 280, 1447–1451.

Weil, D., Blanchard, S., Kaplan, J., Guilford, P., Gibson, F., Walsh, J., Mburu, P., Varela, A., Levilliers, J., Weston, M. D., Kelley, P. M., Kimberling, W. J., Wagenaar, M., Levi-Acobas, F., Larget-Piet, D., Munnich, A., Steel, K. P., Brown, S. D. M. and Petit, C. 1995. Defective myosin VIIA gene responsible for Usher syndrome type 1B. Nature 374, 60–61.

Wu, X. and Hammer, J. A. 3rd. 2000. Making sense of melanosome dynamics in mouse melanocytes. Pigment Cell Res. 13, 241–247.

Yanaihara, N., Nishioka, M., Kohno, T., Otsuka, A., Okamoto, A., Ochiai, K., Tanaka, T. and Yokota, J. 2004. Reduced expression of MYO18B, a candidate tumor-suppressor gene on chromosome arm 22q, in ovarian cancer. Int. J. Cancer 112, 150–154.

Yao, J., Le, T. C., Kos, C. H., Henderson, J. M., Allen, P. G., Denker, B. M. and Pollak, M. R. 2004. Alpha-actinin-4-mediated FSGS: An inherited kidney disease caused by an aggregated and rapidly degraded cytoskeletal protein. PLoS Biol. 2, e167.

Yoshida, H., Cheng, W., Hung, J., Montell, D., Geisbrecht, E., Rosen, D., Liu, J. and Naora, H. 2004. Lessons from border cell migration in the Drosophila ovary: A role for myosin VI in dissemination of human ovarian cancer. Proc. Natl Acad. Sci. USA 101, 8144–8149.

Zheng, L., Sekerkova, G., Vranich, K., Tilney, L. G., Mugnaini, E. and Bartles, J. R. 2000. The deaf jerker mouse has a mutation in the gene encoding the espin actin-bundling proteins of hair cell stereocilia and lacks espins. Cell 102, 377–385.

6
ADF/Cofilin, Actin Dynamics, and Disease

Michael T. Maloney, Andrew W. Kinley, Chi W. Pak, and James R. Bamburg

Introduction

The actin-depolymerizing factor (ADF)/cofilin family of actin assembly regulatory proteins is essential for the survival of all eukaryotes including protists, plants, and animals (reviewed in Bamburg 1999). In multicellular organisms, ADF/cofilin (AC) proteins are highly regulated by complex signaling pathways. Multiple strategies of biochemical regulation are utilized including phosphorylation/dephosphorylation, membrane phosphoinositol-phosphate binding, pH regulation, accessory proteins that can enhance the effects of AC on actin dynamics, and proteins that compete for actin binding, such as tropomyosins (TMs).

AC proteins increase the turnover of actin filaments providing a continuous supply of actin monomers to enhance membrane protrusions and cell motility (Dawe et al. 2003). AC proteins also facilitate the modulation of Golgi membrane dynamics (Rosso et al. 2004), and they can change the membrane delivery and secretory activity of cells. Under certain cellular conditions, AC proteins are targeted to the nucleus where they presumably deliver actin, an important protein in transcriptional complexes (Visa 2005) and chromatin-remodeling complexes (Olave et al. 2002), thus affecting gene expression. Targeting of AC proteins to the outer mitochondrial membrane occurs in response to other signals, where they can open pores for cytochrome c leakage, ultimately leading to the initiation of mitochondrial-dependent apoptosis (Chua et al. 2003).

The turnover of actin filaments is a major energy-utilizing process in cells (Bernstein and Bamburg 2003). Under conditions of cell stress, when ATP levels drop, AC proteins may form nondynamic inclusions with actin that spare ATP (Minamide et al. 2000). In some cells these AC–actin inclusions can reversibly disassemble when ATP levels return to normal; however, in other cells, these inclusions may remain persistent and irreversibly alter cellular function. Due to the rapidity of their formation under conditions of stress and the potential for persistence, these inclusions might initiate, but will certainly aid in the progression of diseases as diverse as Alzheimer's disease (AD), stroke, and ischemic kidney disease. Recent studies suggest that altered expression, regulation, or localization of AC proteins might lead to cognitive impairment, inflammation, infertility, immune deficiencies, loss of contact inhibition and metastases, and other pathophysiological defects. This chapter briefly outlines the normal essential function of AC proteins across phylogeny and then presents the potential role for AC proteins in a broad spectrum of degenerative disorders and diseases starting with examples of AC protein silencing and gene knockouts and finally leading into disorders caused by misregulation.

Overview of Actin Assembly and Its Regulation

Actin, a 42 kDa protein, exists either in an unassembled globular form (G-actin) or in a filamentous polymer (F-actin). Actin is a nucleotide-binding protein found associated with ATP or ADP. ATP-G-actin spontaneously assembles into F-actin *in vitro* under physiological ionic conditions. Spontaneous assembly of actin is prevented *in vivo* by G-actin-sequestering proteins (Sun, Kwiatkowska and Yin 1995), which also play a role in delivering ATP-G-actin to specific sites intracellularly (Goode, Drubin and Lappalainen 1998; Palmgren, Vartainen and Lappalainen 2002; Vartiainen et al. 2002). The presence of F-actin capping/severing proteins (Cooper and Schafer 2000; Maiti and Bamburg 2004), along with ATP-G-actin in discrete subcellular environments, determines sites for actin nucleation and assembly.

After ATP-G-actin assembles into a filament, the ATP is rapidly hydrolyzed to give ADP-P*i*-actin and then in a slower step phosphate is released, creating a nonequilibrium polymer with unequal critical monomer concentrations for assembly at the two filaments ends. The ADP remains bound to actin and is not exchanged for ATP within the filament (Carlier 1989). At steady state, filaments treadmill, i.e., ATP-G-actin adds at the faster-growing barbed end (a.k.a. plus end), which is defined based upon a characteristic arrowhead decoration of F-actin with actin-binding proteolytic pieces of myosin II (Nachmias and Huxley 1970) and ADP-G-actin is lost from the pointed end (a.k.a. minus end) (Fig. 1) (Carlier 1991; Wegner 1982). Actin-ATP hydrolysis in platelets and cultured embryonic neurons consumes ~50% of total cell ATP (Bernstein and Bamburg 2003; Daniel et al. 1986).

Most F-actin is ADP-actin while most G-actin is ATP-actin in complex with binding proteins (Atkinson, Hosford and Molitoris 2004; Rosenblatt, Peluso and Mitchison 1995). Filament length and turnover at steady state are determined by rates of subunit addition and removal, which are in turn regulated by the local concentration of the actin monomer pool. Filament severing creates new pointed and barbed ends, which become sites for disassembly or for nucleated growth, respectively. Indeed, many cellular functions require the simultaneous assembly of actin in one cellular region and disassembly in another. Tight control over actin polymerization and depolymerization in response to differing cellular demands is important in nonmuscle cells where

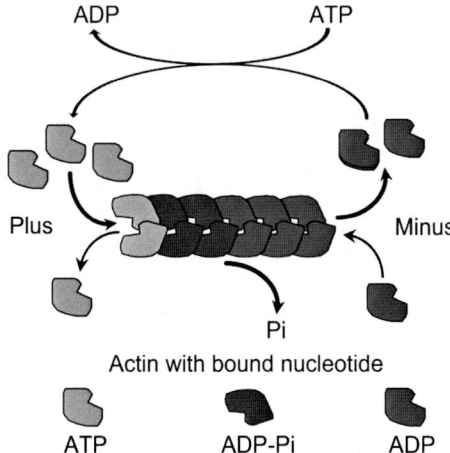

FIG. 1. Actin treadmilling. The major pool of actin monomers in a cell is ATP-actin and this assembles onto free ends of growing filaments as long as it exceeds the critical concentration necessary to sustain growth. ATP is hydrolyzed rapidly after assembly and inorganic phosphate is released more slowly. At steady state shown here, actin filaments undergo treadmilling because the monomer pool exceeds the critical concentration for assembly on the barbed ends of filaments but is below the concentration required for pointed end growth (*See Color Plates*)

much of the actin exists in an unassembled form (Blikstad et al. 1978; Wang 1985). In striated muscle and cardiac muscle, most of the actin is assembled into less dynamic filaments that are part of the sarcomere contractile unit.

Mammals express six different actin genes, two sarcomeric isoforms (α-skeletal and α-cardiac actin), two smooth muscle isoforms (α- and γ-actin), and two cytoskeletal isoforms found in all cells (β- and γ-actin). Each actin isoform has a highly conserved tertiary structure that allows these proteins to assemble into mixed polymers. However, individual affinities for actin association with other cellular proteins are different enough that in cells expressing multiple isoforms of actin, isoforms may sort preferentially, if not exclusively, into different structures.

In nonmuscle cells, the organization of filamentous actin gives rise to various cellular structures, such as filopodia, lamellipodia, and stress fibers (Fig. 2). The physical properties of G-actin, F-actin, and actin-binding proteins (ABPs) contribute to the regulation, organization, and variety of actin structures found in cells. The actin cytoskeleton is directly implicated in a variety of cellular processes such as the establishment and maintenance of cell polarity (Matsudaira 1991), cell motility (Welch et al. 1997), cytokinesis (Fishkind and Wang 1995), cell–cell and cell–matrix interactions (Yamada and Geiger 1997), cell volume and ion channel protein dynamics (Cantiello 1997), mitochondrial-dependent apoptosis and cell aging (Chua et al. 2003; Gourlay and Ayscough 2005), and organelle

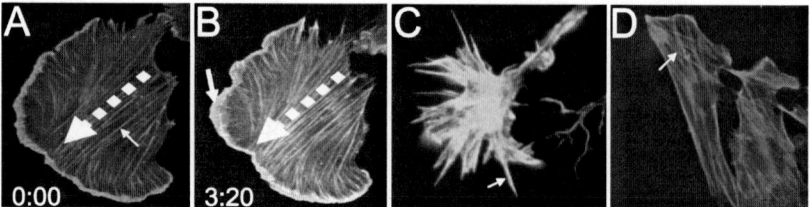

FIG. 2. Major F-actin structures in nonmuscle cells. (**a**) Fluorescent-phalloidin-stained lamellipodium at leading edge of chick cardiac fibroblast with *bold arrow* showing direction of migration and *small arrow* showing a graded polarity actin bundle, which differs from a stress fiber in that polarities of actin filaments within the bundle change between the front and rear of the migrating cell. (**b**) A second image of the cell shown in (**a**), pseudocolored in green, and overlayed on image in (**a**) to show new lamellipodial extension (*arrow: bright green*) that occurred over the 3 min 20 s time period. (**c**) Filopodia (microspikes) on a neuronal growth cone (*arrow*). (**d**) Stress fiber (*arrow*) in a cultured nonpolarized and nonmigrating cell stained with rhodamine-phalloidin. Long smooth membrane on left of cell with *arrow* indicates tension in the filament caused by contraction along the filament bundle (images courtesy of Lubna Tahtamouni) (*See Color Plates*)

and vesicular transport including Golgi vesicle and endosome trafficking (Kubler and Riezman 1993; Langford 1995).

Given actin's central role in cell division and migration, loss-of-function mutations in either actin or a critical actin regulatory protein would be lethal to any multicellular organism early in development. Thus, adult diseases based upon genetic abnormalities in actin or its regulatory proteins are mostly limited to those affecting their tissue-specific expression.

ADF/Cofilin Family of Actin Assembly Regulatory Proteins

Identification and Naming

ADF was initially identified in extracts of embryonic chicken brain and named for its ability to rapidly depolymerize low concentrations of F-actin (Bamburg, Harris and Weeds 1980). Subsequently, many related 13–19 kDa proteins have been isolated from a number of different animals, plants, and protists. Cofilin, named for its ability to form *cofil*amentous structures with act*in* (Nishida, Maekawa and Sakai 1984b), was the first member of this family to be cloned and sequenced (Matsuzaki et al. 1988). Thus, this family of proteins is now referred to as the ADF/cofilin family (Bamburg 1999). Metazoans have multiple genes encoding ADF/cofilin proteins. Mammals have the following: muscle-specific cofilin (cof-2), ubiquitously expressed nonmuscle cofilin (cof-1), and ADF.

Structure of ADF/Cofilin

The primary sequences of the known AC family members contain 118–168 amino acids. NMR and X-ray crystallography have been used to elucidate the three-dimensional structures of porcine destrin (Hatanaka et al. 1996), yeast cofilin (Fedorov et al. 1997), human cofilin (Pope et al. 2004), chick cofilin (Gorbatyuk et al. 2006), plant ADF from *Arabidopsis thaliana*, and *Acanthamoeba actophorin* (Leonard et al. 1997). The entire AC structure (Fig. 3) consists of a four stranded mixed β-sheet surrounded by four α-helices, referred to as the ADF-homology (ADF-H) domain, a sequence found in structurally distinct protein families, including the ADF/cofilins and the Abp1/drebrins (Lappalainen et al. 1998). The predicted structure for the smallest member of the AC family, the 118 amino acid *Toxoplasma gondii* ADF, has the same core with no extra amino acids in any of the loop domains between the β-strands and α-helices.

The ADF-H domain is present in non-AC family members such as yeast Abp-1 (Mabuchi 1983; Drubin et al. 1990) and mammalian drebrin (Lappalainen et al. 1998; Shirao, Kojima and Obata 1992), and is duplicated

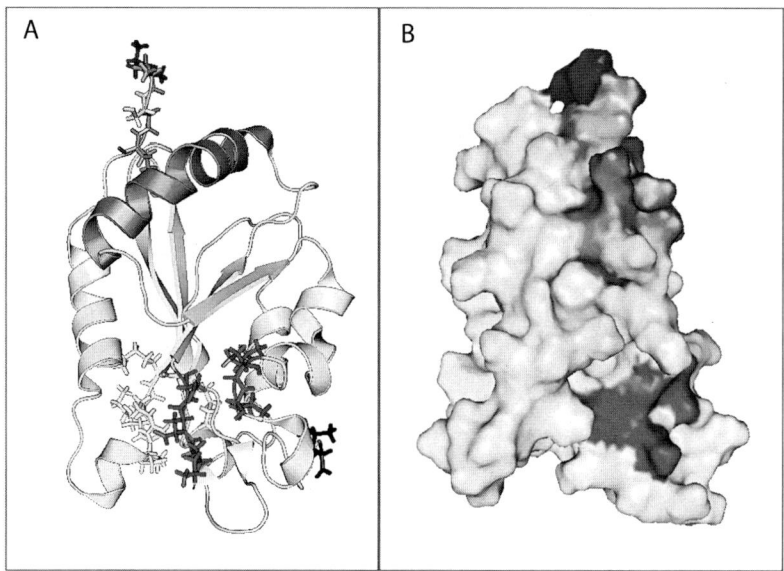

Fig. 3. Ribbon and space-filling model of human cofilin. (a) The ribbon structure of human cofilin is shown looking directly on the actin-binding face with the N-terminus (*dark blue*), the single serine phosphorylation site (*pink*), the nuclear localization signal (*yellow*), the G-actin and upper F-actin subunit major binding helix (*green*), the second lower F-actin subunit-binding residues (*red*), and the C-terminus (*black*). (b) Same color scheme used in (a) is used here for the space-filling model but the molecule is rotated so that the actin-binding surface is on the right side (*See Color Plates*)

in the actin monomer-binding protein twinfilin, triplicated in the severing and/or capping proteins fragmin, severin, and CapG, and repeated six times in gelsolin and villin (Lappalainen et al. 1998). Both N-terminal modification (demethionation and acetylation) and phosphorylation occur on AC family members from metazoans (Agnew, Minamide and Bamburg 1995; Kanamori, Suzuki and Titani 1998). Also, sequence analysis of depactin from starfish oocytes revealed a free proline at the amino terminus suggesting demethionation (Takagi, Konishi and Mabuchi 1988).

Vertebrate members of the AC family contain a nuclear localization sequence (NLS) consisting of the sequence PE(E/D)(I/V)KKRKK (Abe, Nagaoka and Obinata 1993; Nishida et al. 1987). The AC NLS is similar to one identified in the SV40 large T-antigen (Matsuzaki et al. 1988). Microinjection of the cofilin NLS conjugated to bovine serum albumin leads to accumulation of the fusion protein in the nucleus (Abe, Nagaoka and Obinata 1993), whereas mutations in the NLS prevent its nuclear accumulation following heat shock (Iida, Matsumoto and Yahara 1992), suggesting that this region of cofilin is both necessary and sufficient to direct it to the nuclear compartment.

Structural domains of cofilin necessary for targeting to mitochondrial outer membrane have also been characterized (Chua et al. 2003). The regions required include amino acids 15–30, and the C-terminal (CT) domain residues 100–165. A nonfunctional mutant protein with only these domains expressed was targeted to the outer mitochondrial membrane. However, cytochrome c release requires that cofilin maintains functional actin binding.

ADF/Cofilin Expression

Mammalian Genes

Mammals have evolved with three highly conserved ADF/cofilin genes: cofilin-1 (also called *n-cofilin* for nonmuscle cofilin) (Nishida, Maekawa and Sakai 1984b), cofilin-2 (also called *m-cofilin* for muscle cofilin) (Abe, Ohshima and Obinata 1989), and ADF (also called *destrin*) (Moriyama et al. 1990). The human cofilin-2 gene has two different exon 1 domains and different 5′ upstream regulatory sequences but yields mRNAs with identical coding sequence, one (cof-2b) is strongly expressed in skeletal muscle and heart and the other (cof-2a) is expressed at lower levels in other tissues (Thirion et al. 2001).

Developmental and Tissue Expression Pattern

The most comprehensive studies of ADF and cofilin expression during mammalian development have been done in mice. Earliest expression studies were performed by *in situ* hybridization for mRNA (Moriyama et al. 1990). ADF and cofilin mRNAs were both found to be highly expressed in the brain but weakly expressed in liver and muscle. Even though ADF and cofilin mRNAs are often found expressed in the same tissues, the relative amounts of each mRNA differ (Moriyama et al. 1990).

More recent expression studies have been done by immunocytochemistry using antibodies specific to each protein. ADF is expressed at low levels during embryonic development with highest levels found in embryonic heart and the adaxial region of somites of E10.5 embryos (Gurniak, Perlas and Witke 2005). In embryonic stem cells, ADF is present in about 1:5 ratio to cofilin-1. ADF is upregulated after birth in epithelial and endothelial tissues, such as uterus, stomach, and intestine (Gurniak, Perlas and Witke 2005; Vartiainen et al. 2002). Interestingly, ADF expression in the gut depends on the nutrition of newborn mice; upon switching from milk to solid food, ADF is upregulated two- to threefold.

Cofilin-1 is widely expressed during development. Before midgestation (day E8.5), cofilin-1 is highest in the neural fold (Gurniak, Perlas and Witke 2005). By day E9.5, expression persists in the developing nervous system and the neural crest. In E10.5 embryos, cofilin-1 is highest in the myotome cell layer of somites, more ventral than ADF, and with an over-all staining pattern closely resembling that of muscle-specific markers. After day E10.5, cofilin-1 expression is observed in somites and the neural tube, as well as limb buds, becoming more localized by day E11.5. Between day E13.5 and 16.5, cofilin-1 expression becomes more restricted to spe-cific tissues, particularly the central nervous system, trigeminal ganglion, olfactory bulb, lung, thymus, and developing kidney. Cofilin-1 expression is maintained in these tissues after birth, with highest levels in lymphatic organs such as spleen, thymus, and lymph nodes.

Cofilin-2 is not expressed until later in development, especially in differen-tiated muscle (Gurniak, Perlas and Witke 2005; Obinata et al. 1997; Ono et al. 1994; Thirion et al. 2001; Vartiainen et al. 2002). Mature muscle lacks both cofilin-1 and ADF. High levels of cofilin-2 are limited to skeletal muscle, heart, and testis in adult (Ono et al. 1994), although low levels of cofilin-2a mRNA are present elsewhere (Thirion et al. 2001).

To better understand if ADF and cofilin, along with the monomer-sequestering proteins profilin and thymosins, are responsible for maintain-ing the unassembled actin pool in tissues, the amounts of these proteins were quantified and compared with the amount of G-actin in both devel-oping muscle (Obinata et al. 1997) and brain (Devineni et al. 1999) from chick embryos. The amounts of ADF, cofilin, and profilin could account for the monomeric actin pool in later stages of muscle differentiation but not at early stages. However, thymosins are present early on and disappear from muscle as it matures, accounting for the remainder of the monomer-sequestering activity. Thymosins also account for the major G-actin-sequestering activity in brain, with ADF playing a significant role as well, presumably because chick ADF does not display the strong pref-erence for binding ADP-actin over ATP-actin (Chen et al. 2004), as do the mammalian ADF and cofilin homologs (Yeoh at al. 2002; Vartiainen et al. 2002). In a variety of chicken tissues, ADF accounted for 0.1–0.4% of total protein and was expressed in a 0.1–0.2 molar ratio to actin (Bamburg and Bray 1987; Morgan et al. 1993).

Muscle-type cofilin (cof-2) and nonmuscle-type cofilin (cof-1) are both expressed in developing mammalian skeletal and cardiac muscles and are involved in the regulation of actin assembly in developing myotubes. Cosedimentation assays were used to determine that cof-2 precipitated with F-actin *in vitro* more efficiently than cof-1. Also cof-2 inhibited actin–TM interaction more robustly than did cof-1 *in vitro*, suggesting that the two cofilins may play somewhat different roles during myotube development (Nakashima et al. 2005).

Regulation of Expression by the Actin Monomer Pool

The rate of actin synthesis in many different cell types is negatively correlated with the amount of unassembled actin present. This autoregulatory response is posttranscriptional, involves the destabilization of actin mRNA, and depends upon the 3' untranslated region of the mRNA (Serpinskaya et al. 1990; Bershadsky et al. 1995; Reuner et al. 1996; Lyubimova, Bershadsky and Ben-Ze'ev 1997, 1999).

ADF expression is also linked to the actin monomer pool. To increase the actin monomer pool, myoblasts were transfected with a mutant human β-actin (G244D) that does not assemble into normal filaments. Expression of β-actin (G244D) causes a posttranscriptional downregulation of ADF protein and mRNA, whereas cofilin expression is unaffected (Minamide et al. 1997). Also, treatment with latrunculin A, a membrane permeable compound that binds to monomeric actin and prevents its assembly, results in a decrease in the synthesis of ADF and actin, but not cofilin or several other ABPs. Finally, increase of the actin monomer pool causes a robust increase in ADF and cofilin phosphorylation, showing that inactivation precedes the decline in synthesis. Thus, both ADF and cofilin are regulated posttranslationally, but their expression is regulated differently. The regulation of ADF but not cofilin expression by the actin monomer pool may reflect the ability of ADF to modulate the monomer pool size better than cofilin.

Biochemical Analysis of AC–Actin Interaction

Across phylogeny, AC family members have qualitatively similar but quantitatively different activities on actin dynamics (Chen et al. 2004; Ono et al. 2003; Yeoh et al. 2002; Vartiainen et al. 2002). AC severs actin filaments thereby increasing the number of pointed ends from which subunit dissociation can occur (reviewed in Bamburg and Wiggan 2002; Chen, Bernstein and Bamburg

2000), but also creating new free barbed ends for actin assembly (Ghosh et al. 2004; Mouneimne et al. 2004; Zebda et al. 2000). Actin monomer-sequestering proteins such as Srv2/cyclase-associated protein 1 (CAP1) (Mattila et al. 2004; Moriyama and Yahara 2002) and profilin (Buss et al. 1992; Nishida 1985; Pollard, Blanchoin and Mullins 2000) are responsible for catalyzing a nucleotide exchange substituting ATP for the ADP on G-actin dissociated from filaments. Furthermore, actin monomer-sequestering proteins prevent spontaneous nucleation of filaments, keeping an ATP-G-actin pool much higher than the critical concentration required for assembly of purified actin. Profilin also delivers ATP-actin to assembly sites (Goode, Drubin and Lappalainen 1998; Palmgren, Vartainen and Lappalainen 2002; Vartiainen et al. 2003), particularly those utilizing formins for enhancing plus end growth (Kovar 2005).

ADF and cofilin bind to ATP-G-actin (Nishida, Maekawa and Sakai 1984a) but with a 4- to 100-fold weaker affinity than to ADP-G-actin (Carlier et al. 1997; Chen et al. 2004; Vartiainen et al. 2002; Yeoh et al. 2002). Chick ADF has a significantly higher affinity than chick cofilin for monomeric ATP-actin (Chen et al. 2004) and thus contributes to sequestering actin monomer in chick brain (Devineni et al. 1999). There may be less difference in affinity between AC proteins and Ca^{2+}-ADP- or Ca^{2+}-ATP-actin (Chen et al. 2004). Although Ca^{2+} binds to actin with a fivefold greater affinity than Mg^{2+} (Strzelecka-Golaszewska and Drabikowski 1968), Ca^{2+}-actin is not generally considered to be an important physiological species because cells maintain a very low free calcium concentration compared to magnesium. However, in some cells and in some regions of cells where calcium channels may be active, transient pools of calcium-actin may be present, which could be important for local actin dynamics. Additionally, the binding of AC to actin monomers prevents nucleotide exchange (Nishida et al. 1985; Hawkins et al. 1993; Hayden et al. 1993), which can be a rate limiting step for the recycling of actin into the assembly competent ATP-actin pool.

Actin is thought to participate in more protein–protein interactions than any other known protein. To date, only a few crystal structures have been solved of complexes between actin and actin-binding proteins, but these include structures with fragments of gelsolin that contain the ADF-H domain (Burtnick et al. 2004). A hydrophobic pocket that mediates important interactions in five of the existing structures of actin complexes may represent a common actin-binding motif as it remains accessible in F-actin. Such a motif may represent a primary target for many F-actin-binding proteins, reminiscent of the calponin-homology-related proteins and myosin (Dominguez 2004), or of the ADF-homology domain.

To identify the cofilin-binding site on filamentous actin, low-resolution images of cofilin bound to actin were generated using electron cryomicroscopy (Galkin et al. 2002; McGough et al. 1997). Cofilin binds F-actin cooperatively by binding to two adjacent actin subunits. Cofilin binds the upper actin subunits, closest to the minus or pointed end, between actin

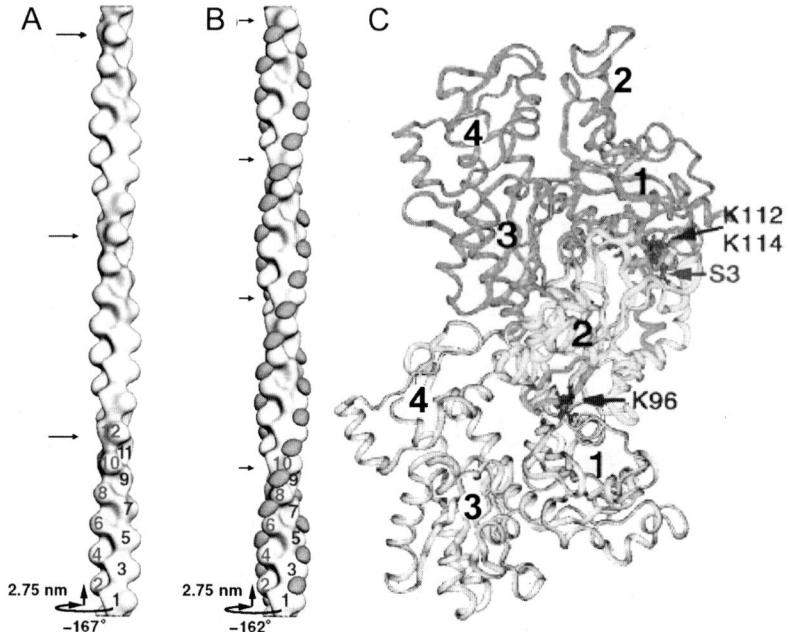

Fig. 4. Effects of AC binding on F-actin conformation. Structural (low resolution) model of naked F-actin (**a**) and F-actin saturated with cofilin (*blue*) (**b**). A higher resolution model (**c**) showing presumed fit of cofilin on two actin subunits of a filament. The subdomains (1–4) of each actin subunit are labeled. Model is based on molecular dynamic simulations for docking of cofilin to F-actin (see Pope et al. 2004) (*See Color Plates*)

subdomains 1 and 3 and interacts with the lower actin subunit at actin subdomain 2 (Fig. 4).

AC binding traps the actin subunits in a normally less occupied rotated conformation, shortening the repeated helical twist (Galkin et al. 2001, 2003; McGough et al. 1997). Upon AC binding, longitudinal interactions between actin subunits are converted to those similar to ones found normally at the filament pointed end, which results in enhanced filament instability (Galkin et al. 2003; Orlova et al. 2004). The twist in F-actin bound to cofilin results in the loss of the phalloidin-binding site (McGough et al. 1997). A pH-dependent binding of a second ADF that enhances a tilt of actin subunits has been inferred from thicker filaments observed by electron cryomicroscopy (Galkin et al. 2001) and used to explain the pH-dependent activity of AC, but no biochemical confirmation of this 2:1 complex has been reported.

AC proteins bind F-actin cooperatively and saturate at a 1:1 AC:actin molar ratio (McGough et al. 1997; Hawkins et al. 1993; Hayden et al. 1993), resulting in shorter and thicker actin filaments (Nishida, Maekawa and Sakai 1984b; McGough et al. 1997). AC proteins preferentially bind to ADP-F-actin over ATP- or ADP-P*i*-F-actin subunits (Carlier et al. 1997; Maciver and Weeds 1994). A detailed biophysical study of cofilin effects on actin filaments has confirmed that cofilin binds cooperatively and increases the rotational motions of subunits within the filament over longer distances than between neighboring subunits (Bobkov et al. 2004; Prochniewicz et al. 2005). Therefore, two distinct structured changes were proposed to occur in F-actin upon cobilin binding (Prochniewicz et al. 2005). A local change affects the structure of actin's C-terminus and is thought to be the one likely to mediate nearest-neighbor cooperative binding and filament severing. The second change is an increase in the torsional flexibility of the filaments, which could explain the increased rate of P*i* release following ATP hydrolysis by the actin subunits (Blanchoin and Pollard 1999), as well as dynamic interactions between other ABPs (e.g., TMs) and AC.

The major actin-binding regions of ADF/cofilin were determined from crosslinking (Sutoh and Mabuchi 1989) and mutagenesis studies (Lappalainen et al. 1997; Moriyama et al. 1992; Yonezawa et al. 1989, 1991). Other point mutations have mapped out residues important for the severing activity and for the monomer-sequestering activity. The major actin monomer-binding regions include the long α-helix extending from residues 111 to 128 of human ADF, especially Lys112 and Lys114 and the N-terminal region (residues 1–20, but especially residues 1–5) (Hatanaka et al. 1996; Lappalainen et al. 1997). Additional residues required for F-actin binding, which are presumed to make contact with a second actin subunit within the filament, are residues 96–98 and 151–152 (Lappalainen et al. 1997; Ono et al. 1991).

In addition to severing, ADF and cofilin enhance the rates of monomer removal (depolymerization) from the pointed ends. Plant ADF1 increases the rate of actin dissociation from the pointed end of F-actin by 25-fold, whereas it does not affect dissociation from the barbed end (Carlier et al. 1997; Maciver et al. 1998). Other AC proteins increase the pointed end off-rate 2- to 40-fold, depending on pH and correcting for their degree of severing (Chen et al. 2004; Yeoh et al. 2002). A schematic model for AC effects on F-actin is shown in Fig. 5.

Regulation of AC

Regulation of AC is critical to proper actin cytoskeletal dynamics and is accomplished in multiple ways. Although single-cell eukaryotes express one AC family member from a single gene, metazoans express multiple isoforms in a developmental- and tissue-dependent manner. Expression of multiple AC family members occurs as a result of alternative splicing in *Caenorhabditis elegans* or gene duplication in vertebrates (Bamburg 1999).

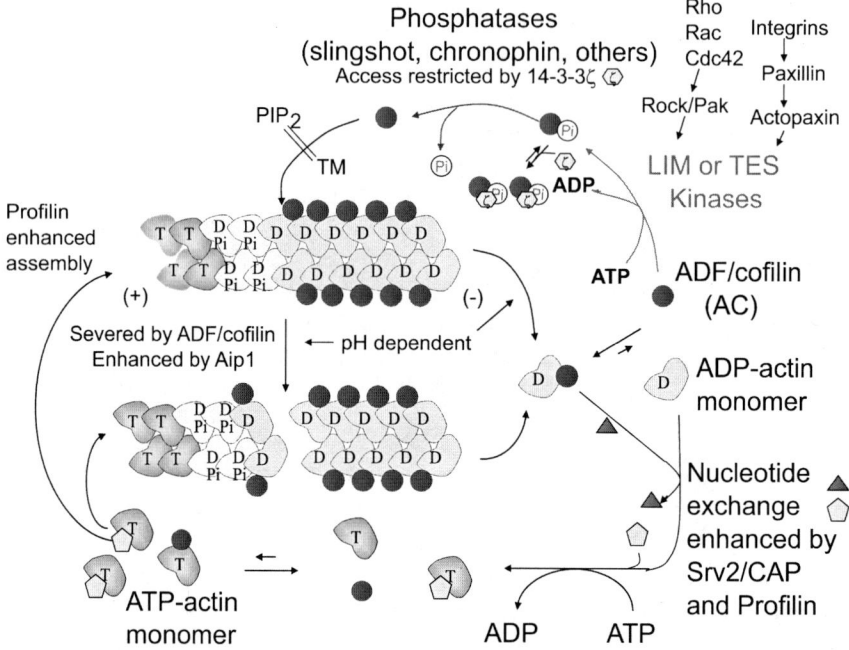

FIG. 5. The actin-dynamizing cycle of ADF/cofilin. Most AC proteins enhance the off-rate of subunits from the filament pointed end and sever filaments. Severing can be enhanced by Aip1. Severing creates more pointed ends from which depolymerization can occur, but also more barbed ends that can promote nucleated growth of filaments if conditions are right. Phosphorylation of metazoan AC proteins by kinases that are regulated downstream of receptors or adhesion molecules inactivates them, whereas dephosphorylation by phosphatases activates them. The phosphorylated AC may be targeted to its site of activation by 14-3-3 scaffolding proteins, which may also protect phosphorylated AC from dephosphorylation by general phosphatases. AC binding to membrane PIP$_2$ inhibits its binding to actin. Many isoforms of TM compete with AC proteins for F-actin binding, although one isoform may cooperate with AC to enhance turnover. Nucleotide exchange on the actin monomer released from the filament is inhibited by AC but can be enhanced by Srv2/CAP and/or profilin, the latter of which helps sequester actin monomers and targets them to the plus ends of F-actin, especially those filaments utilizing formins for enhancing growth (*See Color Plates*)

The AC homolog in *C. elegans* is encoded by the *unc-60* gene, which gives rise to unc-60A and unc-60B through alternative splicing (McKim et al. 1994). Although displaying 38% identity, unc-60A and unc-60B differentially regulate actin dynamics, suggesting the two proteins have separate functions *in vivo* (Ono, Baillie and Benian 1999). Indeed, unc-60B is muscle-specific whereas unc-60A is ubiquitous.

Phosphorylation

Phosphorylation on Ser3 of metazoan ACs generates the inactive phosphoAC (pAC) (Agnew, Minamide and Bamburg 1995; Kanamori, Suzuki and Titani 1998; Morgan et al. 1993; Moriyama, Iida and Yahara 1996; Nebl, Meuer and Samstag 1996). Following phosphorylation, ADF affinity for actin is decreased by 20-fold and most likely results in the observed decrease in actin-depolymerizing activity (Ressad et al. 1998). Phosphorylation of actophorin did not alter its three-dimensional structure, suggesting that the addition of phosphate creates a charge repulsion of the AC with residues in actin at the binding interface (Blanchoin et al. 2000).

Extracellular stimulation can activate actin dynamics by increasing phosphocycling (Meberg et al. 1998; Pandey et al. 2006) or by reducing pAC levels (Aizawa et al. 2001; Bamburg and Wiggan 2002; Chen, Bernstein and Bamburg 2000; Pandey et al. 2006). The kinetics and intensity of kinase/phosphatase activation depends on the pathways stimulated (e.g., Pandey et al. 2006). Examples of actin dynamics changed through net alterations in pAC levels include NGF-stimulated and normal neurite growth (Aizawa et al. 2001; Endo et al. 2003; Meberg and Bamburg 2000; Meberg et al. 1998; Sarmiere and Bamburg 2004), semaphorin 3A-induced collapse of growth cones (Aizawa et al. 2001), and thrombin-stimulated shape change in platelets (Pandey et al. 2006). Cofilin phosphoregulation is also essential for *Drosophila* axon growth (Ng and Luo 2004), fibroblast polarization (Dawe et al. 2003), and directional cell migration (Nishita et al. 2005). Pseudophosphorylated Ser3 to Asp3 or Glu3 AC mutations (S3D or S3E) binds actin weakly (Nagaoka, Abe and Obinata 1996) but sometimes has surprising dominant-negative effects in transfected cells (Dawe et al. 2003; Gehler et al. 2004).

Two related families of ubiquitous kinases, TES and LIM kinases, phosphorylate ACs on Ser3. Testicular kinase (TESK) (Toshima et al. 2001a) is inactivated when bound at focal adhesions but is released by integrin signaling (LaLonde et al. 2005). Lin-11, Isl-1, and Mec-3 kinase (LIMK), a serine/threonine kinase containing LIM and PDZ domains, phosphorylate vertebrate ACs *in vitro* and *in vivo* (Arber et al. 1998; Yang et al. 1998). The LIM domain targets the kinase to the Golgi, whereas the PDZ domain targets the kinase to the plasma membrane or neuronal growth cone (Rosso et al. 2004). LIMK has two isoforms, LIMK1 and LIMK2, each of which is regulated downstream of the Rho family of small GTPases (Amano et al. 2001; Edwards et al. 1999; Nunoue et al. 1995; Sumi, Matsumoto and Nakamura 2001; Sumi et al. 2001). LIM kinases form homodimers (Hiraoka et al. 1996) and, following their activation by phosphorylation on a Thr508 in LIMK1 or Thr505 in LIMK2 (Edwards and Gill 1999; Sumi, Matsumoto and Nakamura 2001), transphosphorylate each other leading to their full activation (Edwards et al. 1999; Hiraoka et al. 1996).

Two brain LIMK isoforms, one of LIMK2 with a six amino acid insert in the kinase domain, and a second of LIMK1 with a 20 amino acid deletion in the kinase domain, are transiently expressed during development. Inactive

LIMK1 behaves as a dominant-negative form (Bernard et al. 1994; Ikebe et al. 1997) and its expression in cells inhibits formation of Rac-induced lamellipodia, whereas expression of constitutively active Rac induces LIMK1 autophosphorylation and phosphorylation of cofilin (Arber et al. 1998; Yang et al. 1998). Disruption of either the LIMK1 or 2 genes is not associated with the development of severe phenotypes (Meng et al. 2002; Takahashi et al. 2002). Although mouse double knockout of LIMK1 and 2 (Meng et al. 2004) exhibits abnormalities in synaptic function, no gross developmental defects are apparent. Some residual phosphorylated AC occurs in these animals, presumably because of compensation by TESK, and other mechanisms of AC regulation (see below) are almost certainly at work.

Because virtually all extracellular regulators of directional cell migration have Rho family GTPases as downstream targets (Song and Poo 2001), ACs are likely integrators of transmembrane signaling that lead to directional migration through localized actin assembly regulation (Meyer and Feldman 2002; Sarmiere and Bamburg 2004). p21-activated kinases (PAKs), which are downstream targets of Rac and Cdc42 (Bagrodia and Cerione 1999), regulate fibroblast motility (Sells, Boyd and Chernoff 1999) and neurite outgrowth (Daniels, Hall and Bokoch 1998). PAK1 and PAK4 bind and phosphorylate LIMK1 within its activation loop and increase LIMK1-mediated phosphorylation of cofilin (Edwards et al. 1999; Dan et al. 2001). PAK2 also mediates cell motility but it can be activated in macrophages by membrane recruitment. Membrane recruitment leads to activation through LIMK association with adapter proteins (e.g., NCK) linked to receptor tyrosine kinases, thereby leading to cofilin phosphorylation (Misra, Sharma and Pizzo 2005). Rho mediates many of its effects on actin dynamics via Rho kinase (ROCK), which, like PAK, activates LIMK1/2 (Edwards et al. 1999; Maekawa et al. 1999; Ohashi et al. 2000; Sumi, Matsumoto and Nakamura 2001). These pathways may be cell type specific in that downstream of RhoA, ROCK may specifically activate LIMK2 and not LIMK1 (Amano et al. 2001; Sumi, Matsumoto and Nakamura 2001). LIMK1 and LIMK2 are also activated by the Cdc42 downstream effector myotonic dystrophy kinase-related Cdc42-binding kinase (MRCKalpha) leading to phosphorylated cofilin (Sumi, Matsumoto and Nakamura 2001).

pAC, a weak substrate for protein phosphatases 1 (PP1), 2A, and 2B (Meberg et al. 1998; Samstag and Nebl 2003), is regulated by more specific phosphatases: the phosphorylation-regulated slingshot (SSH) (Niwa et al. 2002; Nagata-Ohashi et al. 2004; Ohta et al. 2003) and chronophin with unknown regulation (Gohla, Birkenfeld and Bokoch 2005). Three human SSH homologs have been identified each with long (L) and short (S) forms (Niwa et al. 2002; Ohta et al. 2003). These SSH isoforms show distinct subcellular localization and expression profiles (Ohta et al. 2003). SSH overexpression disrupts actin reorganization induced by either LIMK or TESK overexpression, suggesting it counteracts their activity (Niwa et al. 2002). More specifically, SSH-1L, abundant in neuronal growth cones and other dynamic membrane domains (Gehler et al. 2004; Soosairajah et al. 2005), is only active when bound to F-actin through its C-terminal tail and is inactivated by phosphorylation within its N-terminus (Nagata-Ohashi et al. 2004; Soosairajah et al. 2005). Purified SSH-1L and LIMK1 bind directly to one another *in vitro* (Soosairajah et al. 2005). SSH-1L acts as a LIMK1 phosphatase, removing phosphates at Thr508 as well as at autophosphorylation sites. Thus SSH-1L activates AC and inhibits LIMK1. Furthermore, SSH-1L can be inactivated by PAK4, which can activate LIMK1. These protein kinase/protein phosphatase interactions and activities provide a dynamic bidirectional control mechanism governing AC phosphocycling. Currently little is known about mechanisms regulating other SSH isoforms.

In *Drosophila*, loss of the single SSH locus induces an increase in F-actin and phosphorylated cofilin (Niwa et al. 2002). In chick dorsal root ganglia (Endo et al. 2003), and in chemokine-stimulated Jurkat T cells (Nishita et al. 2005), SSH and LIMK1 regulate the activity of cofilin, and subsequently, the changes in the actin cytoskeleton crucial for directed cellular migration. However, neither yeast nor *Dictyostelium* cofilin appears to be regulated by phosphorylation, implying that other mechanisms exist to modulate AC activity. A schematic of the AC phosphocycle and how this is coupled to actin dynamics is shown in Fig. 5. Included also are the sites where regulation is conferred by other proteins, which are discussed below.

Following release of the inorganic phosphate, active AC binds to ADP-F-actin, inducing or stabilizing a twisted form of the filament. AC binding destabilizes F-actin by increasing the twist per actin subunit by 5° (McGough et al. 1997), while weakening the lateral interactions in the filament (Bobkov et al. 2004), the likely cause of a 5- to 20-fold increase in pointed end depolymerization. Cofilin binding to ADP-F-actin is cooperative and mutually exclusive with phalloidin binding. It seems likely, although not proven, that severing occurs at the boundary between the AC saturated and unbound regions. AC bound to monomeric ADP-actin is released from the pointed end of the filament and nucleotide exchange is inhibited as long as the AC remains bound. Although AC binds ADP-actin with a higher affinity than profilin under physiological conditions (Pantaloni and Carlier 1993), profilin-induced nucleotide exchange occurs in the presence of AC (Blanchoin and Pollard 1998; Didry, Carlier and Pantaloni 1998), but exchange by the Srv2/CAP1 protein is probably more relevant *in vivo* (Balcer et al. 2003; Moriyama and Yahara 2002).

pH

At pH below 7.1 *in vitro*, cofilin (Yonezawa, Nishida and Sakai 1985) and ADF (Hawkins et al. 1993; Hayden et al. 1993) have a weaker actin-depolymerizing activity than at higher pH (Chen et al. 2004). However, the ADF–actin complex has a much higher critical concentration for assembly than cofilin actin (Bernstein et al. 2000; Chen et al. 2004; Yeoh et al. 2002). This same behavior occurs *in vivo*, where it has been observed that increasing intracellular pH results in more ADF colocalizing with G-actin. In keeping with the ability of cofilin to associate with and stimulate F-actin increase in cells, the pH shift has little effect on cofilin, which remains mostly F-actin associated (Bernstein et al. 2000).

Phosphatidylinositol-Phosphate Binding

In vitro, phosphatidylinositol (PI) phosphates (PIP and PIP_2) inhibit the interaction between actin and many ABPs including AC (Kusano, Abe and Obinata 1999; Yonezawa et al. 1990, 1991). *In vivo*, epidermal growth factor (EGF) stimulation of adenocarcinoma cells reduces PIP_2 through activation of phospholipase C (PLC) and results in the activation of cofilin, possibly by cofilin release from the membrane (Mouneimne et al. 2004; Wang, Shibasaki and Mizuno 2005). The role of PIP_2 is discussed in more detail in the chapter by dos Remedios and Nosworthy.

14-3-3 Proteins

At least seven isoforms of 14-3-3 proteins occur in mammals, and these affect signaling by modulating localization, activity, or protein–protein interactions of phosphoserine-containing proteins (Fu, Subramanian and Masters 2000). Most phosphoserine-containing proteins will bind a number of different 14-3-3 isoforms, although there is usually some isoform selectivity. Cofilin interacts with 14-3-3ζ and ε, both *in vitro* and *in situ* (Gohla and Bokoch 2002). In addition to its phosphorylatable Ser3, cofilin's interaction with 14-3-3 requires intact Ser23 and Ser24 (Gohla and Bokoch 2002), which are not phosphorylated *in vivo* (Agnew, Minamide and Bamburg 1995; Kanamori, Suzuki and Titani 1998). Overexpression of 14-3-3ζ increases p-cofilin levels (Gohla and Bokoch 2002), suggesting 14-3-3 protects AC from dephosphorylation by some phosphatases *in vivo*. However, 14-3-3 binding does not prevent pAC dephosphorylation by purified SSH-1L *in vitro* or by expressed SSH-1L *in vivo* (Soosairajah et al. 2005).

In addition to interacting directly with AC, 14-3-3 also interacts with the AC kinases, LIMK1, and TESK1 (Birkenfield, Betz and Roth 2003; Toshima et al. 2001a,b). 14-3-3 does not stimulate LIMK phosphorylation of cofilin (Gohla and Bokoch 2002), nor does it inhibit SSH-1L dephosphorylation of AC or LIMK1 (Soosairajah et al. 2005). 14-3-3β inhibits TESK activity and the ability of TESK to be activated by integrin-mediated release from focal adhesions (LaLonde et al. 2005).

Tropomyosins

TMs are α-helical coiled-coil proteins that cooperatively bind along actin filaments and can protect filaments from AC-dependent depolymerization (Bernstein and Bamburg 1982; Ono and Ono 2002) by altering actin structure (McGough 1998). They may also block Arp2/3-mediated branching (Blanchoin, Pollard and Hitchcock-DeGregori 2001). Yeast expressing a mutant actin V159N contain actin filaments that depolymerize slowly because of a failure of actin to undergo a conformational change following inorganic phosphate release that enhances the affinity for cofilin binding

(Belmont and Drubin 1998). Crossing this mutant yeast strain with a knockout of TM1 rescues the normal phenotype, restoring cofilin labeling of actin filaments and strongly suggesting that cofilin and TM work in opposite directions to turnover the actin filament network in yeast. However, the situation in mammals is more complex.

TMs in mammals are transcribed from four different genes (α, β, γ, and δ), which combined with alternative splicing produce >40 different isoforms (Gunning et al. 2005) that are classified into higher (~284 amino acids) and lower (247 amino acids) molecular weight groups. Stable overexpression of the nonmuscle α-TM gene product, TM5NM1, in neuroblastoma cells results in large, spread cells with abundant contractile filaments and an increased but diffusely staining pAC pool (Bryce et al. 2003). In these cells, transient expression of TMBr3, a much weaker F-actin-binding isoform derived from the α-TM gene, decreases stress fibers and active myosin II, and promotes formation of lamellipodia containing AC and TMBr3 (Bryce et al. 2003). These results suggest that some TM isoforms might cooperate with cofilin to turnover actin filaments.

In epithelial cells, TMs aid in controlling apical and basolateral transport (Dalby-Payne, O'Loughlin and Gunning 2003). Since many TM isoforms and AC can compete for binding to F-actin, it is likely that cellular differentiation requires at least two distinct populations of F-actin, one being a more stable TM–actin population and the other a less stable AC–actin population, localized to distinct regions of the cell. However, as noted above, certain TM isoforms may also bind filaments along with AC, perhaps targeting AC to these formerly stable filaments and increasing turnover (Bryce et al. 2003).

TM isoform expression modulates neuronal form and development. Expression of many TM isoforms, including at least ten isoforms in brain, result from different *TM* genes and alternative splicing (Lees-Miller, Goodwin and Helfman 1990; Dufour, Weinberger and Gunning 1998). A subset of TM isoforms is expressed in developing mouse brain in a spatially restricted and temporal manner, suggesting roles in neuritogenesis and axonogenesis by sorting actin filaments into different functional compartments. The spatial and temporal expression patterns of the γ-*TM* gene products in mouse brain (Vrhovski et al. 2003), as well as studies with transgenic mice, support this idea (Schevzov et al. 2005).

TM isoforms are not functionally redundant although some do overlap in their expression and localization. In developing neurons, the proper spatial and temporal localization of TM isoforms is important for the establishment of axonal polarity (Hannan et al. 1995). *TM II* gene disruption in *Drosophila* significantly restructures dendritic fields of the peripheral nervous system (Li and Gao 2003). Cortical neurons from TM5NM1 transgenic mice have longer and more highly branched axons with greatly enlarged growth cones. TM5NM1 is almost exclusively in the axons of newly differentiating neurons but is replaced during outgrowth by TMBr3, an isoform associated with neuronal maturation (Weinberger et al. 1993). Cultured cortical neurons from mice overexpressing the higher molecular weight α-gene product TM3 show an initial inhibition of neurite outgrowth followed by a decrease in dendritic number and length (Schevzov et al. 2005), even though the homozygous transgenic animals develop normally. Of particular interest is the differential sorting of two closely related isoforms: TM5NM1 associated with cultured cell stress fibers and TM5NM2 associated with Golgi short actin filaments (Heimann et al. 1999; Percival et al. 2004). These isoforms differ only in the splicing of a single exon (Gunning et al. 2005). Thus TM isoforms and AC proteins appear to work in concert to protect certain actin filaments and reorganize others during neuronal development.

Actin-Interacting Protein 1

Actin-interacting protein 1 (Aip1; also known as WD-repeat protein 1 or WDR1 in chick and mammals and unc-78 in *C. elegans*) is an AC regulating protein, first identified in yeast (Rodal et al. 1999). It caps barbed ends of AC-bound F-actin, preventing filament annealing (Okada et al. 2002), and it enhances severing activity of AC (Mohri and Ono 2003; Ono, Mohri and Ono 2004). Several Aip1 homologs have been discovered in metazoans (Voegtli, Madrona and Wilson 2003). Chick Aip1 is rapidly upregulated in noise damaged chick cochlea and associates with ADF and actin structures (Oh et al. 2002), as it also does in PC12 cells (Shin et al. 2004). Aip1 supports mammalian mitotic cell rounding (Fujibuchi et al. 2005), and the elimination of Aip1 by siRNA can impair cell migration and cytokinesis (Rogers et al. 2003).

Srv2/CAP

The Srv2/CAP family in yeast has been demonstrated to form a high molecular weight complex linked to actin filaments via Abp1 (Balcer et al. 2003). This Srv2 complex catalytically accelerates cofilin-dependent actin turnover in two ways: by releasing cofilin from ADP-actin monomers and by enhancing profilin-mediated nucleotide exchange on actin monomers. CAPs are a family of highly conserved actin monomer-binding proteins found in all eukaryotes (Bertling et al. 2004; Moriyama and Yahara 2002). Two isoforms, CAP1/CAP2, are expressed in a cell type specific manner and knockdown of CAP1 results in decreased actin filament depolymeriziation and improper cofilin subcellular localization leading to defects in cellular morphology, migration, and endocytosis. Finally, CAP1/ASP56, the human homolog to yeast Srv2/CAP1, was also found to enhance F-actin depolymerization and to increase the rate of G-actin nucleotide exchange (Moriyama and Yahara 2002). Thus CAPs are important regulators of cofilin-mediated actin filament dynamics.

Profilin

Profilin, an actin monomer sequestering protein (reviewed in Sun, Kwiatkowska and Yin 1995), binds to monomeric actin and opens the nucleotide-binding pocket to enhance nucleotide exchange (Nishida, 1985). It was identified as a protein that could cooperate with AC in the turnover of actin filaments by enhancing the exchange of ATP for ADP on AC depolymerized ADP-actin subunits and contribute to the >100 fold increase in filament turnover by AC *in vitro* (Didry, Carlier and Pantaloni 1998). Profilin-actin complexes enhance the growth rate of filament-barbed ends (Gutsche-Perelroizen et al. 1999). Profilin along with AC, Arp2/3 complex, Ena/VASP and a barbed end capping protein were identified as the essential proteins needed to maintain *Listeria* comet-tail motility *in vitro* (Loisel et al. 1999), a model for membrane protrusive activity (Carlier et al. 2003). Within the cell this enhanced assembly is presumably regulated through interactions with barbed end associated assembly factors that have profilin binding domains, such as the Ena/VASP proteins or the Diaphanous related formins (Laurent et al. 1999; Romero et al. 2004).

Functions of ADF/Cofilin That Define Potential Roles in Disease

Cytokinesis

Perturbation of AC activity affects cytokinesis. *Xenopus laevis* expresses two functionally related proteins with 77% sequence identity to chick cofilin and 66% identity to chick ADF, named *Xenopus* ADF/cofilin (XAC) 1 and 2 (Abe et al. 1996). Their activities are very similar to chick ADF (Chen et al. 2004). In the unfertilized *Xenopus* oocyte, most of the XAC is in the inactive phosphory-lated state but activity (dephosphorylation) increases following fertilization and cycles during mitosis. Two spikes of increased XAC activity were seen during M phase, one during prophase and another during telophase. XAC activity was reduced during metaphase and anaphase. This pattern was repeated in consecutive cell cycles (Tanaka, Okubo and Abe 2005). Cytokinesis is blocked either by inhibiting XAC through injecting one blastomere at the two-cell stage with an XAC antibody or by increasing XAC activity by injecting the constitutively active Ser3 to Ala (S3A) mutant XAC S3A. Neither injection affected nuclear division but the XAC S3A injection caused complete regression of the cleavage furrow. Together these results imply that XAC is required for cytokinesis and its regulation by phosphorylation is essential for proper cytokinesis (Abe et al. 1996; Tanaka et al. 2005b).

Mutations in twinstar, the *Drosophila melanogaster* AC homolog, result in aberrant accumulations of F-actin that are specifically observed at the contractile ring (Gunsalus et al. 1995). These accumulations of F-actin do not disassemble and are frequently associated with the failure of cells to progress through cytokinesis. Likewise, treatment of *Drosophila* S2 tissue culture cells with siRNAs specific for *twinstar* produced an excess accumulation of F-actin and prominent actin rings resulting in a disruption in cytokinesis (Somma et al. 2002). Silencing of the AC homolog, *unc-60*, in *C. elegans* with siRNA also disrupted embryonic development and cytokinesis (Ono et al. 2003). Other proteins commonly implicated in cytokinesis defects are often involved in AC and or actin regulation. A mutation of chickadee, the *D. melanogaster* profilin homolog, resulted in severely defective meiotic cytokinesis and failure of spermatocytes to form an actomyosin contractile ring (Giansanti et al. 1998). Temporal regulation of cofilin activity occurs during meiosis via phosphorylation/dephosphorylation. The phosphatase activity of SSH-1 decreases in the early stages of meiosis and is elevated in telophase and cytokinesis, consistent with cofilin dephosphorylation. SSH-1L colocalizes with F-actin and accumulates at the cleavage furrow. Expression of a phosphatase dead SSH-1 results in regression of the cleavage furrow and leads to multinucleated cells (Kaji et al. 2003).

As stated above, changes in ADF/cofilin activity can lead to a lack of actin filament formation or removal creating defects in cytokinesis (Ono et al. 2003; Tanaka et al. 2005; Tanaka, Okubo and Abe 2005). Similar defects in cytokinesis are also observed in cells in which activities of either LIM kinase or slingshot phosphatase are abnormal (Amano et al. 2002; Bernstein et al. 2000; Kaji et al. 2003; Tanaka, Okubo and Abe 2005) and in cells in which both ADF and cofilin expression have been silenced with siRNA (Hotulainen et al. 2005). Rescue of cytokinesis by expression of either ADF or cofilin in these siRNA-treated cells demonstrates their functional overlap in this process. Furthermore, mitotic cell flattening but not rounding is induced in cells by suppression of Aip1 expression, although no changes in cofilin phosphorylation or distribution are observed. Therefore, cell morphological changes during mitotic cell rounding appear to be regulated via a functional cooperativity of Aip1 and cofilin on actin (Fujibuchi et al. 2005).

Cell Polarity and Polarized Migration

Upstream regulators of polarity in both neurons and fibroblasts are part of an evolutionarily conserved complex that includes Par-3 and Par-6, either a Rac or Cdc42 GTPase, an atypical protein kinase C (PKC), a PAK, and PI-3-kinase (reviewed in Doe 2001; Macara 2004a,b; Nichols, Fraser and Heitman 2004; Shi, Jan and Jan 2003; Wodarz 2002). Rac activation is restricted to the leading edge of the lamellipodium in migrating cells by a complex containing α_4-integrin, paxillin, and GIT1 (an Arf-GAP) (Nishiya et al. 2005). Along the sides of the lamellipodium, Arf (and thus Rac) is inactivated by the enhanced GIT1 activity associated with the complex. The Par complex may also work, in part, through Diaphanous-related formins, one of which, mDia1 (Krebs et al. 2001; Palazzo et al. 2001; Watanabe et al. 1997), binds activated Rho GTPase. Par-3 regulates Rac activity through affects on the Rac GTP exchange factor Tiam-1 (Chen and Macara 2005), but can also bind and inactivate LIM kinase 2 directly, leading to enhanced AC activation (Chen and Macara 2006).

There are four well-defined steps involved in cell migration, one of which is the establishment of a single lamellipodium (leading edge) that defines the cell as polar. The protrusion of this membrane followed by establishment of adhesions, contraction of the cell tail and de-adhesion, all require the spatial and temporal regulation of actin dynamics, making cell migration a complex and highly integrated event. Various ABPs influence the arrangement and rate of actin polymerization at the leading edge by regulating the pool of actin monomers and free filament ends (Pollard and Borisy 2003). Microinjection of a cofilin function-blocking antibody into a metastatic carcinoma cell line inhibited both membrane ruffle protrusion and the formation of barbed ends (Chan et al. 2000; Lorenz et al. 2004). In adenocarcinoma cells stimulated to migrate in response to EGF, activation of cofilin occurs by depletion of membrane PIP_2 through the activation of PLCγ, and presumably its release from the membrane (Mouneimne et al. 2004; Wang, Shibasaki and Mizuno 2005). A leading edge in these cells also can be established by local activation of cofilin through photo-uncaging (Ghosh et al. 2004). The active cofilin generates free barbed ends at the membrane necessary for assembly of the branched actin filament network ("dendritic brush") that triggers membrane protrusion through actin assembly (DesMarais et al. 2005).

On a two-dimensional surface, migrating fibroblasts exhibit a polarized shape with a flat, wide lamella that terminates in a lamellipodium consisting of protruding and ruffling membrane with tightly coupled actin disassembly (Cramer 1999). Actin assembly provides the protrusive driving force for lamellipodial and filopodial expansion (Pollard and Borisy 2003; Gehler et al. 2004). When the forward protrusive lamellipodia fail to maintain adhesion and undergo centripetal or retrograde flow, ruffles form. The lamellipodium has two spatially colocalized but kinetically distinct actin networks (Ponti et al. 2004). Random protrusion of the leading edge is driven by actin assembly within 1–3 μm of the membrane; retraction is driven by disassembly (Vallotton et al. 2004). The second colocalized lamellar F-actin network more effectively advances the cell by integrating substrate adhesion and myosin contraction (Ponti et al. 2004) as has been proposed for growth cones (Suter and Forscher 2000). Microinjection of F-actin-stabilizing TM eliminated lamellipodia, but lamella still drove migration (Gupton et al. 2005). However, such cells may be unable to respond to environmental cues because directional migration in response to guidance cues requires dynamic actin filaments (Bentley and Toroian-Raymond 1986; Nishita et al. 2005).

Maintaining polarized migration in fibroblasts requires active ADF/cofilin (Dawe et al. 2003). Inactivating ADF/cofilin through overexpression of LIMK will cause the development of multiple lamellipodia and loss of polarity, but polarity can be rescued by expressing an active, nonphosphorylatable form S3A of ADF/cofilin, suggesting that phosphocycling, which the S3A mutant cannot undergo, is not important to the maintenance of migration but that turnover of actin to maintain protrusion within the leading edge is important. The ability of the S3E mutant of

ADF/cofilin to disrupt polarized migration further suggests that delivery and/or local activation of phospho-ADF/cofilin is important to maintain this process (Dawe et al. 2003). However, the requirements for LIMK and slingshot to alter the direction of polarized migration (Nishita et al. 2005) suggest that the differential activation of ADF/cofilin across the leading edge may be required for execution of signals from extracellular guidance cues.

ADF/cofilin is found throughout the lamellipodium of some nonmigrating fibroblasts all the way to the cell membrane, whereas in the polarized lamellipodium of a rapidly migrating keratocyte it localizes to the middle and rear of the lamellipodium, potentially in a region where ADP-actin filaments are located (Svitkina and Borisy 1999; Pollard and Borisy 2003). However, even in polarized migrating fibroblasts, ADF/cofilin extends to the membrane but the region of most active ADF/cofilin (dephosphorylated) extends only as far into the lamellipodium as that region penetrated by microtubule plus ends (Fig. 6) (Cramer et al., unpublished results). In these cells, active ADF/cofilin surrounds the plus ends of dynamic microtubules but not stable ones, suggesting that microtubules can spatially regulate the activity of ADF/cofilin. Activation of ADF/cofilin and the subsequent severing of older actin filaments could supply the actin monomers necessary for maintaining actin polymerization at the leading edge, or it could maintain a supply of free barbed ends suitable for dendritic nucleation (Ichetovkin, Grant and Condeelis 2002). These data begin to illustrate how ADF/cofilin functions in the complex and coordinated cell migration process.

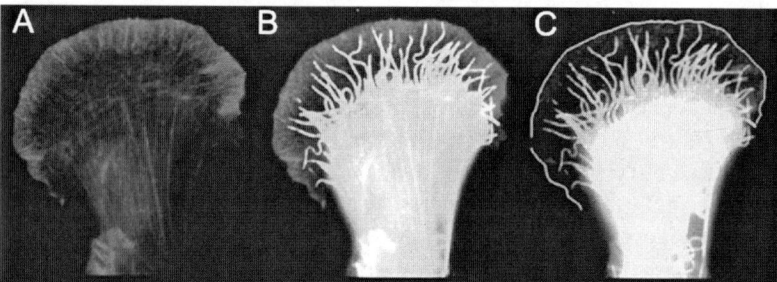

FIG. 6. Microtubules penetrating the lamellipodium of the leading edge of a polarized migrating cell spatially regulate AC activity. (**a**) F-actin in a polarized migrating chick cardiac fibroblast stained with a fluorescent phalloidin. (**b**) Microtubules of same cell stained with antibody to β-tubulin. (**c**) Immunostaining of phosphoAC (*red*) and microtubules (*green*) in leading edge of lamellipodium from the same cell. Total AC staining throughout the lamellipodium is relatively constant (not shown) and thus the compartmentalization of pAC to the leading edge means that the region of most active AC is toward the rear of the lamellipodium where microtubules penetrate. Loss of microtubules or capture of microtubules rapidly alters the zone of active AC (image courtesy of Louise Cramer) (*See Color Plates*)

Golgi Tubule Formation

Textbook models of Golgi function in membrane processing are generally confined to delivery of coatamer protein (COP) II-decorated vesicles from the ER to Golgi, retrograde flow of COP I vesicles from the Golgi to ER, and budding of clathrin-coated vesicles from the trans-Golgi network. Often excluded is the highly dynamic Golgi membrane tubule system, first observed following expression of fluorescent membrane protein chimeras (e.g., Lippincott-Schwartz, Roberts and Hirschberg 2000). Elongated Golgi membrane tubules with high surface/volume ratio sort membrane proteins. These tubules can enhance the rates at which materials transcend the Golgi and are sorted and transferred between compartments (Lippincott-Schwartz, Roberts and Hirschberg 2000), especially between the Golgi and ER membrane systems. Active LIMK1 overexpression in neurons suppresses formation of trans-Golgi-derived tubules, whereas a dominant-negative kinase dead LIMK1 had the opposite effect (Rosso et al. 2004). Overexpressed LIMK1 (or the ΔPDZ-LIMK1 targeted to the Golgi) accelerates axon formation and increases Par-3/Par-6, IGF1 receptors, and NCAM at the growth cone, while inhibiting the export of synaptophysin-containing membrane. Overexpressing the cofilin S3A mutant rescues LIMK1-suppressed Golgi tubule extensions, demonstrating that cofilin is the target of LIMK1. Even in the presence of cofilin (S3A), Golgi tubules are repressed by the actin-stabilizing drug jasplakinolide (see chapter by Braet et al.). Thus, actin filaments are the downstream target of cofilin. These studies suggest that Golgi membrane tubules compete with other delivery systems used for sorting membrane proteins. Altering membrane vesicle targeting can influence biogenesis of cellular membrane compartments, secretion, and development.

Within the region of the Golgi, associated with its vesicles, are short actin filaments, some of which contain the TM isoform TM5NM2 (Percival et al. 2004). Treatment of cells with Brefeldin A (BFA; an inhibitor of ER to Golgi transport) or nocodazole (a microtubule assembly inhibitor) caused Golgi disappearance and dispersion of TM5NM2. Cell treatment with the actin-depolymerizing drug cytochalasin D (CytoD) caused Golgi fragmentation (Rosso et al. 2004), but TM5NM2 remained in a perinuclear region (Percival et al. 2004). One possible interpretation of these results is that there may be more than one actin filament network associated with the Golgi, one containing TM5NM2 that is stable to CytoD (possibly with a very slow turnover), and another that is CytoD sensitive, which regulates Golgi assembly and association with the microtubule system. This latter population may be the one affected by AC.

The essential gene *GAB1* encodes a subunit of the glycosylphosphatidyinositol (GPI) transamidase, a resident ER protein required for synthesis of GPI-linked proteins (Grimme et al. 2004). Interestingly, depletion of the Gab1 protein, or another member of the GPI transamidase complex, Gpi8, causes accumulation of cofilin-decorated actin bars that remain closely associated with perinuclear ER. Morphologically similar bar-like structures containing actin have been observed in nemaline myopathy (NM) (see below) but their origin is still a mystery. However, these results demonstrate a close association between cofilin-dependent actin dynamic processes and the normal functioning of the Golgi and ER in membrane protein trafficking.

Although the specific roles of actin and cofilin in ER and Golgi membrane dynamics are still unknown, recent studies are beginning to shed light on this area. Brefeldin A normally causes a complete resorption of the Golgi into the ER by disrupting ER to Golgi transport without blocking the retrograde transport. In the presence of active LIMK1, which dampens Golgi tubule extensions (Rosso et al. 2004), or in the presence of cofilin S3E, an inactive pseudophosphorylated form, the resorption of the Golgi by the ER is strongly inhibited (Caceras, A., personal communication). Thus, cofilin activity is essential for the retrograde transport from the Golgi to ER. These studies demonstrate a functional interaction between the ER network and the actin cytoskeleton as well as between the Golgi and actin filaments.

Mitochondrial-Dependent Apoptosis

Cofilin plays an important role in the initiation of mitochondrial-dependent apoptosis (Chua et al. 2003). Treatment of neuroblastoma cells with the apoptotic agents, staurosporine or etoposide, induced a translocation of cofilin to the outer mitochondrial membrane measurable within 30 min and peaking by 2 h after treatment. Suppression of cofilin expression by siRNA resulted in cell protection from apoptosis induced by staurosporine or etoposide. Dephosphorylation of cofilin on Ser3 was required for its translocation to mitochondria. In addition, expression of a pseudophosphorylated form of cofilin (S3D) did not localize to mitochondria and inhibited staurosporine-induced apoptosis (Chua et al. 2003). In PC12 cells, overexpression of LIMK1 protected cells from serum-induced apoptosis by inhibiting both caspase 3 and JNK activation (Yang et al. 2004a). These data clearly show that cofilin translocation to mitochondria is an early event in apoptosis and inhibition of cofilin activation by LIMK1, as well as its presumed translocation to mitochondria, can protect cells from caspase-mediated death.

Regulation of apoptosis is essential for normal development of metazoans. Among the regulators of mammalian apoptosis are the six protein members of the Bag family. Ablation of the *Bag1* gene in mice leads to massive cell death of both hematopoietic and neuronal stem cells (Gotz et al. 2005). Bag1 is a known cochaperone for heat shock protein (hsp) 70 (Takayama et al. 1997) and it forms a complex with Akt and B-Raf in normal cells that regulates apoptosis, in part through maintaining phosphorylation of the proapoptotic Bad at Ser136 and also by inhibiting the expression of members of the inhibitor of apoptosis (IAP) family (Gotz et al. 2005). Surprisingly, Akt activity remained normal in Bag1-deficient cells when assayed against other substrates suggesting that blockage of Bad Ser136 phosphorylation is due to factors other than Akt inhibition. The Raf-ERK signaling pathway in Bag1-deficient cells is also undisturbed, but the distribution of B-Raf and Akt were altered in Bag1 null motoneurons, an easily identifiable cell type in the E10–13 embryo. In wild-type motoneurons, active (phospho) Akt is found primarily associated with mitochondria but in the Bag1 null cells, it is diffusely distributed in the cytoplasm, much the same as B-Raf. As discussed in detail later in this review, the *B-Raf* gene is mutated at high frequency in human cancers. In B-Raf null cells, AC is dephosphorylated and F-actin levels are lower (Pritchard et al. 2004). Extrapolating these findings to the cofilin-dependent apoptotic model, the loss of localized B-Raf activity at mitochondria could result in higher levels of the dephosphorylated (active) AC accumulating around mitochondria. Such a condition might enhance cytochrome *c* leakage to stimulate the apoptotic pathway. Indeed, the normal targeting of B-Raf and Akt to this location might be required for maintaining the local inactivation of proapoptotic factors. Furthermore, if active cofilin accumulates, nuclear translocation of cofilin and actin could impact on the regulation of the IAP genes as well (see below).

Nuclear Import of Actin for Chromatin Remodeling and Transcription

The AC proteins in some single-cell organisms and in all metazoans contain a nuclear localization signal, which is cryptic under most conditions, which can be inferred because AC proteins are usually cytoplasmic. Although the function of nuclear cofilin has yet to be established, cofilin and actin are targeted to the nucleus of rat peritoneal mast cells under conditions that disrupt F-actin (latrunculin B) and during ATP depletion (Pendleton et al. 2003). Nuclear translocation of cofilin following its dephosphorylation has been reported during T-cell proliferation and production of interleukin (IL) 2 (Samstag et al. 1994). Furthermore, measurements of free and actin-bound cofilin using fluorescence resonance energy transfer methods of fluorescently labeled actin and cofilin suggest that virtually all of the G-actin within the nucleus remains bound to cofilin, whereas much less cofilin-bound G-actin occurs in the cytoplasm (Chhabra and dos Remedios 2005). Therefore, it is possible that many functions of actin within the nucleus may be mediated by a cofilin–actin complex.

Heat shock and 10% DMSO treatment of rat fibroblastic 3Y1 cells induces dephosphorylation of cofilin preceding nuclear accumulation and nuclear actin/cofilin rod formation (Nishida et al. 1987; Ohta et al. 1989; Ono et al. 1993; Abe et al. 1993; Jiang et al. 1997). DMSO at 10% is used as a cryoprotectant when freezing mammalian cells. When frozen mammalian cell stocks are thawed in fixative, many of the cells have nuclear rods (unpublished observation of the authors), suggesting that translocation of cortical actin to the nucleus may be one mechanism by which cryoprotectants aid in cell freezing. By weakening the crosslinked cortical actin network, a more elastic expansion of the cell membrane can occur during freezing. Human diploid fibroblast cells show increased accumulation of nuclear actin and dephosphocofilin, along with a loss of LIMK1 activity as cells age. Nuclear actin accumulation was found to be both more sensitive and an earlier event than senescence-associated β-galactosidase activity (Kwak et al. 2004).

The INO80 ATP-dependent chromatin-remodeling complex from yeast contains conventional actin along with three actin-related proteins (Arp4, 5, and 8) (Shen et al. 2003). Null mutants of the Arp5 and 8 components are also deficient in binding of Arp4 and actin and have an INO80 complex that is compromised for its ATPase, DNA binding, and nucleosome mobilization activities. Thus, there is a direct involvement of Arps and conventional actin in chromatin remodeling in yeast.

Nuclear actin has been shown to play a role in export of unspliced HIV genomic RNA and the rev responsive element HIV mRNA (Kimura et al. 2000; Hofmann et al. 2001), and also in chromatin remodeling. In human cells, chromatin remodeling is regulated in part by the human Brahma-related gene 1 (Brg1), a catalytic subunit of the Swi/Snf chromatin-remodeling complex, which has been implicated in gene regulation and cell proliferation, as well as a candidate tumor suppressor (Hendricks, Shanahan and Lees 2004). Brg1 regulates genes important for T lymphocyte differentiation (Gebuhr et al. 2003), and thus could be the functional linkage with cofilin nuclear targeting in T-cell proliferation. Brg1 has also been demonstrated to regulate the expression of the cdk inhibitor p21 and is necessary for formation of flat cells, growth arrest, and cell senescence (Kang, Cui and Zhao 2004). Furthermore, mammalian Swi/SNF-like BAF complex binds to phosphatidylinositol 4,5-bisphosphate (PIP$_2$) and this interaction is required for the binding of complex to actin pointed ends and branch points (Rando et al. 2002). Actin binds to two distinct regions of Brg1 one of which is sensitive to PIP$_2$ (Rando et al. 2002). The C-terminus of nuclear DNA helicase II (NDH II) also binds to F-actin, implicating the actin nucleoskeleton in RNA processing, transport, or other actin-related processes (Zhang et al. 2002).

Substantial evidence has also accumulated that actin is required for the transcriptional activity of all three eukaryotic RNA polymerases (reviewed in Visa 2005). Some reports also provide evidence for the presence of a nuclear myosin, which might also play a role in the transcriptional process. While not enough is known to determine the mechanisms by which actin and myosin function in transcription, they may function independently of each other in at least some transcriptional processes that involve RNA polymerase I (Philimonenko et al. 2004).

Together, the above studies demonstrate a nuclear function for actin linked to chromatin remodeling, transcription, and processing of mRNAs. AC proteins, with their NLS, are able to target actin to the nucleus in response to agents that alter the assembly of cytoplasmic F-actin as well as in response to specific signaling pathways. Within the nucleus, much of the G-actin remains bound to cofilin but the nuclear functions of this complex have not been elucidated.

ADF and Cofilin Gene Knockout: Nonmammalian Systems

ADF/cofilin silencing and knockout experiments have been studied in a broad array of model systems ranging from protists to higher eukaryotes. These studies have revealed a central role for ADF/cofilin family members in basic cellular functioning.

Saccharomyces cerevisiae

Yeast contains one copy of cofilin *cof-1* that shares 41% amino acid sequence homology to the mammalian cofilin family of proteins (Moon et al. 1993). Systematic mutagenesis of yeast cofilin has illuminated several unique aspects of cofilin function *in vivo*. A single region of yeast cofilin is sufficient and specific for actin monomer binding; however, an additional region is necessary for binding to filaments (Lappalainen et al. 1997). Yeast cofilin in *Saccharomyces cerevisiae* is essential for cell viability (Moon et al. 1993), and the lethal knockout Δ*cof-1* can be rescued by expression of wild-type mammalian ADF or cofilin (Iida et al. 1993), demonstrating the highly conserved activity of these proteins across phylogeny. Δ*Cof-1* mutants can also be rescued by expression of constitutively active porcine cofilin S3A (Moriyama, Iida and Yahara 1996), suggesting that phosphoregulation of yeast cofilin is not necessary for its function in yeast.

Dictyostelium discoideum

The eukaryotic slime mold *Dictyostelium discoideum* expresses two 15 kDa proteins with 42% primary sequence identity to yeast cofilin (Aizawa et al. 1995). These proteins, D-cofilin-1 and D-cofilin-2, are the product of two genes (Aizawa et al. 1995). Both D-cofilin-1 and -2 depolymerize actin filaments in a concentration and pH-dependent manner, reduce the viscosity of F-actin solutions *in vitro*, and are inhibited by phosphatidylinositol-phosphates (Aizawa et al. 1995, 2001). D-Cofilin-1 is expressed in both vegetative and differentiating cells (Aizawa et al. 1995), whereas D-cofilin-2 is transiently expressed only during the aggregation stage of development (Aizawa et al. 2001). Gene disruption of DCOF1 indicates that it is essential for cell proliferation (Aizawa et al. 1995), whereas disruption of DCOF2 caused an increase in actin accumulation at substrate adhesion sites (Aizawa et al. 2001) but did not affect cell survival and growth.

Caenorhabditis elegans

The nematode *unc-60* gene is homologous to the cofilin gene (McKim et al. 1994). The *unc-60* gene produces two alternatively spiced transcripts sharing high sequence homology and later classified as the Unc-60A and Unc-60B proteins, which possess different roles in regulating actin filament dynamics *in vivo* (Ono and Benian 1998). *Unc-60* is essential for development; a recessive deletion that disrupts both coding regions is lethal (McKim et al. 1994). Unc-60A is expressed ubiquitously in most embryonic cells and RNA interference of Unc-60A is embryonic lethal. Unc-60B, an isoform whose expression pattern is equivalent to mammalian cofilin-2, is found highly expressed in body wall muscle (Ono et al. 2003). Unc-60B is specifically required for proper assembly of actin into myofibrils in the body wall muscle of the worm (Ono, Baillie and Benian 1999). Many different mutants of Unc-60B were made that were deficient in different AC activities associated with F-actin turnover. Regardless of these various molecular phenotypes, thin filaments of muscle cells are disorganized and not bundled with myosin into functional contractile units, resulting in paralysis. Thus, precise control of actin filament dynamics by Unc-60B is required for proper myofibril assembly.

ADF/Cofilin Knockouts in Mammals

Silencing of ADF/Cofilin Expression in Cultured Cells

As expected, cells silenced for both ADF and cofilin with siRNA had severe cytokinetic and morphological alterations (Hotulainen et al. 2005). However, expression of either ADF or cofilin functionally rescued the cells, demonstrating a functional overlap for these two proteins for the growth requirements of cells in culture (Hotulainen et al. 2005).

ADF Null Mice

A spontaneous mutation in mice that caused corneal defects and blindness was mapped to the ADF gene, which was completely deleted (Ikeda et al. 2003). These mice, named *corn1*, develop corneal thickening leading to blindness by 28 days after birth due to abnormal proliferation of corneal epithelial cells, which contain increased amounts of F-actin. Although cofilin expression is upregulated in cells in which ADF is normally highly expressed and rescues most ADF functions, the excessive proliferation of corneal epithelium is not prevented by cofilin (discussed below under corneal disease).

Cofilin-2

There are no reported gene knockouts of cofilin-2 in mammals.

Cofilin-1 Null Mice

Cofilin-1 null mouse embryos die at about day E10.5, even though ADF is upregulated in embryonic cells (Gurniak et al. 2005). At day E9.5, cofilin-1 null embryos are not significantly different from controls, indicating that ADF can substitute for cofilin through gastrulation. However, between day E9.5

and E10.5, cofilin-null embryos showed severe defects in neural tube morpho-genesis, delamination, and neural crest cell migration. Furthermore, cells within cultures of neural crest tissue from these cofilin-null embryos failed to polarize or form F-actin bundles that were prevalent in similar cells from control embryos. Defects in somitogenesis probably contributed to the almost complete lack of cofilin-2 expression in the E10.5 embryos. These results suggest that cofilin-1 is necessary for cell polarity and actin remodeling leading to F-actin bundles and that ADF will not substitute for cofilin in this activity.

Potential Early Developmental Disorders Involving ADF/Cofilin

Because the ADF/cofilins are essential proteins for, cytokinesis, cell polarity, polarized migration, Golgi function, and other processes it is very likely that development of any metazoan would not survive significant defects in the function or expression of these proteins unless other isoforms, such as cofilin in the ADF null mouse, were available to compensate. Likewise, defects in components of certain upstream signaling pathways that regulate the activities of AC are also likely to be lethal unless alternatives for their regulation are available. Thus, many developmental defects that result in miscarriage or stillbirth are possible outcomes of AC deficiency. Below, we will explore the similarities in embryonic defects in mice between some known defects in AC upstream regulatory proteins and the cofilin-null mutant. Mutations in at least one of these upstream molecules are known to cause Waardenburg syndrome, a human developmental defect.

ADF/Cofilin Family Proteins in Disease

Overview

Given the central role for actin in so many cellular processes, and given that actin turnover is so dependent on AC, many diseases ranging from bacterial infection to developmental disorders, neurodegenerative diseases, and cancer may have a connection to AC and actin. An increasing amount of evidence, although at times tenuous, has implicated ADF/cofilin or its regulation in some of these maladies. Here, we will focus on diseases in which a function for AC has been suggested through cell or animal models in which aspects of the disease can be reproduced. We will also discuss some areas in which a role for AC has been hypothesized in order to stimulate thought and experiments to address the issue.

Developmental Disorders

Corneal Disease

Corneal disease is the major cause of bilateral blindness, and accounts for 6% of blindness reported in the US (Corneal Disease Panel 1983). The major cause of corneal thickening is thought to be bacterial infection (Cutler 2004). The cornea, the most anterior portion of the lens, is a smooth and transparent structure that functions to focus light on the retina. Changes in the morphology of the corneal epithelial surface are a common cause of visual loss. Improper thickening of the corneal epithelium can induce corneal scarring, an increase in vulnerability to infection, and corneal stromal neovascularization (Dua and Forrester 1990; Tsai, Sun and Tseng 1990; Turgeon et al. 1990). Typically, corneal stem cells intermittently divide, giving rise to transiently amplifying cells found at the basal layer of the corneal epithelium (Lehrer, Sun and Lavker 1998). The ADF null *corn1* mouse (Ikeda et al. 2003) develops a rough and opaque corneal surface and corneal stromal neovascularization (Smith et al. 1996). In *corn1* mice, an anomaly occurs that causes an increase in cell proliferation of corneal epithelial cells, leading to a phenotype similar to those observed in human corneal disease.

The development of the disease has been followed in *corn1* mice. By 1 week after birth, increased corneal epithelial cell proliferation induced a thickening of the cornea. By day 18, neovascularization in the corneal stroma had occurred (Smith et al. 1996). Another mutation in ADF found in the *corn1²ᴶ* mice, results in an amino acid substitution (P106S) in ADF. *Corn1²ᴶ* mice displayed a milder corneal epithelial phenotype and no neovascularization. The Pro106 residue is conserved and, according to the three-dimensional structure for human ADF (Hatanaka et al. 1995), it marks the end of the fourth β-sheet strand that comes before the major actin-binding α-helix (Lappalainen et al. 1997; Vartiainen et al. 2002). Thus, it is likely that aberrant regulation of the actin cytoskeleton underlies the morphological changes observed in *corn1* and *corn1²ᴶ* mice.

Mutations in *corn1* affect the actin cytoskeleton. Both *corn1* and *corn1²ᴶ* mutant mice displayed an increase in stress fiber-like F-actin staining in corneal epithelial cells compared to wild-type mice, in which F-actin staining was primarily in the cell cortex (Ikeda et al. 2003). Consistent with the properties of ADF, deletion or mutation of ADF results in an accumulation of F-actin. These results imply that proper ADF-mediated actin dynamics are essential for normal regulation of corneal epithelial cell proliferation. In addition, these data describe a potentially powerful animal model that can be used to probe the defects in signaling pathways that contribute to corneal disease.

Mouse cofilin levels are elevated in cells of the *corn1* and *corn1²ᴶ* mutants to compensate for the loss of ADF. Although ADF and cofilin are highly homologous and often redundant in function, they each possess unique properties in the regulation of actin dynamics. Cofilin does not maintain as high an actin monomer pool as ADF (Chen et al. 2004), and actin monomer levels are linked to mRNA stability of many gene products (Minamide et al. 1997). In *corn1* and *corn1²ᴶ* mice, the absence of ADF is not fully compensated by expression of cofilin, which cannot maintain the increased monomer pool. Thus, cells have increased filament formation and perhaps altered turnover of certain mRNAs. We predict that among these mRNAs will be that for β-catenin, which shares with β-actin and ADF a regulatory element involved in actin monomer-dependent mRNA turnover (Hung, H., Curthoys, N., and Bamburg, J. R., unpublished results). β-Catenin is involved in both cytoplasmic actin filament structures and is also a transcription factor that promotes cell proliferation.

Developmental Disorders of the Neural Crest

Pax3 is a member of a class of highly conserved developmental transcription factors essential for normal embryonic development in a wide range of organisms (Dahl, Koseki and Balling 1997). Pax3 contains two DNA-binding motifs: a conserved 128 amino acid paired box (Walther et al. 1991), and a homeobox domain, each of which binds different DNA sequences. There are nine Pax genes in humans, in which mutations are associated with diverse human diseases (Chi and Epstein 2002). Waardenburg syndrome, an autosomal dominant disease characterized by defects to tissues derived from migratory neural crest precursors, develops in humans with Pax3 mutations (Baldwin et al. 1992; Tassabehji et al. 1992). Homozygous mutations to Pax3 in mice results in numerous developmental anomalies including spina bifida, loss of skeletal muscles, skeletal abnormalities and defects to numerous neural crest-derived tissues (Mansouri 1998; Wehr and Gruss 1996). Pax3-deficient mice exhibit somite defects that are consistent with a requirement for Pax3 in the process of mesenchymal condensation and/or mesenchymal to epithelial transition (MET; Fig. 7) (Schubert et al. 2001; Wiggan, Shaw and Bamburg 2006).

Proteins whose expression is regulated by Pax3 in an *in vitro* model for MET include the PAKs, specifically PAK1 and PAK2 (Wiggan, Shaw and Bamburg 2006). PAK2 plays a major role in the actin cytoskeletal organization required for MET, probably exerting its affects via cofilin, whose phosphorylation level is altered as PAK2 expression and activity changes. PAK2 null mice have been generated (Chernoff and Rogers 2004) but not yet extensively characterized. However, these animals are also embryonic lethal at about the same day as the cofilin-null mice (E10.5). Together these results suggest a linkage between Pax3 defects in neural crest cell migration, neural tube closure defects, and PAK2 regulation of cofilin. Further studies will be required to solidify this connection, which undoubtedly also will involve several other targets of PAK2 that are important in cytoskeletal organization.

Gametogenesis and Reproductive Problems

Mutations of the genes encoding ADF/cofilin in *Drosophila* result in sterility of both males and females (Chen et al. 2001). In flies, ADF/cofilin is required for proper border cell migration, cell shape changes, and cell rearrangements during ovary development and oogenesis (Gunsalus et al. 1995). A specific RacGap (RotundRacGAP/RacGAP(84C)) is a protein required for both retinal organization and spermatogenesis in *Drosophila*. Partners of RacGAP(84C) include proteins involved in spermatogenesis, including a serine/threonine kinase called Center divider (Cdi), which shares homology with the human LIM kinase and TESK1 and is a cofilin kinase. *Rac1* and *cdi* are both expressed in *Drosophila*

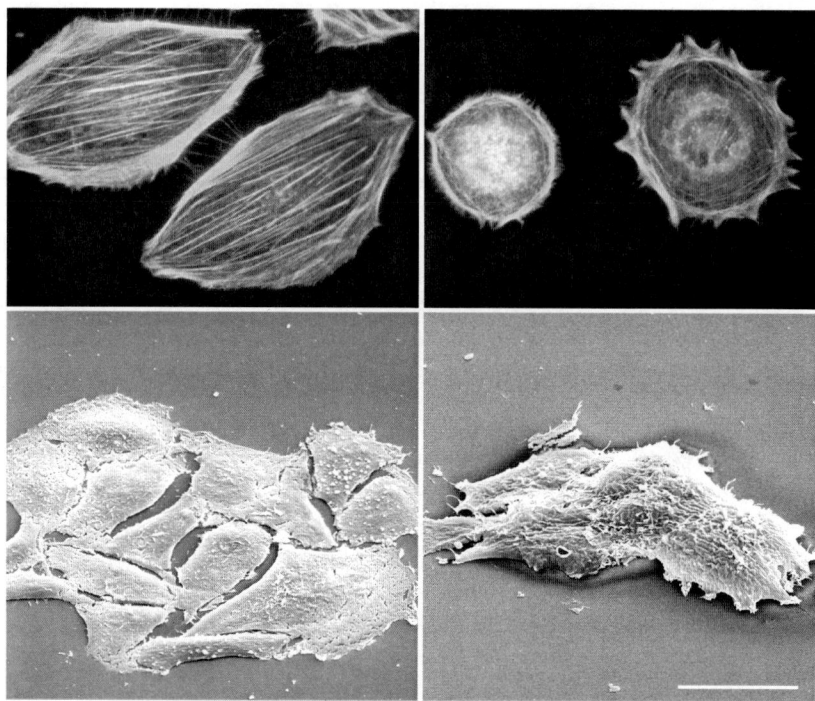

Fig. 7. Mesenchymal to epithelial transition in cultured osteosarcoma cells (Saos2) following expression of Pax3. *Left*: Cells with mesenchymal morphology have actin stress fibers (*green*) anchored at focal adhesions (*red*). Nucleus is stained with DAPI (*blue*). Scanning electron micrograph of similar cells is shown in panel below. Cells are well spread. *Right*: Three days after infection with adenovirus expressing Pax3, the Saos2 cells have undergone a distinctive epithelial morphological change with cortical bands of F-actin (*green*) and loss of focal adhesions. Cells grow in more cuboidal shapes, are taller, and tend to grow more on top of each other as seen in the scanning electron micrograph. Microvillar protrusions are apparent on the apical surface of these cells (image courtesy of O'Neil Wiggan; see Wiggan, Shaw and Bamburg 2006) (*See Color Plates*)

testes and homozygous mutants defective in *Rac1* have decreased fertility, which is exacerbated by introduction of a Cdi loss-of-function mutation. This work demonstrates the existence of a Rac1-Cdi-cofilin pathway regulating spermatogenesis in *Drosophila* (Raymond et al. 2004).

A testis-specific isoform of LIMK2 (tLIMK2) is expressed in differentiated, meiotic stages of mouse spermatogenic cells, suggesting a role in mammalian spermatogenesis (Takahashi, Koshimizu and Nakamura 1998). tLIMK2 is first detected in 20-day postnatal mice with levels increasing during maturation of the testis. tLIMK2 is in differentiated germ cells, but not in spermatogonia or in Sertoli cells, the phagocytic cells in the testis that eliminate apoptotic germ cells during spermatogenesis (Takahashi,

Funakoshi and Nakamura 2003). The RhoB/ROCK/LIMK1 pathway is known to play a critical role in regulation of Sertoli germ cell adherens junction dynamics (Lui, Lee and Cheng 2003).

Tubulobulbar complexes form at the interface between maturing spermatids and Sertoli cells, as well as between two Sertoli cells near the base of the seminiferous epithelium. These complexes are finger-like structures that originate in areas previously occupied by actin filament-associated intercellular adhesion plaques known as ectoplasmic specializations. Actin filaments are also associated with tubulobulbar complexes where they form a network, rather than the tightly packed bundles found in ectoplasmic specializations. These junctional complexes undergo extensive restructuring to facilitate the movement of germ cells across the epithelium. This restructuring is regulated by Rho GTPases, particularly RhoB, with ROCK and LIMK1 activation occurring downstream of RhoB activation (Lui, Lee and Cheng 2003). Cofilin phosphorylation was also detected. A similar pathway was elucidated in *Drosophila* spermatogenesis (Raymond et al. 2004).

Nonmuscle cofilin is concentrated at tubulobulbar complexes and not at ectoplasmic specializations (Guttman et al. 2004). Tubulobulbar complexes may be part of the mechanism by which intercellular adhesion junctions are internalized by Sertoli cells during sperm release (Guttman, Takai and Vogl 2004), and thus disruptions of the actin filament system could impair sperm release. Interestingly, while LIM kinase 2 null mice are viable, males exhibit impaired spermatogenesis and germ cell loss (Takahashi et al. 2002). TESK1 is expressed in testicular germ cells and TESK2 is most highly expressed in Sertoli cells (Toshima et al. 2001a,b). These kinases are regulated through integrin signaling mechanisms involving Rho but not ROCK (Toshima et al. 2001b), and may play additional roles in the adhesion-mediated internalization of adhesion junctions and germ cell release. Together these results demonstrate an essential requirement for ADF/cofilin activity during gametogenesis and suggest that abnormalities in ADF/cofilin regulation could contribute to infertility in humans.

Injection of LIM kinases into *Xenopus* oocytes inhibits oocyte maturation via disruption of the organization, maintenance, and migration of the meiotic spindle precursor, MTOC-TMA (microtubule organizing center and transient microtubule array) (Takahashi et al. 2001). This inhibition is alleviated by coinjecting constitutively active ADF/cofilin, suggesting that ADF/cofilin mediates the effects of LIM kinase. Recent studies have even demonstrated that the actin cytoskeleton is associated with spindle microtubules, and is involved in the poleward flux of tubulin in metaphase kinetochore microtubules (Goode, Drubin and Barnes 2000; Silverman-Gavrila and Forer 2000; Takahashi, Funakoshi and Nakamura 2003), suggesting that AC proteins may have a role in this process as well.

Ovarian muscle contraction in *C. elegans* requires TM and troponin. Both TM and troponin C were associated with actin filaments in the myoepithelial sheath and RNAi knockdown of both induced sterility by inhibiting ovarian contraction. An inactivating mutation in AC suppressed the ovulation defects generated by RNAi-induced defects suggesting that TM and troponin have opposite effects to AC in regulating actomyosin organization for *C. elegans* ovarian contraction (Ono, Mohri and Ono 2004).

Diseases Based on Dendritic Spine Dysgenesis and Synaptic Defects

Although most of the diseases classified under this heading are developmental defects within the hippocampus that affect formation of the important postsynaptic structures, the dendritic spines, they are covered separately because some of the disorders, especially mood or behavioral disorders, can develop in the adult and not all of these have been traced to specific genetic or developmental defects. Indeed, many types of mental retardation have idiopathic causes, e.g., lead poisoning, traumatic head injury, or malnutrition (Ramakers 2002). Studies on biopsied human cortical tissue have shown morphological anomalies in dendrites and dendritic spines that have been linked to various forms of mental retardation and dendritic pathology (Huttenlocher 1970, 1975; Marin-Padilla 1972; Purpura 1974).

Dendritic Spine Development and Function

Dendrite arbor complexity and dendritic spines contribute to the theoretical complexity of postsynaptic signal integration and processing (London and Hausser 2005; Segev and London 2000) and indeed an intimate association between dendrite complexity and cognitive function has been consistently observed *in vivo* (reviewed in Benavides-Piccione et al. 2004; Lewis 2004). Dendritic spines are highly specialized, micron-long protrusions that serve as the major sites of excitatory glutamatergic synapses (>90%). Dendritic spines are found only in selected neuronal populations of vertebrates, especially on hippocampal pyramidal neurons, and in some invertebrates (reviewed in Nimchinsky, Sabatini and Svoboda 2002). Mature dendritic spines are characterized by enlarged, mushroom-shaped heads that are connected to the dendrite shaft by thin necks (Calverley and Jones 1990; Matsuzaki et al. 2004). This characterization represents the classical view of mature dendritic spines, but there is a growing appreciation for the true heterogeneity of dendritic spine shapes even in the adult brain (Jontes and Smith 2000). Through the characterization of heterogeneous spine shapes, three distinct spine types have emerged: mushroom, thin, and stubby (Fig. 8) (Peters and Kaiserman-Abramof 1970). The mushroom spine is the classical mature dendritic spine. The thin spine has a narrower, balloon-shaped head and longer neck than the mushroom spine, although the general shape is similar. The stubby spine is the only spine type to lack a spine neck, as it consists only of a bulbous head attached directly to the dendrite shaft. The dendritic filopodium, though not considered a veritable class of dendritic spine, can be found in young and adult brain and is characterized by a spaghetti-like appearance (Fiala et al. 1998; Vaughn 1989). The heterogeneity in dendritic protrusions may represent

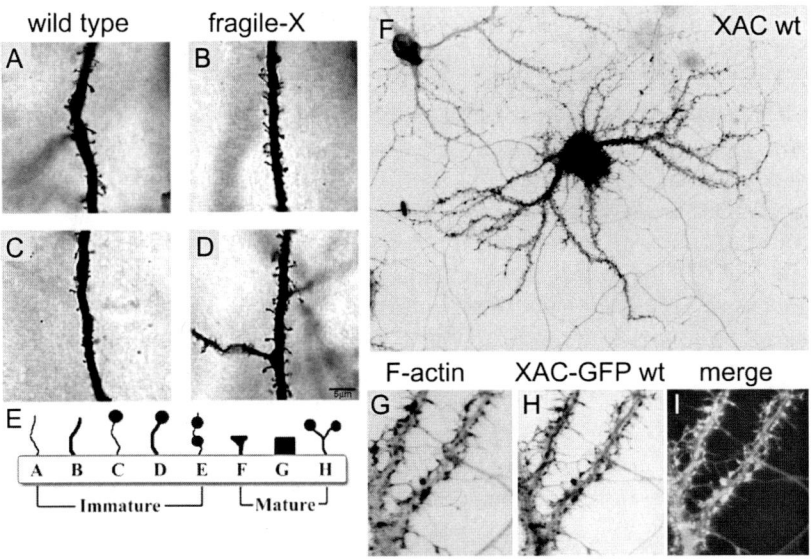

FIG. 8. Morphologies of hippocampal dendritic spines. Spine structures on apical dendrites of neurons from young (**a**) and adult (**c**) wild-type mouse and young (**b**) and adult (**d**) Fragile X mouse (adapted from Galvez and Greenough 2005). Normal spines have various morphologies, but many are compact and have a mushroom or stubby appearance (F and G in panel (**e**)), whereas spines from Fragile X neurons remain more immature in their morphologies (classification scheme adopted from Irwin et al. 2002). (**f**) Inverted fluorescence image of Texas-red phalloidin stained F-actin in cultured rat hippocampal neuron infected with adenovirus expressing *Xenopus* ADF/cofilin–GFP 3 days before fixation at 21 days in culture. (**g–i**) Higher magnification of inverted fluorescence images showing (**g**) location of F-actin (spines are well labeled), and XAC–GFP (**h**) as well as an overlay (F-actin in *red* and XAC–GFP in *green*). Some, but not all, regions containing F-actin also show substantial XAC–GFP (*See Color Plates*)

not only distinct functional classes but also different stages along the shared continuum of spine induction and maturation (Dailey and Smith 1996; Fiala et al. 1998; Harris 1999a,b; Ziv and Smith 1996). Although the predominance of each dendritic protrusion type shifts throughout development, all of these types can be found together in the adult brain (Holtmaat et al. 2005).

Because functional but simple synapses can form directly on the dendritic shaft (Somogyi et al. 1983a,b), the unique geometry of the spine is thought to serve a purpose beyond that of simply supporting synaptic transmission. One such function is thought to be compartmentalization (Axelrod et al. 1976; Kennedy et al. 2005; Koch and Zador 1993; Nimchinsky, Sabatini and Svoboda 2002; Svoboda, Tank and Denk 1996). Compartmentalization includes not only the specific localization of structural protein ensembles and organelles, but also the sequestration of signaling molecules such as Ca^{2+} (Allbritton, Meyer and Stryer 1992; Majewska, Tashiro and Yuste 2000; Majewska et al. 2000; Yuste, Majewska and Holthoff 2000). The localization of structural protein ensembles and organelles requires F-actin, which is highly enriched in the spine (Okamoto et al. 2004). In fact, spine growth and morphology are determined largely if not completely by F-actin assembly and organization, as actin constitutes the predominant cytoskeleton element in the spine (reviewed in Matus 1999, 2000; Zito et al. 2004).

The state and organization of F-actin that determines a particular spine shape invariably also determines the nature and quantity of proteins and organelles localized to the spine. For example, mushroom spines with a large spine head commonly contain a spine apparatus and smooth ER, whereas thin spines are rarely associated with these organelles (Spacek and Harris 1997). In addition, the size of the postsynaptic density (PSD) strongly correlates with spine head volume (Harris and Stevens 1998). The PSD is a dynamic organelle that contains over a hundred different known proteins important for synaptic transmission, modulation, and signaling (Husi et al. 2000; Kennedy 2000; Walikonis et al. 2000). In the hippocampus, the PSD directly juxtaposes a single presynaptic active zone across the synapse (Geinisman et al. 2001). Not surprisingly, spine head volume also correlates positively with synaptic strength, the amplitude of the excitatory postsynaptic potential. Potentiation of synaptic strength is accomplished by increasing both channel conductance and absolute number (Barria, Derkach and Soderling 1997; Barria et al. 1997; Derkach, Barria and Soderling 1999; Heynen et al. 2000; Shi et al. 1999).

The role of RhoA, Rac1, and Cdc42 in dendrite differentiation and dendritic spine growth and maintenance has been extensively characterized (Nakayama and Luo 2000; Negishi and Katoh 2005; Newey et al. 2005; Van Aelst and Cline 2004). In general, Rac1 and Cdc42 act antagonistically to RhoA in nearly every aspect of dendrite differentiation including dendrite initiation, elongation, branching, and dendritic spine initiation and length. Active Rac1 and Cdc42 enhance neurite initiation, elongation, and branching. Active RhoA antagonizes these effects. Active Rac1 and Cdc42 also stimulate dendritic spine initiation and spine extension. Active RhoA inhibits spine initiation and spine extension. Most of what is known about the role of the Rho GTPases in regulating dendrite morphology is from cell culture studies involving diverse neuronal subtypes; however, convincing physiological support of the crucial role of Rho GTPases in normal and diseased cognitive function is provided by several genes that are implicated in the etiology of mental retardation, which can be directly or indirectly linked to these proteins. Furthermore, other Ras superfamily GTPases and their effectors, such as synGAP (Vazquez et al. 2004), may work upstream or in parallel with these Rho GTPases to coordinate spine development and plasticity.

Spines develop from a restructuring of dendritic filopodia (Moeller at al. 2005). Brain-derived neurotrophic factor (BDNF) modulates hippocampal plasticity and hippocampal-dependent memory in cell models and in animals. A single point mutation in BDNF that is associated with poorer episodic memory and abnormal hippocampal activation in human patients was shown to impair activity-dependent secretion of BDNF due to its mislocalization in the secretory pathway (Egan et al. 2003). BDNF induces filopodial elongation on retinal ganglion cell growth cones via activation of cofilin (Gehler et al. 2004), suggesting that regulation of cofilin activity will be necessary for conversion of dendritic filopodia into spines.

In cortical neurons, bone morphogenic protein (BMP) 7 through its binding to the BMP receptor II (BMPRII) specifically affects dendritic morphogenesis (Lee-Hoeflich et al. 2004). LIMK1 colocalizes with BMPRII in the tips of neurites and binds to BMPRII, an interaction that is required for BMP-dependent induction of the dendritic arbor in cortical neurons. The physical interaction of LIMK1 with BMPRII synergizes with the Rho GTPase, Cdc42, to activate LIMK1 catalytic activity (Lee-Hoeflich et al. 2004). The equivalent of the BMPRII receptor in *Drosophila* is called Wishful thinking (Wit), which is required for synapse stabilization (Lee-Hoeflich et al. 2005). In the absence of BMP signaling, synapse disassembly and retraction ensue. BMP is thought to signal through the Smad signaling pathway, but in this system Smad-mediated signaling cannot fully account for the stabilizing activity of the BMP receptor.

Drosophila Lim kinase 1 (DLIMK1)-dependent signaling was identified as a second, parallel pathway conferring added synapse-stabilizing activity of the BMP receptor by binding to a region necessary for synaptic stability but not Smad signaling (Eaton and Davis 2005). Interestingly, DLIMK1 functions presynaptically for synapse stabilization and stabilizes synapses in mutants missing Wit or Smad pathways. DLIMK1 is found near synaptic microtubules and functions independently of AC during synapse stabilization, suggesting that it may have a scaffolding function, in contrast to its AC-dependent phosphorylation function in neurite outgrowth (Eaton and Davis 2005).

Long-Term Depression (LTD) and Long-Term Potentiation (LTP)

The underlying mechanisms regulating F-actin generation required for the induction of LTP are poorly understood. During LTP, actin-rich spines with large synapses are produced. This process appears to require LIMK1 phosphorylation of the ADF cofilin (Lisman 2003). Increased dendritic spine density and/or enlargement accompany the induction of LTP by high-frequency stimulation in acute hippocampal slices from neonatal rats. Induction of LTD by low-frequency stimulation is accompanied by a marked shrinkage of spines, which can be reversed by subsequent high-frequency stimulation. Spine shrinkage is mediated by cofilin, not by PP1, which is essential for LTD, suggesting that different downstream pathways are involved in spine shrinkage and LTD (Zhou, Homma and Poo 2004).

The application of PP1 inhibitors to cultured cortical neurons blocks LTD via both AMPAR- and NMDAR-mediated excitatory postsynaptic currents (EPSCs). Addition of phalloidin or a cofilin inhibitory peptide blocked LTD of NMDAR EPSCs but not AMPAR EPSCs. These findings suggest that the same pattern of afferent activity elicits depression of AMPAR- and NMDAR-mediated synaptic responses by means of distinct triggering and expression mechanisms (Morishita, Marie and Malenka 2005). Together the above results suggest a role for cofilin in synaptic plasticity, an important function of hippocampal synapses in generating memory and learning circuits.

Atypical Dementias with Neurological Symptoms

There are a group of neurodegenerative diseases that affect cortical and subcortical areas of the brain that give rise to dementias with specific neurological symptoms (Kurz 2005). As with the previously described neurodegenerative diseases, AC proteins and actin dynamics could be important to this group of dementias as well. The clinical symptoms reflect the region of the brain affected. Degeneration of the frontal and anterior temporal lobe is often accompanied by behavioral alterations, followed by impaired cognition. Complex visual disturbances accompany posterior cortical atrophy that affects parietal and occipital association cortices. Progressive supranuclear palsy involves the frontal, temporal, and parietal cortex, and parts of the brain stem. Clinical features include a hypokinetic rigid syndrome with nuchal dystonia and vertical gaze palsy. Many other neurodegenerative diseases, such as Huntington's chorea and amyotrophic lateral sclerosis, are often accompanied by dementia causing personality changes and cognitive deterioration (Kurz 2005).

Corticobasal Degeneration

In corticobasal degeneration (CBD), an adult onset progressive neurodegenerative disorder, the focus of pathology includes the frontoparietal cortex and several subcortical nuclei, causing symmetrical rigidity, bradykinesia, myoclonus, and dystonia (Kurz 2005). Changes of proteome profiles were analyzed between three nondemented brains and a CBD brain based on two-dimensional gel electrophoresis. Cofilin-1 (nonmuscle) was one of only two proteins whose expression was upregulated in the CBD brain, whereas six proteins declined in expression (Chen, Ji and Ru 2005). Although it is not possible to draw definitive conclusions from this single sample, developmental changes in cofilin expression or activity could be at the root of the alteration in synaptic function. These changes in cofilin activity could directly alter synapse function through multiple effects: alterations in Golgi dynamics and targeting of neurotransmitter-containing vesicles, changes in spine size and shape, or through formation of intracellular inclusions that block neurite transport.

Williams Syndrome

Williams syndrome (WS) is a rare neurodevelopmental disorder occurring in approximately one in 20,000 to one in 30,000 births (Williams, Barratt-Boyes and Lowe 1961; Beuren, Apitz and Harmjanz 1962). Affected individuals display various diagnostic characteristics, such as craniofacial dysmorphology, dental abnormalities, cardiovascular disorders, hypertension, neonatal hypercalcemia, delayed language and motor development, and abnormal sensitivity to certain sounds (Keating 1997; Bellugi et al. 1999). WS patients display "elfin-like" facial features with a broad forehead, oval ears, and a wide mouth (Preus 1984). Individuals with WS have a distinct cognitive profile showing striking impairments in areas such as general intelligence evident in a low IQ (40–79), poor visuospatial constructive cognition, and attention disorders (Hoogenraad et al. 2004). Surprisingly, WS patients also display areas of relative skill, such as face processing, social engagement, and expressive language (Bellugi et al. 1999).

The genetic cause of WS is a deletion of approximately 1.6 Mb of DNA, termed the WS critical region that accounts for at least 20 individual genes at Chr band 7q11.23. This results in a heterozygous condition for several genes including ones that encode elastin, LIMK1, cytoplasmic linker protein 2 (CYLN2 encodes the cytoplasmic linker protein (CLIP-115) microtubule-binding protein), WBSCR1 (William–Beuren syndrome critical region 1), WBSCR5, and RFC2 (replication factor C subunit 2) (Ewart et al. 1993; Frangiskakis et al. 1996; Meyer-Lindenberg et al. 2005; Osborne et al. 1996; Peoples et al. 1996).

To better understand the relationship between genotype and phenotype in WS patients, two families with a WS phenotype consisting of supravalvular aortic stenosis (SVAS), WS facial features, and impairment in visuospatial constructive cognition were characterized (Frangiskakis et al. 1996). DNA sequence analyses of the 83.6 kb deleted region revealed the deletion of only two genes, *ELN*- and *LIMK1*-encoding elastin, an extracellular matrix protein that provides flexibility to tissues, and LIM kinase 1, respectively. Northern blot and *in situ* hybridization studies showed that *LIMK1* is expressed in distinct regions of fetal and adult brain. In contrast, *ELN* displays negligible expression in fetal and adult brain. In addition, *LIMK1* heterozygosity cosegregated with impaired visual–spatial constructive cognition (Frangiskakis et al. 1996). Also, using genetic linkage and mutational analysis studies, deletions affecting *ELN* have been shown to cause autosomal dominant SVAS, a characteristic of WS (Curran et al. 1993; Ewart et al. 1993, 1994). Together, these data imply that *LIMK1* heterozygosity is responsible for the deficits in visual–spatial constructive cognition observed in WS patients. Therefore, since LIMK1 is an upstream regulator of ADF/cofilin, abnormal regulation of the neuronal actin cytoskeleton might be involved in the WS visual phenotype.

To determine the role of LIMK1 in neuronal development, homozygous LIMK1 null, LIMK2 null, and double null mice were developed (Meng et al. 2002, 2004). Levels of phospho-ADF/cofilin were decreased in the LIMK1$^{-/-}$ mice, but the animals were grossly normal. Cofilin immunostaining of hippocampal neurons from LIMK1$^{-/-}$ mice revealed an absence of growth cones and an increase in actin aggregates along dendrites in mature neurons. Neurons in the hippocampus of the LIMK1 null mice have spines with thicker necks and smaller heads and do show normal LTD, but increased LTP, as well as slight defects in spatial learning performance in the water maze (Meng et al. 2002). Whereas LIMK2 knockout mice (Takahashi et al. 2002) exhibit minimal abnormalities except in the testes, the LIMK1/2 double knockout mice are more severely impaired in both ADF/cofilin phosphorylation and in excitatory synaptic function in the CA1 region of the hippocampus (Meng et al. 2004).

From the above studies, it is clear that LIMK plays important roles in brain development, spine morphology, and synapse stability, but it is not clear that any or all of these effects are mediated by ADF/cofilin or even that kinase activity is required. A recent study suggests that regulation of cofilin by LIMK is indeed important for spine remodeling. Mutations in two tumor suppressor genes of the tuberous sclerosis complex (*TSC1* and *TSC2*) often lead to mental retardation, epilepsy, and autism. Loss of the Tsc1 and Tsc2 proteins in hippocampal neurons triggered enlargement of the somas and spines and altered the properties of the glutaminergic synapses (Tavazoie et al. 2005). These morphological changes were blocked by expression of the cofilin S3A mutant but not wild-type cofilin, suggesting that cofilin phosphorylation by LIMK, which was significantly increased by the loss of Tsc2, is required for these changes.

Another gene found in the WS critical region is *CYLN2*, which encodes the 115 kDa CLIP-115. The CLIP proteins bind to the end of growing microtubules and are believed to function as regulators of microtubule dynamics by promoting persistent growth and nonpersistent shortening (Schuyler and Pellman 2001; Komarova et al. 2002). Mice with haploinsufficiency in CLIP-115 have mild growth deficiency, brain abnormalities, hippocampal dysfunction, and particular deficits in motor coordination (Hoogenraad et al. 2002). A more detailed analysis of brains of WS patients and age matched controls using multimodal neuroimaging showed that hippocampal size was preserved but there were subtle alterations in shape and in resting blood flow and response to visual stimuli. These findings correlated well with the changes observed in both the LIMK null and CLIP-115-deficient mice. Together, these results suggest that the hippocampal abnormalities associated with Williams syndrome arise from both actin filament- and microtubule-dependent processes during development.

X-Linked Nonsyndromic Mental Retardation

Mutations in the *PAK3* gene that generate a protein that is kinase dead, has altered kinase activity, or is defective in GTPase-binding, are associated with an X-linked mental retardation (Bienvenu et al. 2000; Gedeon et al. 2003). Patients with PAK3 mutations also occasionally have neuropsychiatric problems. Although the nature of the defect in neuronal function is unknown and because brain development is grossly normal, PAK3 mutations that result in mental retardation may reflect a requirement for PAK later in development. Such a requirement could be in dendritic spine morphogenesis, for which Rac and PAK signaling are important (Ramakers 2002). PAK3 might be necessary for dendritic development or for the rapid cytoskeletal reorganizations in dendritic spines associated with synaptic plasticity (Park et al. 2003; Penzes et al. 2003).

Of the 11 genes known to cause nonspecific X-linked mental retardation (XMR), three (oligophrenin-1, PAK3, and alpha PIX) are directly linked to the Rho GTPase signaling pathway (reviewed in Ramakers 2002). Oligophrenin-1 is a Rho GTPase activating protein (Rho-GAP) that is absent in a family affected with XMR. It is required for dendritic spine morphogenesis. Reduced oligophrenin-1 levels affect spine length by increasing RhoA and Rho-kinase activities (Govek et al. 2004). Two of the other gene products, PAK3 and alpha PIX, form a complex with the synaptic adaptor protein *G*-protein-coupled receptor kinase-*in*teracting protein *1* (GIT1) (Zhang et al. 2005).

GIT1 is critical for spine and synapse formation. Rac is locally activated in dendritic spines by PIX, a Rac guanine nucleotide exchange factor. PAK1 and PAK3 serve as downstream effectors of Rac in regulating spine and synapse formation. Active PAK promotes the formation of spines and dendritic protrusions, which correlate with an increase in the number of excitatory synapses. These effects are dependent on the kinase activity of PAK. One activity of PAK identified in synapse formation is its phosphorylation of myosin II regulatory light chain (MLC). Activated MLC causes an increase in dendritic spine and synapse formation, whereas inhibiting myosin ATPase activity results in decreased spine and synapse formation. Whether there is a change in cofilin phosphorylation that also helps mediate these effects is currently unknown.

Memory, Learning, Mental Illness

A new model of functional gene pathways involved in neuropsychiatric diseases, such as mood disorders and schizophrenias, suggests these develop as a result of dysfunction of the cytoskeletal system resulting in the disruption of neuronal transmission and transportation (Peter-Ross 2006). In this model, mood disorders are postulated to arise from functional abnormalities in actin microfilaments, whereas schizophrenias are attributed to abnormalities in the microtubules and/or intermediate filament systems. This model is based upon identified genes involved in the disorder or genes residing in chromosomal segments associated with the disorder, or on the probable targets of pharmacological treatments that seem to help patients. This hypothesis-driven approach could lead to more appropriate medications and therapeutic strategies developing from a dialogue between researchers and clinicians in translational molecular psychiatry (Peter-Ross 2006).

Diseases of mRNA Delivery and Translation

Overview of mRNA Delivery and Regulation

Because of their shape, organization, and compartmentalization, neurons face unparalleled challenges in the delivery and maintenance of cellular components required to maintain function. In neurons, transport down axons and dendrites is required to carry components made in the soma to their targets often tens, hundreds, or thousands of cell body diameters away. Thus, although localized synthesis of many cytoskeletal proteins is important in migrating and polarized cells, such as fibroblasts and epithelial cells (Shestakova, Singer and Condeelis 2001), maintaining delivery and local translation of mRNAs in neurons is a more daunting task and defects in these processes develop into neurodegenerative diseases (Bassell and Kelic 2004).

Although evidence in favor of local protein synthesis within axons was presented more than 20 years ago (Koenig and Adams 1982), only recently has it become widely accepted. Locally synthesized proteins include β-actin and ADF (Lee and Hollenbeck 2003), as well as cofilin and other cytoskeletal proteins (Bassell and Kelic 2004; Willis et al. 2005). Neurotrophin treatment of axons results in a very rapid increase in transport of mRNAs into the axon (Willis et al. 2005) as well as local synthesis of actin and cofilin. Indeed synthesis of proteins within growth cones is important for axonal guidance (Campbell and Holt 2001) and regeneration (Verma et al. 2005).

Following synthesis and processing of mRNAs in the nucleus and transport to the cytoplasm, various mRNAs are recognized through sequences, usually in their 3′ untranslated region (3′-UTR), by "zipcode-binding proteins" that target their delivery to cell compartments (Huttelmaier et al. 2005). CRP75, cofilin, and synuclein have 3′-UTR sequences similar to the cis-acting signal of Tau, which could indicate preferential sorting of these messages specifically to axons. Sequences homologous to dendritic target elements could direct the message to dendrites, thus mRNA containing both homologs (e.g., β-actin) are targeted to growth cones of both dendrites and axons (Aranda-Abreu et al. 2005). Defects in mRNA-binding proteins, which regulate mRNA delivery and translation, cause neurodegenerative diseases.

Spinal Muscular Atrophy

Spinal muscular atrophy, an often severe autosomal recessive form of a motor neuron disease, is caused by defects in SMN, part of a multiprotein complex responsible for splicing, targeting, and transport of RNAs, particularly β-actin mRNA, to the axon through interactions with the 3′-UTR zipcode-binding protein (Rossoll et al. 2003; Zhang et al. 2004). Motor neurons isolated from a mouse expressing defective SMN exhibit normal survival but have reduced axon growth, which correlates with reduced levels of β-actin mRNA and β-actin protein in distal axons and growth cones.

Fragile X Syndrome

The most common cause of inherited mental retardation is due to mutations in the Fragile X mental retardation protein (FMRP) (Verkerk et al. 1991). Synaptic abnormalities are thought to be at the basis of mental retardation, brought about through changes in dendritic spines (Greenough et al. 2001). FMRP is an RNA-binding protein implicated in regulation of mRNA translation and/or transport (Darnell et al. 2001; Laggerbauer et al. 2001). In murine fibroblasts Rac1 activation induces relocalization of four FMRP partners to actin ring areas. Rac1-induced actin remodeling is altered in fibroblasts lacking FMRP or carrying a point mutation in the KH1 or KH2 RNA-binding domain. Levels of phosphorylated but not total ADF/cofilin are lower in the absence of wild-type FMRP whereas the levels of PP2Ac catalytic subunit are increased. FMRP binds the 5′-UTR of pp2Acβ mRNA with high affinity, and is a negative regulator of its translation (Castets et al. 2005). Thus, a change in cofilin activity is possibly a contributing factor to the increase in filopodial and balloon-shaped spines observed in brains of patients with Fragile X syndrome (Fig. 8) (McKinney et al. 2005).

Diseases Based on Inclusions

Rods and Aggregates

ADF/cofilin–actin enriched inclusions are a common pathological feature observed in a broad spectrum of neurodegenerative diseases. These aggregates often take the form of rod-shaped bundles of filaments (rods), as irregular aggregates and sheets (inclusions), or as paracrystalline lattices (Hirano bodies). It is unclear whether they share a common origin or are the result of various independent mechanisms ultimately leading to AC–actin aggregation. Actin in rods and actin in stress fibers appears to interact with distinct proteins. Fluorescent antibody staining showed that many proteins associated with stress fibers including myosin, TM, and α-actinin, as well as other cytoskeletal and cytoskeletal-associated proteins, such as filamin, vinculin, vimentin, and tubulin, were not found associated with nuclear or cytoplasmic actin rods (Sanger et al. 1980; Nishida et al. 1987). However, antibodies specific for cofilin labeled both nuclear and cytoplasmic actin rods (Nishida et al. 1987). Thus actin rods and stress fibers have distinct protein compositions.

Pathologists observed rod-like structures in cell nuclei more than 100 years ago (Mann 1894). Rods form in the cytoplasm or nucleus of many types of cultured cells in response to 10% DMSO (Fig. 9) (Fukui 1978; Fukui and Katsumaru 1979), heat shock (Iida, Iida and Yahara 1986; Iida and Yahara 1986; Nishida et al. 1987; Ohta et al. 1989; Iida, Matsumoto and Yahara 1992), osmotic stress (Iida and Yahara 1986; Nishida et al. 1987), and ATP rundown (Bershadsky et al. 1980; Minamide et al. 2000; Ashworth et al. 2003a). In fibroblasts or epithelial cells, rods are generally reversible and do not appear to cause permanent damage to the cell. However, rods form primarily in the axons and dendrites of neurons where clearance is more problematic and their effect on cell function more acute.

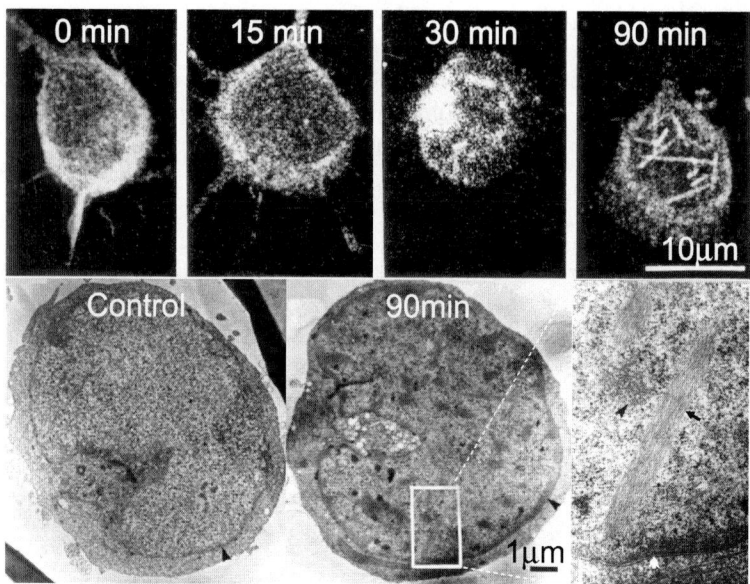

FIG. 9. AC- and actin-containing nuclear rods induced in cultured hippocampal neurons by treatment with 10% DMSO. *Upper panels* show time-course of rod formation after treatment of neurons with 10% DMSO. Rods first appear about 30 min after treatment but reach a maximum by 90 min. Staining in this series is for AC. *Lower panels* show thin section electron micrographs through the nucleus of untreated hippocampal neuron and a neuron treated for 90 min with DMSO. Many rods are observed in the nucleus of the treated neuron, one of which is magnified in the last image. *Arrow* shows rod in longitudinal section and *arrowhead* is a rod in cross-section (courtesy of John R. Jensen and Marcia DeWit)

LIMK2 mutant mice subjected to hyperthermia displayed punctate staining of nonphosphorylated cofilin and actin inclusion in the nucleus within certain populations of germ cells (Takahashi et al. 2002), possibly due to excessive amounts of nonphosphorylated cofilin resulting from the loss of tLIMK2 (Takahashi, Funakoshi and Nakamura 2003). Nuclear actin plays important roles in RNA transcription and splicing, chromosome condensation, and chromatin remodeling (Olave et al. 2002; Rando, Zhao and Crabtree 2000; Visa 2005).

ADF and cofilin are also major components of Hirano bodies (Fig. 10) (Maciver and Harrington 1995). Hirano bodies are unique intracytoplasmic inclusions consisting of a paracrystalline-ordered array of parallel regularly spaced 6–10 nm filaments in orthogonal layers encircled by a less structured actin dense region (Schochet and McCormick 1972; Tomonaga 1974). Hirano bodies were first described in patients with amyotrophic lateral sclerosis and Parkinsonism–dementia complex (ALS-PDC) in Guam (Hirano et al. 1968; Hirano 1994).

Because Hirano bodies are found in aged human brain from individuals with normal cognitive abilities, their link to various ailments in which they

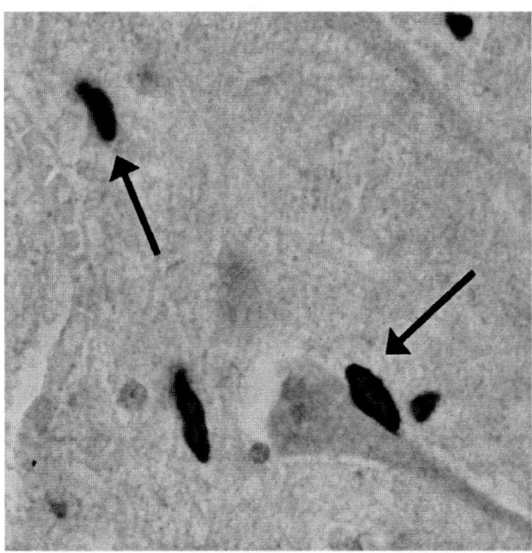

FIG. 10. Hirano bodies in paraffin section of brain from patient with Alzheimer's disease. Section was deparaffinized, stained with hematoxylin and eosin, and then immunocytochemically stained for AC (*brown reaction product*). Hirano bodies (*arrows*) are prevalent throughout the hippocampus and frontal cortex (*See Color Plates*)

occur more frequently is tenuous. These include Alzheimer's disease (Gibson and Tomlinson 1977; Mitake, Ojika and Hirano 1997), in which Alzheimer's patients of the same age as controls displayed a significantly greater number of Hirano bodies (Schmidt, Lee and Trojanowski 1989).

Although Hirano bodies have been found in multiple areas of the brain, they are most frequently found in Sommer's sector of Ammon's horn (Hirano et al. 1968), a region of the brain where Alzheimer's neurofibrillary tangles and Pick bodies are also enriched (Hirano 1994). Since this brain region is involved in the development of new memories, the formation of inclusion bodies here could contribute to the cognitive impairment found in patients of Alzheimer's disease, Parkinsonism–dementia (Hirano et al. 1968), Pick's disease (Schochet, Lampert and Linderberg 1968), amyotrophic lateral sclerosis (Hirano et al. 1968), ataxic Creutzfeldt–Jakob disease (Cartier, Galvez and Gajdusek 1985), scrapie (Field and Narang 1972), Kuru (Field, Mathews and Raine 1969), Papovirus (Hadfield, Martinez and Gilmartin 1974), cancer (Fu, Ward and Young 1975; Gessaga and Anzil 1975), diabetes (Sima and Hinton 1983), and chronic alcoholism (Laas and Hagel 1994). Hirano bodies are not confined to neurons. Other cell types and tissues that exhibit Hirano bodies are astrocytomas, aged extraocular muscle fibers, inflammatory cells of a leptomeningeal vessel (Ho and Allevato 1986), skeletal muscle fibers (Fernandez et al. 1999), and testis (Setoguti, Esumi and Shimizu 1974).

Phalloidin, a probe that recognizes filamentous actin, stains Hirano bodies (Galloway, Perry and Gambetti 1987). Hirano bodies also contain epitopes for microtubule-associated proteins, including tau, and the actin-associated proteins, α-actinin, vinculin, TM, and ADF/cofilin (Galloway, Perry and Gambetti 1987; Peterson et al. 1988; Maciver and Harrington 1995). Thus, it is believed that these structures are primarily composed of actin and ABPs. However, it is unclear in many cases whether the antibodies are staining the Hirano body core or the material surrounding the core, which is important in determining if the epitope is part of the organized crosslinked filament array.

Although the mechanism of Hirano body formation from endogenous proteins is unknown, expression in mammalian cells of a CT fragment of an actin crosslinking protein from *D. discoideum* induces structures morphologically identical to Hirano bodies (Maselli et al. 2002). In epithelial cells, expression of CT-mRFP sequestered actin into Hirano body-like aggregates and prevented accumulation of rods in response to ATP depletion (Gonzalez et al., unpublished results). In neurons, during early stages of CT expression, ATP depletion induced rods in neurites whereas Hirano body-like structures formed within soma. However, actin was more effectively sequestered in Hirano body-like structures by CT-mRFP, which eventually prevented formation of rods in neurites.

Alzheimer's Disease

Proteolytic cleavage of the full-length amyloid precursor protein (APP) by β- and γ-secretases gives rise to amyloid beta (Aβ) peptides from which are formed the extracellular senile plaques characteristic of AD (Price, Sisodia and Gandy 1995; Sisodia and Price 1995; Hardy and Selkoe 2002; Mattson 2004; Tanzi and Bertram 2005). Mutations leading to increased production of the more amyloidogenic $A\beta_{1-42}$ species are linked to early onset familial AD (Chartier-Harlin et al. 1991a,b; Goate et al. 1991; Murrell et al. 1991; Price, Sisodia and Gandy 1995).

Immunostaining of human AD brain for ADF/cofilin identified linear arrays of densely staining rod-like aggregates that were not observed in control human brain (Minamide et al. 2000). In regions surrounding >97% of dense-core amyloid plaques, rod-like staining was observed, but about 45% of the rod-like staining occurred in brain regions that did not contain well formed plaques. A similar cofilin-staining pattern was observed in brains of rapidly perfusion fixed transgenic mice (Tg2576; also called the *Alzheimer mouse*) expressing the Swedish mutation of human APP, but not in control mouse brains fixed identically, strongly suggesting that the formation of these ADF/cofilin staining aggregates are features of the diseased brain and are not postmortem artifacts.

Rod-like inclusions (rods), composed of actin saturated with ADF/cofilin, are induced in both axons and dendrites of cultured hippocampal neurons by ATP depletion, reactive oxygen species, and excess glutamate, common mediators of neuronal stress (Minamide et al. 2000). Survival of cells containing rods did not decrease over 3 days, the longest period followed. Rods that enlarge to occlude the neurite disrupt distal microtubules and eliminate growth cones. Although rods that form initially may be transient and disappear when the stress is removed, they often return within 24 h within a subset of neurites in which mitochondrial activity has been compromised. Because blockage of transport within neurons is one of the earliest defects found to occur in the transgenic mouse model for AD (Stokin et al. 2005), the possible role of rods in the neurodegenerative response has been further investigated.

Soluble forms of synthetic $A\beta_{1-42}$, which is derived naturally from APP by β- and γ-secretase cleavage, induce the formation of rods in 15–19% of cultured E18 rat hippocampal neurons in a time- and concentration-dependent manner (Maloney et al. 2005). About 50% of the maximum neuronal response occurs within 6 h or with as little as 10 nM $A\beta_{1-42}$, as low a concentration as has been reported to induce any physiologically relevant response. Treatment of cultured neurons with $A\beta_{1-42}$ inhibited fast axonal transport (Hiruma et al. 2003). $A\beta_{1-42}$ induces the activation (dephosphorylation) of ADF/cofilin only within that population of neurons that forms rods, suggesting the presence of an Aβ-sensitive hippocampal neuronal population. Treatment of organotypic hippocampal brain slices with Aβ peptides induces rods particularly in neurons within the dentate gyrus (Fig. 11). These studies strongly support a role for ADF/cofilin actin rods in the early blockage of transport associated with the synaptic loss in the mouse model of AD.

The production of Aβ peptides from APP requires its endocytosis from the plasma membrane (Ehehalt et al. 2003). Because blockage of transport within neurons could also lead to accumulation of vesicles that might be involved in either Aβ production or its assembly into more toxic oligomers, neurons were examined for accumulation of vesicles at sites of rods (Maloney et al. 2005). Vesicles containing APP, β-secretase, and presenilin-1, a component of the γ-secretase complex, accumulate at rods. The β-C-terminal fragment of APP,

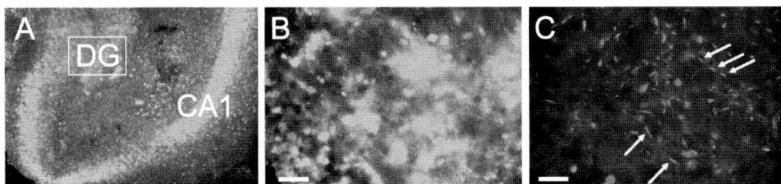

FIG. 11. Alzheimer's $A\beta_{1-42}$ oligomers induce AC–actin rods in cells of organotypic hippocampal slice cultures. A mouse hippocampal organotypic slice cultured 8 days was treated for 1 h by direct application of 2 μM $A\beta_{1-42}$ oligomer over the slice and then for another 47 h after transfer of the peptide to the culture well below the membrane supporting the slice. Slices were then fixed in 4% paraformaldehyde, permabilized in cold methanol, and immunostained for amyloid beta (**a, b**) and AC (**c**). (**b, c**) Higher magnification images of the dentate gyrus (DG) shown boxed in (**a**). Rods are quite prevalent and are often in linear arrays (e.g., *arrows*) similar to their organization in dissociated neuronal cultures (see Maloney et al. 2005) (*See Color Plates*)

the immediate precursor to $A\beta_{1-42}$, or the $A\beta$ peptide itself, also localizes at rods. These results suggest that rods, formed in response to either $A\beta$ or some other stress, block APP transport and provide a site for producing $A\beta$ peptides, which can induce more rods in surrounding neurons, expanding the degenerative zone and eventually resulting in plaque formation. Furthermore, studies in a model system (cultured *Aplysia* neurons), demonstrated that formation of cofilin–actin rods in an axon caused defects in synaptic function (Jang et al. 2005).

If rod formation is detrimental to neurons, why do rods form? Energy rundown and ATP depletion occur in diseased cells and tissues due to the loss of normal mitochondrial function. ATP is the energy currency required for driving many biochemical reactions and it is utilized by kinases to phosphorylate proteins including AC. Actin dynamics relies heavily on ATP hydrolysis and nucleotide exchange to drive the turnover process. ATP rundown leads to an accumulation of ADP actin and activated unphosphorylated AC (Fig. 12). Activated AC binds to ADP actin with high affinity, and this complex assembles

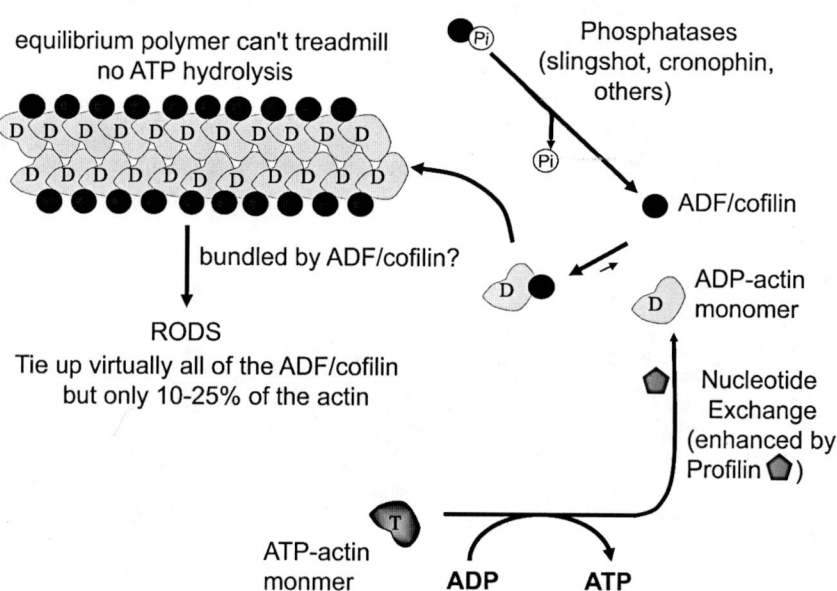

FIG. 12. Proposed mechanism of AC–actin rod formation in cells under stress. As ATP levels decline in stressed cells, the pool of ADP-actin increases and so does the pool of active (dephosphorylated) AC. The higher affinity of AC for ADP-actin leads to complex formation and reassembly into an equilibrium polymer in which the critical concentration for assembly is the same at both ends. Lateral association of filaments is thought to occur as a result of reduced electrostatic repulsion by saturation with AC

into AC-saturated ADP-actin equilibrium polymers (Ressad et al. 1998). Presumably, the slightly basic AC on the filament surface can neutralize the negative charge along F-actin and lead to filament bundling into rods.

Rods may represent a protective mechanism for cells in that by sequestering ADF/cofilin into nondynamic aggregates, the remaining actin will turnover much more slowly, and ATP is spared for more imperative processes. Actin dynamics can account for up to 50% of total ATP consumption in growing neurons (Bernstein and Bamburg 2003) and in resting platelets (Daniel et al. 1986). The size of rods is limited by the amount of ADF/cofilin in the cell. Cells overexpressing XAC form rods that are more abundant and larger in size than rods formed in control cells (Minamide et al. 2000). Actin in cells is more abundant than ADF/cofilin, so rods tie up virtually all of the ADF/cofilin but only a small portion of the actin.

To evaluate the role of rods in sparing cellular ATP, rods were generated in cultured rat E18 hippocampal cells by overexpression of an XAC–GFP (green fluorescent protein) fusion (Bernstein et al. 2006). For a short period (~60 min) immediately after initial rod formation, the loss of mitochondrial membrane potential and ATP in neurites with rods is slower than in neurites without them. Actin and AC in rods form a more stable polymer than elsewhere as demonstrated by the insensitivity of rods to treatment with latrunculin A and by the lack of fluorescence recovery after photobleaching rods containing XAC–GFP. Thus, the formation of rods transiently protects neurites by slowing filament turnover and its associated ATP hydrolysis. In addition to helping to preserve ATP levels, actin rods might protect the cell from apoptosis by sequestering cofilin and preventing its translocation to mitochondria. Additional research is required to build a complete understanding of the signaling pathways that target cofilin to mitochondria in neurons. In addition, it is important to determine if cofilin in rods is prevented from undergoing this translocation.

Stroke

The rapidity of rod formation in cultured hippocampal neurons subjected to transient ATP depletion (Minamide et al. 2000), which appears initially to be neuroprotective if ATP levels recover, strongly suggests that rods are formed during ischemic brain injury and could play a role in synaptic loss that is either temporary or permanent, dependent upon the time-course of recovery. To study this phenomenon in a readily manipulated system in which neurons maintain their normal interconnections, we examined rod formation in organotypic hippocampal slices infected with adenovirus for expression of an AC–GFP (Maloney, Minamide and Bamburg, unpublished results). Prior to becoming hypoxic or anoxic, the brain slices showed a diffuse distribution of the AC–GFP but within 5 min of making the slice anoxic by sandwiching it between coverslips, rods formed throughout the slice, reaching a maximum number, and distribution within 15 min. In slices that were not infected with AC–GFP expressing adenovirus, immunostaining of endogenous AC after 15 min of anoxia demonstrated widespread rod formation, indicating that rods formed during anoxia were not induced because of AC–GFP overexpression. Thus, hippocampal rod formation has kinetics similar to the timing of irreversible ischemic brain injury. Additional studies will be required to determine if rods participate in the transient or permanent loss of synaptic circuitry associated with ischemic brain injury, but synaptic deficits are expected based upon the synaptic disruption caused by rod induction in Aplysia axons (Jang et al. 2005).

Dystonia with Dementia

Dystonia is pathophysiologically defined by the prolonged cocontraction of agonist and antagonist muscle groups causing a distorted or twisted posture or repetitive movements (Yanagisawa and Goto 1971). In addition, dystonia patients have problems recruiting appropriate muscles and experience muscular action in surrounding muscles. As opposed to neurodegeneration, classical dystonia is caused by irregular interneuronal signaling (Berardelli et al. 1998). Neuronal and electrophysiological studies have shown that this disorder is caused by defects in nervous system function, primarily the basal ganglia circuit (Bhatia and Marsden 1994). However, there is a diverse set of neurodegenerative disorders that are classified as subtypes of dystonia, such as hereditodegenerative dystonia in which patients display fundamental brain degeneration (Fahn, Bressman and Marsden 1998). A recent study analyzed the brains of twins that suffered from dystonia with dementia (Gearing et al. 2002). The genetic cause of this dystonia remains unknown: the *TOR1 A* gene, which is responsible for the most severe forms of early onset dystonia, was not mutated, the twins displayed a normal karyotype, and mitochondrial DNA sequencing uncovered no irregularities. The twins had multiple developmental disorders, such as cleft lip and palate, skeletal abnormalities, cataracts, blindness, and deafness. Accumulation of these maladies suggests a problem in embryogenesis or a functional disorder of proteins regulating the actin cytoskeleton. At age 12 the twins suffered from rapidly progressive and dopa-unresponsive generalized dystonia, characterized by motor disorder, progressing to leg dystonia by age 14, followed by loss of fine and, later, gross motor skills by age 15. At age 17, progressive intellectual decline and dementia were observed as the twins began to lose their ability to communicate, and they died within months of each other at ages 21 and 22.

Postmortem brains revealed no major macroscopic defects. Microscopic analysis revealed neither neuronal loss nor neurofibrillary tangles, but eosinophilic, ovoid, or rod-like cytoplasmic inclusions were observed in the neocortex and thalamus that labeled with antibodies to ADF/cofilin. These structures did not stain with actin antibodies. In addition, eosinophilic spherical structures were observed in the striatum, globus pallidus, and substantia nigra. These spherical structures immunostained positively for both actin and ADF/cofilin, but were negative for tau, neurofilaments, glial fibrilary acidic protein (GFAP), α-synuclein, Aβ, and APP. Additionally, the aggregates were elongated (rod-like) in some cases, such as within the globus pallidus and substantia nigra. Oblong aggregations of filaments in neocortical tissue occasionally accompanied by axonal swelling and disruption of the myelin sheath were identified by electron microscopy. Similar eosinophilic, rod-like structures have been observed in normal individuals, patients with neurologic disease, alcoholics, myotonic dystrophy patients, and aging mice (Culebras, Feldman and Merk 1973; Fraser 1969; Kawano and Horoupian 1981; Ono et al. 1987; Pena and Katoh 1989). Unlike one documented case of Meige disease (Kulisevsky et al. 1988), the rod-like structures did not localize to the nucleus.

Although these two cases are phenotypically distinct from all other reported cases of hereditodegenerative dystonia, this study provides evidence for the role of AC and actin containing rod-like aggregates in a human neuronal dysfunction that is different from AD. In addition, these data support a mechanism for AC–actin aggregation as a contributor to the pathology of neurodegenerative disease *in vivo*.

Ischemic Kidney Disease

Both phosphorylated and unphosphorylated AC is localized throughout the cytoplasm of control rat kidney proximal tubule cells (Schwartz et al. 1999). During induction of renal ischemia, ADF was rapidly and progressively dephosphorylated and localized to the apical membrane, sites of actin-rich microvilli. After 25 min of ischemia, AC- and G-actin were localized to intraluminal vesicle and membrane bleb structures (Fig. 13) (Schwartz et al. 1999).

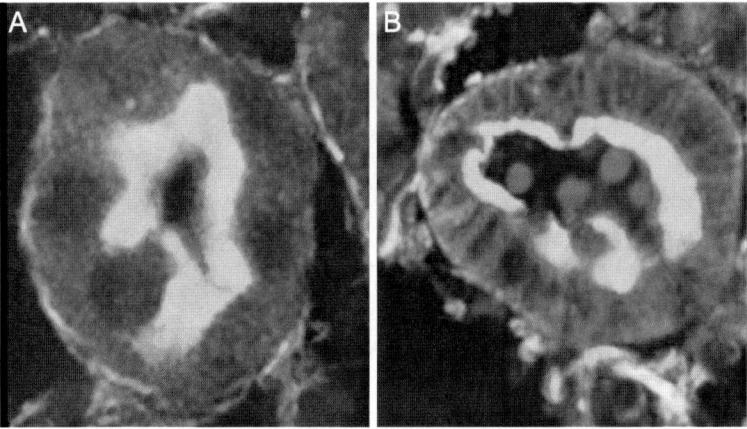

FIG. 13. Images of rat kidney proximal tubules before and after 25 min of ischemia. Sections of the tubules were stained with fluorescent phalloidin (*green*) for F-actin, which is greatly enriched in the apical microvilli lining the lumen. (**a**) Before ischemia, immunostaining for AC (*red*) shows it to be diffuse in the cytoplasm of proximal tubule cells. (**b**) After ischemia, vesicles that are found in the lumen of the kidney, which arise from blebbing of the proximal tubule apical membrane, are enriched in AC as well as G-actin (not shown) (see Schwartz et al. 1999) (*See Color Plates*)

In contrast to control conditions, urine collected during the first 24 h from ischemic rats contained high levels of AC and actin. Following reperfusion, phosphorylated AC levels increased, AC localization returned to the cytoplasm, and disrupted apical microvilli were repaired. These data begin to illustrate an ischemic mechanism involving the dephosphorylation and activation of AC followed by the disruption of the apical cytoskeleton and microvillar destruction.

To study the cellular reorganization and activity of AC during F-actin destabilization and reorganization in response to ischemic cell injury, a cell culture model of proximal tubule cells was utilized. XAC–GFP mutants were overexpressed in a cultured porcine kidney cell line (LLC-PK$_{A4.8}$) (Ashworth et al. 2003a). ATP depletion resulted in a rapid dephosphorylation and activation of both endogenous cofilin and exogenously expressed wild-type XAC(wt). An increase in AC associated with the apical membrane was also observed (Ashworth et al. 2001). These changes in phosphorylation were accompanied with the formation of actin aggregates and rods (Ashworth et al. 2003a,b). Increased F-actin disruption and aggregation was observed in cells expressing greater amounts of XAC(wt)–GFP. Under physiological conditions, XAC(wt) localized only to the cytoplasm, even though F-actin was also found at the basal, lateral, and apical microvilli-containing surfaces. However, following ATP depletion, XAC(wt) and F-actin colocalized in the basal and apical regions of the cell and in dense cytoplasmic actin aggregates. In addition, cells expressing the inactive XAC S3E mutant did not form actin aggregates or actin rods during ATP depletion. These data imply that activated, or nonphosphorylated, AC is central to the changes observed in the actin cytoskeleton in a porcine kidney cell line during ATP depletion.

Whether ATP depletion will induce rod formation or will cause AC to localize to depends preexisting F-actin structures on the relative levels of AC expressed in each cell. Porcine LLC-PK$_{A4.8}$ cells infect readily with recombinant adenovirus. These cells express low levels of AC (<0.1% of soluble protein with 99% as cofilin), whereas actin comprises 2.7% of the soluble protein (Maiti et al. 2002), corresponding to a molar ratio of cofilin:actin of 0.07. Endogenous cofilin undergoes complete dephosphorylation (activation) within 30 min of ATP depletion in these cells. However, these cells form only small aggregates containing actin and cofilin, and do not form rods. Following infection of these cells with adenovirus for expressing XAC, levels of total AC were measured along with the ability of the cells to form rods when ATP depleted. Rods form in the ATP-depleted infected cells when the molar ratio of total cofilin (cofilin + XAC) to actin exceeds >0.5 (between 36 and 48 h after infection), but rod size and number reach a maximum between 72 and 96 h postinfection when total cofilin to actin is >0.7. Thus, rod formation is a function of the AC:actin ratio, explaining why some cell lines readily form rods when ATP is depleted and others do not.

In the microvilli terminal web of proximal tubule cells, the actin filaments are associated with TM. During ischemia the TM is displaced from these filaments, allowing AC access for their depolymerization (Ashworth et al. 2003b). TM also appears as a major urinary protein from the ischemic rat. These results suggest that the competition between TM and AC binding to F-actin is dynamic and the balance in favor of one or the other can be tipped by modulating the pool of active AC through phosphoregulation.

A role for ezrin and myosin in the cellular response to ischemia has also been reported. Ischemia-induced cell injury caused microvillar collapse that was accompanied by ezrin dephosphorylation and dissociation of ezrin from F-actin and the microvillar surface (Chen, Doctor and Mandel 1994; Chen, Cohn and Mandel 1995). In addition, myosin I accumulated with large actin aggregates in the response to ischemic injury (Wagner and Molitoris 1997). Although these data implicate ezrin and myosin I in microvillar structural changes in response to ischemia, the mechanism driving these actin cytoskeletal changes has not been fully elucidated. By understanding the molecular mechanism(s) driving the extreme morphological and functional changes that occur in proximal tubule cells during ischemia, progress can be made toward treating the various clinical consequences resulting from these events.

Musculoskeletal Diseases

Muscular Dystrophy

Although there have been no studies to directly tie in defects in cofilin regulation with muscular dystrophy, the disorganized myofibrillar phenotype of *C. elegans* body wall muscle in which expression of the muscle-type cofilin is defective suggests that cofilin does have a role in myofibril assembly consistent with abnormalities causing muscle disease. In a mouse model of mechanically induced muscle damage (Thirion et al. 2001), the changes of cofilin expression were monitored during the first 10 days of regeneration, with dephosphorylated cofilin-2 being the major isoform at later stages of muscle regeneration. A similar predominance of dephosphorylated cofilin-2 was observed in chronically regenerating dystrophin-deficient muscles of Duchenne muscular dystrophy patients. Therefore, the cofilin-2 isoform may play an important role in normal muscle function and muscle regeneration.

Nemaline Myopathies

Nemaline myopathy (NM) is the most common form of the three congenital myopathies including actin myopathy and intranuclear rod myopathy (Clarkson, Costa and Machesky 2004). Congenital myopathies are neuromuscular disorders characterized by muscle weakness, rod-like Z-line accumulations (nemaline bodies), and myofibrillar disorganization (Corbett et al. 2005). The severities of these myopathies range from neonatal mortality to muscle weakness in adulthood with continual slow progressive degeneration. To date, disease-linked mutations have been identified in five genes encoding the proteins α-actin, α- and β-TM, nebulin, and troponin T. All of the disease-linked mutations occur in protein components of the sarcomeric thin filament.

Fifteen different missense mutations in the human skeletal muscle α-actin gene have been identified to be distributed throughout all six coding exons. These mutations result in 14 amino acid changes some of which involve known functional domains of actin (Nowak et al. 1999). Mutant α-actin isoforms are present within insoluble actin filaments isolated from muscle from two NM patients. Three NM-associated actin mutants (V163L, V163M, and R183G) were transfected into C2C12 myoblasts resulting in abnormal cytoplasmic and intranuclear actin aggregates. Residue 163 is adjacent to the nuclear export signal of actin and both mutations at this residue are known to cause intranuclear rod myopathy (Ilkovski et al. 2004). *In vitro* studies demonstrate that abnormal folding, altered polymerization, and aggregation are common features of NM actin mutant isoforms. Additional NM mutations found in α-actin (L221P, D292V, and P332S) all lay on the surface of the actin monomer contacted by TM during muscle activity (Laing et al. 2004).

The presence of intranuclear or cytoplasmic rods (nemaline rods) characterizes several NMs. Since cofilin has been identified as a component of rods described above for several other diseases, the role of cofilin in nemaline rods is currently being investigated. Genetic defects of cofilin-2 were excluded for one human muscle disorder, chromosome 14-linked distal myopathy, but could possibly be a rare cause of another NM (Thirion et al. 2001). α-Actinin 2 is a major component of rods in muscle biopsies of patients with NMs (Jockusch et al. 1980; Wallgren-Pettersson, Arjomaa and Holmberg 1990). A tissue culture model of various actin mutants that cause NM has been developed (Ilkovski et al. 2004). Overexpression of some α-skeletal actin (ACTA1) mutant chimeras with enhanced GFP in cultured myoblasts results in cytoplasmic actin aggregates. Cofilin has been found to be a component of some of these cytoplasmic aggregates but not of intranuclear aggregates resulting from overexpression of actin V163L, which causes intranuclear rod myopathy (North, K., Domazetovska, A., and Ilkovski, B., personal communication). This suggests that some ACTA1 mutants associated with nemaline myopathies show altered cellular interactions with cofilin. Whether cofilin association with mutant actins contributes to disease progression is unknown and needs to be addressed.

Immune Diseases

T-Cell Activation

Activation of resting T lymphocytes (T cells) leads to clonal growth and expression of their functional repertoires. When T cells encounter their antigen-presenting cell, their initial contacts are strengthened leading to the formation of an immunological synapse, crucial for T-cell activation. Engagement of the

T-cell receptor must be accompanied by activation of an accessory receptor (costimulation) in order to activate T-cell proliferation and interleukin secretion. Costimulation activates the reorganization of the actin cytoskeleton and the directional transport of receptors and lipid microdomains to the immunological synapse. Cofilin has been identified as a component of the costimulatory signaling pathway (Lee, Meuer and Samstag 2000; Samstag et al. 1994). Initial reports suggested cofilin dephosphorylation and nuclear translocation were required for T-cell activation. However, T cells loaded with Penetratin-coupled cofilin peptide sequences (residues 1–14 and 104–115) that inhibited cofilin–actin interaction, but not those loaded with a control peptide from the 104–115 region in which Lys112 and Lys114 were replaced with Gln, blocked formation of the immunological synapse as well as T-cell proliferation and IL secretion when cells were costimulated (Eibert et al. 2004). Thus, cofilin dephosphorylation, which results from the costimulatory signals, leads to enhanced cofilin association with the actin cytoskeleton, and affects the immunological synapse. Whether cofilin targeting to the nucleus is required as well remains unresolved. Commonly used immunosuppressive drugs do not affect cofilin activation by costimulatory signaling (Ambach et al. 2000). Therefore, cofilin might be a useful target for immunomodulatory therapeutics.

Many other autoimmune diseases may be linked through T-cell activation to the disease progression. For example, the respiratory disease asthma, a type of chronic allergic airway disease, is often accompanied by T-cell activation through costimulatory pathways described above (Kroczek and Hamelmann 2005). These diseases, which are among the most difficult to diagnose and treat, affect almost every tissue in the body.

Platelet-Dependent Clotting Disorders

Cofilin is largely phosphorylated in resting platelets. However, thrombin stimulation leads to a reversible dephosphorylation of cofilin peaking at 1–2 min poststimulation and corresponding with or just following the peak of actin assembly. Cofilin rephosphorylation begins at about 2 min and exceeds resting levels by 5–10 min postthrombin stimulation (Falet et al. 2005; Pandey et al. 2006). ROCK activation is responsible for mediating rapid Thr508 phosphorylation and activation of LIMK1, and the F-actin increase during shape change induced in platelets by thrombin stimulation. Two counteracting pathways involving a cofilin phosphatase and LIMK1 are activated during platelet aggregation/secretion regulating cofilin phosphocycling sequentially and independently of integrin $\alpha_{IIb}\beta_3$ engagement (see below). During shape change, cofilin phosphorylation was unaltered, and during aggregation/secretion, cofilin was first rapidly dephosphorylated by an okadaic acid-insensitive phosphatase and then slowly rephosphorylated by LIMK1, although stirring conditions decreased the rephosphorylation step (Falet et al. 2005).

Cofilin is normally rephosphorylated when platelets are stimulated in the presence of wortmannin to block $\alpha_{IIb}\beta_3$ crosslinking and signaling or in platelets isolated from a patient with Glanzmann's thrombasthenia, which express only 2–3% of normal $\alpha_{IIb}\beta_3$ levels. Glanzmann's thrombasthenia is a rare autosomal recessive bleeding disorder, which highlights the vital role played by this receptor in platelet function. Glanzmann's thrombasthenic platelets fail to aggregate due to a lack of surface expression of functional α_{IIb} or β_3 integrins on the platelet surface. Actin assembly and Arp2/3 complex incorporation in the platelet actin cytoskeleton is also decreased when $\alpha_{IIb}\beta_3$ is engaged, suggesting that cofilin is essential for actin dynamics mediated by outside-in signals in activated platelets (Falet et al. 2005). (For a more extensive discussion of ABP and leukocyte function, see chapter by Nunloi.)

Cancer

Although cancer is a genetically heterogeneous disease, two characteristics are believed to be universal in cancer cells: uncontrolled proliferation and increased cell survival (Green and Evan 2002). Cancer is a complex disease that usually involves accumulations of multiple genetic mutations. Results of these mutations are the expression of proteins that promote cell replication and/or inhibit expression of proteins that stop growth. Later mutations result in the expression of proteins that allow the cell to overcome growth inhibitory effects of the environment, and proteins that allow the cell to escape from its surrounding basal lamina and become invasive. Cellular changes that occur during the progression of cancer affect proteins that drive actin dynamics, sometimes modifying their function. Changes in the proteins driving actin dynamics can bring about cellular changes that modulate cell cycle progression and lead to more invasive cancers. As should be evident from what has been presented on cofilin function, cofilin plays important roles in various stages of cancer progression including cellular transformation, immortalization, motility, escape from apoptosis, synthesis and/or secretion of metalloproteases, and angiogenesis.

Cell Transformation

Transformation describes the changes that cells undergo to begin the process toward becoming invasive and metastatic. It usually involves the gain in function of an oncogene involved in upstream signaling for cell growth and/or the loss of tumor suppressor genes that regulate growth. Transformation involves changes in gene expression and signaling, decreased apoptosis, and cell proliferation, and growth control that will be discussed below.

Gene Expression and Signaling

No direct link has been established between the translocation of AC–actin to the nucleus and the expression of specific genes. However there are many reports to suggest that transcription and cytoskeletal reorganization during cell transformation are coordinately regulated.

Serum response factor (SRF) stimulation leads to transcriptional activation and F-actin polymerization in smooth muscle cells and many cell lines. While the exact pathways of SRF-mediated activation of transcription vary in different cell lines, the effects are ultimately mediated through RhoA-actin signaling (Geneste, Copeland and Treisman 2002). Stimulation with SRF also leads to F-actin polymerization through activation of a member of the Vasodilator-stimulated phosphoprotein (Ena/VASP) family and mDia1, a Diaphanous-related formin, which function cooperatively downstream of Rho (Grosse et al. 2003). Members of the formin family function in regulation of actin remodeling and in activation of gene transcription. SRF stimulation leads to activation of formin homology-2-domain containing protein 1 (FHOD1) upstream of transcription and actin stress fiber formation (Westendorf and Koka 2004). Taken together these findings suggest an interconnection between actin dynamics and transcriptional activation.

Cofilin contains a NLS homologous to SV40 large T antigen (Matsuzaki et al. 1988). One mechanism used to target cofilin and actin to the cell nucleus in some cell types is heat shock. Mutations in the cofilin NLS prevent its targeting to the nucleus in response to heat shock even while cofilin functions in cytoplasmic actin-dependent processes remain intact (Iida, Matsumoto and Yahara 1992). Thus, the NLS is required to direct cofilin to the nucleus after heat shock; results also suggest that actin translocation into the nucleus for rod formation is cofilin dependent.

Cyclin D1 is induced in most cells during mid-G1 and controls G1 phase progression through interaction with cyclin-dependent kinases (cdk) 4/6. Its induction is usually the rate limiting step in the activation of cdk4/6 as it integrates signals from receptor tyrosine kinases, integrins, and the actin cytoskeleton to allow cell cycle progression only if growth conditions are optimal. There are two pathways for cyclin D1 induction. One pathway requires prolonged signaling from the extracellular receptor kinase (ERK), whose maintenance in an active state requires integrin–matrix interaction, cytoskeletal tension, and growth factor signaling. In the second pathway, Rac and/or Cdc42 induces cyclin D1 in an ERK-independent manner, but still requires receptor tyrosine kinase and integrin signaling, but it is independent of stress fiber formation and cell tension (Roovers and Assoian 2003). Nuclear activity of LIMK is required to prevent cyclin D1 expression early in G1 either in an ERK-dependent Rho and ROCK pathway or in the Rac/Cdc42-dependent pathway (Roovers et al. 2003). The effect of nuclear LIMK on cyclin D1 expression ultimately regulates the duration of G1 phase and the degree to which G1 phase progression depends on actin stress fiber formation and imposition of cellular tension. Experiments performed in the presence of actin-depolymerizing drugs showed that the LIMK inhibition of cyclin D1 expression was not diminished, suggesting actin assembly was not important. Expression of neither the S3E nor the S3A mutant of cofilin altered the inhibitory effect of LIMK on Rac/Cdc42-dependent cyclin D1 expression, a finding interpreted by the authors to indicate the effect was cofilin independent. We would like to offer another possible interpretation. If the function of cofilin in altering gene expression is through the transport of actin into the nucleus, and the actin functions as part of a complex in which it is not assembled, neither of the cofilin mutants used here would rule out this mechanism. The cofilin S3E mutant would not bind actin and thus would not transport it into the nucleus nor would it prevent endogenous cofilin from doing so. If the cofilin S3A mutant transported actin into the nucleus, the cofilin would remain active and might not release its actin cargo, which is perhaps one function of the nuclear LIMK. Thus, we believe that the nuclear function of cofilin in regulating cyclin D1 expression remains an open question. It is still possible that its function is to release actin from the cofilin so that the actin can form its complexes with chromatin-remodeling factors and RNA polymerases.

The translocation of LIMK1 from the cytoplasm to the nucleus is under regulation of a cdk inhibitory protein, specifically p57^{Kip2} in osteoblasts (Yokoo et al. 2003). The cdk inhibitory proteins regulate progression of the cell cycle and are also important in cell differentiation. Cells with nuclear targeted LIMK1 showed altered cytoplasmic actin filament systems, as would be expected. These findings again support a role for the crosstalk between the cytoskeleton and processes of cell growth and differentiation and they suggest a nuclear role for cofilin.

Binding of receptor-recognized forms of the proteinase inhibitor α_2-macroglobulin (α_2-M) to the 78 kDa glucose-regulated protein (GRP78) on the cell surface of 1-LN human prostate cancer cells induces mitogenic signaling and cellular proliferation. α_2-M exposure induces a two- to threefold increase in phosphorylated PAK2 and a similar increase in its kinase activity toward myelin basic protein. Silencing the expression of the *GRP78* gene greatly attenuated the appearance of phospho-PAK2 in α_2-M-stimulated cells. PAK2 activation was accompanied by increases in phosphorylated LIMK and phosphorylated cofilin. *PAK2* gene silencing greatly reduced phospho-LIMK levels, thus demonstrating activation of PAK2 in 1-LN prostate cancer cells by protein inhibitor α_2-M (Misra, Sharma and Pizzo 2005).

Following transformation by the Rous sarcoma virus (v-Src) tyrosine kinase, cells undergo dramatic changes in the cytoskeletal architecture. Actin stress fibers and focal adhesions are dissolved, and actin-associated adhesive structures called *podosomes* are formed. These changes in the actin cytoskeleton are thought to underlie adhesion-independent growth and increased cell migration, two characteristics of a transformed cell (Pawlak and Helfman 2001). Cortactin is a major Src substrate and is found associated with F-actin in the cell cortex (Wu et al. 1991). Cortactin binds preferentially to ATP and ADP-Pi forms of F-actin, thus enhancing lamellipodial persistence through an antagonistic effect on F-actin turnover by AC (Bryce et al. 2005).

Decreased Apoptosis

Along with aberrant regulation of proliferation, cancerous cells can genetically adapt to protect themselves from programmed cell death or apoptosis. To suppress apoptosis, all somatic cells require constant survival signals.

The development of therapeutics aimed at inducing apoptosis in cancer cells is confounded by the susceptibility of healthy noncancerous cells to also undergo apoptosis. Since apoptosis is a complex signaling event that is often perturbed in cancer cells, molecules that disrupt progression of apoptosis in cancer cells can be therapeutic targets to clinically intervene without affecting healthy cells. Thus, understanding the pathways involved in initiating and inhibiting apoptotic signals is essential for the development of some cancer therapies.

Cells that have initiated apoptosis go through specific morphological changes including chromatin condensation, DNA fragmentation, plasma membrane blebbing, and cell shrinkage. Caspases, a multigene family with 14 known members, are a driving force behind these changes (Kasibhatla and Tseng 2003). Synthesized as inactive precursors, caspases become active following proteolytic cleavage at two or three aspartic acid residues (Kasibhatla and Tseng 2003). Once activated, caspases cleave target substrates and trigger the induction of apoptosis. The activation of the caspase cascade occurs downstream of two pathways. The "extrinsic" pathway is initiated by activation of cell surface receptors responding to apoptotic signals; the "intrinsic" pathway originates at the mitochondria (Kasibhatla and Tseng 2003).

The "intrinsic" apoptosis pathway begins when cytochrome c (or another mitochondrial-derived ligand) is released into the cytosol from the mitochondrial inner membrane space. Once in the cytosol, cytochrome c forms an ATP-dependent complex with Apaf-1 and caspase-9, leading to the activation of caspase-9 and subsequent activation of the execution caspase cascade (Li et al. 1997). Apoptotic signals originating downstream from mitochondria have been well studied; however, signals prior to mitochondrial involvement are less understood. Previously described studies (Chua et al. 2003) suggest that cofilin targeting to mitochondrial outer membrane is both necessary and sufficient for release of cytochrome c to activate this intrinsic pathway. Functional actin binding was not required for cofilin targeting to mitochondria but was required for the release of cytochrome c. Furthermore, expression of a fusion protein between the Bcl-2 mitochondrial-targeting domain and cofilin resulted in its localization to mitochondria where it induced release of cytochrome c into the cytosol. Thus, localization of cofilin to the mitochondria is sufficient to induce the intrinsic apoptotic pathway.

Activation of cofilin in cells is not sufficient to target it to mitochondria and thus its mitochondrial translocation must be regulated through other factors. As a case in point, overexpression of CD44, an adhesion and antiapoptotic molecule, and rearrangement of the actin cytoskeleton are both evident in early stages of colon cancer. In the SW620 human colon cancer cell line, which does not express CD44, cofilin overexpression occurs when the cells were stably transfected with variant isoforms of CD44. The majority of overexpressed cofilin is dephosphorylated, which could bring about directional motility of cells, thus having important implications in the proliferation and motility of colonic epithelial cells in cancer (Subramaniam, Vincent and Jothy 2005).

Overexpression of wild-type LIMK in PC12 cells rendered them more resistant to serum-withdrawal-induced apoptosis (Yang et al. 2004a). This protection could come from maintenance of a phosphocofilin pool, which would decrease mitochondrial translocation and cytochrome *c* release, or it could be through inhibition of JNK activation, which was also observed in these cells. Similarly, SRF is both necessary and sufficient for the enhanced neuronal survival of postnatal cortical neurons to neurotrophic deprivation or campothecin-induced DNA damage (Chang, Poser and Xia 2004). SRF stimulates F-actin polymerization through RhoA-actin signaling (Geneste et al. 2002; Grosse et al. 2003) and may also decrease the ability of cofilin to target actin to mitochondria for cytochrome *c* release (Chua et al. 2003).

Cell Proliferation and Growth Control

Transforming growth factor β (TGFβ) controls cell adhesion, motility, and growth of diverse cell types (Roberts 1998). While the classical TGFβ signaling pathway works via type I and type II receptor subunits to activate Smad signaling effectors (Shi and Massague 2003), TGFβ modulates the actin cytoskeleton in Swiss 3T3 and mouse embryonic fibroblasts through a Rho, ROCK, LIMK2, cofilin pathway mediated by only the TGFβ-type I receptor (Vardouli, Moustakas and Stournaras 2005).

Coordinated changes in the actin cytoskeleton are required for proper progression through the cell cycle, particularly the spatially and temporally regulated process of cytokinesis (Glotzer 2001). ADF/cofilin (Tanaka, Okubo and Abe 2005) and its known regulators, LIMK1 and SSH, have all been shown to function in cytokinesis and undergo alterations in cell cycle regulation (Kaji et al. 2003). HeLa cells, synchronized in prometaphase using nocodazole, display a hyperphosphorylation of LIMK1 and increased activity of LIMK1 toward cofilin (Sumi, Matsumoto and Nakamura 2002). As cells progress into interphase, LIMK1 activity decreases. A separate study in HeLa cells revealed that endogenous LIMK1 and cofilin phosphorylation levels increased during prometaphase and metaphase and returned to basal levels of phosphorylation during telophase and cytokinesis (Amano et al. 2002). Overexpression of LIMK1 increased the levels of phosphocofilin and disrupted cell division resulting in multinucleated cells.

SSH-1 activity is also regulated during mitosis. Endogenous SSH-1 is phosphorylated and displays low phosphatase activity during prometaphase and metaphase; however, SSH-1 is dephosphorylated and active during telophase and cytokinesis (Kaji et al. 2003). Consistent with a role for SSH-1 in the cell cycle during cytokinesis, SSH-1 localizes to the cleavage furrow and midbody. Disruption of proper SSH-1 regulation leads to aberrant changes in the actin cytoskeleton most likely related to changes in phosphocycling of ADF/cofilin. Together, these studies imply that proper phosphoregulation of cofilin by LIMK1 and SSH-1 is important during cell division and cytokinesis.

The large tumor suppressor (LATS) family of proteins is conserved from *Drosophila* to mammals. LATS1 has been found to bind with LIMK1 *in vitro* and to colocalize with LIMK1 at the contractile ring during cytokinesis. LIMK1-mediated cytokinesis defects and phosphorylation of cofilin are both blocked by LATS1 binding. Inactivation of LATS1 disrupts normal cytokinesis resulting in increased numbers of multinucleated cells (Yang et al. 2004b). Thus, LATS1 appears to mediate its tumor suppression activity through AC protein regulation by LIMK1.

Morphological Changes During Transformation

Ras-induced transformation of human diploid fibroblasts is characterized by uncontrolled proliferation and morphological changes resulting from disruption of the actin cytoskeleton. Human fibroblasts are more resistant to Ras-induced transformation than are rodent fibroblasts. Overexpression of activated human Ras (H-Ras) in rat embryo fibroblasts produced decreased expression of TM isoforms and suppression of the ROCK/LIMK/cofilin pathway accompanied by an increased invasive phenotype. These changes do not occur in human fibroblasts in response to H-Ras overexpression. These studies suggest that a human-specific mechanism exists for Ras-mediated transformation and suggests that caution need to be exercised when extrapolating results from rodent cells to human cells (Sukezane et al. 2005a,b).

A mechanism to explain the changes in cell morphology observed in cells expressing v-Src is being explored. Various Src substrates, many of which are actin cytoskeletal proteins, are tyrosine phosphorylated in response to v-Src expression (Erpel and Courtneidge 1995). Similarly, the Ras/MEK and PI3K cell signaling cascades, two pathways implicated in cytoskeletal dynamics, are activated in response to v-Src expression. Because MEK and ERK translocate to focal adhesions following v-Src expression, it is believed that both proteins are involved in the downstream cytoskeletal changes associated with cell transformation (Fincham et al. 2000). Implying a role for MEK in v-Src-mediated transformation, inhibition of MEK prevents the breakdown of stress fibers and focal adhesions and the formation of podosomes after v-Src expression (Pawlak and Helfman 2002a,b).

In agreement with previous reports suggesting Rho acts downstream of v-Src to induce actin cytoskeletal changes (Mayer et al. 1999), overexpression of active mutants of the Rho downstream effectors, ROCK, or LIMK, also inhibited loss of stress fibers in response to v-Src expression (Pawlak and Helfman 2002a,b). Dephosphorylation of cofilin is increased in v-Src expressing cells, suggesting inhibition of the Rho/ROCK/LIMK pathway and the subsequent activation of cofilin (Pawlak and Helfman 2002a,b). Expression of oncogenic H-RasV12 induces the translocation of the cyclin-dependent kinase inhibitor p21(Cip1) from the nucleus to the cytoplasm where it uncouples Rho GTP from stress fiber formation by inhibiting

ROCK (Lee and Helfman 2004). In addition, activation of MEK inhibited expression of ROCK, presenting a possible pathway for how MEK activation downregulates the Rho pathway that normally causes cofilin phosphorylation. Thus, the Rho pathway, which is upstream of cofilin phosphorylation, is a major factor in the actin cytoskeleton changes in response to v-Src expression. Likewise, changes in phosphocofilin levels after v-Src transformation indicate a potential role for cofilin in cytoskeletal modifications. However, establishing a definitive role for cofilin requires repeating these studies in cells in which cofilin expression is silenced.

Cancer cells often acquire increased capacity to migrate and metastasize. As described previously, the establishment of a polarized lamellipodium for directed cell migration requires AC, whether the cell polarizes spontaneously (Dawe et al. 2003) or is responding to chemotactic cues (Chan et al. 2000; Zebda et al. 2000). Directional migration requires LIMK and SSH, implying that the local control of AC through phosphoregulation is required for integration of extracellular signals into a turning response (Nishita et al. 2005). Although there are apparent discrepancies in results presented below as to whether increased levels of total or active cofilin, or increased levels of LIMK1 (and presumably decreased active cofilin levels) promote enhanced cell motility, it is important to reiterate that cofilin phosphocycling rates may be more important than the absolute levels of active cofilin at any point in time. Furthermore, there are undoubtedly optimal levels of cofilin for maintaining the most favorable actin dynamics for persistent cellular migration. For example, the human glioblastoma cell line, U373 MG, displays increased rates of locomotion with increased overproduction of cofilin (Yap et al. 2005). A maximal rate of locomotion, twice as fast as wild-type cells, is reached when cofilin expression levels are 4.5 times that of wild-type cells. Cellular locomotion progressively decreases as cofilin expression increases beyond an optimal level. Such a cofilin-mediated increase in tumor cell locomotion is likely to contribute to cell invasiveness (Fig. 14) (Dang, Bamburg and Ramos 2006).

Overexpression of LIMK1, along with an increase in phosphorylation of cofilin, has been observed in prostate tumors and prostate cancer cell lines (Davila et al. 2003). Decreased expression of LIMK1, using antisense RNA, inhibited the invasiveness of metastatic cells and blocked cell proliferation at the G2/M phase transition (Davila et al. 2003). In addition, ectopic expression of LIMK1 induces a metastatic phenotype in benign prostate epithelial cells (Davila et al. 2003). Invasive breast and prostate cancer cell lines display a higher level of LIMK1 protein expression and kinase activity in comparison to other less invasive cancer cell lines (Yoshioka et al. 2003). Ectopic overexpression of LIMK1 in MCF-7 and MDA-MB0231 human breast cancer cell lines induced an increase in cell motility assayed by an *in vitro* Matrigel invasion assay. The increase in invasiveness was blocked by ROCK and Rho inhibitors, suggesting Rho-mediated activation of LIMK1 is required for the invasive phenotype.

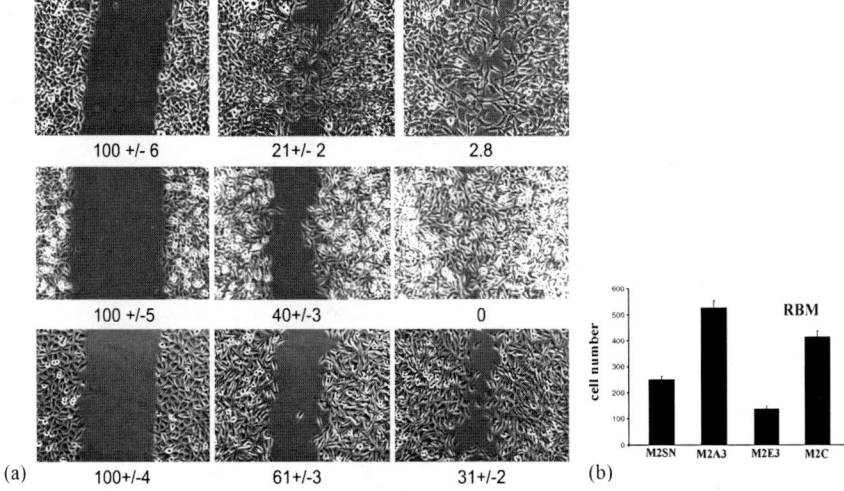

FIG. 14. Migration of melanoma cells is enhanced by expression of active cofilin and inhibited by expression of pseudophosphorylated cofilin. (**a**) Wound healing assay showing closure of wound in control cells (*top panel*), cells expressing XAC S3A (*middle panel*), and cells expressing XAC S3E (*lower panel*). Numbers under panels reflect the relative area of the wound that has remained open. (**b**) Tumor cell reconstituted basement membrane invasion assay on transwell filters. Cell numbers are those moving across the membrane. The invasiveness of the parental line of melanoma cells (M2SN) was increased by expressing the active XAC S3A (M2A3) and decreased by expressing the XAC S3E (M2E3). Overexpressing wild-type XAC (M2C) also enhanced invasiveness (adapted from Dang, Bamburg and Ramos 2006)

Alterations in Secretory Pathways

LIMK1-transfected human breast cancer cells display abnormal dispersed translocation of lysosomes stained for lysosome integral membrane protein (LIMP) II and cathepsin D throughout the cytoplasm. These lysosomes failed to colocalize with the transferrin receptor and LIMPII-positive lysosomes did not colocalize with early endosomes labeled with endocytosed transferrin. LIMK1 overexpression induced a slowing in the receptor-mediated internalization of EGF in comparison with mock-transfected cells. These findings suggest that LIMK1 plays a role in regulating vesicle trafficking of lysosomes and endosomes in invasive tumor cells (Nishimura et al. 2004).

The expression of metalloproteinases MT1-MMP and MMP2 were increased in melanoma cells overexpressing wild-type or cofilin S3A (Dang, Bamburg and Ramos 2006). A 50% reduction of both enzymes was observed in cells expressing the cofilin S3E. Overexpression of active, nonphosphorylatable cofilin S3A was sufficient to induce the expression of MT1-MMP and

MMP2 in β3-integrin-negative M2Tβ3 cells. Interestingly, the invasion (enhanced mobility) of M2Tβ3 cells could be sustained by overexpression of cofilin (S3A). These results suggest that the integrin αvβ3 and cofilin together regulate melanoma cell invasion by enhancing actin dynamics and the secretion of metalloproteinases required for modifying the extracellular matrix. The enhanced secretion could arise from Golgi effects of cofilin on sorting tubules, or on transport, delivery, and/or exocytosis of the enzymes.

Metastasis

Alterations in cell adhesion, often through a change in integrin protein expression, usually precede tumor cell metastasis. A positive relationship between β3-integrin expression and metastatic potential of melanoma cells has been reported (Li et al. 1998). In a more recent study, the relationship between β3-integrin expression, phosphorylation of cofilin, and metalloproteinase production by melanoma cells was examined (Dang, Bamburg and Ramos 2006). The levels of phosphorylated cofilin were tenfold higher in cells expressing integrins αvβ3 than in αvβ3-negative cells when plated on vitronectin for 30 min. However, by 60 min, phosphorylation of cofilin was greater in the β3-negative cells. Expression of the wild-type or nonphosphorylatable cofilin (S3A) increased melanoma cell migration on vitronectin and invasion through a reconstituted basement membrane, whereas expression of a pseudophosphorylated S3E, poorly active cofilin reduced cell invasion. Expression of active cofilin accelerated the phosphorylation of focal adhesion kinase, strongly implicating cofilin as a mediator of cell signaling involved in the motility.

The Rho/ROCK/LIMK pathway is one of the major signaling pathways involved in tumor metastasis and a potential target for antimetastasis therapy (Suyama et al. 2004). Disruption of ROCK causes cells to lose metastatic properties and it is reported that suppression of LIMK2 in human fibrosarcoma cells limits their migration and efficiency to form dense colonies without affecting cell proliferation rate or viability (Suyama et al. 2004). RhoC is overexpressed in metastatic melanoma cells and expression of dominant-negative Rho inhibits metastasis (Clark et al. 2000).

The B-Raf gene is mutated at high frequency in human cancers. Most oncogenic B-Raf mutations have enhanced kinase activity and maintain constitutive MEK and ERK phosphorylation leading to the transformed phenotype. Disruption of B-Raf enhances cell migration by reducing stress fibers and lowering total F-actin (Pritchard et al. 2004) with no change in myosin light chain phosphorylation. However, in B-Raf null cells, AC was dephosphorylated and ROCK expression was reduced. Overexpression of LIMK or MEK restored the stress fibers and resulted in reestablishing the phosphoAC pool and decreasing cell motility. These results suggest that normal B-Raf maintains the actin stress fibers through a MEK/ROCK/LIMK pathway and that its abnormal activity, either too high as in many tumorigenic cell types or too low as in the null mutations, alters remodeling of actin stress fibers to enhance motility.

On the basis of the above studies in which cofilin overactivity seems to enhance cell motility by decreasing actin stress fibers, it is surprising that cofilin-1 is downregulated in highly metastatic cells compared to cells with low metastatic potentials as determined by proteome analysis (Ding et al. 2004). However, it is consistent with the finding of LIMK1 overexpression in highly metastatic tumor cells. Decreased expression of the protein nm23-H1 was also reported in highly metastatic cells. Nm23-HI functions as a metastasis suppressor protein (Steeg et al. 1988). It negatively regulates Tiam1 and inhibits Rac1 activation *in vivo* (Otsuki et al. 2001). Thus, its downregulation could enhance cofilin activity through a decrease in the Rac/PAK/LIMK pathway. Cofilin phosphocycling, i.e., the rate of phosphate turnover, may be more important to the coupling of actin dynamics than is the absolute amount of active cofilin (Meberg et al. 1998; Chen, Bernstein and Bamburg 2000). Thus, even though total cofilin levels decrease and LIMK1 levels increase in metastatic cells, the rate of cofilin phosphocycling may be elevated, which could enhance actin dynamics and cell invasiveness, especially if inhibitory TMs are downregulated.

The loss of high molecular weight TM isoforms occurs in many childhood and adult tumors, particularly those of the central nervous system (Hughes et al. 2003). The morphological changes, associated with cell transformation, in some but not all tumor cell lines can be reversed by exogenous expression of TM1, one of the high molecular weight TMs (Yager et al. 2003). Loss of TM1 expression in breast tumor cells is associated with a resistance to anoikis (Raval et al. 2003), a form of apoptosis that is associated with loss of cell adhesion. Exogenous expression of TM1 in breast tumor cells resensitizes them to anoikis (Bharadwaj et al. 2005). TGFβ induces stress fibers in cultured cells through Smad and p38 Map kinase pathways, but requires the presence of TM1 for stress fiber formation (Bakin et al. 2004). Silencing of the *TM1* gene alters the tumor suppressor function of TGFβ (Varga et al. 2005). Because the high molecular weight TMs compete with AC proteins for actin filament binding (Bryce et al. 2003), the enhanced invasiveness of metastatic tumor cells is probably due to the loss of the AC-inhibitory TMs.

Invasive cancer cells form actin-rich membrane protrusions with matrix degradation activity known as *invadopodia*, which aid in the cells ability to escape from a surrounding basal lamina and pass across the endothelial cell barrier of capillaries. Neural WASP (N-WASP), Arp2/3 complex, and upstream activators including Cdc42 are necessary for invadopodium formation. Silencing cofilin expression interferes with long-lived invadopodia and decreases matrix degradation activity in metastatic carcinoma cells. Thus, invadopodia formed from the N-WASP–Arp2/3 pathway require cofilin for stabilization and maturation (Yamaguchi et al. 2005) as well as for the release of matrix degrading enzymes (Dang, Bamburg and Ramos 2006).

Angiogenesis

Cofilin and the actin cytoskeleton are proposed to function in angiogenesis, the formation of new capillaries. The endothelial cell cytoskeleton is an essential component of the vascular barrier (Garcia and Schaphorst 1995). All mammalian cells need oxygen and nutrients and are typically located within 100–200 μm of blood vessels (Carmeliet and Jain 2000). To grow beyond a certain size, multicellular organisms must generate new sources of oxygen and nutrients through vasculogenesis and angiogenesis requiring proliferation of endothelial and smooth muscle cells. Similarly, to grow and metastasize, tumors require the formation of new blood vessel. An "angiogenic switch" model has been proposed and states that certain triggers, such as metabolic or mechanical stress, genetic mutations, or an immune or inflammatory response, promote angiogenesis (Carmeliet and Jain 2000). One mechanism to explain vascular growth states that endothelial precursors from the vessel wall or bone marrow migrate to the tumor, divide, and form tubules. Inhibitors of angiogenesis have been studied in clinical trials, but the results have not shown promise for these compounds therapeutically (Cristofanilli, Charnsangavej and Hortobagyi 2002). Several studies have begun to determine the effects of angiogenesis inhibitors on the endothelial cell cytoskeleton and the role of ADF/cofilin in angiogenesis.

A proteomics approach was used to determine the proteins that function downstream of antiangiogenic signaling (Keezer et al. 2003). The angiogenic inhibitors used in this study prevent endothelial cell proliferation in vitro and in vivo, prevent endothelial cell migration in vitro, induce endothelial cell apoptosis, and disrupt monocyte and macrophage chemotaxis. Two-dimensional gel electrophoresis was used to compare proteins in endothelial cell extracts prior to or following treatment with inhibitors of angiogenesis. These studies revealed several proteins that displayed distinct intensity changes. Two such proteins were hsp27, a chaperone protein that has actin filament-binding activity, and cofilin. Levels of phosphocofilin increase after treatment with the angiogenesis inhibitors fumagillin, TNP-470, endostatin, EMAP-II, and thrombospondin-1. In addition, cofilin localization, normally found in lamellipodia and cell surface projections, becomes cytoplasmic and diffuse following treatment with fumagillin. Surprisingly, in endothelial cells, phosphocofilin localizes primarily to the nucleus; however, following treatment with fumagillin phosphocofilin localizes to the cytoplasm. The changes in localization of both active and inactive cofilin imply that fumagillin might affect actin turnover via cofilin or actin transport into the nucleus with possible changes in gene regulation.

To determine the effects of angiogenesis inhibitors on the actin cytoskeleton, endothelial cells were treated with fumagillin, thrombospondin-1, and EMAP-II, and then fluorescently labeled phallacidin was used to visualize actin stress fibers. Following treatment with all of the angiogenesis inhibitors, the incidence of cellular projections decreased, the number of focal adhesions increased, and stress fibers increased in density. These data suggest that angiogenesis inhibitors induce greater endothelial cell attachment to the extracellular matrix as a means to inhibit vascular growth. The observed changes in cofilin phosphorylation and localization following antiangiogenic treatments suggest a role for cofilin in the resulting morphological changes to the actin cytoskeleton. Furthermore, inhibition of cofilin activation following ATP depletion of cultured endothelial cells delayed F-actin cytoskeletal destruction, suggesting that endothelial cell damage from ischemia and sepsis that leads to compromised microvascular flow in many organs, is mediated at least in part through cofilin (Suurna et al. 2006).

The growth of preexisiting collateral arteries (arteriogenesis) can be induced in rabbits by femoral artery occlusion (Boengler et al. 2003). A subtractive hybridization screen was used to identify cDNAs encoding new proteins associated with the enhanced arteriogenesis after 24 h. Among the upregulated proteins was cofilin-2, which was confirmed at the protein level and by immunolocalization to the smooth muscle cells of the collaterals, suggesting a role for cofilin-2 in the smooth muscle remodeling associated with building new arteries.

Stretch on the vascular wall contributes to the maintenance of smooth muscle (Albinsson, Nordstrom and Hellstrand 2004). Stretch stimulated Rho activity and phosphorylation of cofilin-2 lead to an enhanced F/G-actin ratio in stretched veins when compared with unstretched ones after 24 h. Cofilin-2 phosphorylation in stretched veins was inhibited by Rho inhibition. The results show that stretch of the vascular wall stimulates increased actin polymerization. The effect is partially, but probably not completely, mediated via Rho kinase (Rock) and cofilin downstream of Rho.

Platelet-derived phospholipids, such as lysophosphatidic acid (LPA), phosphatidic acid (PA), and sphingosine 1-phosphate (S1P), mediate the cellular signals linking hemostasis and angiogenesis through activation of the G-protein-coupled endothelial development gene (Edg) receptors (English, Garcia and Brindley 2001). Signaling cascades driven by platelet-derived phospholipids promote angiogenesis by inducing endothelial cell release from the monolayer, chemotactic motility, cell proliferation, adherens junction assembly, and the formation of new vesicle-like structures (English, Garcia and Brindley 2001). For an extensive discussion of ADF/cofilin and other ABPs and cancer, see the chapter by Van Troys et al.

Inflammation

Increased vascular permeability contributes to the high rate of morbidity and mortality of patients with inflammatory pulmonary diseases. To determine the role of S1P on endothelial cell permeability, bovine and human endothelial monolayers were treated with S1P and barrier protection was enhanced as assayed by a dose-dependent increase of the transmonolayer electrical resistance (TER) (Garcia et al. 2001). In addition, treatment of endothelial cells during or after treatment with thrombin, an edema-inducing agent, inhibits the decrease in TER and reestablishes the integrity of the vascular barrier. Therefore, these data show that S1P enhances the vascular barrier in untreated cells and protects and restores the vascular barrier in cells treated with thrombin.

To determine how S1P mediates the enhancement of TER, the effects of a number of signaling inhibitors were determined. Inhibitors for p38 MAP kinase (SB203580), ERK kinase (MEK) (UO126), phosphatidylinositol-3 kinase

(LY294002), and p60src (PP2) did not affect S1P-mediated vascular barrier protection. However, treatment of endothelial cell monolayers with the tyrosine kinase inhibitors (genistein, herbimycin, and erbstatin) significantly decreased S1P enhancement of TER (Garcia et al. 2001). Immunofluorescent staining of endothelial cells treated with S1P exhibited an increase in cortical F-actin correlating with enhanced cell barrier function. In contrast, endothelial cells treated with thrombin exhibited an increase in stress fibers and a presence of cell gaps consistent with treatment with an edemagenic agent. Reagents to disrupt the actin cytoskeleton and microtubules were used to determine the role of the cytoskeleton in the observed vascular barrier protection following treatment with S1P. In endothelial cells, latrunculin B, an inhibitor of actin polymerization, and cytochalasin B, an actin depolymerization agent, decreased the basal TER and blocked the S1P-mediated enhancement of TER (Garcia et al. 2001). In addition, endothelial cells treated with nocodazole, an inhibitor of microtubule polymerization, showed an S1P-induced enhancement of TER (Garcia et al. 2001). Thus, these data imply a role for actin cytoskeletal dynamics in S1P-mediated vascular barrier protection. However, *in vivo* assays of microvasculature permeability do not always correlate with gaps visualized in endothelial monolayers (Waschke et al. 2004a,b,c, 2005). (Also see chapter by Nunoi.)

Aging and Cellular Senescence

Senescent human diploid fibroblast cells have a significantly reduced activity of LIMK1 and accumulate G-actin and dephosphorylated cofilin in their nucleus (Kwak et al. 2004). Phorbol-ester treatment induced nuclear export of actin and increased DNA synthesis in senescent cells along with changes associated with younger cells such as increased retinoblastoma protein phosphorylation and a decreased expression of p21[WAF1], ERK1/2, and caveolins 1 and 2. Nuclear actin export in young human diploid fibroblast cells was inhibited by leptomycin B and by transfection with nonphosphorylatable mutated S3A cofilin, and both treatments induced the senescent phenotype including G1 arrest. Nuclear actin accumulation was much more sensitive and an earlier event than the well-known, senescence-associated beta-galactosidase activity. These results support a role in cell aging for cofilin-mediated nuclear accumulation of actin, either through cofilin-dependent nuclear translocation or cofilin-inhibition of nuclear export.

Other studies have demonstrated a relationship between cell aging/apoptosis and the ability of the cell to maintain dynamic cytoplasmic actin filaments (reviewed in Gourlay and Ayscough 2005). In yeast, expression of an actin bundling protein Scp1 causes accumulation of actin clumps leading to cell death (Winder, Jess and Ayscough 2003). A human equivalent to this protein, SM22/transgelin, was identified in studies looking for markers of cell senescence (Dumont et al. 2000). In cells in which actin turnover declines in the cytosol, there is a burst of reactive oxygen from mitochondria that is thought to contribute to the damage that ultimately leads to cell death. While the connection between F-actin dynamics and mitochondrial function has not been elucidated, compartmentalization of actin and/or cofilin into the nucleus is one mechanism that would decrease cytoplasmic actin turnover.

Bacterial Invasion and Pathogenesis

Bacterial Uptake

Changes in the host cell actin cytoskeleton are often triggered by an invading microbial pathogen, such as *Salmonella typhinurium*, *Escherichia coli*, *Bordetella pertussis*, and *Neisseria meningitides* (Galan and Zhou 2000; Masuda et al. 2000; Chen et al. 2002; Eugene et al. 2002; Khan et al. 2002). In addition, following infection, many pathogens recruit the host actin cytoskeleton to generate a force to move the organism within the cytoplasm and from cell to cell. Several pathogens are propelled by a polarized assembly of actin "comet" tails, including *Shigella flexneri*, *Listeria monocytogenes*, *Listeria ivanovii*, *Rickettsia conorii*, *Rickettsia rickettsia*, and the vaccinia virus (Frischknecht et al. 1999; Cossart 2000), although vaccinia virus remains extracellular and is propelled around the surface of the cell through comet tail motility of its receptor.

Invasive bacteria induce their uptake and invade nonphagocytic cells by using a variety of bacterial proteins to manipulate the host actin cytoskeleton. Bacteria enter host cells using either the zipper or trigger mechanism (reviewed by Cossart and Sansonetti 2004). The zipper mechanism is employed by *Yersinia pseudotuberculosis* and *Listeria monocytogenes*. Bacterial proteins bind to the host cell through host transmembrane cell adhesion protein receptors. Following membrane extension and actin polymerization, a phagocytic cup forms around the pathogen. Lastly, the phagocytic cup closes and retracts and is accompanied by actin depolymerization (Cossart and Sansonetti 2004). *Shigella* and *Salmonella* use the trigger mechanism. These bacteria secrete multiple proteins that create a pore in the host cell, the type III secretory system, through which bacterial effector proteins are transferred to induce actin cytoskeletal changes required for internalization of the pathogen.

Listeria monocytogenes, a facultative intracellular bacterial pathogen that infects herd animals or humans, causes a food-borne illness often represented by meningitis, encephalitis, or fetal death (Schlech et al. 1983). Recent evidence supports a role for AC in internalization of *L. monocytogenes*. AC and LIMK1 localized to internalin1B (In1B)-induced membrane ruffles and phagocytic cups (Bierne et al. 2001). In addition, overexpression of LIMK1 or dominant-negative LIMK1, to either downregulate or upregulate cofilin activity, respectively, inhibited In1B-induced internalization. Overexpression of the constitutively active cofilin S3A mutant inhibited bacterial entry and In1B-induced membrane ruffles. Overexpression of LIMK1 resulted in uncharacteristic F-actin foci at the phagocytic cup, which inhibited bacterial entry.

Host cell invasion by *Salmonella* is mediated via recruitment of the actin cytoskeleton. *Salmonella* invasion protein A (SipA) enhances actin polymerization near adherent extracellular bacteria leading to internalization of the bacteria. SipA functions through direct binding to F-actin by a central globular domain and by tethering actin monomers in adjacent filaments via two nonglobular arm domains (Lilic et al. 2003). Efficient *Salmonella* internalization involves an initial dephosphorylation of ADF and cofilin followed by phosphorylation, suggesting ADF and cofilin activities are increased briefly (Dai et al. 2004). Expression of a kinase dead form of LIMK1 or catalytically inactive SSH phosphatase, but not constitutively active LIMK1 or wild-type SSH, resulted in decreased invasion. To inhibit regulation of AC by phosphorylation, constitutively active XAC S3A mutants were expressed resulting in decreased *Salmonella* entry. These data suggest that AC activity plays a key role in the actin polymerization/depolymerization process induced by *Salmonella* and that phosphocycling of AC might be involved.

The ability of SipA to disrupt host cell actin dynamics may be mediated via competitive binding with AC and gelsolin (Le Clainche and Drubin 2004). *In vitro* studies have demonstrated the ability of SipA to protect F-actin from depolymerization by preventing AC binding and by displacing bound AC (Dai et al. 2004). SipA also prevents gelsolin severing of F-actin and reanneals severed filaments (McGhie, Hayward and Koronakis 2004). Taken together the ability of pathogenic bacteria to hijack the host cell actin cytoskeleton is regulated through disruption of normal actin dynamics by SipA.

Because of the increase in AIDS patients and the use of immunosuppressive drugs, infection by *Cryptococcus neoformans*, an encapsulated fungal pathogen, has become more prevalent (Mitchell and Perfect 1995). *C. neoformans* enters the host through the respiratory system and spreads to the brain where it can induce potentially fatal meningoencephalitis. To cause meningitis, *C. neoformans* must pass through the tight junctions created by the capillary endothelium (blood brain barrier). The mechanisms used by *C. neoformans* to generate enhanced endothelial cell permeability have been examined (Chen et al. 2003). *C. neoformans*, through the activation of cofilin, induces changes in the actin cytoskeleton. *C. neoformans* adheres to human brain microvascular endothelial cells (HBMEC) and is capable of passing through an HBMEC monolayer, *in vitro*. Treatment of HBMEC with *C. neoformans* induced a drastic change in cell morphology leading to a ruffled apical membrane with significant microvilli-resembling filopodial protrusions and/or apoptotic bodies as observed by transmission electron microscopy. After 16 h of treatment with *C. neoformans*, levels of phosphocofilin decrease, implying an activation of cofilin and subsequent increase in actin dynamics. Coincidentally, stress fibers in HBMEC are destabilized. Lastly, proteolysis of occludin was detected in infected HBMEC, which destabilizes tight junctions and increased permeability (Wachtel et al. 1999).

Bacterial Motility

After uptake, *L. monocytogenes* uses the infected cell's actin cytoskeleton to transport itself from host cell to host cell (Ireton and Cossart 1997). The formation of actin tails, initiated by the bacterial protein ActA, have several similarities to actin cytoskeletal changes at the leading edge of a migrating cell (Cameron et al. 2001). Therefore, bacterial driven actin dynamics has been used as a model system to study actin reorganization and protrusive events at the leading edge of eukaryotic cells. The actin tails are composed of a nonparallel arrangement of short and highly branched actin filaments with the barbed ends oriented toward the bacterium (Tilney, Connelly and Portnoy 1990). For the induction of actin tail formation, *L. monocytogenes* directly or indirectly recruits a variety of actin cytoskeletal host cell proteins, such as Arp2/3 complex, vasodilator-stimulated protein (VASP), Mena, Evl, profilin, cofilin, capping protein, coronin, Rac, and capZ, as well as phosphoinositides (David et al. 1998; Goldberg 2001). Using purified proteins, Loisel et al. (1999) determined the minimum requirements for the formation of *L. monocytogenes*-induced actin tails to be actin, activated Arp2/3 complex, cofilin, and capping protein. Profilin, α-actinin, and VASP are not necessary for tail formation and movement, but when present they increased motility (Loisel et al. 1999).

AC increases the rate of actin turnover and is required to maintain a pool of monomeric G-actin necessary for directional actin polymerization. XAC is required for actin tail depolymerization in the absence of calcium (Fig. 15) (Rosenblatt et al. 1997). Immunodepletion of XAC resulted in longer *L. monocytogenes*-induced actin tails, but did not affect the rate of tail formation. Reintroduction of XAC or chick ADF to immunodepleted extracts returned actin tail size to control lengths. In contrast, addition of the S3E mutant, which mimics the phosphorylated and inactive form of XAC, did not reduce actin tail length. A similar study in diluted platelet extracts reported the requirement of

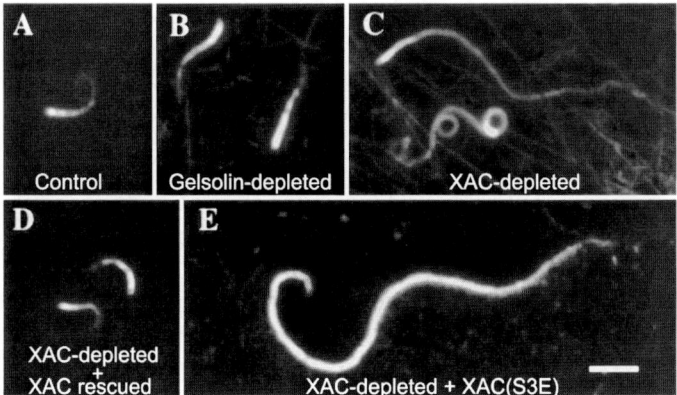

FIG. 15. *Listeria monocytogenes* actin comet tail length is dependent upon AC–actin turnover. Motility assays of *Listeria monocytogenes* in the presence of rhodamine actin and extracts of *Xenopus* eggs treated with (**a**) control IgG, (**b**) IgG to gelsolin, (**c**) IgG to XAC, (**d**) IgG to XAC and then excess wild-type XAC added back, and (**e**) IgG to XAC and then excess XAC S3E added back. The XAC S3E mutant is an inactive recombinant form purified identically to the recombinant XAC (adapted from Rosenblatt et al. 1997)

ADF for an increased rate of *L. monocytogenes* actin driven propulsion that also resulted in shorter actin tails (Carlier et al. 1997). These studies showed that AC is sufficient for actin-depolymerizing activity in *L. monocytogenes*-driven actin tail formation and increased the rate of an actin-based motile process. However, in some cell types, *Listeria* motility may be calcium dependent.

The length of actin filament tails in *L. monocytogenes*-infected Madin–Darby canine kidney cells demonstrates tight correlation with intracellular calcium concentrations (Larson et al. 2005). Reduction of intracellular free Ca^{2+} to 40 nM results in a nearly threefold increase in actin filament tail length. This increase is attributed to a reduction in the filament disassembly rate without a significant change in assembly rate. The calcium sensitive F-actin severing protein gelsolin concentrates in *Listeria* tails when calcium levels are within the normal cellular range but dissociates from the tails when calcium levels drop. *Listeria* tail lengths are also calcium sensitive in *Xenopus* extracts and are calcium insensitive in gelsolin-null mouse embryo fibroblasts. Therefore, *Listeria* tail disassembly can be driven in a calcium-independent manner by AC and in a calcium-dependent manner by gelsolin.

As *L. monocytogenes* is propelled through the cytoplasm, the comet tail can propel the bacterium into extended membranous protrusions that force internalization by adjacent cells. After gaining entry to a new cell via an endocytic event, the cell membranes surrounding the bacterium are lysed and a new phase of division and replication is initiated (Ireton and Cossart 1997). This method of movement is advantageous for *L. monocytogenes* as the bacterium can transport itself from cell to cell without contacting antibody or complement, which could explain why antibodies to *L. monocytogenes* are not protective (MacDonald and Carter 1980).

Rickettsia motility is also driven by actin dynamics but the filaments in the tail are straight and parallel. *Rickettsia* actin tails contain little if any Arp2/3 but do contain Ena/VASP, which depresses Arp2/3 branching (Sechi and Wehland 2004). Although a role for cofilin has not yet been demonstrated in *Rickettsia* motility, it seems likely that it will be required to maintain turnover, as it does for *Listeria* motility. The *Rickettsia* tail may be an actin assembly model for filopodia, whereas the *Listeria* tail is a model for the branched filament network seen in lamellipodia.

Pathogen disruption of the epithelial apical junctional complex (AJC) is important for cellular invasion (Ivanov et al. 2004). This process is characterized by reorganization of apical F-actin into contractile rings and activation of cofilin. In addition Arp2/3, cortactin and activated cofilin were colocalized with these structures. Treatment with blebbistatin, a myosin II inhibitor, disrupted actin reorganization and AJC disassembly. These findings suggest a mechanism of cofilin-dependent depolymerization, Arp2/3-assisted repolymerization, and myosin-driven contraction (Ivanov et al. 2004).

Two members of the ADF/cofilin family (ADF1 and ADF2) are found in the genome of the malaria parasite, *Plasmodium falciparum*. ADF1, which performs a vital role during pathogenic red blood cell stages, lacks residues of the ADF/cofilin consensus sequence implicated in binding to F-actin. Recombinant *P. falciparum* *Pf*ADF1 interacts with monomeric actin and stimulates nucleotide exchange, an opposite effect of other AC proteins. Furthermore, it does not bind to actin polymers. Thus *Pf*ADF1 differs from previously known AC proteins in its biochemical activity and seems to promote turnover by interaction with actin monomers, more akin to profilin (Shuler, Mueller and Matuschewski 2005). For a more detailed discussion of the role of ABPs in pathogen infection, see chapter by Bearer.

Conclusions and Perspectives

The role of AC proteins in actin dynamics has been well studied in a myriad of biological systems. Because actin plays such a central role in most cellular processes, AC proteins do so as well. Regulating actin dynamics is a complex process involving more than just AC proteins. Thus, many of the processes for which we describe a putative role for AC proteins in disease could be disrupted as effectively by abnormalities in other ABPs discussed elsewhere in this volume.

Studies on the function of AC protein and their upstream regulators have led to the discovery of a number of roles for actin in cells that were hitherto unknown. Among these are the release of cytochrome *c* from mitochondria and the role for actin in extending sorting tubules from the Golgi. Roles for actin in the nucleus are just beginning to be understood and more studies are needed to delineate the roles of AC proteins in targeting and function of nuclear actin. Because of these diverse functional roles for AC proteins, therapeutic intervention through their regulation/modification is probably not a realistic goal in many diseases in which AC proteins have been implicated.

One of the unique areas for AC protein involvement in disease is the formation of AC–actin rods induced in neurons by common mediators of neurodegeneration. Rod-related diseases are one area where possible therapeutic intervention might succeed because the protein–protein interface between the AC-saturated filaments that form the rods is different from the interface between AC proteins and the actin filament that is required for the AC-dynamizing functions. Developing reagents to target this interface may be a worthwhile goal for future research.

Acknowledgments. The authors would like to thank Aaron Sholders for assistance with graphics and members of the Bamburg laboratory and our wonderful collaborators for helpful discussions and figures. Support from NIH grants NS40371, NS43115, HL58064, DK69408, NSF grant DGE-0234615, and a grant-in-aid from Sigma Xi are gratefully acknowledged.

References

Abe, H., Nagaoka, R. and Obinata, T. 1993. Cytoplasmic localization and nuclear transport of cofilin in cultured myotubes. Exp. Cell Res. 206, 1–10.

Abe, H., Obinata, T., Minamide, L. S. and Bamburg, J. R. 1996. *Xenopus laevis* actin-depolymerizing factor/cofilin: A phosphorylation-regulated protein essential for development. J. Cell Biol. 871, 885.

Abe, H., Ohshima, S. and Obinata, T. 1989. A cofilin-like protein is involved in the regulation of actin assembly in developing skeletal muscle. J. Biochem. 106, 696–702.

Agnew, B. J., Minamide, L. S. and Bamburg, J. R. 1995. Reactivation of phosphorylated actin depolymerizing factor and identification of the regulatory site. J. Biol. Chem. 270, 17582–17587.

Aizawa, H., Kishi, Y., Kazuko, I., Sameshima, M. and Yahara, I. 2001. Cofilin-2, a novel type of cofilin, is expressed specifically at aggregation stage of *Dictyostelium discoideum* development. Genes Cells 6, 913–921.

Aizawa, H., Sutoh, K., Tsubuki, S., Kawashima, S., Ishii, A. and Yahara, I. 1995. Identification, characterization, and intracellular distribution of cofilin in *Dictyostelium discoideum*. J. Biol. Chem. 270, 10923–10932.

Albinsson, S., Nordstrom, I. and Hellstrand, P. 2004. Stretch of the vascular wall induces smooth muscle differentiation by promoting actin polymerization. J. Biol. Chem. 279, 34849–34855.

Allbritton, N. L., Meyer, T. and Stryer, L. 1992. Range of messenger action of calcium ion and inositol 1,4,5-trisphosphate. Science 258, 1812–1815.

Amano, T., Kaji, N., Ohashi, K. and Mizuno, K. 2002. Mitosis-specific activation of LIM motif-containing protein kinase and roles of cofilin phosphorylation and dephosphorylation in mitosis. J. Biol. Chem. 277, 22093–22102.

Amano, T., Tanabe, K., Eto, T., Narumiya, S. and Mizuno, K. 2001. LIM-kinase 2 induces formation of stress fibres, focal adhesions and membrane blebs, dependent on its activation by Rho-associated kinase-catalysed phosphorylation at threonine-505. Biochem. J. 354, 149–159.

Ambach, A., Saunus, J., Konstandin, M., Wesselborg, S., Meuer, S. C. and Samstag, Y. 2000. The serine phosphatases PP1 and PP2A associate with and activate the actin-binding protein cofilin in human T lymphocytes. Eur. J. Immunol. 30, 3422–3431.

Aranda-Abreu, G. E., Hernandez, M. E., Soto, A. and Manzo, J. 2005. Possible Cis-acting signal that could be involved in the localization of different mRNAs in neuronal axons. Theor. Biol. Med. Model. 2, 33.

Arber, S., Barbayannis, F. A., Hanser, H., Schneider, C., Stanyon, C. A., Bernard, O. and Caroni, P. 1998. Regulation of actin dynamics through phosphorylation of cofilin by LIM-kinase. Nature 393, 805–809.

Ashworth, S. L., Sandoval, R. M., Hosford, M., Bamburg, J. R. and Molitoris, B. A. 2001. Ischemic injury induces ADF relocalization to the apical domain of rat proximal tubule cells. Am. J. Physiol. Renal Physiol. 280, 886–894.

Ashworth, S. L., Southgate, E. L., Sandoval, R. M., Meberg, P. J., Bamburg, J. R. and Molitoris, B. A. 2003a. ADF/cofilin mediates actin cytoskeletal alterations in LLC-PK cells during ATP depletion. Am. J. Physiol. Renal Physiol. 284, 852–862.

Ashworth, S. L., Wean, S. E., Campos, S. B., Temm-Grove, C. J., Southgate, E. L., Vrhovski, B., Gunning, P., Weinberger, R. P. and Molitoris, B. A. 2003b. Renal ischemia induces tropomyosin dissociation-destabilizing microvilli microfilaments. Am. J. Physiol. Renal Physiol. 286, 988–996.

Atkinson, S. J., Hosford, M. A. and Molitoris, B. A. 2004. Mechanisms of actin polymerization in cellular ATP depletion. J. Biol. Chem. 279, 5194–5199.

Axelrod, D., Ravdin, P., Koppel, D. E., Schlessinger, J., Webb, W. W., Elson, E. L. and Podleski, T. R. 1976. Lateral motion of fluorescently labeled acetylcholine receptors in membranes of developing muscle fibers. Proc. Natl Acad. Sci. USA 73, 4594–4598.

Bagrodia, S. and Cerione, R. A. 1999. Pak to the future. Trends Cell Biol. 9, 350–355.

Bakin, A. V., Safina, A., Rimehart, C., Daroqui, C., Darbary, H. and Helfman, D. M. 2004. A critical role of tropomyosins in TGF-beta regulation of the actin cytoskeleton and cell motility in epithelial cells. Mol. Biol. Cell 15, 4682–4694.

Balcer, H. I., Goodman, A. L., Rodal, A. A., Smith, E., Kugler, J., Heuser, J. E. and Goode, B. L. 2003. Coordinated regulation of actin filament turnover by a high-molecular-weight Srv2/CAP complex, cofilin, profilin and Aip1. Curr. Biol. 13, 2159–2169.

Baldwin, C. T., Hoth, C. F., Amos, J. A., da-Silva, E. O. and Milunsky, A. 1992. An exonic mutation in the HuP2 paired domain gene causes Waardenburg's syndrome. Nature 355, 637–638.

Bamburg, J. R. 1999. Proteins of the ADF/cofilin family: Essential regulators of actin dynamics. Annu. Rev. Cell Dev. Biol. 15, 185–230.

Bamburg, J. R. and Bray, D. 1987. Distribution and cellular localization of actin depolymerizing factor. J. Cell Biol. 105, 2817–2825.

Bamburg, J. R., Harris, H. E. and Weeds, A. G. 1980. Partial purification and characterization of an actin depolymerizing factor from brain. FEBS Lett. 121, 178–182.

Bamburg, J. R. and Wiggan, O. P. 2002. ADF/cofilin and actin dynamics in disease. Trends Cell Biol. 12, 598–605.

Barria, A., Derkach, V. and Soderling, T. 1997. Identification of the Ca2+/calmodulin-dependent protein kinase II regulatory phosphorylation site in the alpha-amino-3-hydroxy-5-methyl-4-isoxazole-propionate-type glutamate receptor. J. Biol. Chem. 272, 32727–32730.

Barria, A., Muller, D., Derkach, V., Griffith, L. C. and Soderling, T. R. 1997. Regulatory phosphorylation of AMPA-type glutamate receptors by CaM-KII during long-term potentiation. Science 276, 2042–2045.

Bassell, G. J. and Kelic, S. 2004. Binding proteins for mRNA localization and local translation, and their dysfunction in genetic neurological disease. Curr. Opin. Neurobiol. 14, 574–581.

Bellugi, U., Lichtenberger, L., Mills, D., Galaburda, A. and Korenberg, J. R. 1999. Bridging cognition, the brain and molecular genetics: Evidence from Williams syndrome. Trends Neurosci. 22, 197–207.

Belmont, L. D. and Drubin, D. G. 1998. The yeast V159N actin mutant reveals roles for actin dynamics in vivo. J. Cell Biol. 142, 1289–1299.

Benavides-Piccione, R., Ballesteros-Yanez, I., de Lagran, M. M., Elston, G., Estivill, X., Fillat, C., Defelipe, J. and Dierssen, M. 2004. On dendrites in Down syndrome and DS murine models: A spiny way to learn. Prog. Neurobiol. 74, 111–126.

Bentley, D. and Toroian-Raymond, A. 1986. Disoriented pathfinding by pioneer neurone growth cones deprived of filopodia by cytochalasin treatment. Nature 323, 712–715.

Berardelli, A., Rothwell, J. C., Hallet, M., Thompson, P. D., Manfriedi, M. and Marsden, C. D. 1998. The pathophysiology of primary dystonia. Brain 121, 1195–1212.

Bernard, O., Ganiatsas, S., Kannourakis, G. and Dringer, R. 1994. Kiz-1, a protein with LIM zinc finger and kinase domains, is expressed mainly in neurons. Cell Growth Differ. 5, 1159–1171.

Bernstein, B. W. and Bamburg, J. R. 1982. Tropomyosin binding to F-actin protects the F-actin from disassembly by brain actin-depolymerizing factor (ADF). Cell Motil. 2, 1–8.

Bernstein, B. W. and Bamburg, J. R. 2003. Actin-ATP hydrolysis is a major energy drain for neurons. J. Neurosci. 23, 1–6.

Bernstein, B. W., Chen, H., Boyle, J. A. and Bamburg, J. R. 2006. Formation of actin–ADF/cofilin rods transiently retards decline of mitochondrial potential and ATP in stressed neurons. Am. J. Physiol. Cell Physiol. 291, C828–C839.

Bernstein, B. W., Painter, W. B., Chen, H., Minamide, L. S., Abe, H. and Bamburg, J. R. 2000. Intracellular pH modulation of ADF/cofilin proteins. Cell Motil. Cytoskeleton 47, 319–336.

Bershadsky, A. D., Gelfand, V. I., Svitkina, T. M. and Tint, I. S. 1980. Destruction of microfilament bundles in mouse embryo fibroblasts treated with inhibitors of energy metabolism. Exp. Cell Res. 127, 421–429.

Bershadsky, A. D., Gluck, U., Denisenko, O. N., Sklyarova, T. V., Spector, I. and Ben-Ze'ev, A. 1995. The state of actin assembly regulates actin and vinculin expression by a feedback loop. J. Cell Sci. 108, 1183–1193.

Bertling, E., Hotulainen, P., Matilla, P. K., Tanja, M., Salminen, M. and Lappalainen, P. 2004. Cyclase-associated protein 1 (CAP1) promotes cofilin-induced actin dynamics in mammalian nonmuscle cells. Mol. Biol. Cell 15, 2324–2334.

Beuren, A. J., Apitz, J. and Harmjanz, D. 1962. Supravalvular aortic stenosis in association with mental retardation and a certain facial appearance. Circulation 26, 1235–1240.

Bharadwaj, S., Thanawala, R., Bon, G., Falcioni, R. and Prasad, G. L. 2005. Resensitization of breast cancer cells to anoikis by Tropomyosin-1: Role of Rho kinase-dependent cytoskeleton and adhesion. Oncogene 24, 8291–8303.

Bhatia, K. P. and Marsden, C. D. 1994. The behavioural and motor consequences of focal lesions of the basal ganglia in man. Brain 117, 859–876.

Bienvenu, T., des Portes, V., McDonell, N., Carrie, A., Zemni, R., Couvert, P., Ropers, H. H., Moraine, C., van Bokhoven, H., Fryns, J. P., Allen, K., Walsh, C. A., Boue, J., Kahn, A., Chelly, J. and Beldjord, C. 2000. Missense mutation in PAK3, R67C, causes X-linked nonspecific mental retardation. Am. J. Med. Genet. 93, 294–298.

Bierne, H., Gouin, E., Roux, P., Caroni, P., Yin, H. L. and Cossart, P. 2001. A role for cofilin and LIM kinase in Listeria-induced phagocytosis. J. Cell Biol. 155, 101–112.

Birkenfield, J., Betz, H. and Roth, D. 2003. Identification of cofilin and LIM-domain-containing protein kinase 1 as novel interaction partners of 14-3-3 zeta. Biochem. J. 369, 45–54.

Blanchoin, L. and Pollard, T. D. 1998. Interaction of actin monomers with *Acanthamoeba actophorin* (ADF/cofilin) and profilin. J. Biol. Chem. 273, 25106–25111.

Blanchoin, L. and Pollard, T. D. 1999. Mechanism of interaction of *Acanthamoeba actophorin* (ADF/Cofilin) with actin filaments. J. Biol. Chem. 274, 15538–15546.

Blanchoin, L., Pollard. T. D. and Hitchcock-DeGregori, S. E., 2001. Inhibition of the Arp2/3 complex-nucleated actin polymerization and branch formation by tropomyosin. Curr. Biol. 11, 1300–1304.

Blanchoin, L., Robinson, R. C., Choe, S. and Pollard, T. D. 2000. Phosphorylation of *Acanthamoeba actophorin* (ADF/cofilin) blocks interaction with actin without a change in atomic structure. J. Mol. Biol. 295, 203–11.

Blikstad, I., Markey, F., Carlsson, L., Perssin, T. and Lindberg, U. 1978. Selective assay of monomeric and filamentous actin in cell extracts, using inhibition of deoxyribonuclease I. Cell 15, 935–943.

Bobkov, A. A., Muhlrad, A., Shvetsov, A., Benchaar, S., Scoville, D., Almo, S. C. and Reisler, E. 2004. Cofilin (ADF) affects lateral contacts in F-actin. J. Mol. Biol. 337, 93–104.

Boengler, K., Pipp, F., Broich, K., Fernandez, B., Schaper, W. and Deindl, E. 2003. Identification of differentially expressed genes like cofilin2 in growing collateral arteries. Biochem. Biophys. Res. Commun. 300, 751–756.

Bryce, N. S., Clark, E. S., Leysath, J. L., Currie, J. D., Webb, D. J. and Weaver, A. M. 2005. Cortactin promotes cell motility by enhancing lamellipodial persistence. Curr. Biol. 15, 1276–1285.

Bryce, N. S., Schevzov, G., Ferguson, V., Percival, J. M., Lin, J. J.-C., Matsumura, F., Bamburg, J. R., Jeffrey, P. L., Hardeman, E. C., Gunning, P. and Weinberger, R. P. 2003. Specification of actin filament function and molecular composition by tropomyosin isoforms. Mol. Biol. Cell 14, 1002–1016.

Burtnick, L. D., Urosev, D., Irobi, E., Narayan, K. and Robinson, R. C. 2004. Structure of the N-terminal half of gelsolin bound to actin: Roles in severing, apoptosis and FAF. EMBO J. 23, 2713–2722.

Buss, F., Temm-Grove, C., Henning, S. and Jockusch, B. M. 1992. Distribution of profilin in fibroblasts correlates with the presence of highly dynamic actin filaments. Cell Motil. Cytoskeleton 22, 51–61.

Calverley, R. K. and Jones, D. G. 1990. Contributions of dendritic spines and perforated synapses to synaptic plasticity. Brain Res. Brain Res. Rev. 15, 215–249.

Cameron, L. A., Svitkina, T. M., Vignjevic, D., Theriot, J. A. and Borisy, G. G. 2001. Dendritic organization of actin comet tails. Curr. Biol. 11, 130–135.

Campbell, D. S. and Holt, C. E. 2001. Chemotropic responses of retinal growth cones mediated by rapid local protein synthesis and degradation. Neuron 32, 1013–1026.

Cantiello, H. F. 1997. Role of actin filament organization in cell volume and ion channel regulation. J. Exp. Zool. 279, 425–435.

Carlier, M. F. 1989. Role of nucleotide hydrolysis in the dynamics of actin filaments and microtubules. Int. Rev. Cytol. 115, 139–170.

Carlier, M. F. 1991. Actin: Protein structure and filament dynamics. J. Biol. Chem. 266, 1–4.

Carlier, M. F., Laurent, V. Santolini, J., Melki, R., Didry, D., Xia, G.-X., Hong, Y., Chua, N.-H. and Pantaloni, D. 1997. Actin depolymerizing factor (ADF/cofilin) enhances the rate of filament turnover: Implication in actin-based motility. J. Cell Biol. 136, 1307–1322.

Carlier, M. F., Wiesner, S., Le Clainche, C. and Pantaloni, D. 2003. Actin-based motility as a self-organized system: Mechanism and reconstitution in vitro. Crit. Rev. Biol. 326, 161–170.

Carmeliet, P. and Jain, R. K. 2000. Angiogenesis in cancer and other diseases. Nature 407, 249–257.

Cartier, L., Galvez, S. and Gajdusek, D. C. 1985. Familial clustering of the ataxic form of Creutzfeldt–Jakob disease with Hirano bodies. J. Neurol. Neurosurg. Psychiatry 48, 234–238.

Castets, M., Schaeffer, C., Bechara, E., Schenck, A., Khandjian, E. W., Luche, S., Moine, H., Rabilloud, T., Mandel, J. L. and Bardoni, B. 2005. FMRP interferes with the Rac1 pathway and controls actin cytoskeleton dynamics in murine fibroblasts. Hum. Mol. Genet. 14, 835–844.

Chan, A. Y., Bailly, M., Zebda, N., Segall, J. E. and Condeelis, J. S. 2000. Role of cofilin in epidermal growth factor-stimulated actin polymerization and lamellipod protrusion. J. Cell Biol. 148, 531–542.

Chang, S. H., Poser, S. and Xia, Z. 2004. A novel role for serum response factor in neuronal survival. J. Neurosci. 24, 2277–2285.

Chartier-Harlin, M. C., Crawford, F., Hamandi, K., Mullan, M., Goate, A., Hardy, J., Backhovens, H., Martin, J. J. and Broeckhoven, C. V. 1991a. Screening for the beta-amyloid precursor protein mutation (APP717: Val–Ile) in extended pedigrees with early onset Alzheimer's disease. Neurosci. Lett. 129, 134–135.

Chartier-Harlin, M. C., Crawford, F., Houlden, H., Warren, A., Hughes, D., Fidani, L., Goate, A., Rossor, M., Roques, P., Hardy, J. and Mullan, M. 1991b. Early-onset Alzheimer's disease caused by mutations at codon 717 of the beta-amyloid precursor protein gene. Nature 353, 844–846.

Chen, H., Bernstein, B. W. and Bamburg, J. R. 2000. Regulating actin-filament dynamics in vivo. Trends Biochem. Sci. 25, 19–23.

Chen, H., Bernstein, B. W., Sneider, J. M., Boyle, J. A., Minamide, L. S. and Bamburg, J. R. 2004. In vitro activity differences between proteins of the ADF/cofilin family define two distinct subgroups. Biochemistry 43, 7127–7142.

Chen, Y. H., Chen, S. H.-M., Jong, A., Zhou, Z. Y., Li, W., Suzuki, K. and Huang, S.-H. 2002. Enhanced *Escherichia coli* invasion of human brain microvascular endothelial cells is associated with alternations in cytoskeleton induced by nicotine. Cell. Microbiol. 4, 503–514.

Chen, J., Cohn, J. A. and Mandel, L. J. 1995. Dephosphorylation of ezrin as an early event in renal microvillar breakdown and anoxic injury. Proc. Natl Acad. Sci. USA 92, 7495–7499.

Chen, J., Doctor, R. B. and Mandel, L. J., 1994. Cytoskeletal dissociation of ezrin during renal anoxia: Role in microvillar injury. Am. J. Physiol. 267, 784–795.

Chen, J., Godt, D., Gunsalus, K., Kiss, I., Goldberg, M. and Laski, F. A. 2001. Cofilin/ADF is required for cell motility during Drosophila ovary development and oogenesis. Nat. Cell Biol. 3, 204–209.

Chen, W., Ji, J. and Ru, B. 2005. Proteomic analysis of corticobasal degeneration: A case study of corticobasal degeneration at the proteome level. J. Neuropsychiatry Clin. Neurosci. 17, 364–371.

Chen, X. and Macara, I. G. 2005. Par-3 controls tight junction assembly through the Rac exchange factor Tiam1. Nat. Cell Biol. 7, 262–269.

Chen, X. and Macara, I. G. 2006. Par-3 mediates the inhibition of LIM kinase 2 to regulate cofilin phosphorylation and tight junction assembly. J. Cell Biol. 172, 671–678.

Chen, S. H., Stins, M. F., Huang, S.-H., Chen, Y. H., Kwon-Chung, K. J., Chang, Y., Kim, K. S. Suzuki, K. and Jong, A. Y. 2003. *Cryptococcus neoformans* induces alterations in the cytoskeleton of human brain microvascular endothelial cells. J. Med. Microbiol. 52, 961–970.

Chernoff, N. and Rogers, J. M. 2004. Supernumerary ribs in developmental toxicity bioassays and in human populations: Incidence and biological significance. J. Toxicol. Environ. Health B. Crit. Rev. 7, 437–449.

Chhabra, D. and dos Remedios, C. G. 2005. Cofilin, actin and their complex observed in vivo using fluorescence resonance energy transfer. Biophys. J. 89, 1902–1908.

Chi, N. and Epstein, J. A. 2002. Getting your Pax straight: Pax proteins in development and disease. Trends Genet. 18(1), 41–47.

Chua, B. T., Volbracht, C., Tan, K. O., Li, R., Yu, V. C. and Li, P. 2003. Mitochondrial translocation of cofilin is an early step in apoptosis induction. Nat. Cell Biol. 5, 1083–1089.

Clark, E. A., Golub, T. R., Lander, E. S. and Hyns, R. O. 2000. Genomic analysis of metastasis reveals an essential role for RhoC. Nature 406, 532–535.

Clarkson, E., Costa, C. F. and Machesky, L. M. 2004. Congenital myopathies: Diseases of the actin cytoskeleton. J. Pathol. 204, 407–417.

Cooper, J. A. and Schafer, D. A. 2000. Control of actin assembly and disassembly at filament ends. Curr. Opin. Cell Biol. 12, 97–103.

Corbett, M. A., Akkari, P. A., Domazetovska, A., Cooper, S. T., North, K. N., Laing, N. G., Gunning, P. W. and Hardeman, E. C. 2005. An alphaTropomyosin mutation alters dimer preference in nemaline myopathy. Ann. Neurol. 57, 42–49.

Corneal Disease Panel. 1983. Vision Research – A National Plan: 1983–1987, vol. 2. US Department of Health and Human Services. pp. 1–59.

Cossart, P. 2000. Actin-based motility of pathogens: The Arp2/3 complex is a central player. Cell. Microbiol. 2, 195–205.

Cossart, P. and Sansonetti, P. J. 2004. Bacterial invasion: The paradigms of enteroinvasive pathogens. Science 304, 242–248.

Cramer, L. P. 1999. Role of actin-filament disassembly in lamellipodium protrusion in motile cells revealed using the drug jasplakinolide. Curr. Biol. 9, 1095–1105.

Cristofanilli, M., Charnsangavej, C. and Hortobagyi, G. N. 2002. Angiogenesis modulation in cancer research: Novel clinical approaches. Nat. Rev. Drug Discov. 1, 415–426.

Culebras, A., Feldman, R. G. and Merk, F. B. 1973. Cytoplasmic inclusion bodies within neurons of the thalamus in myotonic dystrophy. A light and electron microscope study. J. Neurol. Sci. 19, 319–329.

Curran, M. E., Atkinson, D. L., Ewart, A. K., Morris, C. A., Leppert, M. F. and Keating, M. T. 1993. The elastin gene is disrupted by a translocation associated with supravalvular aortic stenosis. Cell 73, 159–168.

Cutler, T. J. 2004. Corneal epithelial disease. Vet. Clin. North Am. Equine Pract. 20, 319–343.

Dahl, E., Koseki, H. and Balling. R. 1997. Pax genes and organogenesis. Bioessays 19(9), 755–765.

Dai, S., Sarmiere, P. D., Wiggan, O., Bamburg, J. R. and Zhou, D. 2004. Efficient Salmonella entry requires activity cycles of host ADF and cofilin. Cell. Microbiol. 6(5), 459–471.

Dailey, M. E. and Smith, S. J. 1996. The dynamics of dendritic structure in developing hippocampal slices. J. Neurosci. 16, 2983–2994.

Dalby-Payne, J. R., O'Loughlin, E. V. and Gunning, P. 2003. Polarization of specific tropomyosin isoforms in gastrointestinal epithelial cells and their impact on CFTR at the apical surface. Mol. Biol. Cell 14, 4365–4375.

Dan, C., Kelly, A., Bernard, O. and Minden, A. 2001. Cytoskeletal changes regulated by the PAK4 serine/threonine kinase are mediated by LIM kinase 1 and cofilin. J. Biol. Chem. 276, 32115–32121.

Dang, D., Bamburg, J. R. and Ramos, D. M. 2006. alphavbeta3 integrin and cofilin modulate K1735 melanoma cell invasion. Exp. Cell Res. 312, 468–477.

Daniel, J. L., Molish, I. R., Robkin, L. and Holmsen, H. 1986. Nucleotide exchange between cytosolic ATP and F-actin-bound ADP may be a major energy-utilizing process in unstimulated platelets. Eur. J. Biochem. 156, 677–684.

Daniels, R. H., Hall, P. S. and Bokoch, G. M. 1998. Membrane targeting of p21-activated kinase 1 (PAK1) induces neurite outgrowth from PC12 cells. EMBO J. 17, 754–764.

Darnell, J. C., Jensen, K. B., Jin, P., Brown, V., Warren, S. T. and Darnell, R. B. 2001. Fragile X mental retardation protein targets G quartet mRNAs important for neuronal function. Cell 107, 489–499.

David, V., Gouin, E., Troys, M. V., Grogan, A., Segal, A. W., Ampe, C. and Cossart, P. 1998. Identification of cofilin, coronin, Rac and capZ in actin tails using a Listeria affinity approach. J. Cell Sci. 111, 2877–2884.

Davila, M., Frost, A. R., Grizzle, W. E. and Chakrabarti, R. 2003. LIM kinase 1 is essential for the invasive growth of prostate epithelial cells: Implications in prostate cancer. J. Biol. Chem. 278, 36868–36875.

Dawe, H. R., Minamide, L. S., Bamburg, J. R. and Cramer, L. P. 2003. ADF/cofilin controls cell polarity during fibroblast migration. Curr. Biol. 13, 252–257.

Derkach, V., Barria, A. and Soderling, T. R. 1999. Ca2+/calmodulin-kinase II enhances channel conductance of alpha-amino-3-hydroxy-5-methyl-4-isoxazole-propionate type glutamate receptors. Proc. Natl Acad. Sci. USA 96, 3269–3274.

DesMarais, V., Ghosh, M., Eddy, R. and Condeelis, J. 2005. Cofilin takes the lead. J. Cell Sci. 118, 19–26.

Devineni, N., Minamide, L. S., Niu, M., Safer, D., Verma, R., Bamburg, J. R. and Nachmias, V. T. 1999. A quantitative analysis of G-actin binding proteins and the G-actin pool in developing chick brain. Brain Res. 823, 129–140.

Didry, D., Carlier, M.-F. and Pantaloni, D. 1998. Synergy between actin depolymerizing factor/cofilin and profilin in increasing actin filament turnover. J. Biol. Chem. 273, 25602–25611.

Ding, S. J., Li, Y., Shao, X.-X., Zhou, H., Zeng, R., Tang, Z.-Y. and Xia, X.-C. 2004. Proteome analysis of hepatocellular carcinoma cell strains, MHCC97-H and MHCC97-L, with different metastasis potentials. Proteomics 4, 982–994.

Doe, C. Q. 2001. Cell polarity: The PARty expands. Nat. Cell Biol. 3, 7–9.

Dominguez, R. 2004. Actin-binding proteins – A unifying hypothesis. Trends Biochem. Sci. 29, 572–578.

Drubin, D. G., Mulholland, J., Zhu, Z. M. and Botstein, D. 1990. Homology of a yeast actin-binding protein to signal transduction proteins and myosin-I. Nature 343, 288–290.

Dua, H. S. and Forrester, J. V. 1990. The corneoscleral limbus in human corneal epithelial wound healing. Am. J. Ophthalmol. 110, 646–656.

Dufour, C., Weinberger, R. P. and Gunning, P. 1998. Tropomyosin isoform diversity and neuronal morphogenesis. Immunol. Cell Biol. 76, 424–429.

Dumont, P., Burton, M., Chen, Q. M., Gonos, E. S., Frippiat, C., Mazarati, J. B., Eliaers, F., Remacle, J. and Toussaint, O. 2000. Induction of replicative senescence biomarkers by sublethal oxidative stresses in normal human fibroblast. Free Radic. Biol. Med. 28, 361–373.

Eaton, B. A. and Davis, G. W. 2005. LIM Kinase1 controls synaptic stability downstream of the type II BMP receptor. Neuron 47, 695–708.

Edwards, D. C. and Gill, G. N. 1999. Structural features of LIM kinase that control effects on the actin cytoskeleton. J. Biol. Chem. 274, 11352–11361.

Edwards, D. C., Sanders, L. C., Bokoch, G. M. and Gill, G. N. 1999. Activation of LIM-kinase by Pak1 couples Rac/Cdc42 GTPase signalling to actin cytoskeletal dynamics. Nat. Cell Biol. 1, 253–259.

Egan, M. F., Kojima, M., Callicott, J. H., Goldberg, T. E., Kolachana, B. S., Bertolino, A., Zaitsez, E., Gold, D., Goldman, D., Dean, M., Lu, B. and Weinberger, D. R. 2003. The BDNF val66met polymorphism affects activity-dependent secretion of BDNF and human memory and hippocampal function. Cell 112, 257–269.

Ehehalt, R., Keller, P., Haass, C., Thiele, C. and Simons, K. 2003. Amyloidogenic processing of the Alzheimer beta-amyloid precursor protein depends on lipid rafts. J. Cell Biol. 160, 113–123.

Eibert, S. M., Lee, K.-H., Pipkorn, R., Sester, U., Wabnitz, G. H., Giese, T., Meuer, S. C. and Samstag, Y. 2004. Cofilin peptide homologs interfere with immunological synapse formation and T cell activation. Proc. Natl Acad. Sci. USA 101, 1957–1962.

Endo, M., Ohashi, K., Sasaki, Y., Goshima, Y., Niwa, R., Uemura, T. and Mizuno, K. 2003. Control of growth cone motility and morphology by LIM kinase and Slingshot via phosphorylation and dephosphorylation of cofilin. J. Neurosci. 23, 2527–2537.

English, D., Garcia, D. J. and Brindley, D. N. 2001. Platelet-released phospholipids link haemostasis and angiogenesis. Cardiovasc. Res. 49, 588–599.

Erpel, T. and Courtneidge, S. A. 1995. Src family protein tyrosine kinases and cellular signal transduction pathways. Curr. Opin. Cell Biol. 7, 176–182.

Eugene, E., Hoffmann, I., Pujol, C., Couraud, P.-O., Bourdoelous, S. and Nass, X. 2002. Microvilli-like structures are associated with the internalization of virulent capsulated *Neisseria meningitidis* into vascular endothelial cells. J. Cell Sci. 115, 1231–1241.

Ewart, A. K., Jin, W., Atkinson, D., Morris, C. A. and Keating, M. T. 1994. Supravalvular aortic stenosis associated with a deletion disrupting the elastin gene. J. Clin. Invest. 93, 1071–1077.

Ewart, A. K., Morris, C. A., Atkinson, D., Jin, W., Sternes, K., Spallone, P., Stock, A. D., Leppert, M. and Keating, M. T. 1993. Hemizygosity at the elastin locus in a developmental disorder, Williams syndrome. Nat. Genet. 5, 11–16.

Fahn, S., Bressman, S. B. and Marsden, C. D. 1998. Classification of dystonia. Adv. Neurol. 78, 1–10.

Falet, H., Chang, G., Brohard-Bohn, B., Rendu, F. and Hartwig, J. H. 2005. Integrin alpha(IIb)beta3 signals lead cofilin to accelerate platelet actin dynamics. Am. J. Physiol. Cell Physiol. 289, 819–825.

Fedorov, A. A., Lappalainen, P., Federov, E. V., Drubin, D. G. and Almo, S. C. 1997. Structure determination of yeast cofilin. Nat. Struct. Biol. 4, 366–369.

Fernandez, R., Fernandez, J. M., Cerva, C., Teijiera, S., Teijeiro, A., Dominguez, C. and Navarro, C. 1999. Adult glycogenosis II with paracrystalline mitochondrial inclusions and Hirano bodies in skeletal muscle. Neuromuscul. Disord. 9, 136–143.

Fiala, J. C., Feinberg, M., Popov, V. and Harris, K. M. 1998. Synaptogenesis via dendritic filopodia in developing hippocampal area CA1. J. Neurosci. 18, 8900–8911.

Field, E. J., Mathews, J. D. and Raine, C. S. 1969. Electron microscopic observations on the cerebellar cortex in kuru. J. Neurol. Sci. 8, 209–224.

Field, E. J. and Narang, H. K. 1972. An electron-microscopic study of scrapie in the rat, further observations on "inclusion bodies" and virus-like particles. J. Neurol. Sci. 17, 347–364.

Fincham, V. J., James, M., Frame, M. C. and Winder, S. J. 2000. Active ERK/MAP kinase is targeted to newly forming cell–matrix adhesions by integrin engagement and v-Src. EMBO J. 19, 2911–2923.

Fishkind, D. J. and Wang, Y. L. 1995. New horizons for cytokinesis. Curr. Opin. Cell Biol. 7, 23–31.

Frangiskakis, J. M., Ewart, A. K., Morris, C. A., Mervis. C. B., Bertrand, J., Robinson, B. F., Klein, B. P., Ensing, G. J., Everett, L. A., Green, E. D., Prochel, C., Gutowski, N. J., Noble, M., Atkinson, D. L., Odelberg, S. J. and Keating, T. M. 1996. LIM-kinase1 hemizygosity implicated in impaired visuospatial constructive cognition. Cell 86, 59–69.

Fraser, H. 1969. Eosinophilic bodies in some neurones in the thalamus of ageing mice. J. Pathol. 98, 201–204.

Frischknecht, F., Cudmore, S., Moreau, V., Reckmann, I., Rottger, S. and Way, M. 1999. Tyrosine phosphorylation is required for actin-based motility of vaccinia but not Listeria or Shigella. Curr. Biol. 9, 89–92.

Fu, H., Subramanian, R. R. and Masters, S. C. 2000. 14-3-3 proteins: Structure, function and regulation. Annu. Rev. Pharmacol. Toxicol. 40, 617–647.

Fu, Y., Ward, J. and Young, H. F. 1975. Unusual, rod-shaped cytoplasmic inclusions (Hirano bodies) in a cerebellar hemangioblastoma. Acta Neuropathol. 31, 129–135.

Fujibuchi, T., Abe, Y., Takeuchi, T., Imai, Y., Kamei, Y., Murase, R., Ueda, N., Shigemoto, K., Yamamoto, H. and Kito, K. 2005. AIP1/WDR1 supports mitotic cell rounding. Biochem. Biophys. Res. Commun. 327, 268–275.

Fukui, Y. 1978. Intranuclear actin bundles induced by dimethyl sulfoxide in interphase nucleus of Dictyostelium. J. Cell Biol. 76, 146–157.

Fukui, Y. and Katsumaru, H. 1979. Nuclear actin bundles in Amoeba, Dictyostelium and human HeLa cells induced by dimethyl sulfoxide. Exp. Cell Res. 120, 451–455.

Galan, J. E. and Zhou, D. 2000. Striking a balance: Modulation of the actin cytoskeleton by Salmonella. Proc. Natl Acad. Sci. USA 97, 8754–8761.

Galkin, V. E., Orlova, A., Lukoyanova, N., Wriggers, W. and Egelman, E. H. 2001. Actin depolymerizing factor stabilizes an existing state of F-actin and can change the tilt of F-actin subunits. J. Cell Biol. 153, 75–86.

Galkin, V. E., Orlova, A., VanLoock, M. S., Shvetsov, A., Reisler, E., Egelman, E. H. 2003. ADF/cofilin use an intrinsic mode of F-actin instability to disrupt actin filaments. J. Cell Biol. 163, 1057–1066.

Galkin, V. E., VanLoock, M. S., Orlova, A. and Egelman, E. H. 2002. A new internal mode in F-actin helps explain the remarkable evolutionary conservation of actin's sequence and structure. Curr. Biol. 12, 570–575.

Galloway, P. G., Perry, G. and Gambetti, P. 1987. Hirano body filaments contain actin and actin-associated proteins. J. Neuropathol. Exp. Neurol. 46, 185–199.

Galvez, R. and Greenough, W. T. 2005. Sequence of abnormal dendritic spine development in primary somatosensory cortex of a mouse model of the fragile X mental retardation syndrome. Am. J. Med. Genet. 135, 155–160.

Garcia, J. G., Liu, F., Verin, A. D., Birukova, A., Dechert, M. A., Gerthoffer, W. T., Bamburg, J. R. and English, D. 2001. Sphingosine 1-phosphate promotes endothelial cell barrier integrity by Edg-dependent cytoskeletal rearrangement. J. Clin. Invest. 108, 689–701.

Garcia, J. G. and Schaphorst, K. L. 1995. Regulation of endothelial cell gap formation and paracellular permeability. J. Investig. Med. 43, 117–126.

Gearing, M., Juncos, J. L., Procaccio, V., Gutekunst, C.-A., Marino-Rodriguez, E. M., Gyure, K. A., Ono, S., Santoianni, R., Kraweicki, M. S., Wallace, D. C. and Wainer, B. H. 2002. Aggregation of actin and cofilin in identical twins with juvenile-onset dystonia. Ann. Neurol. 52, 465–476.

Gebuhr, T. C., Kovalev, G. I., Bultman, S., Godfrey, V., Su, L. and Magnuson, T. 2003. The role of Brg1, a catalytic subunit of mammalian chromatin-remodeling complexes, in T cell development. J. Exp. Med. 198, 1937–1949.

Gedeon, A. K., Nelson, J., Gecz, J. and Mulley, J. C. 2003. X-linked mild non-syndromic mental retardation with neuropsychiatric problems and the missense mutation A365E in PAK3. Am. J. Med. Genet. A 120, 509–517.

Gehler, S., Shaw, A. E., Sarmiere, P. D., Bamburg, J. R. and Letourneau, P. C. 2004. Brain-derived neurotrophic factor regulation of retinal growth cone filopodial dynamics is mediated through actin depolymerizing factor/cofilin. J. Neurosci. 24, 10741–10749.

Geinisman, Y., Berry, R. W., Disterhoft, J. F., Power, J. M. and Van der Zee, E. A. 2001. Associative learning elicits the formation of multiple-synapse boutons. J. Neurosci. 21, 5568–5573.

Geneste, O., Copeland, J. W. and Treisman, R. 2002. LIM kinase and Diaphanous cooperate to regulate serum response factor and actin dynamics. J. Cell Biol. 157, L831–L838.

Gessaga, E. C. and Anzil, A. P. 1975. Rod-shaped filamentous inclusions and other ultrastructural features in a cerebellar astrocytoma. Acta Neuropathol. 33, 119–127.

Ghosh, M., Song, X., Mouneimne, G., Sidani, M., Lawrence, D. S. and Condeelis, J. S. 2004. Cofilin promotes actin polymerization and defines the direction of cell motility. Science 304, 743–746.

Giansanti, M. G., Bonaccorsi, S., Williams, B., Williams, W. V., Santolamazza, C., Goldberg, M. L. and Gatti, M. 1998. Cooperative interactions between the central spindle and the contractile ring during Drosophila cytokinesis. Genes Dev. 12, 396–410.

Gibson, P. H. and Tomlinson, B. E. 1977. Numbers of Hirano bodies in the hippocampus of normal and demented people with Alzheimer's disease. J. Neurol. Sci. 33, 199–206.

Glotzer, M. 2001. Animal cell cytokinesis. Annu. Rev. Cell Dev. Biol. 17, 351–386.

Goate, A., Chartier-Harlin, M. C., Mullan, M., Brown, J., Crawford, F., Fidani, L., Giuffra, L., Haynes, A., Irving, N., James, L., Mant, R., Newton, P., Rooke, R., Roquis, P., Talbot, C., Pericak-Vance, M., Roses, A., Williamson, R., Rossor, M., Owen, M., Owen, M. and Hardy, J. 1991. Segregation of a missense mutation in the amyloid precursor protein gene with familial Alzheimer's disease. Nature 349, 704–706.

Gohla, A., Birkenfeld, J. and Bokoch, G. M. 2005. Chronophin, a novel HAD-type serine protein phosphatase, regulates cofilin-dependent actin dynamics. Nat. Cell Biol. 7, 21–29.

Gohla, A. and Bokoch, G. M. 2002. 14-3-3 regulates actin dynamics by stabilizing phosphorylated cofilin. Curr. Biol. 12, 1704–1710.

Goldberg, M. B. 2001. Actin-based motility of intracellular microbial pathogens. Microbiol. Mol. Biol. Rev. 65, 595–626.

Goode, B. L., Drubin, D. G. and Barnes, G. 2000. Functional cooperation between the microtubule and actin cytoskeletons. Curr. Opin. Cell Biol. 12, 63–71.

Goode, B. L., Drubin, D. G. and Lappalainen, P. 1998. Regulation of the cortical actin cytoskeleton in budding yeast by twinfilin, a ubiquitous actin monomer-sequestering protein. J. Cell Biol. 142, 723–733.

Gorbatyuk, V. Y., Nosworthy, N. J., Robson, S. A., Bains, N. P. S., Maciejewski, M. W., dos Remedios, C. G. and King, G. F. 2006. Mapping of a novel phosphoinositide binding site on chick cofilin explains how PIP2 regulated the cofilin–actin interaction. Mol. Cell 24, 511–522.

Gotz, R., Wiese, S., Takayama, S., Camarero, G. C., Rossoll, W., Schweizer, U., Troppmair, J., Jablonka, S., Holtmann, B., Reed, J. C., Rapp, U. R. and Sendtner, M. 2005. Bag1 is essential for differentiation and survival of hematopoietic and neuronal cells. Nat. Neurosci. 8, 1169–1178.

Gourlay, C. W. and Ayscough, K. R. 2005. A role for actin in aging and apoptosis. Biochem. Soc. Trans. 33, 1260–1264.

Govek, E. E., Newey, S. E., Akerman, C. J., Cross, J. R., Van der Veken, L. and Van Aelst, L. 2004. The X-linked mental retardation protein oligophrenin-1 is required for dendritic spine morphogenesis. Nat. Neurosci. 7, 364–372.

Green, D. R. and Evan, G. I. 2002. A matter of life and death. Cancer Cell 1, 19–30.

Greenough, W. T., Klintsova, A. Y., Irwin, S. A., Galvez, R., Bates, K. E. and Weiler, I. J. 2001. Synaptic regulation of protein synthesis and the fragile X protein. Proc. Natl Acad. Sci. USA 98, 7101–7106.

Grimme, S. J., Gao, X.-D., Martin, P. S., Tu, K., Tcheperegine, S. E., Corrado, K., Farewell, A. E., Orlean, P. and Bi, E. 2004. Deficiencies in the endoplasmic reticulum (ER)-membrane protein Gab1p perturb transfer of glycosylphosphatidylinositol to proteins and cause perinuclear ER-associated actin bar formation. Mol. Biol. Cell 15, 2758–2770.

Grosse, R., Copeland, J. W., Newsome, T. P., Way, M. and Treisman, R. 2003. A role for VASP in RhoA-Diaphanous signalling to actin dynamics and SRF activity. EMBO J. 22, 3050–3061.

Gunning, P. W., Schevzov, G., Kee, A. J. and Hardeman, E. C. 2005. Tropomyosin isoforms: Divining rods for actin cytoskeleton function. Trends Cell Biol. 15, 333–341.

Gunsalus, K. C., Bonaccorsi, S., Williams, E., Verni, F., Gatti, M. and Goldberg, M. L. 1995. Mutations in twinstar, a Drosophila gene encoding a cofilin/ADF homologue, result in defects in centrosome migration and cytokinesis. J. Cell Biol. 131, 1243–1259.

Gupton, S. L., Anderson, K. L., Kole, T. P., Fischer, R. S., Ponti, A., Hitchcock-DeGregori, S. E., Danuser, G., Fowler, V. M., Wirtz, D., Hanein, D. and Waterman-Storer, C. M. 2005. Cell migration without a lamellipodium: Translation of actin dynamics into cell movement mediated by tropomyosin. J. Cell Biol. 168, 619–631.

Gurniak, C. B., Perlas, E. and Witke, W. 2005. The actin depolymerizing factor n-cofilin is essential for neural tube morphogenesis and neural crest cell migration. Dev. Biol. 278, 231–241.

Gutsche-Perelroizen, I., Lepault, J., Ott, A. and Carlier, M. F. 1999. Filament assembly from profilin-actin. J. Biol. Chem. 274, 6234–6243.

Guttman, J. A., Obinata, T., Shima, J., Griswold, M. and Vogl, A. W. 2004. Nonmuscle cofilin is a component of tubulobulbar complexes in the testis. Biol. Reprod. 70, 805–812.

Guttman, J. A., Takai, Y. and Vogl, A. W. 2004. Evidence that tubulobulbar complexes in the seminiferous epithelium are involved with internalization of adhesion junctions. Biol. Reprod. 71, 548–559.

Hadfield, M. G., Martinez, A. J. and Gilmartin, R. C. 1974. Progressive multifocal leukoencephalopathy with paramyxovirus-like structures, Hirano bodies and neurogibrillary tangles. Acta Neuropathol. (Berl.) 27(4), 277–288.

Hannan, A. J., Schevzov, G., Gunning, P., Jeffrey, P. L. and Weinberger, R. P. 1995. Intracellular localization of tropomyosin mRNA and protein is associated with development of neuronal polarity. Mol. Cell. Neurosci. 6, 397–412.

Hardy, J. and Selkoe, D. J. 2002. The amyloid hypothesis of Alzheimer's disease: Progress and problems on the road to therapeutics. Science 297, 353–356.

Harris, K. M. 1999a. Calcium from internal stores modifies dendritic spine shape. Proc. Natl Acad. Sci. USA 96, 12213–12215.

Harris, K. M. 1999b. Structure, development, and plasticity of dendritic spines. Curr. Opin. Neurobiol. 9, 343–348.

Harris, K. M. and Stevens, J. K. 1998. Dendritic spines of rat cerebellar Purkinje cells: Serial electron microscopy with reference to their biophysical characteristics. J. Neurosci. 8, 4455–4469.

Hatanaka, K., Li, X. A., Masuda, K., Yutani, C. and Yamamoto, A. 1995. Immunohistochemical localization of C-reactive protein-binding sites in human atherosclerotic aortic lesions by a modified streptavidin–biotin-staining method. Pathol. Int. 45, 635–641.

Hatanaka, H., Ogura, K., Moriyama, K., Ichikawa, S., Yahara, I. and Inagaki, F. 1996. Tertiary structure of destrin and structural similarity between two actin-regulating protein families. Cell 85, 1047–1055.

Hawkins, M., Pope, B., Maciver, S. K. and Weeds, A. G. 1993. Human actin depolymerizing factor mediates a pH-sensitive destruction of actin filaments. Biochemistry 32, 9985–9993.

Hayden, S. M., Miller, P. S., Brauweiler, A. and Bamburg, J. R. 1993. Analysis of the interactions of actin depolymerizing factor with G- and F-actin. Biochemistry 32, 9994–10004.

Heimann, K., Percival, J. M., Weinberger, R., Gunning, P. and Stow, J. L. 1999. Specific isoforms of actin-binding proteins on distinct populations of Golgi-derived vesicles. J. Biol. Chem. 274, 10743–10750.

Hendricks, K. B., Shanahan, F. and Lees, E. 2004. Role for BRG1 in cell cycle control and tumor suppression. Mol. Cell. Biol. 24, 362–376.

Heynen, A. J., Quinlan, E. M., Bae, D. C. and Bear, M. F. 2000. Bidirectional, activity-dependent regulation of glutamate receptors in the adult hippocampus in vivo. Neuron 28, 527–536.

Hirano, A. 1994. Hirano bodies and related neuronal inclusions. Neuropathol. Appl. Neurobiol. 20, 3–11.

Hirano, A., Dembitzer, H. M., Kurland, L. T. and Zimmerman, H. M. 1968. The fine structure of some intraganglionic alterations. Neurofibrillary tangles, granulovacuolar bodies and "rod-like" structures as seen in Guam amyotrophic lateral sclerosis and Parkinsonism–dementia complex. J. Neuropathol. Exp. Neurol. 27, 167–182.

Hiraoka, J., Okano, I., Higuchi, O., Yang, N. and Mizuno, K. 1996. Self-association of LIM-kinase 1 mediated by the interaction between an N-terminal LIM domain and a C-terminal kinase domain. FEBS Lett. 399, 117–121.

Hiruma, H., Katakura, T., Takahashi, S., Ichikawa, T. and Kawakami, T. 2003. Glutamate and amyloid beta-protein rapidly inhibit fast axonal transport in cultured rat hippocampal neurons by different mechanisms. J. Neurosci. 23, 8967–8977.

Ho, K. L. and Allevato, P. A. 1986. Hirano body in an inflammatory cell of leptomeningeal vessel infected by fungus Paecilomyces. Acta Neuropathol. 71, 159–162.

Hofmann, W., Reichart, B., Ewald, A., Müller, E., Schmitt, I., Stauber, R. H., Lottspeich, F., Jockusch, B. M., Scheer, U., Hauber, J. and Dabauvalle, M.-C. 2001. Cofactor requirements for nuclear export of Rev response element (RRE)- and constitutive transport element (CTE)-containing retroviral RNAs. An unexpected role for actin. J. Cell Biol. 152, 895–910.

Holtmaat, A. J., Trachtenberg, J. T., Wilbrecht, L., Shepherd, G. M., Zhang, X., Knott, G. W. and Svoboda, K. 2005. Transient and persistent dendritic spines in the neocortex in vivo. Neuron 45, 279–291.

Hoogenraad, C. C., Akhmanova, A., Galjart, N. and De Zeeuw, C. I. 2004. LIMK1 and CLIP-115: Linking cytoskeletal defects to Williams syndrome. Bioessays 26, 141–150.

Hoogenraad, C. C., Koekkoek, B., Akhmanova, A., Krugers, H., Dortland, B., Miedema, M., van Alphen, A., Kistler, W. M., Jaegle, M., Doutsourakis, M., Van Camp, N., Verhoye, M., van der Linder, A., Kaverina, I., Grosveld, F., De Zeeuw, C. I. and Galjart, N. 2002. Targeted mutation of Cyln2 in the Williams syndrome critical region links CLIP-115 haploinsufficiency to neurodevelopmental abnormalities in mice. Nat. Genet. 32, 116–127.

Hotulainen, P., Paunola, E., Vartiainen, M. K. and Lappalainen, P. 2005. Actin-depolymerizing factor and cofilin-1 play overlapping roles in promoting rapid F-actin depolymerization in mammalian nonmuscle cells. Mol. Biol. Cell 16, 649–664.

Hughes, J. A., Cooke-Yarborough, C. M., Chadwick, N. C., Schevzov, G., Arbuckle, S. M., Gunning, P. and Weinberger, R. P. 2003. High-molecular-weight tropomyosins localize to the contractile rings of dividing CNS cells but are absent from malignant pediatric and adult CNS tumors. Glia 42, 25–35.

Husi, H., Ward, M. A., Choudhary, J. S., Blackstock, W. P. and Grant, S. G. 2000. Proteomic analysis of NMDA receptor-adhesion protein signaling complexes. Nat. Neurosci. 3, 661–669.

Huttelmaier, S., Zenklusen, D., Lederer, M., Dictenberg, J., Lorenz, M., Meng, X., Bassell, G. J., Condeelis, J. and Singer, R. H. 2005. Spatial regulation of beta-actin translation by Src-dependent phosphorylation of ZBP1. Nature 438, 512–515.

Huttenlocher, P. R. 1970. Dendritic development and mental defect. Neurology 20, 381.

Huttenlocher, P. R. 1975. Synaptic and dendritic development and mental defect. UCLA Forum Med. Sci. 18, 123–140.

Ichetovkin, I., Grant, W. and Condeelis, J. 2002. Cofilin produces newly polymerized actin filaments that are preferred for dendritic nucleation by the Arp2/3 complex. Curr. Biol. 12, 79–84.

Iida, K., Iida, H. and Yahara, I. 1986. Heat shock induction of intranuclear actin rods in cultured mammalian cells. Exp. Cell Res. 165, 207–215.

Iida, K., Matsumoto, S. and Yahara, I. 1992. The KKRKK sequence is involved in heat shock-induced nuclear translocation of the 18-kDa actin-binding protein, cofilin. Cell Struct. Funct. 17, 39–46.

Iida, K., Moriyama, K., Matsumoto, S., Kawasaki, H., Nixhida, E. and Yahara, I. 1993. Isolation of a yeast essential gene, COF1, that encodes a homologue of mammalian cofilin, a low-M(r) actin-binding and depolymerizing protein. Gene 124, 115–120.

Iida, K. and Yahara, I. 1986. Reversible induction of actin rods in mouse C3H-2K cells by incubation in salt buffers and by treatment with non-ionic detergents. Exp Cell Res. 164, 492–506.

Ikebe, C., Ohashi, K., Fujimori, T., Bernard, O., Noda, T., Robvertson, E. J. and Mizuno, K. 1997. Mouse LIM-kinase 2 gene: cDNA cloning, genomic organization, and tissue-specific expression of two alternatively initiated transcripts. Genomics 46, 504–508.

Ikeda, S., Cunningham, L. A., Boggers, D., Hobson, C. D., Sundberg, J. P., Naggert, J. K., Smith, R. S. and Nishina, P. M. 2003. Aberrant actin cytoskeleton leads to accelerated proliferation of corneal epithelial cells in mice deficient for destrin (actin depolymerizing factor). Hum. Mol. Genet. 12, 1029–1037.

Ilkovski, B., Nowak, K. J., Domazetovska, A., Maxwell, A. L., Clement, S., Davies, K. E., Laing, N. G., North, K. N. and Cooper, S. T. 2004. Evidence for a dominant-negative effect in ACTA1 nemaline myopathy caused by abnormal folding, aggregation and altered polymerization of mutant actin isoforms. Hum. Mol. Genet. 13, 1727–1743.

Ireton, K. and Cossart, P. 1997. Host–pathogen interactions during entry and actin-based movement of *Listeria monocytogenes*. Annu. Rev. Genet. 31, 113–138.

Irwin, S. A., Idupulapati, M., Gilbert, M. E., Harris, J. B., Chakravarti, A. B., Rogers, E. J., Crisostomo, R. A., Larsen, B. P., Mehta, A., Alcantara, C. J., Patel, B., Swain, R. A., Weiler, I. J., Oostra, B. A. and Greenough, W. T. 2002. Dendritic spine and densritic field characteristics of layer V pyramidal neurons in the visual cortex of fragile-X knockout mice. Am. J. Med. Genet. 111, 140–146.

Ivanov, A. I., McCall, I. C., Parkos, C. A. and Nusrat, A. 2004. Role for actin filament turnover and a myosin II motor in cytoskeleton-driven disassembly of the epithelial apical junctional complex. Mol. Biol. Cell 15, 2639–2651.

Jang, D. H., Han, J. H., Lee, S. H., Lee, Y. S., Park, H., Lee, S. H., Kim, H. and Kaang, B. K. 2005. Cofilin expression induces cofilin–actin rod formation and disrupts synaptic structure and function in Aplysia synapses. Proc. Natl Acad. Sci. USA 102, 16072–16077.

Jiang, C. J., Weeds, A. J. and Hussey, P. J. 1997. The maize actin-depolymerizing factor, ZmADF3, redistributes to the growing tip of elongating root hairs and can be induced to translocate into the nucleus with actin. Plant J. 12, 1035–1043.

Jockusch, B. M., Veldman, H., Griffiths, G. W., van Oost, B. A. and Jennekens, F. G. 1980. Immunofluorescence microscopy of a myopathy. alpha-Actinin is a major constituent of nemaline rods. Exp. Cell Res. 127, 409–420.

Jontes, J. D. and Smith, S. J. 2000. Filopodia, spines, and the generation of synaptic diversity. Neuron 27, 11–14.

Kaji, N., Ohashi, K., Shuin, M., Niwa, R., Uemura, T. and Mizuno, K. 2003. Cell cycle-associated changes in Slingshot phosphatase activity and roles in cytokinesis in animal cells. J. Biol. Chem. 278, 33450–30455.

Kanamori, T., Suzuki, M and Titani, K. 1998. Complete amino acid sequences and phosphorylation sites, determined by Edman degradation and mass spectrometry, of rat parotid destrin- and cofilin-like proteins. Arch. Oral Biol. 43, 955–967.

Kang, H., Cui, K. and Zhao, K. 2004. BRG1 controls the activity of the retinoblastoma protein via regulation of p21CIP1/WAF1/SDI. Mol. Cell. Biol. 24, 1188–1199.

Kasibhatla, S. and Tseng, B. 2003. Why target apoptosis in cancer treatment? Mol. Cancer Ther. 2, 573–580.

Kawano, N. and Horoupian, D. S. 1981. Intracytoplasmic rod-like inclusions in caudate nucleus. Neuropathol. Appl. Neurobiol. 7, 307–314.

Keating, M. T. 1997. On the trail of genetic culprits in Williams syndrome. Cardiovasc. Res. 36, 134–137.

Keezer, S. M., Ivie, S. E., Krutzsch, H. C., Tandle, A., Libutti, S. K. and Roberts, D. D. 2003. Angiogenesis inhibitors target the endothelial cell cytoskeleton through altered regulation of heat shock protein 27 and cofilin. Cancer Res. 63, 6405–6412.

Kennedy, M. B. 2000. Signal-processing machines at the postsynaptic density. Science 290, 750–754.

Kennedy, M. B., Beale, H. C., Carlisle, H. J. and Washburn, L. R. 2005. Integration of biochemical signalling in spines. Nat. Rev. Neurosci. 6, 423–434.

Khan, N. A., Wang, Y., Kim, K. J., Chung, J. W., Wass, C. A. and Kim, K. S. 2002. Cytotoxic necrotizing factor-1 contributes to *Escherichia coli* K1 invasion of the central nervous system. J. Biol. Chem. 277, 15607–15612.

Kimura, T., Hashimoto, I., Yamamoto, A., Nixhikawa, M. and Fujisawa, J.-I. 2000. Rev-dependent association of the intron-containing HIV-1 gag mRNA with the nuclear actin bundles and the inhibition of its nucleocytoplasmic transport by latrunculin-B. Genes Cells 5, 289–307.

Koch, C. and Zador, A. 1993. The function of dendritic spines: Devices subserving biochemical rather than electrical compartmentalization. J. Neurosci. 13, 413–422.

Koenig, E. and Adams, P. 1982. Local protein synthesizing activity in axonal fields regenerating in vitro. J. Neurochem. 39, 386–400.

Komarova, Y. A., Akhmanova, A. S., Kojima, S.-I., Galjart, N. and Borisy, G. G. 2002. Cytoplasmic linker proteins promote microtubule rescue in vivo. J. Cell Biol. 159, 589–599.

Kovar, D. R. 2005. Molecular details of formin-mediated actin assembly. Curr. Opin. Cell Biol. 18, 1–7.

Krebs, A., Rothkegel, M., Klar, M. and Jockusch, B. M. 2001. Characterization of functional domains of mDia1, a link between the small GTPase Rho and the actin cytoskeleton. J. Cell Sci. 114, 3663–3672.

Kroczek, R. and Hamelmann, E. 2005. T-cell costimulatory molecules: Optimal targets for the treatment of allergic airway disease with monoclonal antibodies. J. Allergy Clin. Immunol. 116, 906–909.

Kubler, E. and Riezman, H. 1993. Actin and fimbrin are required for the internalization step of endocytosis in yeast. EMBO J. 12, 2855–2862.

Kulisevsky, J., Marti, M. J., Ferrer, I. and Tolosa, E. 1988. Meige syndrome: Neuropathology of a case. Mov. Disord. 3, 170–175.

Kurz, A. F. 2005. Uncommon neurodegenerative causes of dementia. Int. Psychogeriatr. 17(Suppl. 1), 35–49.

Kusano, K.-I., Abe, H. and Obinata, T. 1999. Detection of a sequence involved in actin-binding and phosphoinositide-binding in the N-terminal side of cofilin. Mol. Cell. Biochem. 190, 133–141.

Kwak, I. H., Kim, H. S., Choi, O. R., Ryu, S. and Lim, K. 2004. Nuclear accumulation of globular actin as a cellular senescence marker. Cancer. Res. 64, 572–580.

Laas, R. and Hagel, C. 1994. Hirano bodies and chronic alcoholism. Neuropathol. Appl. Neurobiol. 20, 12–21.

Laggerbauer, B., Ostareck, D., Keidel, E. M., Ostareck-Lederer, A. and Fischer, U. 2001. Evidence that fragile X mental retardation protein is a negative regulator of translation. Hum. Mol. Genet. 10, 329–338.

Laing, N. G., Clarke, N. F., Dye, D. E., Liyanage, K., Walker, K. R., Kobayashi, Y., Shimakawa, S., Hagiwara, T., Ouvrier, R., Sparrow, J. C., Nishino, I., North, K. N. and Nonaka, I. 2004. Actin mutations are one cause of congenital fibre type disproportion. Ann. Neurol. 56, 689–694.

LaLonde, D. P., Brown, M. C., Bouverat, B. P. and Turner, C. E. 2005. Actopaxin interacts with TESK1 to regulate cell spreading on fibronectin. J. Biol. Chem. 280, 21680–21688.

Langford, G. M. 1995. Actin- and microtubule-dependent organelle motors: Interrelationships between the two motility systems. Curr. Opin. Cell Biol. 7, 82–88.

Lappalainen, P., Fedorov, E. V., Fedorov, A. A., Almo, S. C. and Drubin, D. G. 1997. Essential functions and actin-binding surfaces of yeast cofilin revealed by systematic mutagenesis. EMBO J. 16, 5520–5530.

Lappalainen, P., Kessels, M. M., Cope, M. J. T. V. and Drubin, D. G. 1998. The ADF homology (ADF-H) domain: A highly exploited actin-binding module. Mol. Biol. Cell 9, 1951–1959.

Larson, L., Arnaudeau, S., Gibson, B., Li, W., Krause, R., Hao, B., Bamburg, J. R., Lew, D. P., Demaurex, N. and Southwick, F. 2005. Gelsolin mediates calcium-dependent disassembly of Listeria actin tails. Proc. Natl Acad. Sci. USA 102, 1921–1926.

Laurent, V., Loisel, T. P., Harbeck, B., Wehman, A., Grobe, L., Jockusch, B. M., Wehland, J., Gertler, F. B. and Carlier, M. F. 1999. Role of proteins of the Ena/VASP family in actin-based motility of Listeria monocytogenes. J. Cell Biol. 144, 1245–1258.

Le Clainche, C. and Drubin, D. G. 2004. Actin lessons from pathogens. Mol. Cell 13, 453–454.

Lee, S. and Helfman, D. M. 2004. Cytoplasmic p21Cip1 is involved in Ras-induced inhibition of the ROCK/LIMK/cofilin pathway. J. Biol. Chem. 279, 1885–1891.

Lee, S. K and Hollenbeck, P. J. 2003. Organization and translation of mRNA in sympathetic axons. J. Cell Sci. 116, 4467–4478.

Lee, K. H., Meuer, S. C. and Samstag, Y. 2000. Cofilin: A missing link between T cell co-stimulation and rearrangement of the actin cytoskeleton. Eur. J. Immunol. 30, 892–899.

Lee-Hoeflich, S. T., Causing, C. G., Podkowa, M., Zhao, X., Wrana, J. L. and Attisano, L. 2004. Activation of LIMK1 by binding to the BMP receptor, BMPRII, regulates BMP-dependent dendritogenesis. EMBO J. 23, 4792–4801.

Lee-Hoeflich, S. T., Zhao, X., Mehra, A. and Attisano, L. 2005. The Drosophila type II receptor, Wishful thinking, binds BMP and myoglianin to activate multiple TGF beta family signaling pathways. FEBS Lett. 579, 4615–4621.

Lees-Miller, J. P., Goodwin, L. O. and Helfman, D. M. 1990. Three novel brain tropomyosin isoforms are expressed from the rat alpha-tropomyosin gene through the use of alternative promoters and alternative RNA processing. Mol. Cell. Biol. 10, 1729–1742.

Lehrer, M. S., Sun, T. T. and Lavker, R. M. 1998. Strategies of epithelial repair: Modulation of stem cell and transit amplifying cell proliferation. J. Cell Sci. 111, 2867–2875.

Leonard, S. A., Gittis, A. G., Petrella, E. C., Pollard, T. D. and Lattman, E. E. 1997. Crystal structure of the actin-binding protein actophorin from Acanthamoeba. Nat. Struct. Biol. 4, 369–373.

Lewis, M. H. 2004. Environmental complexity and central nervous system development and function. Ment. Retard. Dev. Disabil. Res. Rev. 10, 91–95.

Li, X., Chen, B., Blystone, S. D., McHugh, K. P., Ross, F. P. and Ramos, D. M. 1998. Differential expression of alphav integrins in K1735 melanoma cells. Invasion Metastasis 18, 1–14.

Li, W. and Gao, F. B. 2003. Actin filament-stabilizing protein tropomyosin regulates the size of dendritic fields. J. Neurosci. 23, 6171–6175.

Li, P., Nijhawan, D., Budihardjo, I., Srinivasula, S. M., Ahmad, M., Alnemri, E. S. and Wang, X. 1997. Cytochrome c and dATP-dependent formation of Apaf-1/caspase-9 complex initiates an apoptotic protease cascade. Cell 91, 479–489.

Lilic, M., Galkin, V. E., Orlova, A., VanLoock, M. S., Egleman, E. H. and Stebbins, C. E. 2003. Salmonella SipA polymerizes actin by stapling filaments with nonglobular protein arms. Science 301, 1918–1921.

Lippincott-Schwartz, J., Roberts, T. H. and Hirschberg, K. 2000. Secretory protein trafficking and organelle dynamics in living cells. Annu. Rev. Cell Dev. Biol. 16, 557–589.

Lisman, J. 2003. Actin's actions in LTP-induced synapse growth. Neuron 38, 361–362.

Loisel, T. P., Boujemaa, R., Pantaloni, D. and Carlier, M.-F. 1999. Reconstitution of actin-based motility of Listeria and Shigella using pure proteins. Nature 401, 313–316.

London, M. and Hausser, M. 2005. Dendritic computation. Annu. Rev. Neurosci. 28, 503–532.

Lorenz, M., DesMarais, V., Macaluso, F., Singer, R. H. and Condeelis, J. 2004. Measurement of barbed ends, actin polymerization, and motility in live carcinoma cells after growth factor stimulation. Cell Motil. Cytoskeleton 57, 207–217.

Lui, W. Y., Lee, W. M. and Cheng, C. Y. 2003. Sertoli-germ cell adherens junction dynamics in the testis are regulated by RhoB GTPase via the ROCK/LIMK signaling pathway. Biol. Reprod. 68, 2189–2206.

Lyubimova, A., Bershadsky, A. D. and Ben-Ze'ev, A. 1997. Autoregulation of actin synthesis responds to monomeric actin levels. J. Cell Biochem. 65, 469–478.

Lyubimova, A., Bershadsky, A. D. and Ben-Ze'ev, A. 1999. Autoregulation of actin synthesis requires the 3'-UTR of actin mRNA and protects cells from actin overproduction. J. Cell Biochem. 76, 1–12.

Mabuchi, I. 1983. An actin-depolymerizing protein (depactin) from starfish oocytes: Properties and interaction with actin. J. Cell Biol. 97, 1612–1621.

Macara, I. G. 2004a. Par proteins: Partners in polarization. Curr. Biol. 14, 160–162.

Macara, I. G. 2004b. Parsing the polarity code. Nat. Rev. Mol. Cell Biol. 5, 220–231.

MacDonald, T. T. and Carter, P. B. 1980. Cell-mediated immunity to intestinal infection. Infect. Immun. 28, 516–523.

Maciver, S. K. and Harrington, C. R. 1995. Two actin binding proteins, actin depolymerizing factor and cofilin, are associated with Hirano bodies. Neuroreport 6, 1985–1988.

Maciver, S. K., Pope, B. J., Whytock, S. and Weeds, A. G. 1998. The effect of two actin depolymerizing factors (ADF/cofilins) on actin filament turnover: pH sensitivity of F-actin binding by human ADF, but not of Acanthamoeba actophorin. Eur. J. Biochem. 256, 388–397.

Maciver, S. K. and Weeds, A. G. 1994. Actophorin preferentially binds monomeric ADP-actin over ATP-bound actin: Consequences for cell locomotion. FEBS Lett. 347, 251–256.

Maekawa, M., Ishizaki, T., Boku, S., Watanabe, N., Fujita, A., Iwamatsu, A., Obinata, T., Ohashi, K., Mizuno, K. and Narumiya, S. 1999. Signaling from Rho to the actin cytoskeleton through protein kinases ROCK and LIM-kinase. Science 285, 895–898.

Maiti, S. and Bamburg, J. R. 2004. Actin capping and severing proteins. In Encyclopedia of Biological Chemistry, vol. 1. W. J. Lennarz and M. D. Lane (Editors). Elsevier, Amsterdam. pp. 19–26.

Maiti, S., Boyle, J. A., Minamide, L. S., Gungabissoon, R. A. and Bamburg. J. R. 2002. Isolation and characterization of ADF/cofilin–actin rods from cultured cells. Mol. Biol. Cell 13, 37.

Majewska, A., Brown, E., Ross, J. and Yuste, R. 2000. Mechanisms of calcium decay kinetics in hippocampal spines: Role of spine calcium pumps and calcium diffusion through the spine neck in biochemical compartmentalization. J. Neurosci. 20, 1722–1734.

Majewska, A., Tashiro, A. and Yuste, R. 2000. Regulation of spine calcium dynamics by rapid spine motility. J. Neurosci. 20, 8262–8268.

Maloney, M. T., Minamide, L. S., Kinley, A. W., Boyle, J. A. and Bamburg, J. R. 2005. Beta-secretase-cleaved amyloid precursor protein accumulates at actin inclusions induced in neurons by stress or amyloid beta: A feedforward mechanism for Alzheimer's disease. J. Neurosci. 25, 11313–11321.

Mann, G. 1894. Histochemical changes induced in sympathetic, motor, and sensory nerve cells by functional activity. J. Anat. Physiol. London 19, 100–108.

Mansouri, A. 1998. The role of Pax3 and Pax7 in development and cancer. Crit. Rev. Oncog. 9, 141–149.

Marin-Padilla, M. 1972. Structural abnormalities of the cerebral cortex in human chromosomal aberrations: A Golgi study. Brain Res. 44, 625–629.

Maselli, A. G., Davis, R., Thomson, S. A. M., Davis, R. C. and Fechheimer, M. 2002. Formation of Hirano bodies in Dictyostelium and mammalian cells induced by expression of a modified form of an actin-crosslinking protein. J. Cell Sci. 115, 1939–1949.

Masuda, M., Betancourt, L., Matsuzawa, T., Kashimoto, T., Takao, T., Shimonishi, Y. and Horiguchi, Y. 2000. Activation of rho through a cross-link with polyamines catalyzed by Bordetella dermonecrotizing toxin. EMBO J. 19, 521–530.

Matsudaira, P. 1991. Modular organization of actin crosslinking proteins. Trends Biochem. Sci. 16, 87–92.

Matsuzaki, M., Honkura, N., Ellis-Davies, G. C. and Kasai, H. 2004. Structural basis of long-term potentiation in single dendritic spines. Nature 429, 761–766.

Matsuzaki, F., Matsumoto, S., Yahara, I., Yonezawa, N., Nishida, E and Sakai, H. 1988. Cloning and characterization of porcine brain cofilin cDNA. Cofilin contains the nuclear transport signal sequence. J. Biol. Chem. 263, 11564–11568.

Mattila, P. K., Quintero-Monzon, O., Kugler, J., Moseley, J. B., Almo, S. C., Lappalainen, P. and Goode, B. L. 2004. A high-affinity interaction with ADP-actin monomers underlies the mechanism and in vivo function of Srv2/cyclase-associated protein. Mol. Biol. Cell 15, 5158–5171.

Mattson, M. P. 2004. Pathways towards and away from Alzheimer's disease. Nature 430, 631–639.

Matus, A. 1999. Postsynaptic actin and neuronal plasticity. Curr. Opin. Neurobiol. 9, 561–565.

Matus, A. 2000. Actin-based plasticity in dendritic spines. Science 290, 754–758.

Mayer, T., Meyer, M., Janning, A., Schiedel, A. C. and Barnekow, A. 1999. A mutant form of the rho protein can restore stress fibers and adhesion plaques in v-src transformed fibroblasts. Oncogene 18, 2117–2128.

McGhie, E. J., Hayward, R. D. and Koronakis, V. 2004. Control of actin turnover by a salmonella invasion protein. Mol. Cell 13, 497–510.

McGough, A. 1998. F-actin-binding proteins. Curr. Opin. Struct. Biol. 8, 166–176.

McGough, A., Pope, B., Chiu, W. and Weeds, A. 1997. Cofilin changes the twist of F-actin: Implications for actin filament dynamics and cellular function. J. Cell Biol. 138, 771–781.

McKim, K. S., Matheson, C., Marra, M. A., Wakarchuk, M. F. and Baillie, D. L. 1994. The *Caenorhabditis elegans* unc-60 gene encodes proteins homologous to a family of actin-binding proteins. Mol. Gen. Genet. 242, 346–357.

McKinney, B. C., Grossman, A. W., Elisseou, N. M. and Greenough, W. T. 2005. Dendritic spine abnormalities in the occipital cortex of C57BL/6 Fmr1 knockout mice. Am. J. Med. Genet. B. Neuropsychiatr. Genet. 136, 98–102.

Meberg, P. J. and Bamburg, J. R. 2000. Increase in neurite outgrowth mediated by overexpression of actin depolymerizing factor. J. Neurosci. 20, 2459–2469.

Meberg, P. J., Ono, S., Minamide, L. S., Takahashi, M. and Bamburg, J. R. 1998. Actin depolymerizing factor and cofilin phosphorylation dynamics: Response to signals that regulate neurite extension. Cell Motil. Cytoskeleton 39, 172–190.

Meng, Y., Takahashi, H., Meng, J., Zhang, Y., Lu, G., Asrar, S., Nakamura, T. and Jia, Z. 2004. Regulation of ADF/cofilin phosphorylation and synaptic function by LIM-kinase. Neuropharmacology 47, 746–754.

Meng, Y., Zhang, Y., Tregoubov, V., Janus, C., Cruz, L., Jackson, M., Lu, W.-Y., MacDonald, J. F., Wang, J. Y., Falls, D. L. and Jia, Z. 2002. Abnormal spine morphology and enhanced LTP in LIMK-1 knockout mice. Neuron 35, 121–133.

Meyer, G. and Feldman, E. L. 2002. Signaling mechanisms that regulate actin-based motility processes in the nervous system. Neurochemistry 83, 490–503.

Meyer-Lindenberg, A., Mervis, C. B., Sarpal, D., Koch, P., Steele, S., Kohn, P., Marenco, S., Morris, C. A., Das, S., Kippenhan, S., Mattay, V. S., Weinberger, D. R. and Berman, K. F. 2005. Functional, structural and metabolic abnormalities of the hippocampal formation in Williams syndrome. J. Clin. Invest. 115, 1888–1895.

Minamide, L. S., Painter, W. B., Schevzov, G., Gunning, P. and Bamburg, J. R. 1997. Differential regulation of actin depolymerizing factor and cofilin in response to alterations in the actin monomer pool. J. Biol. Chem. 272, 8303–8309.

Minamide, L. S., Striegl, A. M., Boyle, J. A., Meberg, P. J. and Bamburg, J. R. 2000. Neurodegenerative stimuli induce persistent ADF/cofilin–actin rods that disrupt distal neurite function. Nat. Cell Biol. 2, 628–636.

Misra, U. K., Sharma, T. and Pizzo, S. V. 2005. Ligation of cell surface-associated glucose-regulated protein 78 by receptor-recognized forms of alpha 2-macroglobulin: Activation of p21-activated protein kinase-2-dependent signaling in murine peritoneal macrophages. J. Immunol. 175, 2525–2533.

Mitake, S., Ojika, K. and Hirano, A. 1997. Hirano bodies and Alzheimer's disease. Kaohsiung J. Med. Sci. 13, 10–18.

Mitchell, T. G. and Perfect, J. R. 1995. Cryptococcosis in the era of AIDS-100 years after the discovery of *Cryptococcus neoformans*. Clin. Microbiol. Rev. 8, 515–548.

Moeller, M. L., Shi, Y., Reichardt, L. F. and Ethell, I. M. 2005. EphB receptors regulate dendritic spine morphogenesis through the recruitment/phosphorylation of FAK and RhoA activation. J. Biol. Chem. 281, 1587–1598.

Mohri, K. and Ono, S. 2003. Actin filament disassembling activity of *Caenorhabditis elegans* actin-interacting protein 1 (UNC-78) is dependent on filament binding by a specific ADF/cofilin isoform. J. Cell Sci. 116, 4107–4108.

Moon, A. L., Janmey, P. A., Louie, K. A. and Drubin, D. G. 1993. Cofilin is an essential component of the yeast cortical cytoskeleton. J. Cell Biol. 120, 421–435.

Morgan, T. E., Lockerbie, R. O., Minamide, L. S., Browning, M. D. and Bamburg, J. R. 1993. Isolation and characterization of a regulated form of actin depolymerizing factor. J. Cell Biol. 122, 623–633.

Morishita, W., Marie, H. and Malenka, R. C. 2005. Distinct triggering and expression mechanisms underlie LTD of AMPA and NMDA synaptic responses. Nat. Neurosci. 8, 1043–1050.

Moriyama, K., Iida, K. and Yahara, I. 1996. Phosphorylation of Ser-3 of cofilin regulates its essential function on actin. Genes Cells 1, 73–86.

Moriyama, K., Nishida, E., Yonezawa, N., Sakai, H., Matsumoto, S., Iida, K. and Yahara, I. 1990. Destrin, a mammalian actin-depolymerizing protein, is closely related to cofilin. Cloning and expression of porcine brain destrin cDNA. J. Biol. Chem. 265, 5768–5773.

Moriyama, K. and Yahara, I. 2002. Human CAP1 is a key factor in the recycling of cofilin and actin for rapid actin turnover. J. Cell Sci. 115, 1591–1601.

Moriyama, K., Yonezawa, N., Sakai, H., Yahara, I. and Nishida, E. 1992. Mutational analysis of an actin-binding site of cofilin and characterization of chimeric proteins between cofilin and destrin. J. Biol. Chem. 267, 7240–7244.

Mouneimne, G., Soon, L., DesMarais, V., Sidani, M., Song, X., Yip, S. C., Ghosh, M., Eddy, R., Backer, J. M. and Condeelis, J. 2004. Phospholipase C and cofilin are required for carcinoma cell directionality in response to EGF stimulation. J. Cell Biol. 166, 697–708.

Murrell, J., Farlow, M., Ghetti, B. and Benson, M. D. 1991. A mutation in the amyloid precursor protein associated with hereditary Alzheimer's disease. Science 254, 97–99.

Nachmias, V. T. and Huxley, H. E. 1970. Electron microscope observations of actomyosin and actin preparations from Physarum polycephalum, and on their interaction with heavy meromyosin subfragment I from muscle myosin. J. Mol. Biol. 50, 83–90.

Nagaoka, R., Abe, H. and Obinata, T. 1996. Site-directed mutagenesis of the phosphorylation site of cofilin: Its role in cofilin–actin interaction and cytoplasmic localization. Cell Motil. Cytoskeleton 35, 200–209.

Nagata-Ohashi, K., Ohta, Y., Goto, K., Chiba, S., Mori, R., Nishita, M., Ohashi, K., Kousaka, K., Iwamatsu, A., Niwa, R., Uemura, T. and Mizuno, K. 2004. A pathway of neuregulin-induced activation of cofilin-phosphatase Slingshot and cofilin in lamellipodia. J. Cell Biol. 165, 465–471.

Nakashima, K., Sato, N., Nakagaki, T., Abe, H., Ono, S. and Obinata, T. 2005. Two mouse cofilin isoforms, muscle-type (MCF) and non-muscle type (NMCF), interact with F-actin with different efficiencies. J. Biochem. 138, 519–526.

Nakayama, A. Y. and Luo, L. 2000. Intracellular signaling pathways that regulate dendritic spine morphogenesis. Hippocampus 10, 582–586.

Nebl, G., Meuer, S. C. and Samstag, Y. 1996. Dephosphorylation of serine 3 regulates nuclear translocation of cofilin. J. Biol. Chem. 271, 26276–26280.

Negishi, M. and Katoh, H. 2005. Rho family GTPases and dendrite plasticity. Neuroscientist 11, 187–191.

Newey, S. E., Velamoor, V., Govek, E. E. and Van Aelst, L. 2005. Rho GTPases, dendritic structure, and mental retardation. J. Neurobiol. 64, 58–74.

Ng, J. and Luo, L. 2004. Rho GTPases regulate axon growth through convergent and divergent signaling pathways. Neuron 44, 779–793.

Nichols, C. B., Fraser, J. A. and Heitman, J. 2004. PAK kinases Ste20 and Pak1 govern cell polarity at different stages of mating in Cryptococcus neoformans. Mol. Biol. Cell 15, 4476–4489.

Nimchinsky, E. A., Sabatini, B. L. and Svoboda, K. 2002. Structure and function of dendritic spines. Annu. Rev. Physiol. 64, 313–353.

Nishida, E. 1985. Opposite effects of cofilin and profilin from porcine brain on rate of exchange of actin-bound adenosine 5′-triphosphate. Biochemistry 24, 1160–1164.

Nishida, E., Iida, K., Yonezawa, N., Koyasu, S., Yahara, I. and Sakai, H. 1987. Cofilin is a component of intranuclear and cytoplasmic actin rods induced in cultured cells. Proc. Natl Acad. Sci. USA 84, 5262–5266.

Nishida, E., Maekawa, S. and Sakai, H. 1984a. Characterization of the action of porcine brain profilin on actin polymerization. J. Biochem. 95, 399–404.

Nishida, E., Maekawa, S. and Sakai, H. 1984b. Cofilin, a protein in porcine brain that binds to actin filaments and inhibits their interactions with myosin and tropomyosin. Biochemistry 23, 5307–5313.

Nishida, E., Muneyuki, E., Maekawa, S., Ohta, Y. and Sakai, H. 1985. An actin-depolymerizing protein (destrin) from porcine kidney. Its action on F-actin containing or lacking tropomyosin. Biochemistry 24, 6624–6630.

Nishimura, Y., Yoshioka, K., Bernard, O., Himeno, M. and Itoh, K. 2004. LIM kinase 1: Evidence for a role in the regulation of intracellular vesicle trafficking of lysosomes and endosomes in human breast cancer cells. Eur. J. Cell Biol. 83, 369–380.

Nishita, M., Tomizawa, C., Yamamoto, M., Horita, Y., Ohashi, K. and Mizuno, K. 2005. Spatial and temporal regulation of cofilin activity by LIM kinase and Slingshot is critical for directional cell migration. J. Cell Biol. 171, 349–359.

Nishiya, N., Kiosses, W. B., Han, J. and Ginsberg, M. H. 2005. An alpha4 integrin–paxillin–Arf–GAP complex restricts Rac activation to the leading edge of migrating cells. Nat. Cell Biol. 7, 343–352.

Niwa, R., Nagata-Ohashi, K., Takeichi, M., Mizuno, K. and Uemura, T. 2002. Control of actin reorganization by Slingshot, a family of phosphatases that dephosphorylate ADF/cofilin. Cell 108, 233–246.

Nowak, K. J., Wattanasirichaigoon, D., Goebel, H. H., Wilce, M., Pelin, K., Donner, K., Jacob, R. L., Hubner, C., Oexle, K., Anderson, J. R., Verity, C. M., North, K. N., Iannaccone, S. T., Muller, C. R., Nurnberg, P., Muntoni, F., Sewry, C., Hughes, I., Sutphen, R., Lacson, A. G., Swoboda, K. J., Vigneron, J., Wallgren-Pettersson, C., Beggs, A. H. and Laing, N. G. 1999. Mutations in the skeletal muscle alpha-actin gene in patients with actin myopathy and nemaline myopathy. Nat. Genet. 23, 208–212.

Nunoue, K., Ohashi, K., Okano, I. and Mizuno, K. 1995. LIMK-1 and LIMK-2, two members of a LIM motif-containing protein kinase family. Oncogene 11, 701–710.

Obinata, T., Nagaoka-Yasuda, R., Ono, S., Kusano, K., Mohri, K., Ohtaka, Y., Yamashiro, S., Okada, K. and Abe, H. 1997. Low molecular-weight G-actin binding proteins involved in the regulation of actin assembly during myofibrillogenesis. Cell Struct. Funct. 22, 181–189.

Oh, S. H., Adler, H. J., Raphael, Y. and Lomax, M. I. 2002. WDR1 colocalizes with ADF and actin in the normal and noise-damaged chick cochlea. J. Comp. Neurol. 448, 399–409.

Ohashi, K., Nagata, K., Maekawa, M., Ishizaki, T., Narumiya, S. and Mizuno, K. 2000. Rho-associated kinase ROCK activates LIM-kinase 1 by phosphorylation at threonine 508 within the activation loop. J. Biol. Chem. 275, 3577–3582.

Ohta, Y., Kousaka, K., Nagata-Ohashi, K., Ohashi, K., Murramoto, A., Shima, Y., Niwa, R., Uemura, T. and Mizuno, K. 2003. Differential activities, subcellular distribution and tissue expression patterns of three members of Slingshot family phosphatases that dephosphorylate cofilin. Genes Cells 8, 811–824.

Ohta, Y., Nishida, E., Sakai, H. and Miyamoto, E. 1989. Dephosphorylation of cofilin accompanies heat shock-induced nuclear accumulation of cofilin. J. Biol. Chem. 264, 16143–16148.

Okada, K., Blanchoin, L., Abe, H., Chen, H., Pollard, T. D. and Bamburg, J. R. 2002. Xenopus actin-interacting protein 1 (XAip1) enhances cofilin fragmentation of filaments by capping filament ends. J. Biol. Chem. 277, 43011–43016.

Okamoto, K., Nagai, T., Miyawaki, A. and Hayashi, Y. 2004. Rapid and persistent modulation of actin dynamics regulates postsynaptic reorganization underlying bidirectional plasticity. Nat. Neurosci. 7, 1104–1112.

Olave, I., Wang, W., Xue, Y., Kou, A. and Crabtree, G. R. 2002. Identification of a polymorphic, neuron-specific chromatin remodeling complex. Genes Dev. 16, 2509–2517.

Ono, S., Abe, H., Nagaoka, R. and Obinata, T. 1993. Colocalization of ADF and cofilin in intranuclear actin rods of cultured muscle cells. J. Muscle Res. Cell Motil. 14, 195–204.

Ono, S., Baillie, D. L. and Benian, G. M. 1999. UNC-60B, an ADF/cofilin family protein, is required for proper assembly of actin into myofibrils in *Caenorhabditis elegans* body wall muscle. J. Cell Biol. 145, 491–502.

Ono, S. and Benian, G. M. 1998. Two *Caenorhabditis elegans* actin depolymerizing factor/cofilin proteins, encoded by the unc-60 gene, differentially regulate actin filament dynamics. J. Biol. Chem. 273, 3778–3783.

Ono, S., Inoue, K., Mannen, T., Kanda, F., Jinnai, K. K. and Takahashi, K. 1987. Neuropathological changes of the brain in myotonic dystrophy-some new observations. J. Neurol. Sci. 81, 301–320.

Ono, S., McGough, A., Pope, B. J., Tolbert, V. T., Bui, A., Pohl, J., Benian, G. M., Gernert, K. M. and Weeds, A. G. 1991. The C-terminal tail of UNC-60B (actin depolymerizing factor/cofilin) is critical for maintaining its stable association with F-actin and is implicated in the second actin-binding site. J. Biol. Chem. 276, 5952–5958.

Ono, S., Minami, N., Abe, H. and Obinata, T. 1994. Characterization of a novel cofilin isoform that is predominantly expressed in mammalian skeletal muscle. J. Biol. Chem. 269, 15280–15286.

Ono, S., Mohri, K. and Ono, K. 2004. Microscopic evidence that actin-interacting protein 1 actively disassembles actin-depolymerizing factor/Cofilin-bound actin filaments. J. Biol. Chem. 279, 14207–14212.

Ono, S. and Ono, K. 2002. Tropomyosin inhibits ADF/cofilin-dependent actin filament dynamics. J. Cell Biol. 156, 1065–1076.

Ono, K., Parast, M., Alberico, C., Benian, G. M. and Shoichiro, O. 2003. Specific requirement for two ADF/cofilin isoforms in distinct actin-dependent processes in *Caenorhabditis elegans*. J. Cell Sci. 116, 2073–2085.

Orlova, A., Shvetsov, A., Galkin, V. E., Kudryashov, D. S., Rubenstein, P. A., Egelman, E. H. and Reisler, E. 2004. Actin-destabilizing factors disrupt filaments by means of a time reversal of polymerization. Proc. Natl Acad. Sci. USA 101, 17664–17668.

Osborne, L. R., Martindale, D., Scherer, S. W., Shi, X.-M., Huizenga, J., Heng, H. H. Q., Costa, T., Pober, B., Lew, L., Brinkman, J., Rommens, J., Koop, B. and Tsui, L.-C. 1996. Identification of genes from a 500-kb region at 7q11.23 that is commonly deleted in Williams syndrome patients. Genomics 36, 328–336.

Otsuki, Y., Tanaka, M., Yoshii, S., Kawazoe, N., Nakaya, K. and Sugimura, H. 2001. Tumor metastasis suppressor nm23H1 regulates Rac1 GTPase by interaction with Tiam1. Proc. Natl Acad. Sci. USA 98, 4385–4390.

Palazzo, A. F., Cook, T. A., Alberts, A. S. and Gundersen, G. G. 2001. mDia mediates Rho-regulated formation and orientation of stable microtubules. Nat. Cell Biol. 3, 723–729.

Palmgren, S., Vartainen, M. and Lappalainen, P. 2002. Twinfilin, a molecular mailman for actin monomers. J. Cell Sci. 115, 881–886.

Pandey, D., Goyal, P., Bamburg, J. R. and Siess, W. 2006. Regulation of LIM-kinase 1 and cofilin in thrombin-stimulated platelets. Blood 107, 575–583.

Pantaloni, D. and Carlier, M. F. 1993. How profilin promotes actin filament assembly in the presence of thymosin beta 4. Cell 75, 1007–1014.

Park, E., Na, M., Choi, J., Kim, S., Lee, J. R., Yoon, J., Park, D., Sheng, M. and Kim, E. 2003. The Shank family of postsynaptic density proteins interacts with and promotes synaptic accumulation of the beta PIX guanine nucleotide exchange factor for Rac1 and Cdc42. J. Biol. Chem. 278, 19220–19229.

Pawlak, G. and Helfman, D. M. 2001. Cytoskeletal changes in cell transformation and tumorigenesis. Curr. Opin. Genet. Dev. 11, 41–47.

Pawlak, G. and Helfman, D. M. 2002a. Post-transcriptional down-regulation of ROCKI/Rho-kinase through an MEK-dependent pathway leads to cytoskeleton disruption in Ras-transformed fibroblasts. Mol. Biol. Cell 13, 336–347.

Pawlak, G. and Helfman, D. M. 2002b. MEK mediates v-Src-induced disruption of the actin cytoskeleton via inactivation of the Rho–ROCK–LIM kinase pathway. J. Biol. Chem. 277, 26927–26933.

Pena, C. E. and Katoh, A. 1989. Intracytoplasmic eosinophilic inclusions in the neurons of the central nervous system. Acta Neuropathol. 79, 73–77.

Pendleton, A., Pope, B., Weeds, A and Koffer, A. 2003. Latrunculin B or ATP depletion induces cofilin-dependent translocation of actin into nuclei of mast cells. J. Biol. Chem. 278, 14394–14340.

Penzes, P., Beeser, A., Chernoff, J., Schiller, M. R., Eipper, B. A., Mains, R. E. and Huganir, R. L. 2003. Rapid induction of dendritic spine morphogenesis by trans-synaptic ephrinB-EphB receptor activation of the Rho-GEF kalirin. Neuron 37, 263–274.

Peoples, R., Perez-Jurado, L., Wang, Y. K., Kaplan, P. and Francke, U. 1996. The gene for replication factor C subunit 2 (RFC2) is within the 7q11.23 Williams syndrome deletion. Am. J. Hum. Genet. 58, 1370–1373.

Percival, J. M., Hughes, J. A., Brown, D. L., Schevzov, G., Heimann, K., Vrhovski, B., Bryce, N., Stow, J. L. and Gunning, P. W. 2004. Targeting of a tropomyosin isoform to short microfilaments associated with the Golgi complex. Mol. Biol. Cell 15, 268–280.

Peter-Ross, E. 2006. A new hypothesis with models for the genes and etiopathobiologies of mood disorders and schizophrenias. Mol. Psychiatry (in press).

Peters, A. and Kaiserman-Abramof, I. R. 1970. The small pyramidal neuron of the rat cerebral cortex. The perikaryon, dendrites and spines. Am. J. Anat. 127, 321–355.

Peterson, C., Kress, Y., Valle, R. and Goldman, J. E. 1988. High molecular weight microtubule-associated proteins bind to actin lattices (Hirano bodies). Acta Neuropathol. 77, 168–174.

Philimonenko, V. V., Zhao, J., Iben, S., Dingova, H., Kysela, K., Kahle, M., Zentgraf, H., Hofmann, W. A., de Lanerolle, P., Hozak, P. and Grummt, I. 2004. Nuclear actin and myosin I are required for RNA polymerase I transcription. Nat. Cell Biol. 6, 1165–1172.

Pollard, T. D., Blanchoin, L. and Mullins, R. D. 2000. Molecular mechanisms controlling actin filament dynamics in nonmuscle cells. Annu. Rev. Biophys. Biomol. Struct. 29, 545–576.

Pollard, T. D. and Borisy, G. G. 2003. Cellular motility driven by assembly and disassembly of actin filaments. Cell 112, 453–465.

Ponti, A., Machacek, M., Gupton, S. L., Waterman-Storer, C. M. and Danuser, G. 2004. Two distinct actin networks drive the protrusion of migrating cells. Science 305, 1782–1786.

Pope, B. J., Zierler-Gould, K. M., Kühne, R., Weeds, A. G. and Ball, L. J. 2004. Solution structure of human cofilin: Actin binding, pH sensitivity, and relationship to actin-depolymerizing factor. J. Biol. Chem. 279, 4840–4848.

Preus, M. 1984. The Williams syndrome: Objective definition and diagnosis. Clin. Genet. 25, 422–428.

Price, D. L., Sisodia, S. S. and Gandy, S. E. 1995. Amyloid beta amyloidosis in Alzheimer's disease. Curr. Opin. Neurol. 8, 268–274.

Pritchard, C. A., Hayes, L., Wojnowski, L., Zimmer, A., Marais, R. M. and Norman, J. C. 2004. B-Raf acts via the ROCKII/LIMK/cofilin pathway to maintain actin stress fibers in fibroblasts. Mol. Cell. Biol. 24, 5937–5952.

Prochniewicz, E., Janson, N., Thomas, D. D. and De la Cruz, E. M. 2005. Cofilin increases the torsional flexibility and dynamics of actin filaments. J. Mol. Biol. 353, 990–1000.

Purpura, D. P. 1974. Dendritic spine "dysgenesis" and mental retardation. Science 186, 1126–1128.

Ramakers, G. J. 2002. Rho proteins, mental retardation and the cellular basis of cognition. Trends Neurosci. 25, 191–199.

Rando, O. J., Zhao, K. and Crabtree, G. R. 2000. Searching for a function for nuclear actin. Trends Cell Biol. 10, 92–97.

Rando, O. J., Zhao, K., Janmey, P. and Crabtree, G. R. 2002. Phosphatidylinositol-dependent actin filament binding by the SWI/SNF-like BAF chromatin remodeling complex. Proc. Natl Acad. Sci. USA 99, 2824–2829.

Raval, G. N., Bharadwaj, S., Levine, E. A., Willingham, M. C., Geary, R. L., Kute, T. and Prasad, G. L. 2003. Loss of expression of tropomyosin-1, a novel class II tumor suppressor that induces anoikis, in primary breast tumors. Oncogene 22, 6194–6203.

Raymond, K., Bergeret, E., Avet-Rochex, A., Griffin-Shea, R. and Fauvarque, M. O. 2004. A screen for modifiers of RacGAP(84C) gain-of-function in the Drosophila eye revealed the LIM kinase Cdi/TESK1 as a downstream effector of Rac1 during spermatogenesis. J. Cell Sci. 117, 2777–2789.

Ressad, F., Didry, D., Xia, G.-X., Hong, Y., Chua, N.-H., Pantaloni, D. and Carlier, M.-F. 1998. Kinetic analysis of the interaction of actin-depolymerizing factor (ADF)/cofilin with G- and F-actins. Comparison of plant and human ADFs and effect of phosphorylation. J. Biol. Chem. 273, 20894–20902.

Reuner, K. H., Dunker, P., van der Does, A., Wiederhold, M., Just, I., Aktories, K. and Katz, N. 1996. Regulation of actin synthesis in rat hepatocytes by cytoskeletal rearrangements. Eur. J. Cell Biol. 69, 189–196.

Roberts, A. B. 1998. Molecular and cell biology of TGF-beta. Miner. Electrolyte Metab. 24, 111–119.

Rodal, A. A., Tetreault, J. W., Lappalainen, P., Drubin, D. G. and Amberg, D. C. 1999. Aip1p interacts with cofilin to disassemble actin filaments. J. Cell Biol. 145, 1251–1264.

Rogers, S. L., Wiedemann, U., Stuurman, N. and Vale, R. D. 2003. Molecular requirements for actin-based lamella formation in Drosophila S2 cells. J. Cell Biol. 162, 1079–1088.

Romero, S., Le Clainche, C., Didry, D., Egile, C., Pantaloni, D. and Carlier, M. F. 2004. Formin is a processive motor that requires profilin to accelerate actin assembly and associated ATP hydrolysis. Cell 119, 419–429.

Roovers, K. and Assoian, R. K. 2003. Effects of rho kinase and actin stress fibers on sustained extracellular signal-regulated kinase activity and activation of G(1) phase cyclin-dependent kinases. Mol. Cell. Biol. 23, 4283–4294.

Roovers, K., Klein, E. A., Castagnino, P. and Assoian, R. K. 2003. Nuclear translocation of LIM kinase mediates Rho–Rho kinase regulation of cyclin D1 expression. Dev. Cell 5, 273–284.

Rosenblatt, J., Agnew, B. J., Abe, H., Bamburg, J. R. and Mitchison, T. 1997. Xenopus actin depolymerizing factor/cofilin (XAC) is responsible for the turnover of actin filaments in Listeria monocytogenes tails. J. Cell Biol. 136, 1323–1332.

Rosenblatt, J., Peluso, P. and Mitchison, T. J. 1995. The bulk of unpolymerized actin in Xenopus egg extracts is ATP-bound. Mol. Biol. Cell 6, 227–236.

Rosso, S., Peretti, D., Bollati, F., Sumi, T., Nakamura, T., Quioraga, S., Ferreira, A. and Cáceres, A. 2004. LIMK1 regulates Golgi dynamics, traffic of Golgi-derived vesicles, and process extension in primary cultured neurons. Mol. Biol. Cell 15, 3433–3449.

Rossoll, W., Jablonka, S., Andreassi, C., Kröning, A.-K., Karle, K., Monani, U. R. and Sendtner, M. 2003. Smn, the spinal muscular atrophy-determining gene product, modulates axon growth and localization of beta-actin mRNA in growth cones of motoneurons. J. Cell Biol. 163, 801–812.

Samstag, Y., Eckerskorn, C., Wesselborg, S., Genning, S., Wallich, R. and Meuer, S. C. 1994. Costimulatory signals for human T-cell activation induce nuclear translocation of pp19/cofilin. Proc. Natl Acad. Sci. USA 91, 4494–4498.

Samstag, Y. and Nebl, G. 2003. Interaction of cofilin with the serine phosphatases PP1 and PP2A in normal and neoplastic human T lymphocytes. Adv. Enzyme Regul. 43, 197–211.

Sanger, J. W., Sanger, J. M., Dreis, T. E. and Jockusch, B. M. 1980. Reversible translocation of cytoplasmic actin into the nucleus caused by dimethyl sulfoxide. Proc. Natl Acad. Sci. USA 77, 5268–5272.

Sarmiere, P. D. and Bamburg, J. R. 2004. Regulation of the neuronal actin cytoskeleton by ADF/cofilin. J. Neurobiol. 58, 103–117.

Schevzov, G., Bryce, N. S., Almonte-Baldonado, R., Joya, J., Lin, J. J., Hardeman, E., Weinberger, R. and Gunning, P. 2005. Specific features of neuronal size and shape are regulated by tropomyosin isoforms. Mol. Biol. Cell 16, 3425–3437.

Schlech, W. F. III, Lavigne, P. M., Bortolussi, R. A., Allen, A. C., Haldane, E. V., Wort, A. J., Hightower, A. W., Johnson, S. E., King, S. H., Nicholls, E. S. and Broome, C. V. 1983. Epidemic listeriosis-evidence for transmission by food. N. Engl. J. Med. 308, 203–206.

Schmidt, M. L., Lee, V. M. and Trojanowski, J. Q. 1989. Analysis of epitopes shared by Hirano bodies and neurofilament proteins in normal and Alzheimer's disease hippocampus. Lab. Invest. 60, 513–522.

Color Plates

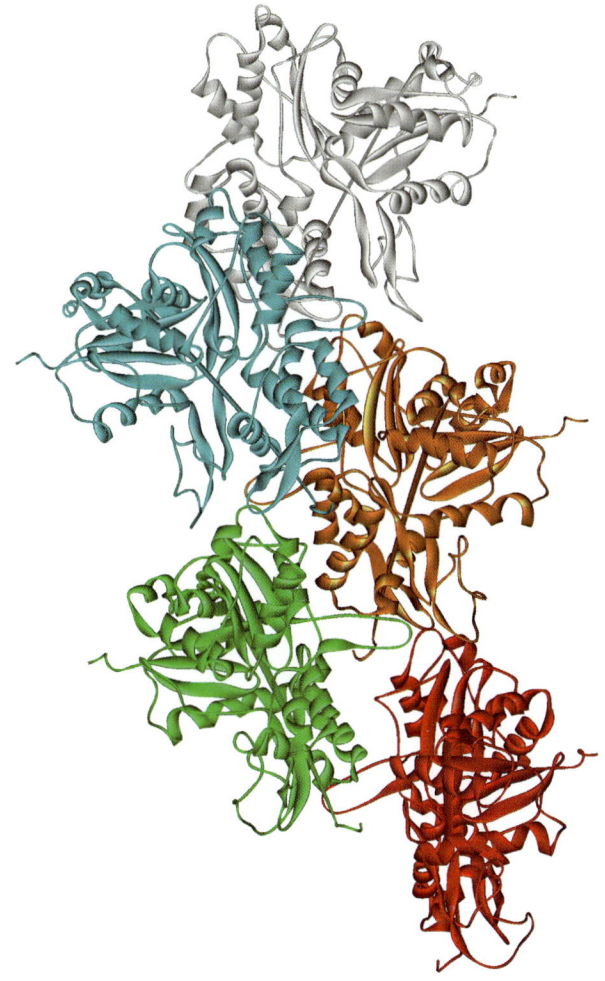

FIG. 1-3. A model of the actin filament (Lorenz et al. 1993). The single-start left-handed helix starts from the monomer in red going to green, orange, cyan then gray. The two right-handed helices start for the first helix from the monomer in red going to orange then gray, while the second helix is from the monomer in green going to cyan. The gray and cyan monomers are localized at the barbed end of the filament; the red and green monomers are localized at the pointed end of the actin filament. Reproduced from dos Remedios et al. 2003 with permission from the American Physiological Society

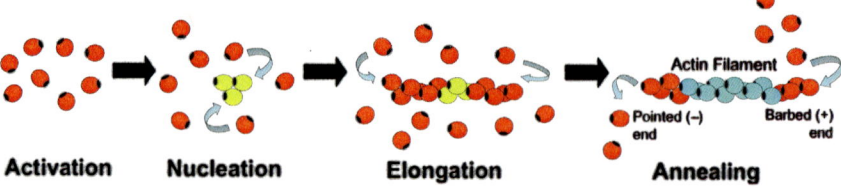

FIG. 1-4. The polymerization of actin consists of four steps – activation, nucleation, elongation, and annealing. See text for details. Image courtesy of Ms. Lan Kim Nguyen (University of Sydney)

Furthermore, globular actin is diffuse and faintly stained. Scale bars, 5 μm. (**c–e**). Transmission electron micrographs of whole mount formaldehyde prefixed and detergent-extracted hepatic endothelial cells. (**c**) Low magnification showing the nuclear area (*N*) and surrounding extracted cytoplasm. Note that the sieve plates are well defined by a darker border (*arrow*) and that inside the sieve plates fenestrae (*arrowhead*) can be observed. Scale bar, 5 μm. (**d**) High magnification micrograph of a hepatic endothelial cell treated with 100 nM dihydrohalichondramide for 30 min, showing a fenestrae-forming center (*asterisk*) (FFC) to which rows (*arrow*) of fenestrae (as indicated by *artificial colors*) with increasing diameter are connected. Scale bar, 200 nm. (**e**) Low magnification image illustrating a hepatic endothelial cell treated with 100 nM dihydrohalichondramide for 120 min, showing huge fenestrated areas (*arrowhead*), sieve plates (*arrow*). Compare with (**c**) for the difference. Scale bar, 5 μm. (**f**) Effect of the different antiactin-binding drugs tested on the number of fenestrae per micrometer squared (nF μm^{-2}) in time. From this graph, we can conclude that all agents increase the number of fenestrae, although at a different rate and different maximum

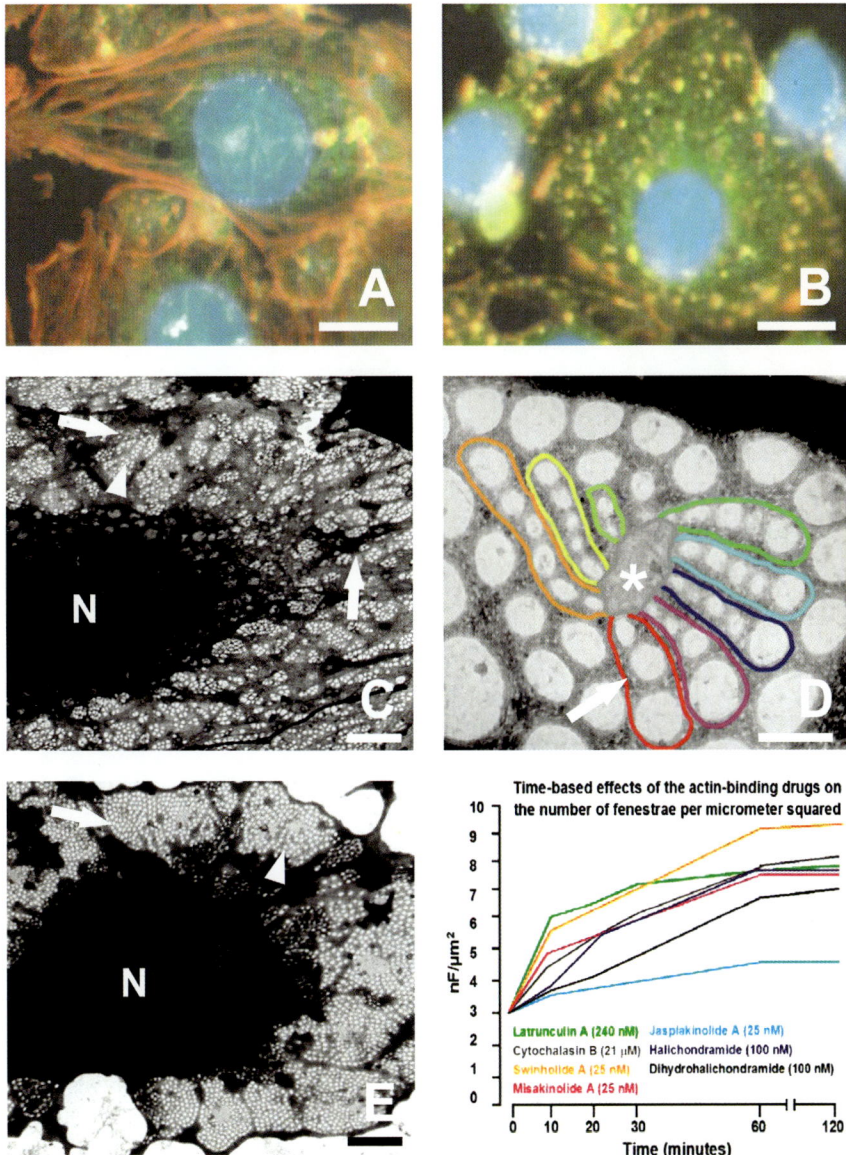

Fig. 3-2. Actin-mediated changes and numerical fenestrae dynamics in hepatic endothelial cells. (**a–b**) Fluorescence micrographs showing the effect of dihydrohalichondramide on actin organization in hepatic endothelial cells when compared to control cells, monitored with rhodamine-phalloidin (filamentous-actin/red) and fluorescein-DNase I staining (globular actin/green). *Blue color* represents the nuclei stained with DAPI. (**a**) Filamentous-actin distribution in control cells shows the presence of cytoplasmic stress fibers and peripheral bands of actin bundles that line the cell margin. Globular actin is mainly localized in the perinuclear region. (**b**) Hepatic endothelial cells treated with 100 nM of dihydrohalichondramide for 10 min show loss of filamentous-actin bundles and appearance of brightly stained filamentous-actin patches.

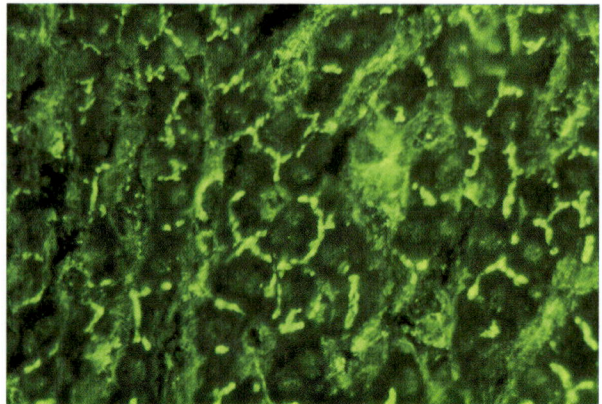

FIG. 4-2. Immunofluorescence photomicrograph of antiactin serum on frozen section of mouse liver, showing the typical "polygonal" pattern given by reactivity of serum with submembranous actin in hepatocytes

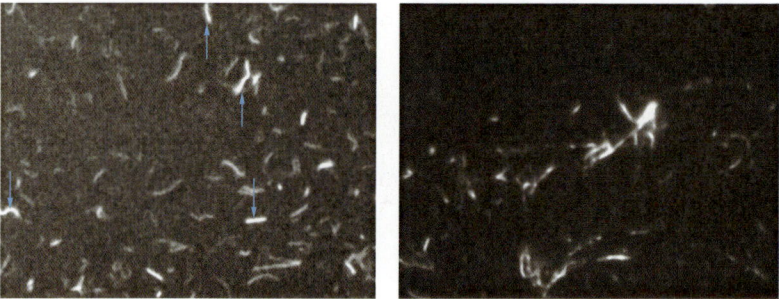

FIG. 4-4. *In vitro* motility images. (a) Control image showing motile actin filaments (indicated by *arrows*); (b) after washing with IgG in M buffer, actin filaments are immobilized and clumped

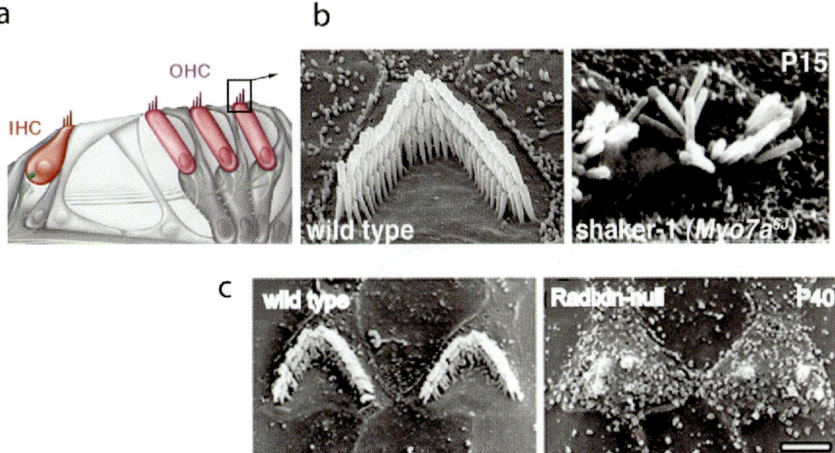

FIG. 5-4. Actin-binding proteins and structural organization of the sensory stereocilia. (a) Mammalian organ of Corti, the sensory element of the inner ear, contains inner and outer hair cells (IHC and OHC), which form numerous mechanosensory stereocilia on their surface. In OHCs of wild-type mice stereocilia form well-organized parallel arrays (b and c, visualized using scanning electron microscopy). In myosin7-mutant mice (Shaker-1, b) or mice lacking actin-bundling protein radixin (c) stereocilia are disorganized and reduced in number. Parts a and b are reproduced with permission from the Company of Biologists (El-Amraoui and Petit 2005), and part c with permission from the Rockefeller University Press (Kitajiri et al. 2004)

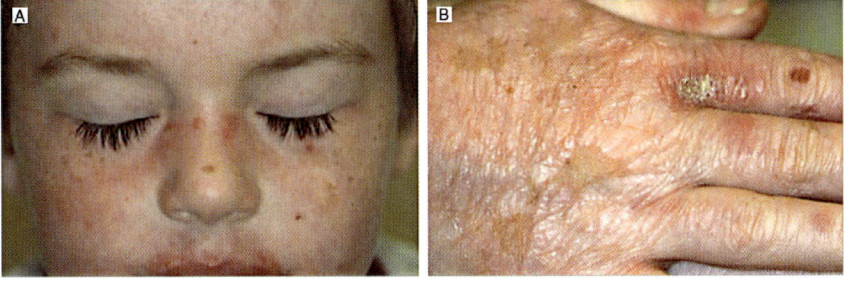

FIG. 5-5. Mutations in actin-ECM crosslinker kindlin-1 result in Kindler syndrome, a skin-blistering disease. A Kindler syndrome patient exhibits healing blisters on the nose and facial telangiectasia (red blotches created by dilated capillaries) (a) and fine wrinkling of the skin and thickened brownish patches of the outer layer of skin (keratoses) on the forearms and hands (b). Reproduced with permission from Burch et al. (2006). Copyright AMA

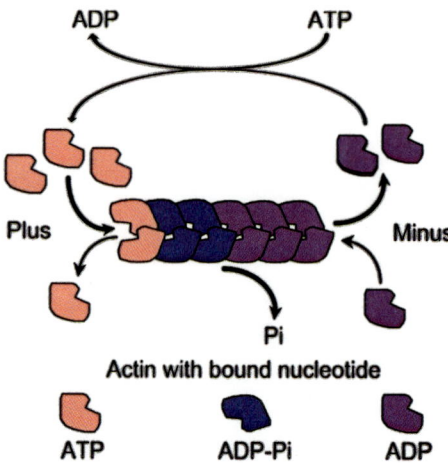

FIG. 6-1. Actin treadmilling. The major pool of actin monomers in a cell is ATP-actin and this assembles onto free ends of growing filaments as long as it exceeds the critical concentration necessary to sustain growth. ATP is hydrolyzed rapidly after assembly and inorganic phosphate is released more slowly. At steady state shown here, actin filaments treadmill because the monomer pool exceeds the critical concentration for assembly on the barbed ends of filaments but is below the concentration required for pointed end growth

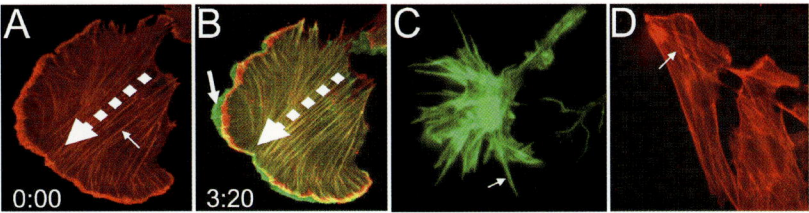

FIG. 6-2. Major F-actin structures in nonmuscle cells. (**a**) Fluorescent-phalloidin-stained lamellipodium at leading edge of chick cardiac fibroblast with *bold arrow* showing direction of migration and *small arrow* showing a graded polarity actin bundle, which differs from a stress fiber in that polarities of actin filaments within the bundle change between the front and rear of the migrating cell. (**b**) A second image of the cell shown in (**a**), pseudocolored in green, and overlayed on image in (**a**) to show new lamellipodial extension (*arrow: bright green*) that occurred over the 3 min 20 s time period. (**c**) Filopodia (microspikes) on a neuronal growth cone (*arrow*). (**d**) Stress fiber (*arrow*) in a cultured nonpolarized and nonmigrating cell stained with rhodamine-phalloidin. Long smooth membrane on left of cell with *arrow* indicates tension in the filament caused by contraction along the filament bundle (images courtesy of Lubna Tahtamouni)

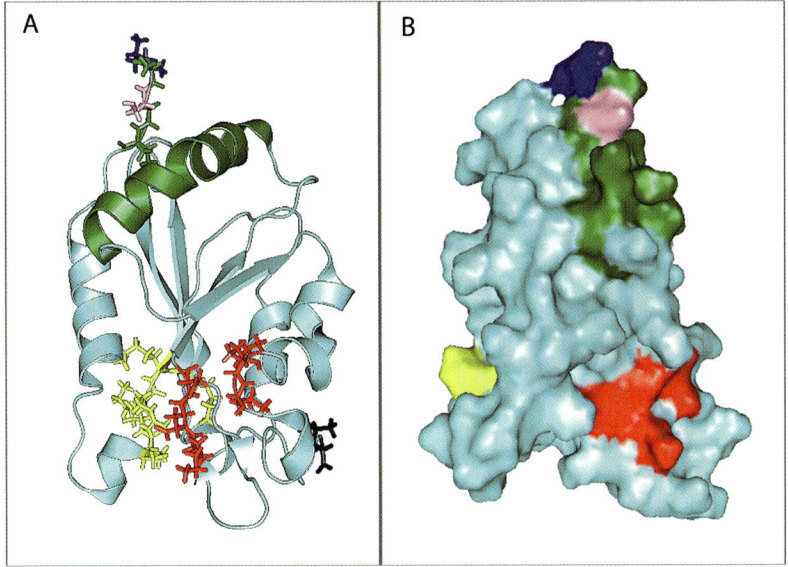

FIG. 6-3. Ribbon and space-filling model of human cofilin. (**a**) The ribbon structure of human cofilin is shown looking directly on the actin-binding face with the N-terminus (*dark blue*), the single serine phosphorylation site (*pink*), the nuclear localization signal (*yellow*), the G-actin and upper F-actin subunit major binding helix (*green*), the second lower F-actin subunit-binding residues (*red*), and the C-terminus (*black*). (**b**) Same color scheme used in (**a**) is used here for the space-filling model but the molecule is rotated so that the actin-binding surface is on the right side

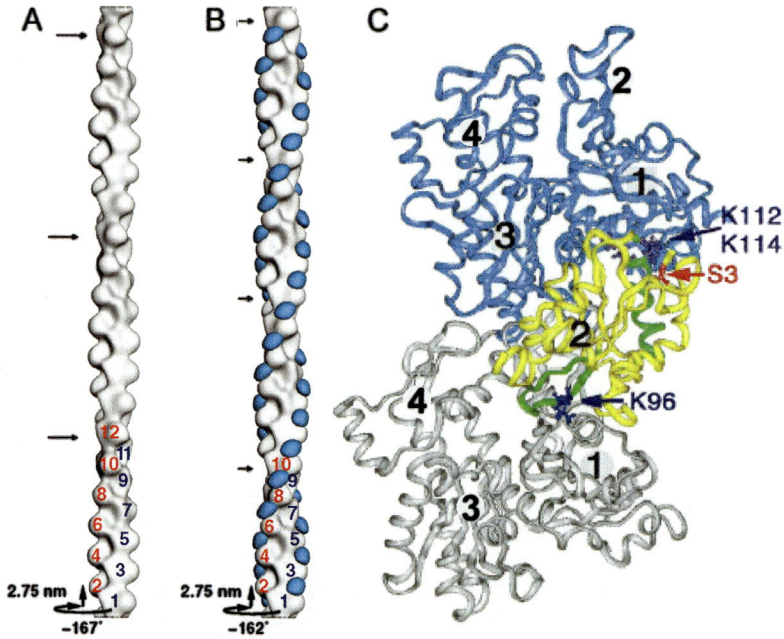

FIG. 6-4. Effects of AC binding on F-actin conformation. Structural (low resolution) model of naked F-actin (**a**) and F-actin saturated with cofilin (*blue*) (**b**). A higher resolution model (**c**) showing presumed fit of cofilin on two actin subunits of a filament. The subdomains (1–4) of each actin subunit are labeled. Model is based on molecular dynamic simulations for docking of cofilin to F-actin (see Pope et al. 2004)

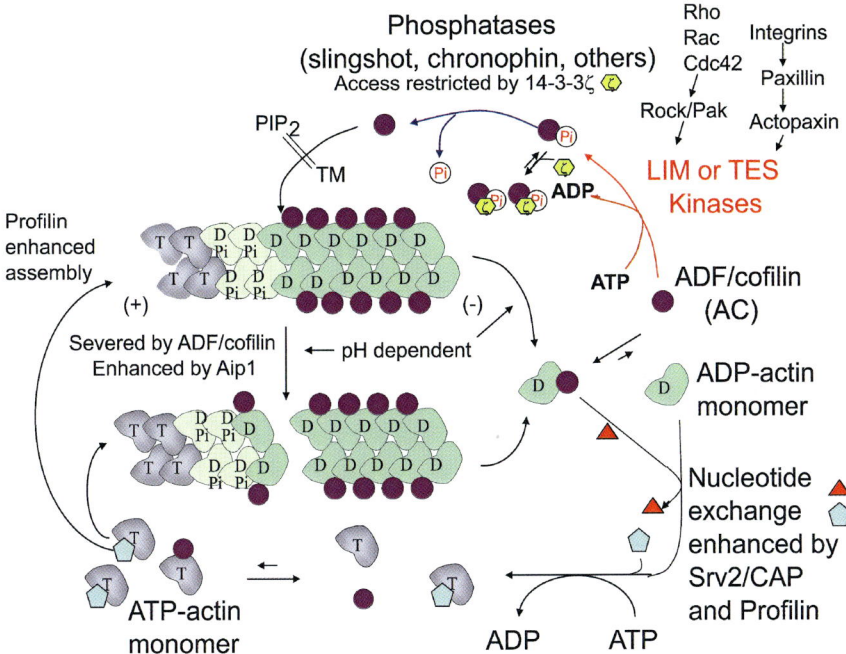

Fig. 6-5. The actin-dynamizing cycle of ADF/cofilin. Most AC proteins enhance the off-rate of subunits from the filament pointed end and sever filaments. Severing can be enhanced by Aip1. Severing creates more pointed ends from which depolymerization can occur, but also more barbed ends that can promote nucleated growth of filaments if conditions are right. Phosphorylation of metazoan AC proteins by kinases that are regulated downstream of receptors or adhesion molecules inactivates them, whereas dephosphorylation by phosphatases activates them. The phosphorylated AC may be targeted to its site of activation by 14-3-3 scaffolding proteins, which may also protect phosphorylated AC from dephosphorylation by general phosphatases. AC binding to membrane PIP$_2$ inhibits its binding to actin. Many isoforms of TM compete with AC proteins for F-actin binding, although one isoform may cooperate with AC to enhance turnover. Nucleotide exchange on the actin monomer released from the filament is inhibited by AC but can be enhanced by Srv2/CAP and/or profilin, the latter of which helps sequester actin monomers and targets them to the plus ends of F-actin, especially those filaments utilizing formins for enhancing growth

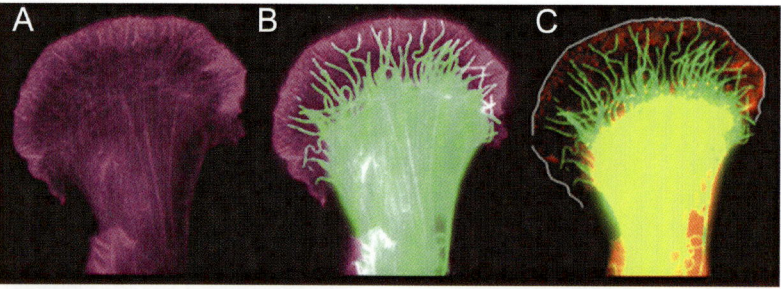

FIG. 6-6. Microtubules penetrating the lamellipodium of the leading edge of a polarized migrating cell spatially regulate AC activity. (**a**) F-actin in a polarized migrating chick cardiac fibroblast stained with a fluorescent phalloidin. (**b**) Microtubules of same cell stained with antibody to β-tubulin. (**c**) Immunostaining of phosphoAC (*red*) and microtubules (*green*) in leading edge of lamellipodium from the same cell. Total AC staining throughout the lamellipodium is relatively constant (not shown) and thus the compartmentalization of pAC to the leading edge means that the region of most active AC is toward the rear of the lamellipodium where microtubules penetrate. Loss of microtubules or capture of microtubules rapidly alters zone of active AC (image courtesy of Louise Cramer)

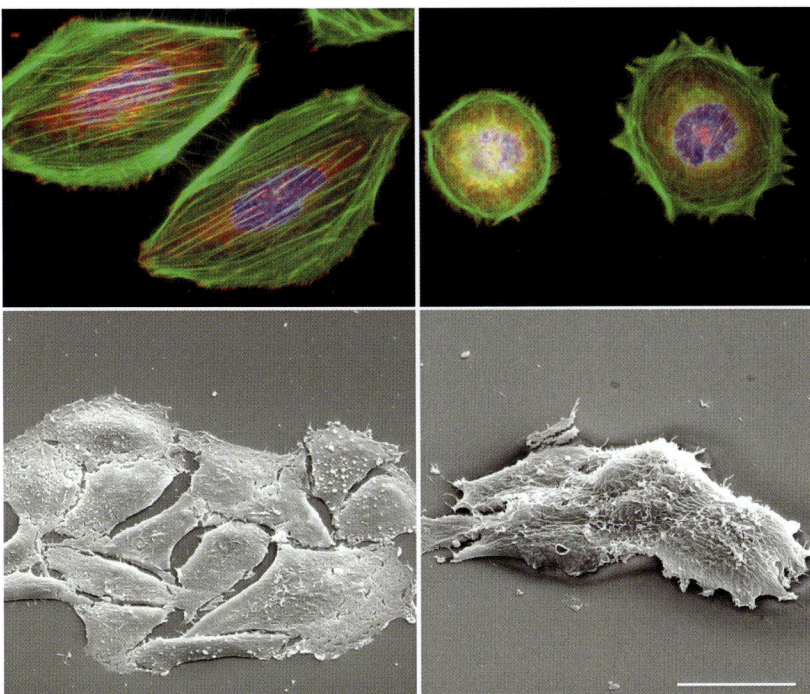

Fig. 6-7. Mesenchymal to epithelial transition in cultured osteosarcoma cells (Saos2) following expression of Pax3. *Left*: Cells with mesenchymal morphology have actin stress fibers (*green*) anchored at focal adhesions (*red*). Nucleus is stained with DAPI (*blue*). Scanning electron micrograph of similar cells is shown in panel below. Cells are well spread. *Right*: Three days after infection with adenovirus expressing Pax3, the Saos2 cells have undergone a distinctive epithelial morphological change with cortical bands of F-actin (*green*) and loss of focal adhesions. Cells grow in more cuboidal shapes, are taller, and tend to grow more on top of each other as seen in the scanning electron micrograph. Microvillar protrusions are apparent on the apical surface of these cells (image courtesy of O'Neil Wiggan; see Wiggan, Shaw and Bamburg 2006)

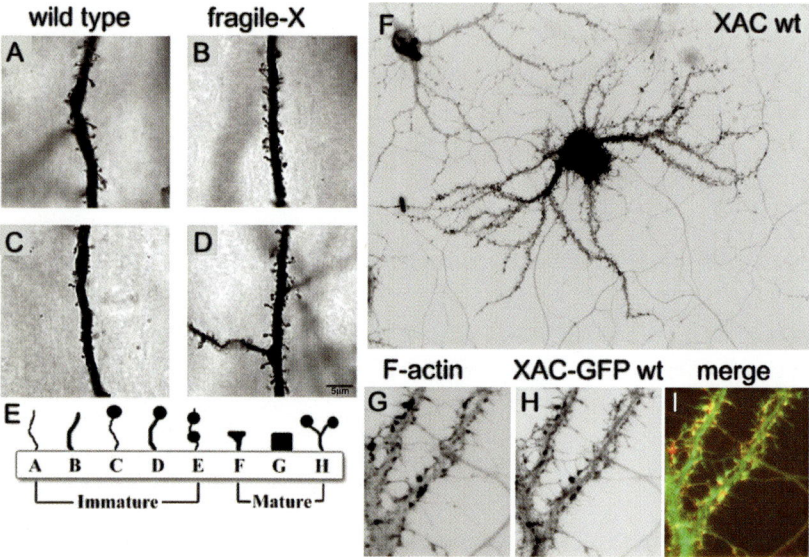

FIG. 6-8. Morphologies of hippocampal dendritic spines. Spine structures on apical dendrites of neurons from young (**a**) and adult (**c**) wild-type mouse and young (**b**) and adult (**d**) Fragile X mouse (adapted from Galvez and Greenough 2005). Normal spines have various morphologies, but many are compact and have a mushroom or stubby appearance (F and G in panel (**e**)), whereas spines from Fragile X neurons remain more immature in their morphologies (classification scheme adopted from Irwin et al. 2002). (**f**) Inverted fluorescence image of Texas-red phalloidin stained F-actin in cultured rat hippocampal neuron infected with adenovirus expressing *Xenopus* ADF/cofilin–GFP 3 days before fixation at 21 days in culture. (**g–i**) Higher magnification of inverted fluorescence images showing (**g**) location of F-actin (spines are well labeled), and XAC–GFP (**h**) as well as an overlay (F-actin in *red* and XAC–GFP in *green*). Some, but not all, regions containing F-actin also show substantial XAC–GFP

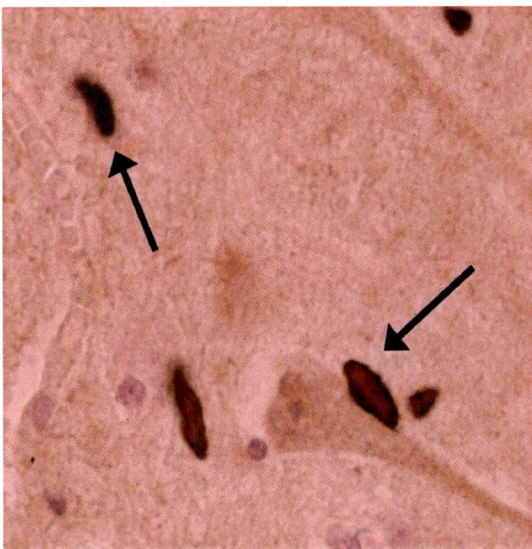

FIG. 6-10. Hirano bodies in paraffin section of brain from patient with Alzheimer's disease. Section was deparaffinized, stained with hematoxylin and eosin, and then immunocytochemically stained for AC (*brown reaction product*). Hirano bodies (*arrows*) are prevalent throughout the hippocampus and frontal cortex

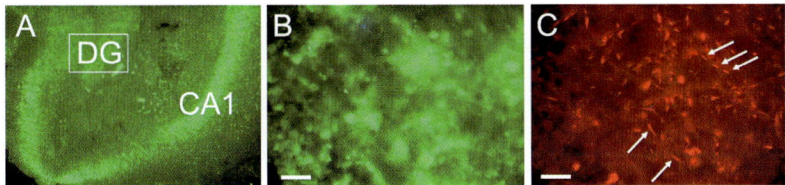

FIG. 6-11. Alzheimer's $A\beta_{1-42}$ oligomers induce AC–actin rods in cells of organotypic hippocampal slice cultures. A mouse hippocampal organotypic slice cultured 8 days was treated for 1 h by direct application of 2 μM $A\beta_{1-42}$ oligomer over the slice and then for another 47 h after transfer of the peptide to the culture well below the membrane supporting the slice. Slices were then fixed in 4% paraformaldehyde, permabilized in cold methanol, and immunostained for amyloid beta (**a, b**) and AC (**c**). (**b, c**) Higher magnification images of the dentate gyrus (DG) shown boxed in (**a**). Rods are quite prevalent and are often in linear arrays (e.g., *arrows*) similar to their organization in dissociated neuronal cultures (see Maloney et al. 2005)

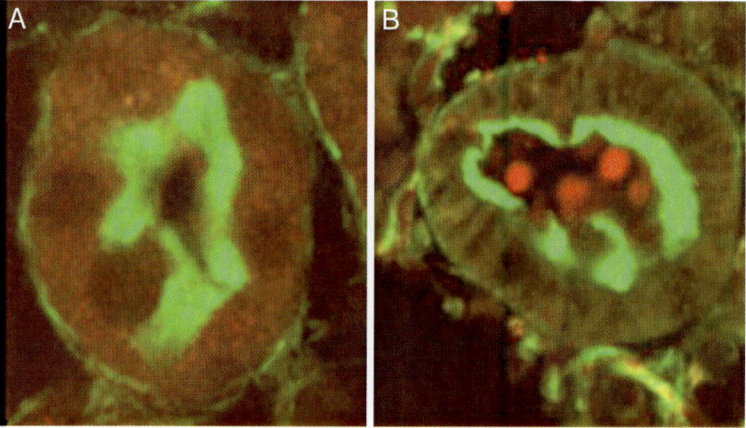

FIG. 6-13. Images of rat kidney proximal tubules before and after 25 min of ischemia. Sections of the tubules were stained with fluorescent phalloidin (*green*) for F-actin, which is greatly enriched in the apical microvilli lining the lumen. (**a**) Before ischemia, immunostaining for AC (*red*) shows it to be diffuse in the cytoplasm of proximal tubule cells. (**b**) After ischemia, vesicles that are found in the lumen of the kidney, which arise from blebbing of the proximal tubule apical membrane, are enriched in AC as well as G-actin (not shown) (see Schwartz et al. 1999)

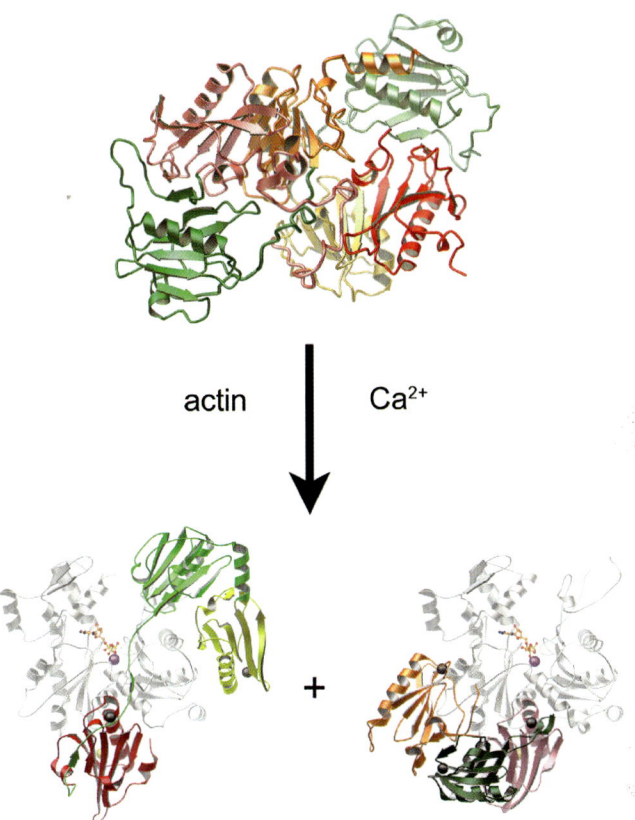

FIG. 7-1. Activation of gelsolin. Inactive gelsolin folds into a compact globular structure in which the actin binding sites are masked (PDB ID 1D0N, Burtnick et al. 1997). The individual domains are colored: (G1) red, (G2) light green, (G3) yellow, (G4) pink, (G5) dark green, and (G6) orange. The helical extension at the C-terminus of G6 constitutes part of the tail latch that reaches across to contact the long helix of G2. A structure for intact activated gelsolin is not yet available, but the structures of the activated individual N-terminal (PDB ID 1RGI, Burtnick et al. 2004) and C-terminal (PDB ID 1H1V, Choe et al. 2002) halves of gelsolin, each bound to an actin monomer, are known. The gelsolin domains are colored as above, and actin is presented in *gray*, with its bound ATP shown in *ball-and-stick format*. Bound calcium ions are represented with *spheres*. The two complex structures can be used to construct models of a gelsolin cap on the barbed end of an actin filament (Burtnick et al. 2004)

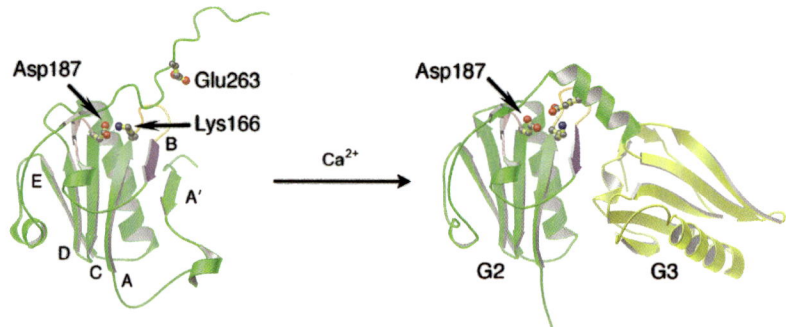

FIG. 7-2. The roles of Asp187 in resting and activated gelsolin. Prior to activation, Asp187 in domain G2 of plasma gelsolin forms a salt bridge to Lys166, which contributes to the stability of the core β-sheet in G2 (excised from PDB ID 1D0N, Burtnick et al. 1997). Strand B of this sheet (*colored purple*) contains the Arg172–Ala173 peptide bond that is susceptible to cleavage by furin in the D187N and D187Y mutant gelsolins. As a part of the structural rearrangements that occur on activation of gelsolin, the A′ strand is peeled from the edge of the core sheet in G2 and a new contact surface is created between G2 and G3 (excised from PDB ID 1RGI, Burtnick et al. 2004). The net result is that the Arg172–Ala173 peptide bond remains obscured to proteolytic attack. In the mutant gelsolins, the lost charge at position 187 may adversely influence the efficiency of the transition between inactive and activated states, enabling furin an opportunity to approach and cut the target bond, initiating the cascade of events that produces the 71-residue amyloidogenic peptides

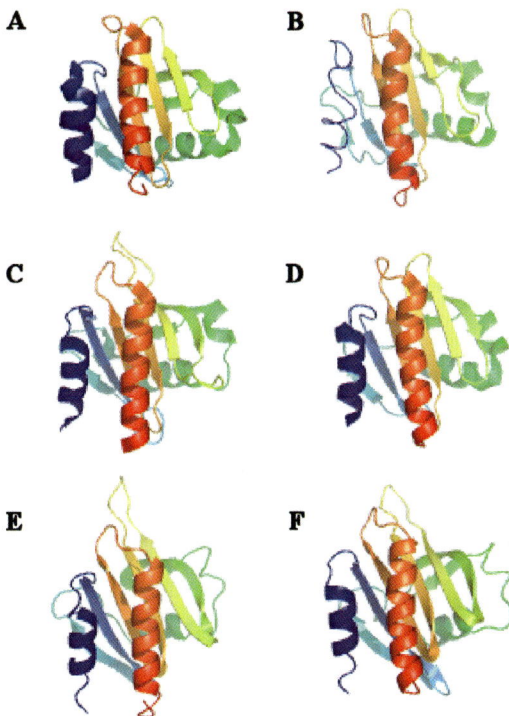

FIG. 8-2. Ribbon representation of various isoforms of profilin. (**a**) *Acanthamoeba castellanii* profilin II (pdbID: 1F2K). (**b**) Allergen profilins HEVB8 and BETV2 (pdbID: 1G5U). (**c**) Human profilin 2 (pdbID: 1D1J). (**d**) Yeast profilin 1(pdbID: 1K0K). (**e**) Human profilin 1 (pdbID: 1FIK). (**f**) Bovine profilin 1 (pdbID: 1PNE). The graphics were obtained using PyMol DeLano Scientific LLC

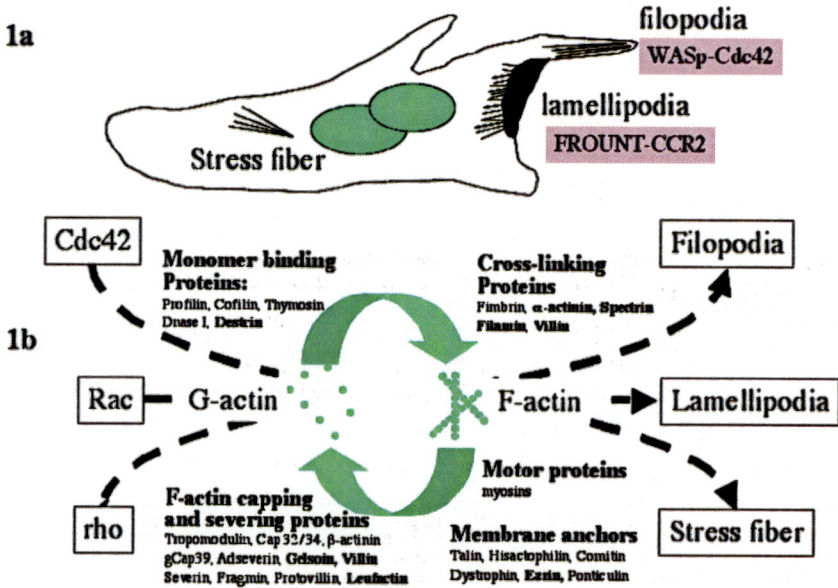

1a

filopodia
WASp-Cdc42

lamellipodia
FROUNT-CCR2

Stress fiber

1b

Cdc42

Monomer binding
Proteins:
Profilin, Cofilin, Thymosin
Dnase I, Destrin

Cross-linking
Proteins
Fimbrin, α-actinin, Spectrin
Filamin, Villin

Filopodia

Rac — G-actin

F-actin → Lamellipodia

Motor proteins
myosins

F-actin capping
and severing proteins
Tropomodulin, Cap 32/34, β-actinin
gCap39, Adseverin, Gelsoin, Villin
Severin, Fragmin, Protovillin, Leufactin

Membrane anchors
Talin, Hisactophilin, Comitin
Dystrophin, Ezrin, Ponticulin

rho

Stress fiber

FIG. 11-1. The structural organization of the microfilament network on cell membranes
(**a**) as well as the equilibrium between monomeric G-actin and filamentous F-actin
(**b**) are regulated by actin-binding proteins (ABPs) and small GTP binding proteins.
They either bind to monomeric actin thus inhibiting polymerization, or they cap,
sever, anchor, crosslink or move actin filaments by binding to the ends or along the
filaments. Active forms of each small GTP-binding protein display specific interactions with the microfilament network and form cell filopodia, lamelliopodia, and
stress-fiber formations (**a**, **b**)

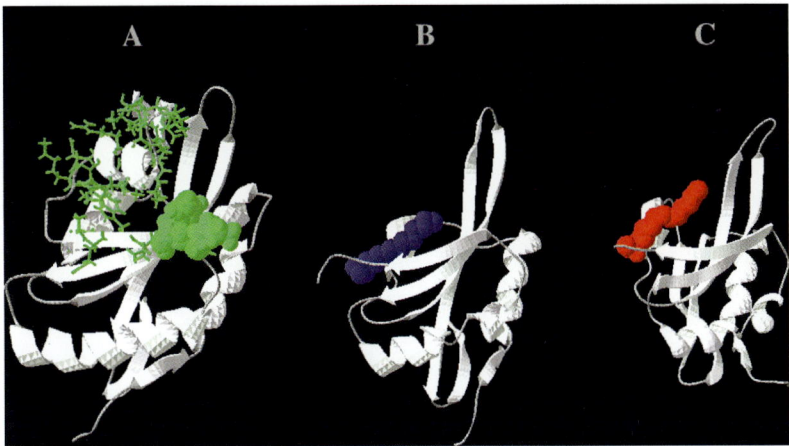

FIG. 12-1. (**a**) NMR structure of chick cofilin (PDB file 1TVJ) with residues perturbed by the binding of PIP2. K132 and H133 are shown spaced-filled and were shown to be direct contact points. (**b**) Crystal structure of *Arabidopsis thaliana* ADF1 (PDB file 1F7S) showing single molecule of cocrystallized LDAO. (**c**) Crystal structure of fission yeast cofilin (PDB file 2I2Q) showing single molecule of cocrystallized LDAO

Schochet, S. S. Jr., Lampert, P. W. and Linderberg, R. 1968. Fine structure of the Pick and Hirano bodies in a case of Pick's disease. Acta Neuropathol. 11, 330–337.

Schochet, S. S. Jr. and McCormick, W. F. 1972. Ultrastructure of Hirano bodies. Acta Neuropathol. 21, 50–60.

Schubert, F. R., Tremblay, P., Mansouri, A., Faisst, A. M., Kammandel, B., Lumsden, A., Gruss, P. and Dietrich, S. 2001. Early mesodermal phenotypes in splotch suggest a role for Pax3 in the formation of epithelial somites. Dev. Dyn. 222, 506–521.

Schuyler, S. C. and Pellman, D. 2001. Microtubule "plus-end-tracking proteins": The end is just the beginning. Cell 105, 421–424.

Schwartz, N., Hosford, M., Sandoval, R. M., Wagner, M. C., Atkinson, S. J., Bamburg, J. and Molitoris, B. A. 1999. Ischemia activates actin depolymerizing factor: Role in proximal tubule microvillar actin alterations. Am. J. Physiol. 276, 544–551.

Sechi, A. S. and Wehland, J. 2004. ENA/VASP proteins: Multifunctional regulators of actin cytoskeleton dynamics. Front. Biosci. 9, 1294–1310.

Segev, I. and London, M. 2000. Untangling dendrites with quantitative models. Science 290, 744–750.

Sells, M. A., Boyd, J. T. and Chernoff, J. 1999. p21-activated kinase 1 (Pak1) regulates cell motility in mammalian fibroblasts. J. Cell Biol. 145, 837–849.

Serpinskaya, A. S., Denisenko, O. N., Gelfand, V. I. and Bershadsky, A. D. 1990. Stimulation of actin synthesis in phalloidin-treated cells. Evidence for autoregulatory control. FEBS Lett. 277, 11–14.

Setoguti, T., Esumi, H., Shimizu, T. 1974. Specific organization of intracytoplasmic filaments in the dog testicular interstitial cell. Cell Tissue Res. 148, 493–497.

Shen, X., Ranallo, R., Choi, E. and Wu, C. 2003. Involvement of actin-related proteins in ATP-dependent chromatin remodeling. Mol. Cell 12, 147–155.

Shestakova, E. A., Singer, R. H. and Condeelis, J. 2001. The physiological significance of beta -actin mRNA localization in determining cell polarity and directional motility. Proc. Natl Acad. Sci. USA 98, 7045–7050.

Shi, S. H., Hayashi, Y., Petralia, R. S., Zaman, S. H., Wenthold, R. J., Svoboda, K. and Malinow, R. 1999. Rapid spine delivery and redistribution of AMPA receptors after synaptic NMDA receptor activation. Science 284, 1811–1816.

Shi, S. H., Jan, L. Y. and Jan, Y. N. 2003. Hippocampal neuronal polarity specified by spatially localized mPar3/mPar6 and PI 3-kinase activity. Cell 112, 63–75.

Shi, Y. and Massague, J. 2003. Mechanisms of TGF-beta signaling from cell membrane to the nucleus. Cell 113, 685–700.

Shin, D. H., Lee, E., Chung, Y. H., Mun, G. G., Park, J., Lomax, M. I. and Oh, S. H. H. 2004. Subcellular localization of WD40 repeat 1 protein in PC12 rat pheochromocytoma cells. Neuorsci. Lett. 367, 399–403.

Shirao, T., Kojima, N. and Obata, K. 1992. Cloning of drebrin A and induction of neurite-like processes in drebrin-transfected cells. Neuroreport 3, 109–112.

Shuler, H., Mueller, A. K. and Matuschewski, K. 2005. A Plasmodium actin-depolymerizing factor that binds exclusively to actin monomers. Mol. Biol. Cell 16, 4013–4023.

Silverman-Gavrila, R. V. and Forer, A. 2000. Evidence that actin and myosin are involved in the poleward flux of tubulin in metaphase kinetochore microtubules of crane-fly spermatocytes. J. Cell Sci. 113, 597–609.

Sima, A. A. and Hinton, D. 1983. Hirano-bodies in the distal symmetric polyneuropathy of the spontaneously diabetic BB-Wistar rat. Acta Neurol. Scand. 68, 107–112.

Sisodia, S. S. and Price, D. L. 1995. Role of the beta-amyloid protein in Alzheimer's disease. FASEB J. 9, 366–370.

Smith, R. S., Hawes, N. L., Kuhlmann, S. D., Heckenlively, J. R., Chang, B., Roderick, T. H. and Sundberg, J. P. 1996. Corn1: A mouse model for corneal surface disease and neovascularization. Invest. Ophthalmol. Vis. Sci. 37, 397–404.

Somma, M. P., Fasulo, B., Cenci, G., Cundari, E. and Gatti, M. 2002. Molecular dissection of cytokinesis by RNA interference in Drosophila cultured cells. Mol. Biol. Cell 13, 2448–2460.

Somogyi, P., Freund, T. F., Wu, J. Y. and Smith, A. D. 1983a. The section-Golgi impregnation procedure. 2. Immunocytochemical demonstration of glutamate decarboxylase in Golgi-impregnated neurons and in their afferent synaptic boutons in the visual cortex of the cat. Neuroscience 9, 475–490.

Somogyi, P., Kisvarday, Z. F., Martin, K. A. and Whitteridge, D. 1983b. Synaptic connections of morphologically identified and physiologically characterized large basket cells in the striate cortex of cat. Neuroscience 10, 261–294.

Song, H. and Poo, M. 2001. The cell biology of neuronal navigation. Nat. Cell Biol. 3, 81–88.

Soosairajah, J., Maiti, S., Wiggan, O., Sarmiere, P., Moussi, N., Sarcevic, B., Sampath, R., Bamburg, J. R. and Bernard, O. 2005. Interplay between components of a novel LIM kinase-slingshot phosphatase complex regulates cofilin. EMBO J. 24, 473–486.

Spacek, J. and Harris, K. M. 1997. Three-dimensional organization of smooth endoplasmic reticulum in hippocampal CA1 dendrites and dendritic spines of the immature and mature rat. J. Neurosci. 17, 190–203.

Steeg, P. S., Bevilacqua, G., Pozzatti, R., Liotta, L. A. and Sobel, M. E. 1988. Altered expression of NM23, a gene associated with low tumor metastatic potential, during adenovirus 2 Ela inhibition of experimental metastasis. Cancer Res. 48, 6550–6554.

Stokin, G. B., Lillo, C., Falzone, T. L., Brusch, R. G., Rockenstein, E., Mount, S. L., Raman, R., Davies, P., Masliah, E., Williams, D. S. and Goldstein, L. S. 2005. Axonopathy and transport deficits early in the pathogenesis of Alzheimer's disease. Science 307, 1282–1288.

Strzelecka-Golaszewska, H. and Drabikowski, W. 1968. Studies on the exchange of G-actin-bound calcium with bivalent cations. Biochim. Biophys. Acta 162, 581–595.

Subramaniam, V., Vincent, I. R. and Jothy, S. 2005. Upregulation and dephosphorylation of cofilin: Modulation by CD44 variant isoform in human colon cancer cells. Exp. Mol. Pathol. 79, 187–193.

Sukezane, T., Oneyama, C., Kakumoto, K., Shibutani, K., Hanafusa, H. and Akagi, T. 2005a. Human diploid fibroblasts are resistant to MEK/ERK-mediated disruption of the actin cytoskeleton and invasiveness stimulated by Ras. Oncogene 24, 8216.

Sukezane, T., Oneyama, C., Kakumoto, K., Shibutani, K., Hanafusa, H. and Akagi, T. 2005b. Human diploid fibroblasts are resistant to MEK/ERK-mediated disruption of the actin cytoskeleton and invasiveness stimulated by Ras. Oncogene 24, 5648–5655.

Sumi, T., Matsumoto, K. and Nakamura, T. 2001. Specific activation of LIM kinase 2 via phosphorylation of threonine 505 by ROCK, a Rho-dependent protein kinase. J. Biol. Chem. 276(1), 670–676.

Sumi, T., Matsumoto, K. and Nakamura, T. 2002. Mitosis-dependent phosphorylation and activation of LIM-kinase 1. Biochem. Biophys. Res. Commun. 290, 1315–1320.

Sumi, T., Matsumoto, K., Shibuya, A. and Nakamura, T. 2001. Activation of LIM kinases by myotonic dystrophy kinase-related Cdc42-binding kinase alpha. J. Biol. Chem. 276, 23092–23096.

Sun, H. Q., Kwiatkowska, K. and Yin, H. L. 1995. Actin monomer binding proteins. Curr. Opin. Cell Biol. 7, 202–210.

Suter, D. M. and Forscher, P. 2000. Substrate-cytoskeletal coupling as a mechanism for the regulation of growth cone motility and guidance. J. Neurobiol. 44, 97–113.

Sutoh, K. and Mabuchi, I. 1989. End-label fingerprintings show that an N-terminal segment of depactin participates in interaction with actin. Biochemistry 28, 102–106.

Suurna, M. V., Ashworth, S. L., Hosford, M., Sandoval, R. M., Wean, S. E., Shah, B. M., Bamburg, J. R. and Molitoris, B. A. 2006. Cofilin mediates ATP depletion-induced endothelial cell actin alterations. Am. J. Physiol. Renal Physiol. 290, 1398–1407.

Suyama, E., Wadhwa, R., Kawasaki, H., Yaguchi, T., Kaul, S. C., Nakajima, M. and Taira, K. 2004. LIM kinase-2 targeting as a possible anti-metastasis therapy. J. Gene Med. 6, 357–363.

Svitkina, T. M. and Borisy, G. G. 1999. Arp2/3 complex and actin depolymerizing factor/cofilin in dendritic organization and treadmilling of actin filament array in lamellipodia. J. Cell Biol. 145, 1009–1026.

Svoboda, K., Tank, D. W. and Denk, W. 1996. Direct measurement of coupling between dendritic spines and shafts. Science 272, 716–719.

Takagi, T., Konishi, K. and Mabuchi, I. 1988. Amino acid sequence of starfish oocyte depactin. J. Biol. Chem. 263, 3097–3102.

Takahashi, H., Funakoshi, H. and Nakamura, T. 2003. LIM-kinase as a regulator of actin dynamics in spermatogenesis. Cytogenet. Genome. Res. 103, 290–298.

Takahashi, T., Koshimizu, U., Abe, H., Obinata, T. and Nakamura, T. 2001. Functional involvement of Xenopus LIM kinases in progression of oocyte maturation. Dev. Biol. 229, 554–567.

Takahashi, H., Koshimizu, U., Miyazaki, J.-I. and Nakamura, T. 2002. Impaired spermatogenic ability of testicular germ cells in mice deficient in the LIM-kinase 2 gene. Dev. Biol. 241, 259–272.

Takahashi, H., Koshimizu, U. and Nakamura, T. 1998. A novel transcript encoding truncated LIM kinase 2 is specifically expressed in male germ cells undergoing meiosis. Biochem. Biophys. Res. Commun. 249, 138–145.

Takayama, S., Bimston, D. N., Matsuzawa, S., Freeman, B. C., Aime-Sempe, C., Xie, Z., Morimoto, R. I. and Reed, J. C. 1997. BAG-1 modulates the chaperone activity of Hsp70/Hsc70. EMBO J. 16, 4887–4896.

Tanaka, K., Nishio, R., Haneda, K. and Abe, H. 2005. Functional involvement of Xenopus homologue of ADF/cofilin phosphatase, slingshot (XSSH), in the gastrulation movement. Zool. Sci. 22, 955–969.

Tanaka, K., Okubo, Y. and Abe, H. 2005. Involvement of slingshot in the Rho mediated dephosphorylation of ADF/cofilin during Xenopus cleavage. Zool. Sci. 22, 971–984.

Tanzi, R. E and Bertram, L. 2005. Twenty years of the Alzheimer's disease amyloid hypothesis: A genetic perspective. Cell 120, 545–555.

Tassabehji, M., Read, A. P., Newton, V. E., Harris, R., Balling, R., Gruss, P. and Strachan, T. 1992. Waardenburg's syndrome patients have mutations in the human homologue of the Pax-3 paired box gene. Nature 355, 635–636.

Tavazoie, S. F., Alvarez, V. A., Ridenour, D. A., Kwiatkowski, D. J. and Sabatini, B. L. 2005. Regulation of neuronal morphology and function by the tumor suppressors Tsc1 and Tsc2. Nat. Neurosci. 8, 1727–1734.

Thirion, C., Stucka, R., Mendel, B., Gruhler, A., Jaksch, M., Nowak, K. J., Binz, N., Laing, N. G., Lochmuller, H. 2001. Characterization of human muscle type cofilin (CFL2) in normal and regenerating muscle. Eur J. Biochem. 263, 3473–3482.

Tilney, L. G., Connelly, P. S. and Portnoy, D. A. 1990. Actin filament nucleation by the bacterial pathogen, *Listeria monocytogenes*. J. Cell Biol. 111, 2979–2988.

Tomonaga, M. 1974. Ultrastructure of Hirano bodies. Acta Neuropathol. 28, 365–366.

Toshima, J., Toshima, J. Y., Amano, T., Yang, N., Narumiya, S. and Mizuno, K. 2001b. Cofilin phosphorylation by protein kinase testicular protein kinase 1 and its role in integrin-mediated actin reorganization and focal adhesion formation. Mol. Biol. Cell 12, 1131–1145.

Toshima, J., Toshima, J. Y., Takeuchi, K., Mori, R. and Mizuno, K. 2001a. Cofilin phosphorylation and actin reorganization activities of testicular protein kinase 2 and its predominant expression in testicular Sertoli cells. J. Biol. Chem. 276, 31449–31458.

Tsai, R. J., Sun, T. T. and Tseng, S. C. 1990. Comparison of limbal and conjunctival autograft transplantation in corneal surface reconstruction in rabbits. Ophthalmology 97, 446–455.

Turgeon, P. W., Nauheim, R. C., Roat, M. I., Stopak, S. S. and Thoft, R. A. 1990. Indications for keratoepithelioplasty. Arch. Ophthalmol. 108, 233–236.

Vallotton, P., Gupton, S. L., Waterman-Storer, C. M., Danuser, G. 2004. Simultaneous mapping of filamentous actin flow and turnover in migrating cells by quantitative fluorescent speckle microscopy. Proc. Natl Acad. Sci. USA 101, 9660–9665.

Van Aelst, L. and Cline, H. T. 2004. Rho GTPases and activity-dependent dendrite development. Curr. Opin. Neurobiol. 14, 297–304.

Vardouli, L., Moustakas, A. and Stournaras, C. 2005. LIM-kinase 2 and cofilin phosphorylation mediate actin cytoskeleton reorganization induced by transforming growth factor-beta. J. Biol. Chem. 280, 11448–11457.

Varga, A. E., Stourman, N. V., Zheng, Q., Safina, A. F., Quan, L., Li, X., Sossey-Alaoui, K. and Bakin, A. V. 2005. Silencing of the Tropomyosin-1 gene by DNA methylation alters tumor suppressor function of TGF-beta. Oncogene 24, 5043–5052.

Vartiainen, M. K., Mustonen, T., Matilla, P. K., Ojala, P. J., Thesleff, I., Partanen, J. and Lappalainen, P. 2002. The three mouse actin-depolymerizing factor/cofilins evolved to fulfill cell-type-specific requirements for actin dynamics. Mol. Biol. Cell 13, 183–194.

Vartiainen, M. K., Sarkkinen, E. M., Matilainen, T., Salminen, M. and Lappalainen, P. 2003. Mammals have two twinfilin isoforms whose subcellular localizations and tissue distributions are differentially regulated. J. Biol. Chem. 278, 34347–34355.

Vaughn, J. E. 1989. Fine structure of synaptogenesis in the vertebrate central nervous system. Synapse 3, 255–285.

Vazquez, L. E., Chen, H. J., Sokolova, I., Knuesel, I. and Kennedy, M. B. 2004. SynGAP regulates spine formation. J. Neurosci. 24, 8862–8872.

Verkerk, A. J., Pieretti, M., Sutcliffe, J. S., Fu, Y. H., Kuhl, D. P., Pizzuti, A., Reiner, O., Richards, S., Victoria, M. F., Zhang, F. P., Eussen, B. E., van Ommen, G.-J. B., Blonden, L. A. J., Riggins, G. J., Chastain, J. L., Kunst, C. B., Galjaard, H., Caskey, C. T., Nelson, D. L., Oostra, B. A. and Warran, S. T. 1991. Identification of a gene (FMR-1) containing a CGG repeat coincident with a breakpoint cluster region exhibiting length variation in fragile X syndrome. Cell 65, 905–914.

Verma, P., Chierzi, S., Codd, A. M., Campbell, D. S., Meyer, R. L., Holt, C. E. and Fawcett, J. W. 2005. Axonal protein synthesis and degradation are necessary for efficient growth cone regeneration. J. Neurosci. 25, 331–342.

Visa, N., 2005. Actin in transcription. Actin is required for transcription by all three RNA polymerases in the eukaryotic cell nucleus. EMBO. Rep. 6, 218–219.

Voegtli, W. C., Madrona, A. Y. and Wilson, D. K. 2003. The structure of Aip1p, a WD repeat protein that regulates cofilin-mediated actin depolymerization. J. Biol. Chem. 278, 34373–34379.

Vrhovski, B., Schevzov, G., Dingle, S., Lessard, J. L., Gunning, P. and Weinberger, R. P. 2003. Tropomyosin isoforms from the gamma gene differing at the C-terminus are spatially and developmentally regulated in the brain. J. Neurosci. Res. 72, 373–383.

Wachtel, M., Frei, K., Ihler, E., Fontana, A., Winterhalter, K. and Gloor, S. M. 1999. Occludin proteolysis and increased permeability in endothelial cells through tyrosine phosphatase inhibition. J. Cell Sci. 112, 4347–4356.

Wagner, M. C. and Molitoris, B. A. 1997. ATP depletion alters myosin I beta cellular location in LLC-PK1 cells. Am. J. Physiol. 272, 1680–1690.

Walikonis, R. S., Jensen, O. N., Mann, M., Provance, D. W. Jr., Mercer, J. A. and Kennedy, M. B. 2000. Identification of proteins in the postsynaptic density fraction by mass spectrometry. J. Neurosci. 20, 4069–4080.

Wallgren-Pettersson, C., Arjomaa, P. and Holmberg, C. 1990. Alpha-actinin and myosin light chains in congenital nemaline myopathy. Pediatr. Neurol. 6, 171–174.

Walther, C., Guenet, J. L., Simon, D., Deutsch, U., Jostes, B., Goulding, M. D., Plachov, D., Balling, R. and Gruss, P. 1991. Pax: A murine multigene family of paired box-containing genes. Genomics 11, 424–434.

Wang, Y. L. 1985. Exchange of actin subunits at the leading edge of living fibroblasts: Possible role of treadmilling. J. Cell Biol. 101, 597–602.

Wang, Y., Shibasaki, F. and Mizuno, K. 2005. Calcium signal-induced cofilin dephosphorylation is mediated by Slingshot via calcineurin. J. Biol. Chem. 280, 12683–12689.

Waschke, J., Baumgartner, W., Adamson, R. H., Zeng, M., Aktories, K., Barth, H., Wilde, C., Curry, F. E. and Drenckhahn, D. 2004a. Requirement of Rac activity for maintenance of capillary endothelial barrier properties. Am. J. Physiol. Heart Circ. Physiol. 286, 394–401.

Waschke, J., Curry, F. E., Adamson, R. H. and Drenckhahn, D. 2005. Regulation of actin dynamics is critical for endothelial barrier functions. Am. J. Physiol. Heart Circ. Physiol. 288, 1296–1305.

Waschke, J., Drenckhahn, D., Adamson, R. H. and Curry, F. E. 2004b. Role of adhesion and contraction in Rac 1-regulated endothelial barrier function in vivo and in vitro. Am. J. Physiol. Heart Circ. Physiol. 287, 704–711.

Waschke, J., Drenckhahn, D., Adamson, R. H. and Curry, F. E. 2004c. cAMP protects endothelial barrier functions by preventing Rac-1 inhibition. Am. J. Physiol. Heart Circ. Physiol. 287, 2427–2433.

Watanabe, N., Madaule, P., Reid, T., Ishizaki, T., Watanabe, G., Kakizuka, A., Saito, Y., Nakao, K., Jockusch, B. M. and Narumiya, S. 1997. p140mDia, a mammalian homolog of Drosophila diaphanous, is a target protein for Rho small GTPase and is a ligand for profilin. EMBO J. 16, 3044–3056.

Wegner, A. 1982. Treadmilling of actin at physiological salt concentrations. An analysis of the critical concentrations of actin filaments. J. Mol. Biol. 161, 607–615.

Wehr, R. and Gruss, P. 1996. Pax and vertebrate development. Int. J. Dev. Biol. 40, 369–377.

Weinberger, R. P., Henke, R. C., Tolhurst, O., Jeffrey, P. L. and Gunning, P. 1993. Induction of neuron-specific tropomyosin mRNAs by nerve growth factor is dependent on morphological differentiation. J. Cell Biol. 120, 205–215.

Welch, M. D., Mallavarapu, A., Rosenblatt, J. and Mitchison, T. J. 1997. Actin dynamics in vivo. Curr. Opin. Cell Biol. 9, 54–61.

Westendorf, J. J. and Koka, S. 2004. Identification of FHOD1-binding proteins and mechanisms of FHOD1-regulated actin dynamics. J. Cell. Biochem. 92, 29–41.

Wiggan, O., Shaw, A. E. and Bamburg, J. R. 2006. Essential requirement for Rho family GTPase signaling in Pax3 induced mesenchymal–epithelial transition. Cell Signal. 18, 1501–1514.

Williams, J. C., Barratt-Boyes, B. G. and Lowe, J. B. 1961. Supravalvular aortic stenosis. Circulation 24, 1311–1318.

Willis, D., Li, K. W., Zheng, J. Q., Chang, J. H., Smit, A., Kelly, T., Merianda, T. T., Sylvester, J., van Minnen, J. and Twiss, J. L. 2005. Differential transport and local translation of cytoskeletal, injury-response, and neurodegeneration protein mRNAs in axons. J. Neurosci. 25, 778–791.

Winder, S. J., Jess, T. and Ayscough, K. R. 2003. SCP1 encodes an actin-bundling protein in yeast. Biochem. J. 375, 287–295.

Wodarz, A. 2002. Establishing cell polarity in development. Nat. Cell Biol. 4, 39–44.

Wu, H., Reynolds, A. B., Kanner, S. B., Vines, R. R. and Parsons, J. T. 1991. Identification and characterization of a novel cytoskeleton-associated pp160src substrate. Mol. Cell. Biol. 11, 5113–5124.

Yager, M. L., Hughes, J. A., Lovicu, F. J., Gunning, P. W., Weinberger, R. P. and O'Neill, G. M. 2003. Functional analysis of the actin-binding protein, tropomyosin 1, in neuroblastoma. Br. J. Cancer 89, 860–863.

Yamada, K. M. and Geiger, B. 1997. Molecular interactions in cell adhesion complexes. Curr. Opin. Cell Biol. 9, 76–85.

Yamaguchi, H., Lorenz, M., Kempiak, S., Sarmiento, C., Coniglio, S., Symons, M., Segall, J., Eddy, R., Miki, H., Takenawa, T. and Condeelis, J. 2005. Molecular mechanisms of invadopodium formation: The role of the N-WASP–Arp2/3 complex pathway and cofilin. J. Cell Biol. 168, 441–452.

Yanagisawa, N. and Goto, A. 1971. Dystonia musculorum deformans. Analysis with electromyography. J. Neurol. Sci. 13, 39–65.

Yang, N., Higuchi, O., Ohashi, K., Nagata, K., Wada, A., Kangawa, K., Nishida, E. and Mizuno, K. 1998. Cofilin phosphorylation by LIM-kinase 1 and its role in Rac-mediated actin reorganization. Nature 393, 809–812.

Yang, E., Kim, H., Shin, J.-S., Yoon, S.-J. and Choi, I.-H. 2004a. Overexpression of LIM kinase 1 renders resistance to apoptosis in PC12 cells by inhibition of caspase activation. Cell. Mol. Neurobiol. 24, 181–192.

Yang, E. J., Yoon, J.-H., Min, D. S. and Chung, K. C. 2004b. LIM kinase 1 activates cAMP-responsive element-binding protein during the neuronal differentiation of immortalized hippocampal progenitor cells. J. Biol. Chem. 279, 8903–8910.

Yap, C. T., Simpson, T. I., Pratt, T., Price, D. J. and Maciver, S. K. 2005. The motility of glioblastoma tumour cells is modulated by intracellular cofilin expression in a concentration-dependent manner. Cell Motil. Cytoskeleton 60, 153–165.

Yeoh, S., Pope, B., Mannherz, H. G. and Weeds, A. 2002. Determining the differences in actin binding by human ADF and cofilin. J. Mol. Biol. 315, 911–925.

Yokoo, T., Toyoshima, H., Miura, M., Wang, Y., Iida, K. T., Suzuki, H., Sone, H., Shimano, H., Gotoda, T., Nishimori, S., Tanaka, K. and Yamada, N. 2003.

p57Kip2 regulates actin dynamics by binding and translocating LIM-kinase 1 to the nucleus. J. Biol. Chem. 278, 52919–52923.

Yonezawa, N., Nishida, E., Iida, K., Yahara, I. and Sakai, H. 1990. Inhibition of the interactions of cofilin, destrin, and deoxyribonuclease I with actin by phospho-inositides. J. Biol. Chem. 265, 8382–8386.

Yonezawa, N., Nishida, E., Iida, K., Yahara, I. and Sakai, H. 1991. Inhibition of actin polymerization by a synthetic dodecapeptide patterned on the sequence around the actin-binding site of cofilin. J. Biol. Chem. 266, 10485–10489.

Yonezawa, N., Nishida, E., Ohba, M., Seki, M., Kumagai, H. and Sakai, H. 1989. An actin-interacting heptapeptide in the cofilin sequence. Eur. J. Biochem. 183, 235–238.

Yonezawa, N., Nishida, E. and Sakai, H. 1985. pH control of actin polymerization by cofilin. J. Biol. Chem. 260, 14410–14412.

Yoshioka, K., Foletta, V., Bernard, O. and Itoh, K. 2003. A role for LIM kinase in cancer invasion. Proc. Natl Acad. Sci. USA 100, 7247–7252.

Yuste, R., Majewska, A. and Holthoff, K. 2000. From form to function: Calcium compartmentalization in dendritic spines. Nat. Neurosci. 3, 653–659.

Zebda, N., Bernard, O., Bailly, M., Welti, S., Lawrence, D. S. and Condeelis, J. S. 2000. Phosphorylation of ADF/cofilin abolishes EGF-induced actin nucleation at the leading edge and subsequent lamellipod extension. J. Cell Biol. 151, 1119–1128.

Zhang, S., Buder, K., Burkhardt, C., Schlott, B., Görlach, M. and Grosse, F. 2002. Nuclear DNA helicase II/RNA helicase A binds to filamentous actin. J. Biol. Chem. 277, 843–853.

Zhang, S., Köhler, C., Hemmerich, P. and Grosse, F. 2004. Nuclear DNA helicase II (RNA helicase A) binds to an F-actin containing shell that surrounds the nucleolus. Exp. Cell Res. 293, 248–258.

Zhang, H., Webb, D. J., Asmussen, H., Niu, S. and Horwitz, A. F. 2005. A GIT1/PIX/Rac/PAK signaling module regulates spine morphogenesis and synapse formation through MLC. J. Neurosci. 25, 3379–3388.

Zhou, Q., Homma, K. J. and Poo, M. M. 2004. Shrinkage of dendritic spines associated with long-term depression of hippocampal synapses. Neuron 44, 749–757.

Zito, K., Knott, G., Shepherd, G. M., Shenolikar, S. and Svoboda, K. 2004. Induction of spine growth and synapse formation by regulation of the spine actin cytoskeleton. Neuron 44, 321–334.

Ziv, N. E. and Smith, S. J. 1996. Evidence for a role of dendritic filopodia in synaptogenesis and spine formation. Neuron 17, 91–102.

7
Gelsolin and Disease

Leslie D. Burtnick and Robert C. Robinson

Introduction

Gelsolin is a multifunctional regulator of actin filament formation and disassembly (reviewed in Sun et al. 1999; Kwiatkowski 1999; McGough et al. 2003; Silacci et al. 2004). It can bind to the side of an F-actin filament, sever that filament in a nonhydrolytic manner, and cap the fast growing end of one of the resultant fragments. Uncapping can be induced through the binding of specific phospholipids (see chapter by dos Remedios), for example, phosphatidylinositol 4,5-bisphosphate (PIP_2), to expose the fast growing end of a prenucleated filament.

Gelsolin operates both within cells and extracellularly, and a single gene on human chromosome 9 encodes both the cytoplasmic and secreted variants (Kwiatkowski et al. 1986). Alternative transcription initiation sites and selective RNA processing lead to distinct mRNA messages that, in turn, produce unique protein products. Plasma gelsolin, representative of the secreted form, consists of a single 755-amino acid polypeptide chain and differs in sequence from its cytoplasmic variant only in possessing a 25-amino acid N-terminal extension. In addition, extracellular gelsolin undergoes oxidation to form a disulfide bond that links Cys188–Cys201 (Wen et al. 1996).

Structure

Gelsolin is composed of six domains, named sequentially from the N-terminus as G1–G6. These appear to be the result of gene triplication followed by gene duplication, giving rise to a modular protein in which the C-terminal half, G4–G6, repeats the architecture of the N-terminal half, G1–G3 (Kwiatkowski et al. 1986; Way and Weeds 1988; Burtnick et al. 1997; Koepf et al. 1998). In the Ca^{2+}-free structure, the actin monomer binding sites on G1 and G4 and the filament side-binding site on G2 are hidden within the globular packing arrangement of the domains (reviewed in Pope, Gooch and Weeds 1997;

Burtnick, Robinson and Choe 2001). Activation is a multistep process (Kinosian et al. 1998; Robinson et al. 1999; Lueck et al. 2000; Choe et al. 2002; Kiselar et al. 2003; Burtnick et al. 2004) that involves release of the noncovalent attachments between the two halves, for example between the C-terminal tail helix and the long helix of G2, and large-scale rearrangement of the relative positions of the domains (Fig. 1).

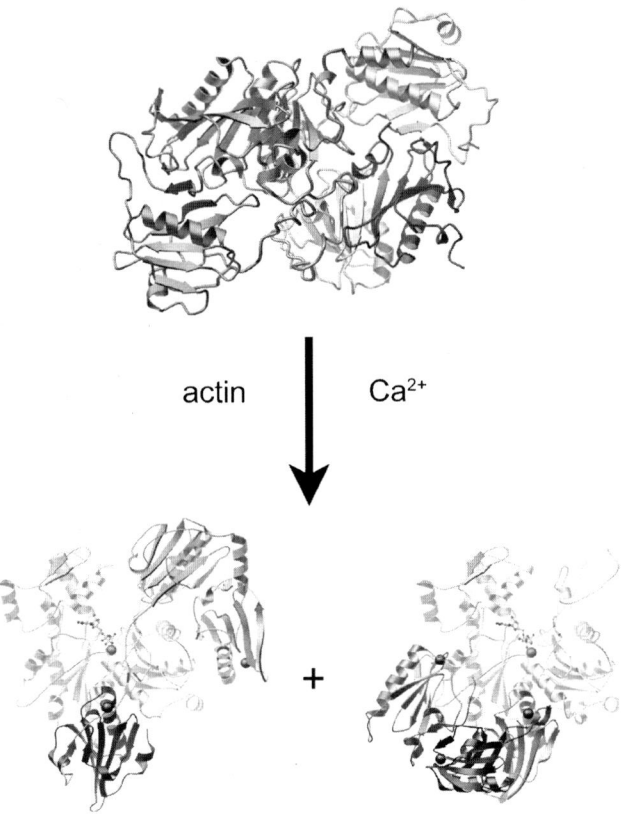

FIG. 1. Activation of gelsolin. Inactive gelsolin folds into a compact globular structure in which the actin binding sites are masked (PDB ID 1D0N, Burtnick et al. 1997). The individual domains are colored: (G1) red, (G2) light green, (G3) yellow, (G4) pink, (G5) dark green, and (G6) orange. The helical extension at the C-terminus of G6 constitutes part of the tail latch that reaches across to contact the long helix of G2. A structure for intact activated gelsolin is not yet available, but the structures of the activated individual N-terminal (PDB ID 1RGI, Burtnick et al. 2004) and C-terminal (PDB ID 1H1V, Choe et al. 2002) halves of gelsolin, each bound to an actin monomer, are known. The gelsolin domains are colored as above, and actin is presented in *gray*, with its bound ATP shown in *ball-and-stick format*. Bound calcium ions are represented with *spheres*. The two complex structures can be used to construct models of a gelsolin cap on the barbed end of an actin filament (Burtnick et al. 2004) (*See Color Plates*)

Gelsolin Genes

Given the wide tissue distribution of gelsolin and its intimate involvement with dynamic remodeling of actin architectures, it is intriguing that deletion of gelsolin genes in mice is not lethal and does not preclude apparently normal development (Witke et al. 1995). To date, the simplest explanation of these data invokes compensatory actions by other members of the gelsolin superfamily that may be present in the affected tissues, even if these molecules function at reduced activity levels. That the gelsolin knockout mice are not completely normal is demonstrated by impaired functioning of platelets, neutrophils, and fibroblasts when subjected to stresses that would require actin filament reorganization (Witke et al. 1995). Gelsolin null female mice exhibit defects in mammary gland morphogenesis subsequent to the onset of puberty (Crowley et al. 2000).

Hippocampal neurons in gelsolin null mice extend filopodia at normal rates, but exhibit impaired retraction rates when compared to wild-type controls, suggestive of a role for gelsolin in initiation of the retraction phase of filopodial activity (Lu et al. 1997). In addition, elimination of gelsolin from mice produces osteoclasts that are hypomotile, with age-related consequences that include increased bone thickness and mechanically stronger bones (Chellaiah et al. 2000).

Gelsolin-Related Amyloidosis

The strong influence of gelsolin on the ability of actin to perform its myriad of crucial functions obfuscates discovery of causal relationships between gelsolin and specific diseases. Only in the case of gelsolin-related amyloidosis is a disease state the clear outcome of a defect in gelsolin, a point mutation that results in expression of either an Asn or a Tyr in place of the normal Asp at position 187 in the amino acid sequence (Maury et al. 2000). Several other relationships have been established that ultimately tie together the ability of gelsolin to regulate actin form and function with a disease state. We categorize some of these relationships in the discussions below.

Gelsolin is one of more than a dozen distinct proteins that can undergo abnormal processing leading to deposition of amyloid plaques in tissues or in the extracellular matrix. The pathological connections between such proteins and those involved in Alzheimer's disease and prion disease are an ongoing area of intense investigation (Sacchettini and Kelly 2002).

Familial amyloidosis of the Finnish type (FAF) was the first heritable disease identified to have a gelsolin-based cause (Maury, Alli and Baumann 1990). FAF is a systemic polynueropathy that is dominantly inherited. Manifestations of the disease generally appear in middle age and include skin changes, as well as corneal lattice and cranial dystrophies. A point mutation in the gelsolin gene, converting a guanine at position 654 to adenine, results in the production of Asn187 in the expressed protein in place of the wild-type Asp187. Subsequently, in a family of Danish origin, an essentially indistinguishable amyloid disease was shown to be the result of a mutation of Asp187 to a Tyr residue, the consequence of a point mutation of guanine-654 in the gelsolin gene to a thymine (Maury et al. 2000).

During transit through the *trans*-Golgi network, these mutant gelsolins transiently encounter furin, which hydrolyzes the peptide bond between Arg172 and Ala173 (Chen et al. 2001). The resultant 68-kDa C-terminal fragment of gelsolin is secreted into the circulation and can be detected in cerebrospinal fluid. This large fragment can undergo subsequent cleavage by β-gelsolinase, which may be a matrix metalloproteinase (Page et al. 2005), ultimately to produce an amyloidogenic 71-residue fragment of gelsolin, Ala173–Met243. Furin attack on wild-type gelsolin is only detectable in cells in which furin is overexpressed (Kangas et al. 1996).

The susceptibility of the Arg172–Ala173 peptide bond in the mutant gelsolins is speculated to be the result of removal of the negative charge from the side chain position at 187, which is expected to contribute to the binding of a type-2 calcium ion in domain G2 during the activation of gelsolin (Zapun et al. 2000; Kazmirski et al. 2002; Choe et al. 2002; Huff et al. 2003). Subsequent elucidation of the crystal structure of the N-terminal half of wild-type gelsolin bound to actin in the presence of Ca^{2+} necessitated elaboration of this hypothesis, as no metal ion was detected in the type-2 metal ion-binding site in domain G2 (Burtnick et al. 2004). We suggest that the binding of Ca^{2+} to the relevant site in G2 is a transient event that mediates the transition between the inactive and active states of gelsolin (Fig. 2).

Mutation of Asp187 abolishes the transient binding of a calcium ion during this transition and could kinetically hang the protein in between the inactive conformation, in which the susceptible bond is protected within the core β-sheet of domain G2, and the activated state, in which the same bond is protected at the newly created interface between domains G2 and G3 (Fig. 2). With gelsolin stranded, at least temporarily, in a conformation between these two extremes, the scissile bond dangles in a precariously exposed condition. This explanation presumes an initial calcium-binding event to be a requirement for breaking apart the edge-on contacts between the core β-sheets of domains G1 and G3 observed in Ca^{2+}-free gelsolin (Burtnick et al. 1997; Robinson et al. 1999), and a calcium-release event that accompanies the formation of the G2–G3 interface observed in the activated state (Burtnick et al. 2004). The proposition also infers that the bond should be accessible to furin for a brief time during activation of wild-type gelsolin. This is consistent with the observation that overexpression of furin does lead to detectable levels of a 68-kDa cleavage product of wild-type gelsolin during transit through the *trans*-Golgi network (Kangas et al. 1996).

As inhibitors of furin and furin-like proteins are known to inhibit processing of FAF variant gelsolins to yield the 68-kDa fragment that is the source of the amyloidogenic peptide, it may be possible to alleviate the progression

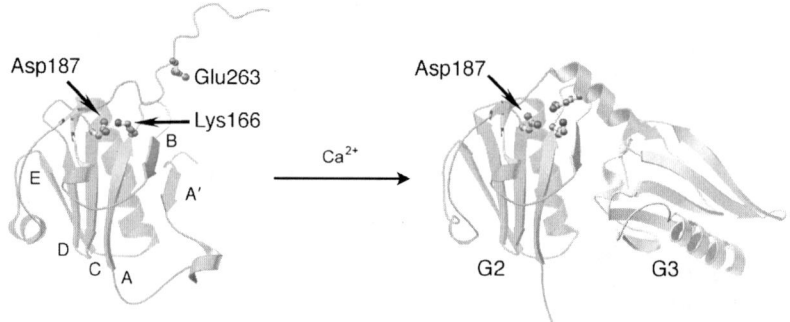

FIG. 2. The roles of Asp187 in resting and activated gelsolin. Prior to activation, Asp187 in domain G2 of plasma gelsolin forms a salt bridge to Lys166, which contributes to the stability of the core β-sheet in G2 (excised from PDB ID 1D0N, Burtnick et al. 1997). Strand B of this sheet (*colored purple*) contains the Arg172–Ala173 peptide bond that is susceptible to cleavage by furin in the D187N and D187Y mutant gelsolins. As a part of the structural rearrangements that occur on activation of gelsolin, the A′ strand is peeled from the edge of the core sheet in G2 and a new contact surface is created between G2 and G3 (excised from PDB ID 1RGI, Burtnick et al. 2004). The net result is that the Arg172–Ala173 peptide bond remains obscured to proteolytic attack. In the mutant gelsolins, the lost charge at position 187 may adversely influence the efficiency of the transition between inactive and activated states, enabling furin an opportunity to approach and cut the target bond, initiating the cascade of events that produces the 71-residue amyloidogenic peptides (*See Color Plates*)

of the gelsolin-related amyloidosis (Chen et al. 2001). However, the experimental data were collected using cultured BHK cells and the probably serious consequences of inhibiting furin on its normal secretory targets (Thomas 2002) were not addressed.

As a likely outcome of the similarities in the folding and aggregation properties of polypeptide sequences that form amyloid deposits, regardless of the protein of origin, gelsolin interacts with the soluble form of amyloid-beta protein (Aβ), which is the generator of amyloid plaques in the brains of patients with Alzheimer's disease (Ray et al. 2000). Interaction with gelsolin inhibits fibrillization of Aβ, and even shows evidence of defibrillizing preformed fibrils in a time-dependent manner. In this way, gelsolin displays activity as an antiamyloidogenic protein in plasma and cerebrospinal fluid.

Gelsolin in the Extracellular Actin Scavenging System

In an extracellular environment such as blood plasma, conditions of ionic strength, ionic composition, temperature, and pH favor the polymerized form of actin. To combat potential circulatory problems, particularly in the microvasculature, that might ensue should actin escape the confines of the cell as a result of lysis due to injury or disease, an extracellular actin scavenging system (EASS) has evolved to rapidly clear actin from the circulation. In blood plasma, the EASS consists of two proteins: the secreted variant of gelsolin and a vitamin D transport protein (DBP; also known as Gc-globulin). Gelsolin severs filamentous actin into short oligomers and DBP sequesters actin monomer units in the form of 1:1 actin–DBP complexes that are removed quickly from the circulatory system by the liver. Both gelsolin and DBP are present at moderately high levels in the plasma of all vertebrates, each at approximately 300 mg L^{-1} of plasma (reviewed in Lind et al. 1986; Lee and Galbraith 1992).

The circulating levels of EASS proteins are reduced dramatically as a result of disease (Van Baelen, Bouillon and De Moor 1980) or multiple trauma situations (e.g., automobile accidents; Dahl et al. 1998, 1999; Mounzer et al. 1999). There is a strong correlation between reduction of circulating EASS proteins and development in trauma victims of multiple organ dysfunction syndrome (MODS), including adult acute respiratory distress syndrome (ARDS). Assays for actin-bound and actin-free gelsolin and DBP in the circulation (Safer 1989; Dahl et al. 1999) can be used to assess the severity of diseases such as hepatitis and malaria, and assist in assessment of the point at which organ transplant becomes critical (reviewed in Lee and Galbraith 1992).

It is hoped that infusion of gelsolin into victims of disease or trauma who suffer a depleted EASS can extend the protection afforded by that system. For example, ARDS is a common complication in patients who have suffered a variety of insults. Sera from patients with ARDS contain F-actin and are toxic to sheep pulmonary endothelial cells, but toxicity can be moderated by pretreatment of the sera with gelsolin (Erukhimov et al. 2000). Treatment of ARDS victims with elevated oxygen levels may lead to enhanced lung damage. In mice subjected to 95% oxygen, depletion of free gelsolin levels in bronchoalveolar fluid is concurrent with hyperoxic lung injury. Repletion with recombinant plasma gelsolin significantly reduces the acute inflammatory response to hyperoxia compared to treatment with serum albumin (Christofidou-Solomidou et al. 2002). Similarly, infusion of recombinant gelsolin into rats attenuates burn injury in the lungs (Rothenbach et al. 2004). Also, gelsolin is able to dramatically reduce the viscosity of sputum from sufferers of cystic fibrosis, probably through its actin-severing activity (Vasconcellos et al. 1994).

Plasma gelsolin may have a second scavenging role in addition to its actin-related one. Toxic shock due to release of lipopolysaccharide (LPS) endotoxins from the outer walls of Gram-negative bacteria and the resultant stimulation of complex inflammatory pathways for which no effective treatment exits can cause death. Plasma gelsolin, *in vitro*, binds LPS molecules from various bacteria with high affinity and offers a possible treatment (Bucki et al. 2005). One consequence of gelsolin binding endotoxin *in vivo* may be related to toxic shock through interference with the gelsolin-dependent ability of actin filament systems to be dynamic in nature. A similar consequence involving *Salmonella* invasion protein A (Sip A) has been documented (McGhie, Hayward and Koronakis 2004). Sip A blocks actin turnover and promotes actin polymerization, at least in part, through inhibition of the severing activity of gelsolin.

Gelsolin in Cancer and Apoptosis

Gelsolin expression is curtailed in many cancer cell lines and tumors, and transfection to restore gelsolin levels suggests a tumor suppressive role for gelsolin (see chapter by Nieuw et al.). Such cases include cancers of the human breast (Asch et al. 1996; Mielnicki et al. 1999), prostate (Lee et al. 1999), bladder (Tanaka et al. 1995), colon (Furuuchi et al. 1996; Klampfer et al. 2004), lung (Sagawa et al. 2003), endometrium and ovaries (Afify and Werness 1998), and stomach (Moriya et al. 1994). Yet, a negative prognostic correlation between gelsolin expression and patient survival exists at least in some breast (Thor et al. 2001) and lung cancers (Shieh et al. 1999; Yang et al. 2006). Based on observed enhancement of motility in fibroblast cell lines that overexpress gelsolin (Cunningham, Stossel and Kwiatkowski 1991), these negative correlations could be an outcome of enhanced tumor cell motility leading to migration and metastasis. However, gelsolin overexpression in a mouse melanoma cell line seems to suppress motility and metastasis (Fujita et al. 2001). The effects of gelsolin on motility, metastasis, tumorigenicity, and mortality are not yet clear and may prove to depend not only on the cell type involved, but also on multiple factors that are still uncertain.

Cancer cells generally exhibit reduced apoptotic activity, which contributes to the lack of clarity on the relationship of gelsolin to cancer, as gelsolin is associated with both promotional and inhibitory activities in apoptosis. Caspase-3, one of the executioner group of aspartate-specific cysteine proteases activated during apoptosis triggered by caspase-9, cleaves the Asp352–Gly353 peptide bond in human cytoplasmic gelsolin (Kothakota et al. 1997). This locus is in the peptide linker that connects domains G3 and G4, so the result is to liberate the two halves of gelsolin, G1–G3 and G4–G6, to operate independently of each other. The N-terminal half of gelsolin displays microfilament-severing activity that is independent of control by Ca^{2+} and is thought to be responsible for disassembly of actin networks observed on microinjection of G1–G3, but not of G4–G6, into cells. Transient overexpression of G1–G3 induces apoptosis, while gelsolin-deficient cells display a retarded onset of apoptosis (Kothakota et al. 1997). In this way, gelsolin acts within the apoptotic scheme as a promoter of programmed cell death.

In contrast, in Jurkat cells (a human T-cell line), overexpression of gelsolin is associated with a counter-apoptotic activity (Ohtsu et al. 1997).

Such activity may be cell-specific as overexpression of gelsolin fails to limit lymphocyte apoptosis (Posey et al. 2000). Nevertheless, antiapoptotic activity of gelsolin has been confirmed elsewhere in that gelsolin blocks actin-mediated voltage-dependant anion channels and prevents mitochondrial membrane potential loss and cytochrome C release in response to excessive influx of Ca^{2+} (Kusano et al. 2000; Koya et al. 2000). Part of an explanation for apparently contradictory roles of gelsolin may come from its position in or near a variety of signaling pathways, such as those involving phospholipids (Sagawa et al. 2003). PIP_2 binds to gelsolin and can advance to form a ternary complex involving caspase-3 to inhibit apoptotic progression (Azuma et al. 2000).

While naturally occurring mutations of gelsolin are not known in regard to cancer or apoptosis, epigenetic factors may come into play. Posttranslational N-myristoylation of the C-terminal caspase-3 cleavage product of gelsolin in COS-1 cells is a requirement for its antiapoptotic activity (Sakurai and Utsumi 2006). The actin cytoskeleton certainly provides for a connection among aging, cancer, and apoptosis (Gourlay and Ayscough 2005), so it is understandable that cytoskeletal regulators, such as gelsolin, play a part. Susceptibility to cancer and resistance to apoptotic signals are both age-related and have been linked to increased expression of gelsolin (Ahn et al. 2003).

Concluding Remarks

Mutations of the secreted form of gelsolin have a clear role in the development of at least one inherited set of conditions, the gelsolin-related amyloidoses. The function of secreted gelsolin as an actin-scavenger in extracellular fluids also has a predictable influence on the progress of a group of syndromes and diseases. But, given the intimate relationship between actin and cytoplasmic gelsolin in such dynamic processes as cell proliferation, cell motility, and cell death, it is not surprising that there remains much to learn about the role of gelsolin in human disease. This convoluted issue is further complicated by the effects on major signaling pathways of the interaction of gelsolin with chemical messages that regulate a host of cellular and tissue responses. Ongoing investigations in many laboratories have revealed connections that had not been anticipated. As our understanding of these connections deepens, so will our successes in understanding the progress and control of diseases, including cancer, and of aging and death.

Acknowledgments. LDB thanks the Canadian Institutes for Health Research and the Heart and Stroke Foundation of British Columbia and Yukon for grant support, and RCR thanks A*STAR for support.

References

Afify, A. M. and Werness, B. A. 1998. Decreased expression of the actin-binding protein gelsolin in endometrial and ovarian adeno-carcinomas. Appl. Immunohistochem. 6, 30–34.

Ahn, J. S., Jang, I. -S., Kim, D. -I., Cho, K. A., Park, Y. H., Kim, K., Kwak, C. S. and Park, S. C. 2003. Aging-associated increase of gelsolin for apoptosis resistance. Biochem. Biophys. Res. Commun. 312, 1335–1341.

Asch, H. L., Head, K., Dong, Y., Natoli, F., Winston, J. S., Connolly, J. L. and Asch, B. B. 1996. Widespread loss of gelsolin in breast cancers of humans, mice and rats. Cancer Res. 56, 4841–4845.

Azuma, T., Koths, K., Flanagan, L. and Kwiatkowski, D. 2000. Gelsolin in complex with phosphatidylinositol 4,5-bisphosphate inhibits caspase-3 and -9 to retard apoptotic progression. J. Biol. Chem. 275, 3761–3766.

Bucki, R., Georges, P. C., Espinassous, Q., Funaki, M., Pastore, J. J., Chaby, R. and Janmey, P. A. 2005. Inactivation of endotoxin by human plasma gelsolin. Biochemistry 44, 9590–9597.

Burtnick, L. D., Koepf, E. K., Grimes, J. M., Jones, E. Y., Stuart, D. I., McLaughlin, P. J. and Robinson, R. C. 1997. The crystal structure of plasma gelsolin: Implications for actin severing, capping and nucleation. Cell 90, 661–670.

Burtnick, L. D., Robinson, R. C. and Choe, S. 2001. Structure and function of gelsolin. Results Probl. Cell Differ. 32, 201–211.

Burtnick, L. D., Urosev, D., Irobi, E., Narayan, K. and Robinson, R. C. 2004. Structure of the N-terminal half of gelsolin bound to actin: Roles in severing, apoptosis and FAF. EMBO J. 23, 2713–2722.

Chellaiah, M., Kizer, N., Silva, M., Alvarez, U., Kwiatkowski, D. and Hruska, K. A. 2000. Gelsolin deficiency blocks podosome assembly and produces increased bone mass and strength. J. Cell Biol. 148, 665–678.

Chen, C. -D., Huff, M., Matteson, J., Page, L., Phillips, R., Kelly, J. W. and Balch, W. E. 2001. Furin initiates gelsolin familial amyloidosis in the Golgi through a defect in Ca^{2+} stabilization. EMBO J. 20, 6277–6287.

Choe, H., Burtnick, L. D., Mejillano, M., Yin, H. L., Robinson, R. C. and Choe, S. 2002. The calcium activation of gelsolin: Insights from the 3 Å structure of the G4–G6/actin complex. J. Mol. Biol. 324, 691–702.

Christofidou-Solomidou, M., Scherpereel, A., Solomides, C. C., Christie, J. D., Stossel, T. P., Goelz, S. and DiNubile, M. J. 2002. Recombinant plasma gelsolin diminishes the acute inflammatory response to hyperoxia in mice. J. Investig. Med. 50, 54–60.

Crowley, M. R., Head, K. L., Kwiatkowski, D., Asch, H. and Asch, B. 2000. The mouse mammary gland requires the actin-binding protein gelsolin for proper ductal morphogenesis. Dev. Biol. 225, 407–423.

Cunningham, C. C., Stossel, T. P. and Kwiatkowski, D. J. 1991. Enhanced mobility in NIH 3T3 fibroblasts that overexpress gelsolin. Science 251, 1233–1236.

Dahl, B., Schiodt, F. V., Ott, P., Gvozdenovic, R., Yin, H. L. and Lee, W. M. 1999. Plasma gelsolin is reduced in trauma patients. Shock 12, 102–104.

Dahl, B., Schiodt, F. V., Ott, P., Kiaer, T., Bondesen, S. and Tygstrup, N. 1998. Gc-Globulin in the early course of multi-trauma. Crit. Care Med. 26, 285–289.

Erukhimov, J. A., Tang, Z. -L., Johnson, B. A., Donahoe, M. P., Razzack, J. A., Gibson, K. F., Lee, W. M., Wasserloos, K. J., Watkins, S. A. and Pitt, B. R. 2000. Actin-containing sera from patients with adult respiratory distress syndrome are toxic to sheep pulmonary endothelial cells. Am. J. Respir. Crit. Care Med. 162, 288–294.

Fujita, H., Okada, F., Hamada, J. -I., Hosokawa, M., Moriuchi, T., Koya, R. C. and Kuzumaki, N. 2001. Gelsolin functions as a metastasis suppressor in B16–BL6 mouse melanoma cells and requirement of the carboxyl-terminus for its effect. Int. J. Cancer 93, 773–780.

Furuuchi, K., Fujita, H., Tanaka, M., Shinohara, N., Senmaru, N., Ogiso, Y., Moriya, S., Hamada, M., Kato, H. and Kuzumaki, N. 1996. Gelsolin as a suppressor of malignant phenotype in human colon cancer. Tumor Target 2, 277–283.

Gourlay, C. W. and Ayscough, K. R. 2005. The actin cytoskeleton: A key regulator of apoptosis and ageing? Nat. Rev. Mol. Cell Biol. 6, 583–589.

Huff, M. E., Page, L. J., Balch, W. E. and Kelly, J. W. 2003. Gelsolin domain 2 Ca^{2+} affinity determines susceptibility to furin proteolysis and familial amyloidosis of Finnish type. J. Mol. Biol. 334, 119–127.

Kangas, H., Paunio, T., Kalkkinen, N., Jalanko, A. and Peltonen, L. 1996. *In vitro* expression analysis shows that the secretory form of gelsolin is the sole source of amyloid in gelsolin-related amyloidosis. Hum. Mol. Genet. 5, 1237–1243.

Kazmirski, S. L., Isaacson, R. L., An, C., Buckle, A., Johnson, C. M., Daggett, V. and Fersht, A. R. 2002. Loss of a metal-binding site in gelsolin leads to familial amyloidosis-Finnish type. Nat. Struct. Biol. 9, 112–116.

Kinosian, H. J., Newman, J., Lincoln, B., Selden, L. A., Gershman, L. C. and Estes, J. E. 1998. Ca^{2+} regulation of gelsolin activity: Binding and severing of F-actin. Biophys. J. 75, 3101–3109.

Kiselar, J. G., Janmey, P. A., Almo, S. and Chance, M. R. 2003. Visualizing the Ca^{2+}-dependent activation of gelsolin by using synchrotron footprinting. Proc. Natl Acad. Sci. USA 100, 3942–3947.

Klampfer, L., Huang, J., Sasazuki, T., Shirasawa, S. and Augenlicht, L. 2004. Oncogenic Ras promotes butyrate-induced apoptosis through inhibition of gelsolin expression. J. Biol. Chem. 279, 36680–36688.

Koepf, E. K., Hewitt, J., Vo, H., MacGillivray, R. T. A. and Burtnick, L. D. 1998. *Equus caballus* gelsolin: cDNA sequence and protein structural implications. Eur. J. Biochem. 251, 613–621.

Kothakota, S., Azuma, T., Reinhard, C., Klippel, A., Tang, J., Chu, K., McGarry, T. J., Kirschner, M. W., Koths, K., Kwiatkowski, D. J. and Williams, L. T. 1997. Caspase-3-generated fragment of gelsolin: Effector of morphological change in apoptosis. Science 278, 294–298.

Koya, R. C., Fujita, H., Shimizu, S., Ohtsu, M., Takimoto, M., Tsujimoto, Y. and Kuzumaki, N. 2000. Gelsolin inhibits apoptosis by blocking mitochondrial membrane potential loss and cytochrome c release. J. Biol. Chem. 275, 15343–15349.

Kusano, H., Shimizu, S., Koya, R. C., Fujita, H., Kamada, S., Kuzumaki, N. and Tsujimoto, Y. 2000. Human gelsolin prevents apoptosis by inhibiting apoptotic mitochondrial changes via closing VDAC. Oncogene 19, 4807–4814.

Kwiatkowski, D. J. 1999. Functions of gelsolin: Motility, signaling, apoptosis, cancer. Curr. Opin. Cell Biol. 11, 103–108.

Kwiatkowski, D. J., Stossel, T. P., Orkin, S. H., Mole, J. E., Colten, H. and Yin, H. L. 1986. Plasma and cytoplasmic gelsolins are encoded by a single gene and contain a duplicated actin binding domain. Nature 323, 455–458.

Lee, H. -K., Driscoll, D., Asch, H., Asch, B. and Zhang, P. J. 1999. Downregulated gelsolin expression in hyperplastic and neoplastic lesions of the prostate. Prostate 40, 14–19.

Lee, W. M. and Galbraith, R. M. 1992. The extracellular actin-scavenger system and actin toxicity. N. Engl. J. Med. 326, 1335–1341.

Lind, S. E., Smith, C. J., Janmey, P. A. and Stossel, T. P. 1986. Role of plasma gelsolin and the vitamin D-binding protein in clearing actin from the circulation. J. Clin. Invest. 78, 736–742.

Lu, M., Witke, W., Kwiatkowski, D. J. and Kosik, K. S. 1997. Delayed retraction of filopodia in gelsolin null mice. J. Cell Biol. 138, 1279–1287.

Lueck, A., Yin, H. L., Kwiatkowski, D. J. and Allen, P. G. 2000. Calcium regulation of gelsolin and adseverin: A natural test of the helix latch hypothesis. Biochemistry 39, 5274–5279.

Maury, C. P. J., Alli, K. and Baumann, M. 1990. Finnish hereditary amyloidosis. Amino acid sequence homology between the amyloid fibril protein and human plasma gelsolin. FEBS Lett. 260, 85–87.

Maury, C. P. J., Liljeström, M., Boysen, G., Törnroth, T., de la Chapelle, A. and Nurmiaho-Lassila, E. -L. 2000. Danish type gelsolin related amyloidosis: 654G-T mutation is associated with a disease pathogenetically and clinically similar to that caused by the 654G-A mutation (familial amyloidosis of the Finnish type). J. Clin. Pathol. 53, 95–99.

McGhie, E. J., Hayward, R. D. and Koronakis, V. 2004. Control of actin turnover by a salmonella invasion protein. Mol. Cell 13, 497–510.

McGough, A., Staiger, C. J., Mina, J. -K. and Simonetti, K. D. 2003. The gelsolin family of actin regulatory proteins: Modular structures, versatile functions. FEBS Lett. 552, 75–81.

Mielnicki, L. M., Ying, A. M., Head, K. L., Asch, H. and Asch, B. 1999. Epigenetic regulation of gelsolin expression in human breast cancer cells. Exp. Cell Res. 249, 161–176.

Moriya, S., Yanagihara, K., Fujita, H. and Kuzumaki, N. 1994. Differential expression of hsp90, gelsolin and GST-p in human gastric carcinoma cell lines. Int. J. Oncol. 5, 1347–1351.

Mounzer, K. C., Moncure, M., Smith, Y. R. and DiNubile, M. J. 1999. Relationship of admission plasma gelsolin levels to clinical outcomes in patients after major trauma. Am. J. Respir. Crit. Care Med. 160, 1673–1681.

Ohtsu, M., Sakai, N., Fujita, H., Kashiwagi, M., Gasa, S., Shimizu, S., Eguchi, Y., Tsujimoto, Y., Sakiyama, Y., Kobayashi, K. and Kuzumaki, N. 1997. Inhibition of apoptosis by the actin-regulatory protein gelsolin. EMBO J. 16, 4650–4656.

Page, L. J., Suk, J. Y., Huff, M. E., Lim, H. -J., Venable, J., Yates, J., III, Kelly, J. W., and Balch, W. 2005. Metalloendoprotease cleavage triggers gelsolin amyloidogenesis. EMBO J. 24, 4124–4132.

Pope, B. J., Gooch, J. T. and Weeds, A. G. 1997. Probing the effects of calcium on gelsolin. Biochemistry 36, 15848–15855.

Posey, S. C., Martelli, M. P., Azuma, T., Kwiatkowski, D. J. and Bierer, B. E. 2000. Failure of gelsolin overexpression to regulate lymphocyte apoptosis. Blood 95, 3483–3488.

Ray, I., Chauhan, A., Wegiel, J. and Chauhan, V. P. S. 2000. Gelsolin inhibits the fibrillization of amyloid beta-protein, and also defibrillizes its preformed fibrils. Brain Res. 853, 344–351.

Robinson, R. C., Mejillano, M., Le, V. P., Burtnick, L. D., Yin, H. L. and Choe, S. 1999. Domain movement in gelsolin: A calcium-activated switch. Science 286, 1939–1942.

Rothenbach, P. A., Dahl, B., Schwartz, J. J., O'Keefe, G. E., Yamamoto, M., Lee, W. M., Horton, J. W., Yin, H. L. and Turnage, R. H. 2004. Recombinant plasma gelsolin infusion attenuates burn-induced pulmonary microvascular dysfunction. J. Appl. Physiol. 96, 25–31.

Sacchettini, J. C. and Kelly, J. W. 2002. Therapeutic strategies for human amyloid diseases. Nat. Rev. Drug Discov. 1, 267–275.

Safer, D. 1989. An electrophoretic procedure for detecting proteins that bind actin monomers. Anal. Biochem. 178, 32–37.

Sagawa, N., Fujita, H., Banno, Y., Nozawa, Y., Katoh, H. and Kuzumaki, N. 2003. Gelsolin suppresses tumourigenicity through inhibiting PKC activation in human lung cancer cell line, PC10. Br. J. Cancer 88, 606–612.

Sakurai, N. and Utsumi, T. 2006. Posttranslational N-myristoylation is required for the anti-apoptotic activity of human tGelsolin, the C-terminal caspase cleavage product of human gelsolin. J. Biol. Chem. 281, 14288–14295.

Shieh, D. G. J., Sugarbaker, D. J., Herndon, J. and Kwiatdowski, D. J. 1999. Motility as a prognostic factor in non-small lung cancer: Role of gelsolin expression. Cancer 85, 47–57.

Silacci, P., Mazzolai, L., Gauci, C., Stergiopulos, N., Yin, H. L. and Hayoz, D. 2004. Gelsolin superfamily proteins: Key regulators of cellular functions. Cell. Mol. Life Sci. 61, 2614–2623.

Sun, H. Q., Yamamoto, M., Mejillano, M. and Yin, H. L. 1999. Gelsolin, a multifunctional actin regulatory protein. J. Biol. Chem. 274, 33179–33182.

Tanaka, M., Müllauer, L., Ogiso, Y., Fujita, H., Moriya, S., Furuuchi, K., Harabayashi, T., Shinohara, N., Koyanagi, T. and Kuzumaki, N. 1995. Gelsolin: A candidate for suppressor of human bladder cancer. Cancer Res. 55, 3228–3232.

Thomas, G. 2002. Furin at the cutting edge: From protein traffic to embryogenesis and disease. Nat. Rev. Mol. Cell Biol. 3, 753–766.

Thor, A. D., Edgerton, S. M., Liu, S., Moore, D. H., II. and Kwiatkowski, D. J. 2001. Gelsolin as a negative prognostic factor and effector of motility in erbB-2-positive epidermal growth factor receptor-positive breast cancers. Clin. Cancer Res. 7, 2415–2424.

Van Baelen, H., Bouillon, R. and De Moor, P. 1980. Vitamin D-binding protein (Gc-globulin) binds actin. J. Biol. Chem. 255, 2270–2272.

Vasconcellos, C. A., Allen, P. G., Wohl, M. E., Drazen, J. M., Janmey, P. A. and Stossel, T. P. 1994. Reduction in viscosity of cystic fibrosis sputum in vitro by gelsolin. Science 263, 969–971.

Way, M. and Weeds, A. G. 1988. Nucleotide sequence of pig plasma gelsolin. Comparison of protein sequence with human gelsolin and other actin-severing proteins shows strong homologies and evidence for large internal repeats. J. Mol. Biol. 203, 1127–1133.

Wen, D., Corina, K., Chow, E., Miller, S., Janmey, P. and Pepinsky, R. 1996. The plasma and cytoplasmic forms of human gelsolin differ in disulfide structure. Biochemistry 35, 9700–9709.

Witke, W., Sharpe, A. H., Hartwig, J. H., Azuma, T., Stossel, T. P. and Kwiatkowski, D. J. 1995. Hemostatic, inflammatory, and fibroblast responses are blunted in mice lacking gelsolin. Cell 81, 41–51.

Yang, J., Ramnath, N., Moysich, K. B., Asch, H. L., Swede, H., Alrawi, S. J., Huberman, J., Geradts, J., Brooks, J. S. J. and Tan, D. 2006. Prognostic significance of MCM2, Ki-67 and gelsolin in non-small cell lung cancer. BMC Cancer 6, 203.

Zapun, A., Grammatyka, S., Deral, G. and Vernet, T. 2000. Calcium-dependent conformational stability of modules 1 and 2 of human gelsolin. Biochem. J. 350 (Part 3), 873–888.

8
Profilin

Pierre D. J. Moens

Introduction

Profilin, a small ubiquitous nonmuscle protein of 12–14 kDa, is found in eukaryotic cells (Carlsson et al. 1977; Reichstein and Korn 1979) including plants (Valenta et al. 1992), and viruses (Machesky et al. 1994) (see chapter by Bearer). It is expressed in all eukaryotic organisms studied to date. Profilin is essential for the normal development and cytokinesis of *Dictyostelium* Amoeba (Haugwitz et al. 1994). These authors showed that in profilin-null mutants cell motility was significantly reduced and development was blocked prior to fruiting body formation. Furthermore, these cells could not be grown in shaking culture under normal conditions.

In 2001, profilin 1 was also shown to be essential for cell survival and cell division in mice (Witke et al. 2001). Specifically, profilin 1 double knockout (pfn1[ko/ko]) embryos died as early as the two-cell stage, and pfn1[ko/ko] blastocysts were not detectable. Although profilin has been intensely studied since its discovery nearly 20 years ago, its *in vivo* functions are still poorly understood. Only recently has research highlighted the role of profilin in diseases such as cancer (Janke et al. 2000; Roy and Jacobson 2004; also see chapter by Van Troys) and Parkinson's disease (Basso et al. 2004), although its involvement as an allergen has been known since the early 1990s (Valenta et al. 1991a,b). In this chapter, the role of profilin in these disorders will be reviewed and discussed.

Profilin and Its Ligands

Profilin is known to form complexes with different ligands including (1) actin monomers; (2) polyphosphoinositides (PPIs) (see chapter by dos Remedios); (3) the p85α subunit of phosphatidylinositol 3-kinase (PI3-kinase); and (4) poly-L-proline sequences including proteins such as dynamin I, clathrin, synapsin, a member of the NSF/sec18 family of proteins, Rho-associated coiled-coil kinase, and the Rac-associated protein, NAP1.

Actin

Binding of actin to profilin was first reported in 1977 by Carlsson and coworkers (1977) and this remains the function that has probably received the most attention in the last decade. Actin binds vertebrate profilin 1 tightly with a dissociation constant ranging between 0.2 and 0.1 μM under physiological ionic conditions (Perelroizen, Carlier and Pantaloni 1995). Profilin acts as a nucleotide exchange factor, replacing the ADP in G-actin with ATP and promoting actin polymerization at the barbed end of the filament (Goldschmidt-Clermont et al. 1992; Pantaloni and Carlier 1993; Vinson et al. 1998).

Profilin can facilitate actin polymerization by transporting monomers to the fast-growing ends of filaments. However, in the current model of F-actin, profilin and actin compete for the same binding site. Therefore, to add monomers to the growing end of the filament, profilin must either move to a new site on actin or completely dissociate (dos Remedios et al. 2003). In agreement with the complete dissociation, Gutsche-Perelroizen et al. (1999) did not detect the incorporation of profilin into actin filaments.

Polyphosphoinositides

It is well known that $PI(4,5)P_2$ micelles dissociate the actin:profilin complex (Lassing and Lindberg 1988) and that platelet profilin binds with high affinity to small clusters of $PI(4,5)P_2$ molecules (Goldschmidt-Clermont et al. 1990). The profilin association/dissociation constant to $PI(4,5)P_2$ micelles was determined using gel filtration. However, dissociation constants varying by >200-fold have been reported (0.13 μM (Ostrander, Gorman and Carman 1995) to 35 μM (Lu et al. 1996)).

Lu et al. (1996) proposed that the putative utility of $PI(4,5)P_2$–profilin association is twofold. First, profilin is sequestered from actin monomers, thus preventing actin–profilin complex formation which is characterized as a precursor of actin filament formation (Markey, Persson and Lindberg 1981). Second, this binding protects $PI(4,5)P_2$ from being hydrolyzed by phospholipase C-γ1, thus implicating profilin in the regulation of inositol phosphate production (Goldschmidt-Clermont et al. 1990).

$PI(3,4)P_2$ and $PI(3,4,5)P_3$, the lipid products of phosphoinositide 3-kinase (PI 3-kinase), have greater affinity for profilin than $PI(4,5)P_2$ (Lu et al. 1996). These D-3 phosphoinositides modulate the turnover of $PI(4,5)P_2$ by counteracting the inhibitory effect of profilin on phospholipase C-γ1 and by activating phospholipase C-γ1 (Lu et al. 1996; Bae et al. 1998). Therefore, Lu et al. proposed that the D-3 phosphoinositides produced in response to agonist stimulation spontaneously take over profilin from $PI(4,5)P_2$ resulting in local actin assembly.

PIP-Binding Site and Mutants of Profilin

The exact binding site of the phosphoinositides on profilin has not yet been elucidated. Yu et al. (1992) proposed that a sequence required for high-affinity phosphoinositide binding identified in gelsolin may represent a consensus sequence for binding phosphoinositide in a variety of proteins. This sequence corresponds to the region spanning residues 126–136 of human profilin (Fig. 1). Schutt et al. (1993) proposed that the four positively charged residues in the C-terminal helix of profilin might be involved in the interaction with phosphoinositides.

In 1995, Sohn et al. (1995) showed that a mutation of the Arg88 residue of human profilin reduces the profilin inhibition of phospholipase C-γ1, implicating this residue in the binding of PI(4,5)P$_2$ to profilin. Several reports (Sohn et al. 1995; Lambrechts et al. 2002; Skare and Karlsson 2002) used site-directed mutagenesis to localize the binding site for PI(4,5)P$_2$ on profilin (Table 1). While some of the mutations reduced the lipid-binding capacity of the profilin, others seem to have the opposite effect and increased its affinity for PI(4,5)P$_2$ (see Table 1). For two of the mutations (W3N and the deletion of residues P96 and T97: P96ΔT97Δ), the increase in affinity is about twofold over the wild-type isoform (Skare and Karlsson 2002).

In 2002, Lambrechts et al. (2002) and Skare and Karlsson (2002) suggested that there are two binding regions for PI(4,5)P$_2$ on human and mammalian profilin 1 (Fig. 1). One is close to the binding site of poly(L-proline) and spans residues 126–136. The other includes residue 88 and overlaps with the

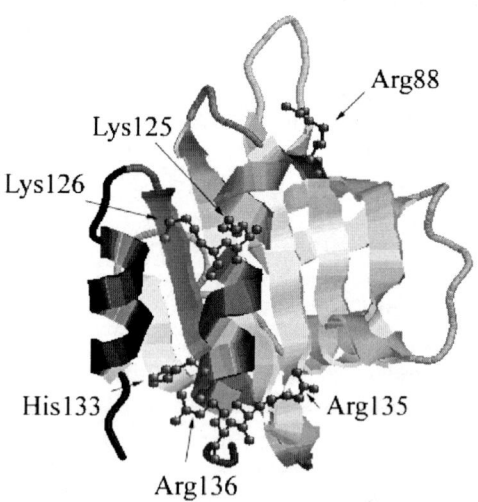

FIG. 1. Representation of the crystal structure of human profilin (Fedorov, Pollard and Almo 1994). The residues thought to be involved in phosphoinositides binding are highlighted and represented as "balls and sticks" models

TABLE 1. Effect of profilin mutations on its lipid binding capacity

Profilin mutants	Change in PI(4,5)P$_2$ affinity	PI(4,5)P$_2$ affinity relative to wild-type	References
W3N	↑	2.1	Skare and Karlsson (2002)
D8A	↑	–	Sohn et al. (1995)
K69N	↓	0.75	Skare and Karlsson (2002)
R88L	↓	–	Sohn et al. (1995)
K90E	↓	0.65	Skare and Karlsson (2002)
P96ΔT97Δ	↑	1.91	
K125N	↓	0.71	
R136D	↓	–	

actin-binding surface. Therefore, binding of PPI to the region spanning residues 126–136 could allow for the formation of a ternary complex of actin:profilin:PPI, while binding of PPI to the region including residue 88 results in the dissociation of the profilin:actin complex.

The binding stoichiometry of PI(4,5)P$_2$ ranges from 5:1 to 10:1 (Goldschmidt-Clermont et al. 1991; Lu et al. 1996). Goldschmidt-Clermont et al. (1990) suggested that this difference in binding stoichiometry could be due to steric hindrance among profilin molecules binding to micelles. Such hindrance would result in an overestimation of the number of PI(4,5)P$_2$ molecules bound per profilin. They also suggested that profilin aggregates PI(4,5)P$_2$ into small patches. We recently demonstrated that the affinity of profilin for submicellar concentration of PPI is significantly lower than for PPI in micelles or small vesicles (Moens and Bagatolli 2007). These results support the hypothesis that profilin binds to small patches of PPI and not to single molecules of PPI lipid.

Upon binding, the α-helical content varies from 1.4% to 17.5% depending on whether profilin is associated with PI(3,4)P$_2$ or PI(4,5)P$_2$, respectively (Lu et al. 1996). Profilin binding to PI(3,4,5)P$_3$ has an intermediate α-helical content of 11.5% (Lu et al. 1996). Because the interaction between profilin and PPI is probably mostly electrostatic, charged interactions could stabilize the association between profilin and PPI (Machesky, Goldschmidt-Clermont and Pollard 1990). Since basic/aromatic regions of proteins can reversibly sequester PI(4,5)P$_2$ by electrostatic interactions (Gambhir et al. 2004; Haleva, Ben-Tal and Diamant 2004; Wang et al. 2004), profilin interaction with PI(4,5)P$_2$ could result in the sequestration of 5–10 molecules of PI(4,5)P$_2$. Such interactions could also account for the higher affinity of profilin for PI(3,4,5)P$_3$ (Lu et al. 1996). Indeed, PPIs with a higher negative charge should be sequestered even more strongly (Wang et al. 2004). It is not clear why the affinity of profilin for PI(3,4)P$_2$ is higher than that for PI(3,4,5)P$_3$, for example, such difference could be explained by changes in its α-helical content. However, the mechanism by which a small molecule like profilin sequesters 5–10 PPI molecules is still to be elucidated.

Profilin Membrane Fraction

In cells, profilin partitions between the plasma membrane and the cytosol in response to membrane-associated PI(4,5)P$_2$ levels. In resting *Saccharomyces cerevisiae*, about 80% is associated with plasma membrane phospholipids (Ostrander, Gorman and Carman 1995). When membrane PI(4,5)P$_2$ levels are modified, either by inositol starvation of *ino1* cells or glucose starvation of respiratory-deficient cells, the proportion of profilin associated with the membrane varies accordingly (Ostrander, Gorman and Carman 1995).

In resting platelets, Hartwig et al. (1989) showed that only 36% of the total profilin is associated with the cell membrane (see chapter by Nunoi). Upon activation by thrombin, these authors noted an increase of the fraction of profilin associated with the membrane to between 41% and 52%. Taken together, these data suggest that membrane-associated profilin is regulated by membrane PI(4,5)P$_2$ levels but that this regulation probably varies with the type and function of cells investigated.

The p85α Subunit of Phosphatidylinositol 3-Kinase (PI3-Kinase)

Profilin/p85α complex stimulates the activity of the kinase and inhibits actin polymerization (Singh et al. 1996a; Bhargavi, Chari and Singh 1998). PI-3 kinase activity is regulated by profilin in a concentration-dependent manner, with maximal stimulation at 7.5 μM profilin. This activation seems to be due to the direct interaction of profilin with the p85 subunit of PI-3 kinase and cannot be attributed solely to the interaction of profilin with phosphoinositides. The binding of profilin to the p85 subunit does not change the affinities of the enzyme for ATP or PI(4,5)P$_2$; however, it increases the V_{max} of PI-3 kinase suggesting that profilin facilitates the accessibility of PI-3 kinase to substrates (Singh et al. 1996a). Finally, binding of profilin to the p85 subunit does not affect the interaction between actin and profilin supporting the presence of two distinct binding sites on profilin (Singh et al. 1996a).

Poly-L-Proline Sequences and Proline-Rich Motif (PRM) Proteins

These sequences are present in proteins such as those from the vasodilator-stimulated phosphoprotein (VASP) family and could regulate actin polymerization in response to external stimuli (Reinhard et al. 1995). VASP was the first natural PRM-containing ligand identified. Since then, a variety of other proteins have been identified as potential ligands, most of them binding to the poly-L-proline-binding domain of profilin (Witke 2004). These proteins involve profilin in endocytosis (Witke et al. 1998; Witke 2004), membrane vesicle trafficking (Witke et al. 1998), focal contacts (Gertler et al. 1996; Parast and Otey 2000), axon pathfinding (Gertler et al. 1996), platelet activation

(Reinhard et al. 1995), receptor clustering and dendritic spines (Mammoto et al. 1998), presynaptic vesicle trafficking (Wang et al. 1999), glutamate receptor function (Miyagi et al. 2002), nuclear export (Stuven, Hartmann and Gorlich 2003), mRNA splicing and spinal muscular atrophy (Giesemann et al. 1999), and the Rac/Rho or cdc42 signaling pathway (Ramesh et al. 1997; Watanabe et al. 1997; Miki, Suetsugu and Takenawa 1998; Suetsugu, Miki and Takenawa 1998; Witke et al. 1998; Boettner et al. 2000; Yayoshi-Yamamoto, Taniuchi and Watanabe 2000; Camera et al. 2003).

Profilin Isoforms and Functions

Profilin 1

Until now, five isoforms of profilin have been identified (Carlsson et al. 1977; Honore et al. 1993; Di Nardo et al. 2000; Hu et al. 2001; Obermann et al. 2005). Although the overall fold and shape of crystallized profilin isoforms as well as their function are remarkably conserved (Fig. 2), their

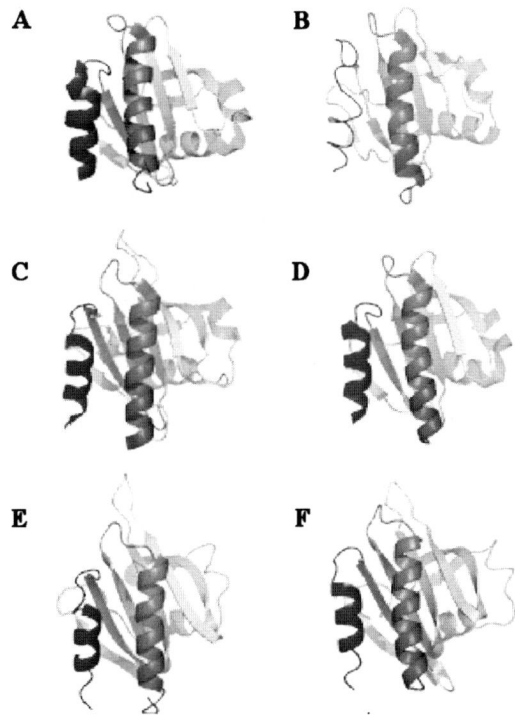

FIG. 2. Ribbon representation of various isoforms of profilin. (**a**) *Acanthamoeba castellanii* profilin II (pdbID: 1F2K). (**b**) Allergen profilins HEVB8 and BETV2 (pdbID: 1G5U). (**c**) Human profilin 2 (pdbID: 1D1J). (**d**) Yeast profilin 1(pdbID: 1K0K). (**e**) Human profilin 1 (pdbID: 1FIK). (**f**) Bovine profilin 1 (pdbID: 1PNE). The graphics were obtained using PyMol DeLano Scientific LLC (*See Color Plates*)

sequence identity is only moderate. Indeed, profilin 1 shares only 37 and 30% amino acid identity with profilin 3 and 4, respectively (Witke 2004; Obermann et al. 2005). Despite this, Rothkegel et al. (1996) demonstrated that profilin from plants, which share only 22% sequence identity with profilin from bovine thymus, can functionally substitute for the endogenous mammalian profilin. Also, defects in cell shape, cytokinesis, and development of *Dictyostelium discoideum* profilin-minus cells can be rescued by introducing the pollen-specific profilins 1 or 2 from maize (Karakesisoglou et al. 1996). However, isoforms in the same organisms differ sufficiently that they cannot complement each other, since deletion of the profilin 2 isoforms in mice results in severe neurological problems despite the presence of profilin 1 (Witke 2004).

Profilin 1 is ubiquitously expressed except in skeletal muscle (Witke et al. 1998). It is diffusely distributed throughout the cytoplasm and as dot-like structures in the nucleus (Giesemann et al. 1999). Profilin 1 is actively involved in the regulation and re/organization of the cytoskeleton and has been involved in the regulation of nonmuscle cell motility.

Profilin 2

Profilin 2 is expressed in motorneurons in the anterior horn of the spinal cord in mice and in some interneurons (Giesemann et al. 1999). It is also abundant in brain (Witke et al. 1998). In mouse, profilin 2 is further subdivided in profilin IIA and profilin IIB (Di Nardo et al. 2000). Profilin IIB is expressed mostly in kidney and embryonic stem cells and results from mRNA alternative splicing. Profilins IIA and IIB share the first 107 amino acids but have distinct C-terminal domains. These differences result in a lower affinity of profilin IIB for actin and poly-L-proline and the binding of profilin IIB to tubulin.

The profilin IIA isoforms have been shown to associate with dynamin 1 and a role for profilin IIA in membrane trafficking and endocytosis has been proposed (Witke et al. 1998) as well as a role in cell cycle control (Di Nardo et al. 2000).

Profilin 3 and Profilin 4

Profilins 3 and 4 are two isoforms expressed in the testis (Hu et al. 2001; Braun et al. 2002; Obermann et al. 2005). These two isoforms are expressed at different stages of the spermatogenesis cycle and are differentially regulated during the postnatal testicular development. These testis-specific profilins may interact with the actin cytoskeleton of developing male germs cells at distinct point locations. Their disturbance may cause abnormalities in the morphology of sperm head and in the function of acrosome in infertile men (Obermann et al. 2005).

Profilin as an Allergen

Up to 15% of the populations in industrialized countries suffer from allergies of the immediate type (Type I). Patients with pollen allergy also frequently experience allergic symptoms on ingestion of plant-derived foods (fruits, vegetables, and spices). Pollen-food allergies are attributed to the crossreactivity of IgE antibodies to conserved plant allergens. These allergens are expressed in related and unrelated plant species and different plant tissues. Conserved plant allergens are proteins that fulfill important biological functions and therefore are conserved in their sequence and/or structure. As illustrated above, profilins have very conserved shapes and folds (Fig. 2) resulting in conserved functions. Profilin was first identified in birch pollen (Bet v 2) in 1991(Valenta et al. 1991b).

Profilins represent crossreactive allergens for almost 20% of all pollen allergic patients (Valenta et al. 1992). This crossreactivity probably results from the high amino acid sequence identity of the plant profilins, i.e., between 70% and 85% while the nonallergenic profilins from other eukaryotes have lower sequence identities, i.e., between 30% and 40% (Radauer and Breiteneder 2006). Despite the lower sequence identity between plant and human profilin, the shape/structure conservation between these profilins could be responsible for the weaker crossreactivity of human profilin with human IgE. This crossreactivity could explain the exacerbation and prolongation of the initial sensitization of allergic patient to plant profilins (Fedorov et al. 1997).

Profilin and Cancer

In 2000, Janke et al. (2000) reported that profilin 1 is expressed in smaller amounts in several tumorigenic breast epithelial cell lines compared to nontumorigenic ones. They also showed that overexpression of profilin in breast cancer cells suppressed the tumorigenicity of these cells. They suggest that the effect of profilin on tumorigenicity is not directly linked to its role in actin organization, but rather to its role in signal transduction to the actin cytoskeleton. Downregulation of profilin expression in migratory tumor cells was also reported for astrocytic tumor cells after stimulation with extracellular S100A4 (a calcium-binding protein) (Belot et al. 2002).

Janke et al. (2000) also reported an increased spreading of breast cancer cells when profilin is overexpressed. In contrast, Roy and Jacobson (2004) showed that a moderate overexpression of profilin in breast cancer cells was enough to impair the spreading of these cells. They suggested that the difference in cell behavior might be due to different effect of profilin on actin assembly depending on profilin concentration in the cells.

Profilin–Actin Interaction and Cancer

In 2004, Wittenmayer et al. (2004) demonstrated the importance of the profilin–actin-binding site on its tumor suppressor activity. These authors also
produced a profilin mutant PFN1/R88L with a ~3–4-fold reduced affinity for
$PI(4,5)P_2$. Since they could not see a difference between the overexpression of
this mutant and the wild-type, they concluded that only the actin-binding site
was involved in the reduced tumorigenicity. However, they may not have chosen the best mutant since the mutated residue Arg88 is part of the PI-binding
region that overlaps the actin-binding region (Skare and Karlsson 2002).
Indeed, the PFN1/R88L mutant has also a ~3-fold (from 0.71×10^{-6} to $2.03
\times 10^{-6}$ M) reduced affinity for actin. Therefore, it is probable that the 3–4-fold
reduction in affinity for both PI and actin is not sufficient to prevent the
effect of profilin overexpression on tumorigenicity.

Roy and Jacobson (2004) showed that the overexpression of profilin in breast
cancer cell had a dramatic effect on the organization of actin in the transfected
cells. The increased profilin concentration resulted in the disappearance of cytoplasmic actin filaments together with an increase in the G-actin sequestering
effect and the formation of a thick cortical rim of actin filaments at the periphery of the cells. These authors also reported an increase adhesive strength of the
cancer cells determined by the resistance to detachment in presence of trypsin.
However, the mechanisms responsible for this increased resistance to detachment and its relation to increased profilin expression is still unclear.

Profilin–Polyphosphoinositides Interaction and Cancer

Many human cancers have mutations affecting the PI-3 kinase signaling
pathway (Vivanco and Sawyers 2002). As a consequence, the levels of
$PI(3,4,5)P_3$ in cancer cells are elevated resulting in the activation of the protein kinase Akt and thus tumor progression (Vivanco and Sawyers 2002).
Importantly, Stocker et al. (2002) demonstrated that a mutation reducing the
Akt affinity for $PI(3,4,5)P_3$ is sufficient to rescue the lethality in *Drosophila
melanogaster* that had an increased level of $PI(3,4,5)P_3$.

Since the level of $PI(3,4,5)P_3$ is critical for the development of cancer, the
sequestration of $PI(3,4,5)P_3$ by a molecule or protein can prevent the interaction between Akt and $PI(3,4,5)P_3$. Because of the higher affinity of profilin
for $PI(3,4,5)P_3$ compared to $PI(4,5)P_2$ (Lu et al. 1996) it is possible that the
overexpression of profilin results in the preferential sequestration of
$PI(3,4,5)P_3$, preventing the activation of Akt, therefore reducing the tumorigenicity of these cells.

Profilin–PRM-Containing Protein Interaction and Cancer

In recent years, the number of proteins identified that bind profilin through
its poly-L-proline-binding site has increased dramatically (see above). Among

these are proteins such as the three mouse homologs of the *Drosophila* gene *diaphanous* product and the profilin ligand WASp (Wiskott–Aldrich syndrome protein). These proteins are involved in the Rac/Rho signaling pathway and are implicated in actin nucleation (Witke 2004). Actin sequestered by profilin can be mobilized by diaphanous and hence promote actin polymerization (Li and Higgs 2003). Also, profilin seems to enhance the activation of WASp by cdc42 (Yang et al. 2000). This activation induces actin nucleation resulting in the formation of actin filaments.

In 2001, Vemuri and Singh reported that profilin is phosphorylated on the C-terminal Ser 137 by protein kinase C (PKC), a key enzyme of the PI3-kinase signaling pathway (Singh et al. 1996b; Vemuri and Singh 2001). Later, Sathish et al. (2004) demonstrated that the phosphorylation of profilin regulates its interaction with actin and poly-L-proline by increasing its affinity for G-actin and decreasing its affinity for poly-L-proline.

Several profilin-interacting PRM-containing proteins are important for eliciting actin reorganization in response to signals. It is therefore possible that profilin phosphorylation is one of the mechanisms to modulate profilin's interaction with various PRM proteins. This can in turn affect actin assembly and hence, cell migration and metastasis.

Profilin in Other Disorders and Diseases

Mesangial Proliferative Glomerulonephritis

Mesangial cell proliferation occurs in many types of human glomerular diseases. It is closely associated with intracellular signaling as well as rearrangement of the actin-based cytoskeleton (Tamura et al. 2000a). Since profilin is known to be involved in the reorganization of the cytoskeleton, its expression was followed in a rat model of mesangial proliferative glomerulonephritis (Tamura et al. 1996, 2000a,b). Profilin was constitutively expressed at a very low level in normal rat kidney but showed a marked increase in both mRNA and protein level, 1 day after disease induction, with a peak on day 4 and a subsequent decrease in expression after 14 days (Tamura et al. 2000a). Profilin was expressed at the plasma membrane and the rough endoplasmic reticulum of mesangial cells but was also present in the extracellular space around mesangial cells (Tamura et al. 2000a). The elevated levels of profilin in this space could arise from the release of profilin from dying cells. Tamura et al. (2000b) reported that the extracellular pool of profilin was able to stimulate the AP-1 transcription factor and DNA synthesis *in vitro*. These authors proposed that these effects were mediated through the binding of profilin to a single class of high-affinity receptors with a K_D of 30–60 nM which are neither PI(4,5)P$_2$, VASP, nor actin. However, the molecular identities of these receptors are still undetermined.

Alzheimer's Disease

Presenilin I and II have been identified as being the causal genes for early-onset familial Alzheimer's disease (Rogaev et al. 1995; Sherrington et al. 1995). In 1996, Kondo et al. (1996) identified several upregulated genes in an attempt to clarify the unidentified function(s) of presenilin I and II. Seventeen genes were identified including profilin however; the significance of this upregulation is still unclear but it may be related to the high motility of reactive glial cells in the affected regions of Alzheimer's disease brains.

Fragile X Syndrome

Fragile X syndrome is an inherited mental retardation, due to the loss of Fragile X mental retardation protein (FMRP) coded by the FMR1 gene on the long arm of the X chromosome (Chiurazzi, Neri and Oostra 2003). The *Drosophila* homolog of the FMRP (dFMRP) binds the mRNA of the *Drosophila* homolog of profilin and negatively regulates its expression (Reeve et al. 2005). In the lateral neurons located ventrolaterally (LNv) and Mushroom Bodies neurons, Reeve et al. found that upregulation of profilin mimics the phenotypes caused by loss of *dfmr1* and can rescue the overexpression of the dFMRP in the LNv neurons. Also, a reduction in profilin expression can rescue the phenotypes observed in *dfmr1* gene mutants. These experiments show that profilin and dFMRP contribute to the correct patterning of brain neurons in *Drosophila* through a possible feedback mechanism, whereby dFMRP represses excessive profilin translation. These authors suggest that misregulation of the actin network in neurons is an important factor in Fragile X syndrome.

Miller–Dieker Syndrome

This is a rare congenital disorder characterized by facial abnormalities and lissencephaly (Kwiatkowski et al. 1990). Lissencephaly results in the brain having a smooth surface with the patient frequently exhibiting epilepsy and mental retardation. Lissencephaly is recognized to be due to neuronal migrational defect during brain development and it has been shown that these patients have defects in both inhibitory interneuron migration and migration of excitatory projection neurons (Pancoast, Dobyns and Golden 2005).

A profilin gene was found to be part of the deletion of the 17p13 locus on chromosome 17 in some patients with Miller–Dieker syndrome. However, other patients had smaller deletions that did not involve the profilin locus. Therefore, profilin may contribute to the clinical phenotype of the Miller–Dieker syndrome but does not play a major role (Kwiatkowski et al. 1990).

Wiskott–Aldrich Syndrome

The Wiskott–Aldrich syndrome is an X chromosome-linked disorder characterized by impaired humoral and cellular immunity, increased susceptibility to lymphoid malignancy, thrombocytopenia with reduced mean platelet volume, and eczema (Aldrich, Steinberg and Campbell 1954; Rosen, Cooper and Wedgwood 1995). Patient T cells have striking restricted defect in their proliferative response (Molina et al. 1993) and the lymphocytes from affected individuals have defective cytoskeletal structure (Gallego et al. 1997). The diseased gene encodes a protein, WASp, rich in proline (Derry, Ochs and Francke 1994a,b) which was shown to bind profilin (Suetsugu, Miki and Takenawa 1998). WASp plays an important role in the integration of signaling pathways and the cytoskeleton. Together, WASp-interacting protein (WIP) (Ramesh et al. 1997; Stewart, Tian and Nelson 1999) and profilin regulate localized actin polymerization via interactions with PI(4,5)P$_2$ and/or SH3 (src Homology-3) domain proteins. Although profilin does not seem to be directly involved in the syndrome, perturbations of the binding between WASp, WIP, and profilin are likely to affect the cytoskeleton structure.

β-Thalassemia

β-Thalassemia is an autosomal recessive genetic disorder resulting in inadequate production of normal hemoglobin. In homozygous β-Thalassemia, Alekperova, Orudzhev and Javadov (2004) found significant changes in the platelet membrane fraction of patients. They reported significant decreases in several proteins including profilin. The significance of this finding and the role of profilin in β-Thalassemia are still to be clarified.

Parkinson's Disease

A recent proteomic analysis of human substantia nigra from patients with Parkinson's disease revealed that profilin was one of the proteins with elevated expression levels (Basso et al. 2004). These authors suggest that the overexpression of profilin could be linked to modifications of neurofilaments at the substantia nigra pars compacta level which take place during the neurodegenerative process of Parkinson's disease.

Conclusions and Perspectives

With its numerous potential ligands, its involvement in signaling, trafficking, receptor activity, nuclear activities, and cytoskeleton organization, profilin can be considered a key element which controls molecular interactions. Although profilin is not necessarily the direct cause of many diseases, because it has such widespread array of functions, mutations of the protein or mutations affecting the interaction of profilin with its ligands will undoubtedly alter important cellular processes. This chapter has only "skimmed" the surface of the functions and controls exerted by profilin in cells and many questions are still unresolved. The identification of all the molecular interactions of profilin with its ligands together with a better understanding of the regulations of these interactions is a challenge that lies ahead. Combined proteomics and system biology approaches will probably prove to be invaluable tools if we are to understand the complex and diverse functions of profilin.

References

Aldrich, R. A., Steinberg, A. G. and Campbell, D. C. 1954. Pedigree demonstrating a sex-linked recessive condition characterized by draining ears, eczematoid dermatitis and bloody diarrhea. Pediatrics 13, 133–139.

Alekperova, G. A., Orudzhev, A. G. and Javadov, S. A. 2004. Analysis of erythrocyte and platelet membrane proteins in various forms of beta-Thalassemia. Biochemistry 69, 748–753.

Bae, Y. S., Cantley, L. G., Chen, C. S., Kim, S. R., Kwon, K. S. and Rhee, S. G. 1998. Activation of phospholipase C-gamma by phosphatidylinositol 3,4,5-trisphosphate. J. Biol. Chem. 273, 4465–4469.

Basso, M., Giraudo, S., Corpillo, D., Bergamasco, B., Lopiano, L. and Fasano, M. 2004. Proteome analysis of human substantia nigra in Parkinson's disease. Proteomics 4, 3943–3952.

Belot, N., Pochet, R., Heizmann, C. W., Kiss, R. and Decaestecker, C. 2002. Extracellular S100A4 stimulates the migration rate of astrocytic tumor cells by modifying the organization of their actin cytoskeleton. Biochim. Biophys. Acta 1600, 74–83.

Bhargavi, V., Chari, V. B. and Singh, S. S. 1998. Phosphatidylinositol 3-kinase binds to profilin through the p85 alpha subunit and regulates cytoskeletal assembly. Biochem. Mol. Biol. Int. 46, 241–248.

Boettner, B., Govek, E. E., Cross, J. and Van Aelst, L. 2000. The junctional multidomain protein AF-6 is a binding partner of the Rap1A GTPase and associates with the actin cytoskeletal regulator profilin. Proc. Natl Acad. Sci. USA 97, 9064–9069.

Braun, A., Aszodi, A., Hellebrand, H., Berna, A., Fassler, R. and Brandau, O. 2002. Genomic organization of profilin-III and evidence for a transcript expressed exclusively in testis. Gene 283, 219–225.

Camera, P., da Silva, J. S., Griffiths, G., Giuffrida, M. G., Ferrara, L., Schubert, V., Imarisio, S., Silengo, L., Dotti, C. G. and Di Cunto, F. 2003. Citron-N is a neuronal Rho-associated protein involved in Golgi organization through actin cytoskeleton regulation. Nat. Cell Biol. 5, 1071–1078.

Carlsson, L., Nystrom, L. E., Sundkvist, I., Markey, F. and Lindberg, U. 1977. Actin polymerizability is influenced by profilin, a low molecular weight protein in non-muscle cells. J. Mol. Biol. 115, 465–483.

Chiurazzi, P., Neri, G. and Oostra, B. A. 2003. Understanding the biological underpinnings of fragile X syndrome. Curr. Opin. Pediatr. 15, 559–566.

Derry, J. M., Ochs, H. D. and Francke, U. 1994a. Isolation of a novel gene mutated in Wiskott–Aldrich syndrome. Cell 78, 635–644.

Derry, J. M., Ochs, H. D. and Francke, U. 1994b. Isolation of a novel gene mutated in Wiskott–Aldrich syndrome. Cell 79, following 922.

Di Nardo, A., Gareus, R., Kwiatkowski, D. and Witke, W. 2000. Alternative splicing of the mouse profilin II gene generates functionally different profilin isoforms. J. Cell Sci. 113, 3795–3803.

dos Remedios CG, Chhabra D, Kekic M, Dedova IV, Tsubakihara M, et al. 2003. Actin binding proteins: regulation of cytoskeletal microfilaments. Physiol. Rev. 83, 433–473.

Fedorov, A. A., Ball, T., Mahoney, N. M., Valenta, R. and Almo, S. C. 1997. The molecular basis for allergen cross-reactivity: Crystal structure and IgE-epitope mapping of birch pollen profilin. Structure 5, 33–45.

Fedorov, A. A., Pollard, T. D. and Almo, S. C. 1994. Purification, characterization and crystallization of human platelet profilin expressed in Escherichia coli. J. Mol. Biol. 241, 480–482.

Gallego, M. D., Santamaria, M., Pena, J. and Molina, I. J. 1997. Defective actin reorganization and polymerization of Wiskott–Aldrich T cells in response to CD3-mediated stimulation. Blood 90, 3089–3097.

Gambhir, A., Hangyas-Mihalyne, G., Zaitseva, I., Cafiso, D. S., Wang, J., Murray, D., Pentyala, S. N., Smith, S. O. and McLaughlin, S. 2004. Electrostatic sequestration of PIP2 on phospholipid membranes by basic/aromatic regions of proteins. Biophys. J. 86, 2188–2207.

Gertler, F. B., Niebuhr, K., Reinhard, M., Wehland, J. and Soriano, P. 1996. Mena, a relative of VASP and Drosophila enabled, is implicated in the control of microfilament dynamics. Cell 87, 227–239.

Giesemann, T., Rathke-Hartlieb, S., Rothkegel, M., Bartsch, J. W., Buchmeier, S., Jockusch, B. M. and Jockusch, H. 1999. A role for polyproline motifs in the spinal muscular atrophy protein SMN. Profilins bind to and colocalize with smn in nuclear gems. J. Biol. Chem. 274, 37908–37914.

Goldschmidt-Clermont, P. J., Furman, M. I., Wachsstock, D., Safer, D., Nachmias, V. T. and Pollard, T. D. 1992. The control of actin nucleotide exchange by thymosin beta 4 and profilin. A potential regulatory mechanism for actin polymerization in cells. Mol. Biol. Cell 3, 1015–1024.

Goldschmidt-Clermont, P. J., Kim, J. W., Machesky, L. M., Rhee, S. G. and Pollard, T. D. 1991. Regulation of phospholipase C-gamma 1 by profilin and tyrosine phosphorylation. Science 251, 1231–1233.

Goldschmidt-Clermont, P. J., Machesky, L. M., Baldassare, J. J. and Pollard, T. D. 1990. The actin-binding protein profilin binds to PIP2 and inhibits its hydrolysis by phospholipase C. Science 247, 1575–1578.

Gutsche-Perelroizen, I., Lepault, J., Ott, A., Carlier, M. F. 1999. Filament assembly from profilin-actin. J. Biol. Chem. 274, 6234–6243.

Haleva, E., Ben-Tal, N. and Diamant, H. 2004. Increased concentration of polyvalent phospholipids in the adsorption domain of a charged protein. Biophys. J. 86, 2165–2178.

Hartwig, J. H., Chambers, K. A., Hopcia, K. L. and Kwiatkowski, D. J. 1989. Association of profilin with filament-free regions of human leukocyte and platelet membranes and reversible membrane binding during platelet activation. J. Cell Biol. 109, 1571–1579.

Haugwitz, M., Noegel, A. A., Karakesisoglou, J. and Schleicher, M. 1994. Dictyostelium amoebae that lack G-actin-sequestering profilins show defects in F-actin content, cytokinesis, and development. Cell 79, 303–314.

Honore, B., Madsen, P., Andersen, A. H. and Leffers, H. 1993. Cloning and expression of a novel human profilin variant, profilin II. FEBS Lett. 330, 151–155.

Hu, E., Chen, Z., Fredrickson, T. and Zhu, Y. 2001. Molecular cloning and characterization of profilin-3: A novel cytoskeleton-associated gene expressed in rat kidney and testes. Exp. Nephrol. 9, 265–274.

Janke, J., Schluter, K., Jandrig, B., Theile, M., Kolble, K., Arnold, W., Grinstein, E., Schwartz, A., Estevez-Schwarz, L., Schlag, P. M., Jockusch, B. M. and Scherneck, S. 2000. Suppression of tumorigenicity in breast cancer cells by the microfilament protein profilin 1. J. Exp. Med. 191, 1675–1686.

Karakesisoglou, I., Schleicher, M., Gibbon, B. C. and Staiger, C. J. 1996. Plant profilins rescue the aberrant phenotype of profilin-deficient Dictyostelium cells. Cell Motil. Cytoskeleton 34, 36–47.

Kondo, T., Shirasawa, T., Itoyama, Y. and Mori, H. 1996. Embryonic genes expressed in Alzheimer's disease brains. Neurosci. Lett. 209, 157–160.

Kwiatkowski, D. J., Aklog, L., Ledbetter, D. H. and Morton, C. C. 1990. Identification of the functional profilin gene, its localization to chromosome subband 17p13.3, and demonstration of its deletion in some patients with Miller–Dieker syndrome. Am. J. Hum. Genet. 46, 559–567.

Lambrechts, A., Jonckheere, V., Dewitte, D., Vandekerckhove, J. and Ampe, C. 2002. Mutational analysis of human profilin I reveals a second PI(4,5)-P2 binding site neighbouring the poly(L-proline) binding site. BMC Biochem. 3, 12.

Lassing, I. and Lindberg, U. 1988. Specificity of the interaction between phosphatidylinositol 4,5-bisphosphate and the profilin:actin complex. J. Cell Biochem. 37, 255–267.

Li, F. and Higgs, H. N. 2003. The mouse formin mDia1 is a potent actin nucleation factor regulated by autoinhibition. Curr. Biol. 13, 1335–1340.

Lu, P. J., Shieh, W. R., Rhee, S. G., Yin, H. L. and Chen, C. S. 1996. Lipid products of phosphoinositide 3-kinase bind human profilin with high affinity. Biochemistry 35, 14027–14034.

Machesky, L. M., Cole, N. B., Moss, B. and Pollard, T. D. 1994. *Vaccinia* virus expresses a novel profilin with a higher affinity for polyphosphoinositides than actin. Biochemistry 33, 10815–10824.

Machesky, L. M., Goldschmidt-Clermont, P. J. and Pollard, T. D. 1990. The affinities of human platelet and *Acanthamoeba* profilin isoforms for polyphosphoinositides account for their relative abilities to inhibit phospholipase C. Cell Regul. 1, 937–950.

Mammoto, A., Sasaki, T., Asakura, T., Hotta, I., Imamura, H., Takahashi, K., Matsuura, Y., Shirao, T. and Takai, Y. 1998. Interactions of drebrin and gephyrin with profilin. Biochem. Biophys. Res. Commun. 243, 86–89.

Markey, F., Persson, T. and Lindberg, U. 1981. Characterization of platelet extracts before and after stimulation with respect to the possible role of profilactin as microfilament precursor. Cell 23, 145–153.

Miki, H., Suetsugu, S. and Takenawa, T. 1998. WAVE, a novel WASp-family protein involved in actin reorganization induced by Rac. EMBO J. 17, 6932–6941.

Miyagi, Y., Yamashita, T., Fukaya, M., Sonoda, T., Okuno, T., Yamada, K., Watanabe, M., Nagashima, Y., Aoki, I., Okuda, K., Mishina, M. and Kawamoto, S. 2002. Delphilin: A novel PDZ and formin homology domain-containing protein that synaptically colocalizes and interacts with glutamate receptor delta 2 subunit. J. Neurosci. 22, 803–814.

Moens, P. D., Bagatolli, L. A. 2007. Profilin binding to sub-micellar concentrations of phosphatidylinositol (4,5) bisphosphate and phosphatidylinositol (3,4,5) trisphosphate. Biochim. Biophys. Acta 1768, 439–449.

Molina, I. J., Sancho, J., Terhorst, C., Rosen, F. S., and Remold'Donnell, E. 1993. T cells of patients with the Wiskott–Aldrich syndrome have a restricted defect in proliferative responses. J. Immunol. 151, 4383–4390.

Obermann, H., Raabe, I., Balvers, M., Brunswig, B., Schulze, W. and Kirchhoff, C. 2005. Novel testis-expressed profilin IV associated with acrosome biogenesis and spermatid elongation. Mol. Hum. Reprod. 11, 53–64.

Ostrander, D. B., Gorman, J. A. and Carman, G. M. 1995. Regulation of profilin localization in Saccharomyces cerevisiae by phosphoinositide metabolism. J. Biol. Chem. 270, 27045–27050.

Pancoast, M., Dobyns, W. and Golden, J. A. 2005. Interneuron deficits in patients with the Miller–Dieker syndrome. Acta Neuropathol. 109, 400–404.

Pantaloni, D. and Carlier, M. F. 1993. How profilin promotes actin filament assembly in the presence of thymosin beta 4. Cell 75, 1007–1014.

Parast, M. M. and Otey, C. A. 2000. Characterization of palladin, a novel protein localized to stress fibers and cell adhesions. J. Cell Biol. 150, 643–656.

Perelroizen, I., Carlier, M. F. and Pantaloni, D. 1995. Binding of divalent cation and nucleotide to G-actin in the presence of profilin. J. Biol. Chem. 270, 1501–1508.

Radauer, C. and Breiteneder, H. 2006. Pollen allergens are restricted to few protein families and show distinct patterns of species distribution. J. Allergy Clin. Immunol. 117, 141–147.

Ramesh, N., Anton, I. M., Hartwig, J. H. and Geha, R. S. 1997. WIP, a protein associated with Wiskott–Aldrich syndrome protein, induces actin polymerization and redistribution in lymphoid cells. Proc. Natl Acad. Sci. USA 94, 14671–14676.

Reeve, S. P., Bassetto, L., Genova, G. K., Kleyner, Y., Leyssen, M., Jackson, F. R. and Hassan, B. A. 2005. The Drosophila fragile X mental retardation protein controls actin dynamics by directly regulating profilin in the brain. Curr. Biol. 15, 1156–1163.

Reichstein, E. and Korn, E. D. 1979. Acanthamoeba profilin. A protein of low molecular weight from Acanthamoeba castellanii that inhibits actin nucleation. J. Biol. Chem. 254, 6174–6179.

Reinhard, M., Giehl, K., Abel, K., Haffner, C., Jarchau, T., Hoppe, V., Jockusch, B. M. and Walter, U. 1995. The proline-rich focal adhesion and microfilament protein VASP is a ligand for profilins. EMBO J. 14, 1583–1589.

dos Remedios, C. G., Chhabra, D., Kekic, M., Dedova, I. V., Tsubakihara, M., Berry, D. A. and Nosworthy, N. J. 2003. Actin binding proteins: Regulation of cytoskeletal microfilaments. Physiol. Rev. 83, 433–473.

Rogaev, E. I., Sherrington, R., Rogaeva, E. A., Levesque, G., Ikeda, M., Liang, Y., Chi, H., Lin, C., Holman, K., Tsuda, T., Mar, L., Sorbi, S., Nacmias, B., Piacentini, S., Amaducci, L., Chumakov, I., Cohen, D., Lannfelt, L., Fraser, P. E., Rommens, J. M. and St George-Hyslop, P. 1995. Familial Alzheimer's disease in kindreds with missense mutations in a gene on chromosome 1 related to the Alzheimer's disease type 3 gene. Nature 376, 775–778.

Rosen, F. S., Cooper, M. D. and Wedgwood, R. J. 1995. The primary immunodeficiencies. N. Engl. J. Med. 333, 431–440.

Rothkegel, M., Mayboroda, O., Rohde, M., Wucherpfennig, C., Valenta, R. and Jockusch, B. M. 1996. Plant and animal profilins are functionally equivalent and stabilize microfilaments in living animal cells. J. Cell Sci. 109, 83–90.

Roy, P. and Jacobson, K. 2004. Overexpression of profilin reduces the migration of invasive breast cancer cells. Cell Motil. Cytoskeleton 57, 84–95.

Sathish, K., Padma, B., Munugalavadla, V., Bhargavi, V., Radhika, K. V., Wasia, R., Sairam, M. and Singh, S. S. 2004. Phosphorylation of profilin regulates its interaction with actin and poly (L-proline). Cell Signal. 16, 589–596.

Schutt, C. E., Myslik, J. C., Rozycki, M. D., Goonesekere, N. C. and Lindberg, U. 1993. The structure of crystalline profilin-beta-actin. Nature 365, 810–816.

Sherrington, R., Rogaev, E. I., Liang, Y., Rogaeva, E. A., Levesque, G., Ikeda, M., Chi, H., Lin, C., Li, G., Holman, K. et al. 1995. Cloning of a gene bearing missense mutations in early-onset familial Alzheimer's disease. Nature 375, 754–760.

Singh, S. S., Chauhan, A., Murakami, N. and Chauhan, V. P. 1996a. Profilin and gelsolin stimulate phosphatidylinositol 3-kinase activity. Biochemistry 35, 16544–16549.

Singh, S. S., Chauhan, A., Murakami, N., Styles, J., Elzinga, M. and Chauhan, V. P. 1996b. Phosphoinositide-dependent in vitro phosphorylation of profilin by protein kinase C. Phospholipid specificity and localization of the phosphorylation site. Recept. Signal Transduct. 6, 77–86.

Skare, P. and Karlsson, R. 2002. Evidence for two interaction regions for phosphatidylinositol(4,5)-bisphosphate on mammalian profilin I. FEBS Lett. 522, 119–124.

Sohn, R. H., Chen, J., Koblan, K. S., Bray, P. F. and Goldschmidt-Clermont, P. J. 1995. Localization of a binding site for phosphatidylinositol 4,5-bisphosphate on human profilin. J. Biol. Chem. 270, 21114–21120.

Stewart, D. M., Tian, L. and Nelson, D. L. 1999. Mutations that cause the Wiskott–Aldrich syndrome impair the interaction of Wiskott–Aldrich syndrome protein (WASP) with WASP interacting protein. J. Immunol. 162, 5019–5024.

Stocker, H., Andjelkovic, M., Oldham, S., Laffargue, M., Wymann, M. P., Hemmings, B. A. and Hafen, E. 2002. Living with lethal PIP3 levels: Viability of flies lacking PTEN restored by a PH domain mutation in Akt/PKB. Science 295, 2088–2091.

Stuven, T., Hartmann, E. and Gorlich, D. 2003. Exportin 6: A novel nuclear export receptor that is specific for profilin.actin complexes. EMBO J. 22, 5928–5940.

Suetsugu, S., Miki, H. and Takenawa, T. 1998. The essential role of profilin in the assembly of actin for microspike formation. EMBO J. 17, 6516–6526.

Tamura, M., Tanaka, H., Hirano, T., Ueta, Y., Higashi, K. and Hirano, H. 1996. Enhanced glomerular profilin gene and protein expression in experimental mesangial proliferative glomerulonephritis. Biochem. Biophys. Res. Commun. 222, 683–687.

Tamura, M., Tanaka, H., Yashiro, A., Osajima, A., Okazaki, M., Kudo, H., Doi, Y., Fujimoto, S., Higashi, K., Nakashima, Y. and Hirano, H. 2000a. Expression of profilin, an actin-binding protein, in rat experimental glomerulonephritis and its upregulation by basic fibroblast growth factor in cultured rat mesangial cells. J. Am. Soc. Nephrol. 11, 423–433.

Tamura, M., Yanagihara, N., Tanaka, H., Osajima, A., Hirano, T., Higashi, K., Yamada, K. M., Nakashima, Y. and Hirano, H. 2000b. Activation of DNA synthesis and AP-1 by profilin, an actin-binding protein, via binding to a cell surface receptor in cultured rat mesangial cells. J. Am. Soc. Nephrol. 11, 1620–1630.

Valenta, R., Duchene, M., Breitenbach, M., Pettenburger, K., Koller, L., Rumpold, H., Scheiner, O. and Kraft, D. 1991a. A low molecular weight allergen of white birch (*Betula verrucosa*) is highly homologous to human profilin. Int. Arch. Allergy Appl. Immunol. 94, 368–370.

Valenta, R., Duchene, M., Ebner, C., Valent, P., Sillaber, C., Deviller, P., Ferreira, F., Tejkl, M., Edelmann, H. and Kraft, D. 1992. Profilins constitute a novel family of functional plant pan-allergens. J. Exp. Med. 175, 377–385.

Valenta, R., Duchene, M., Pettenburger, K., Sillaber, C., Valent, P., Bettelheim, P., Breitenbach, M., Rumpold, H., Kraft, D. and Scheiner, O. 1991b. Identification of profilin as a novel pollen allergen: IgE autoreactivity in sensitized individuals. Science 253, 557–560.

Vemuri, B. and Singh, S. S. 2001. Protein kinase C isozyme-specific phosphorylation of profilin. Cell Signal. 13, 433–439.

Vinson, V. K., De La Cruz, E. M., Higgs, H. N. and Pollard, T. D. 1998. Interactions of *Acanthamoeba* profilin with actin and nucleotides bound to actin. Biochemistry 37, 10871–10880.

Vivanco, I. and Sawyers, C. L. 2002. The phosphatidylinositol 3-kinase AKT pathway in human cancer. Nat. Rev. Cancer 2, 489–501.

Wang, J., Gambhir, A., McLaughlin, S. and Murray, D. 2004. A computational model for the electrostatic sequestration of PI(4,5)P2 by membrane-adsorbed basic peptides. Biophys. J. 86, 1969–1986.

Wang, X., Kibschull, M., Laue, M. M., Lichte, B., Petrasch-Parwez, E. and Kilimann, M. W. 1999. Aczonin, a 550-kD putative scaffolding protein of presynaptic active zones, shares homology regions with Rim and Bassoon and binds profilin. J. Cell Biol. 147, 151–162.

Watanabe, N., Madaule, P., Reid, T., Ishizaki, T., Watanabe, G., Kakizuka, A., Saito, Y., Nakao, K., Jockusch, B. M. and Narumiya, S. 1997. p140mDia, a mammalian homolog of *Drosophila* diaphanous, is a target protein for Rho small GTPase and is a ligand for profilin. EMBO J. 16, 3044–3056.

Witke, W. 2004. The role of profilin complexes in cell motility and other cellular processes. Trends Cell Biol. 14, 461–469.

Witke, W., Podtelejnikov, A. V., Di Nardo, A., Sutherland, J. D., Gurniak, C. B., Dotti, C. and Mann, M. 1998. In mouse brain profilin I and profilin II associate with regulators of the endocytic pathway and actin assembly. EMBO J. 17, 967–976.

Witke, W., Sutherland, J. D., Sharpe, A., Arai, M. and Kwiatkowski, D. J. 2001. Profilin I is essential for cell survival and cell division in early mouse development. Proc. Natl Acad. Sci. USA 98, 3832–3836.

Wittenmayer, N., Jandrig, B., Rothkegel, M., Schluter, K., Arnold, W., Haensch, W., Scherneck, S. and Jockusch, B. M. 2004. Tumor suppressor activity of profilin requires a functional actin binding site. Mol. Biol. Cell 15, 1600–1608.

Yang, C., Huang, M., DeBiasio, J., Pring, M., Joyce, M., Miki, H., Takenawa, T. and Zigmond, S. H. 2000. Profilin enhances Cdc42-induced nucleation of actin polymerization. J. Cell Biol. 150, 1001–1012.

Yayoshi-Yamamoto, S., Taniuchi, I. and Watanabe, T. 2000. FRL, a novel formin-related protein, binds to Rac and regulates cell motility and survival of macrophages. Mol. Cell Biol. 20, 6872–6881.

Yu, F. X., Sun, H. Q., Janmey, P. A. and Yin, H. L. 1992. Identification of a polyphosphoinositide-binding sequence in an actin monomer-binding domain of gelsolin. J. Biol. Chem. 267, 14616–14621.

9
The Roles of Thymosin β_4 in Cell Migration and Cell-to-Cell Signaling in Disease

Joshua K. Au, Mira Krendel, Daniel Safer, and Enrique M. De La Cruz

Summary

The β-thymosins are a highly conserved family of strongly polar ~5 kDa polypeptides that are widely distributed in metazoan cells (Fig. 1). Thymosin β_4, the most abundant and best-characterized β-thymosin, binds monomeric actin in a stable 1:1 complex and acts as an actin "buffer," preventing spontaneous polymerization but supplying high concentrations of free actin monomers for rapid filament elongation when cells are stimulated by extracellular cues. Several biological regulatory effects are attributed to $T\beta_4$ and oxidized $T\beta_4$. Among these are the induction of angiogenesis, tumor metastasis and the inhibition of inflammation. Correspondingly, several therapeutic applications for $T\beta_4$ have been proposed.

Discovery and Properties of Thymosins

The family of acidic, heat-stable polypeptides (Fig. 1) collectively termed "thymosin" was identified and isolated from calf thymus in 1965 during a search for molecules mediating the immunoregulatory activities of the thymus gland (Goldstein, Slater and White 1966). The thymosins are divided into three subfamilies according to their isoelectric points; α-thymosins have isoelectric points below pH 5.0, β-thymosins between pH 5.0 and 7.0, and γ-thymosins above pH 7.0. The numerical subscript denotes the chronological order of the polypeptides as they were isolated and sequenced (Goldstein et al. 1977).

Thymosin β_4 ($T\beta_4$), the most abundant and thoroughly studied mammalian β-thymosin, was first isolated from the partially purified thymosin preparation termed "thymosin fraction 5" since the purification preparation involved five chromatographic separations (Low, Hu and Goldstein 1981). Amino acid sequencing indicated that $T\beta_4$ has a molecular weight of 4,964 Da and is slightly acidic, with an isoelectric point of 5.1 (Low, Hu and

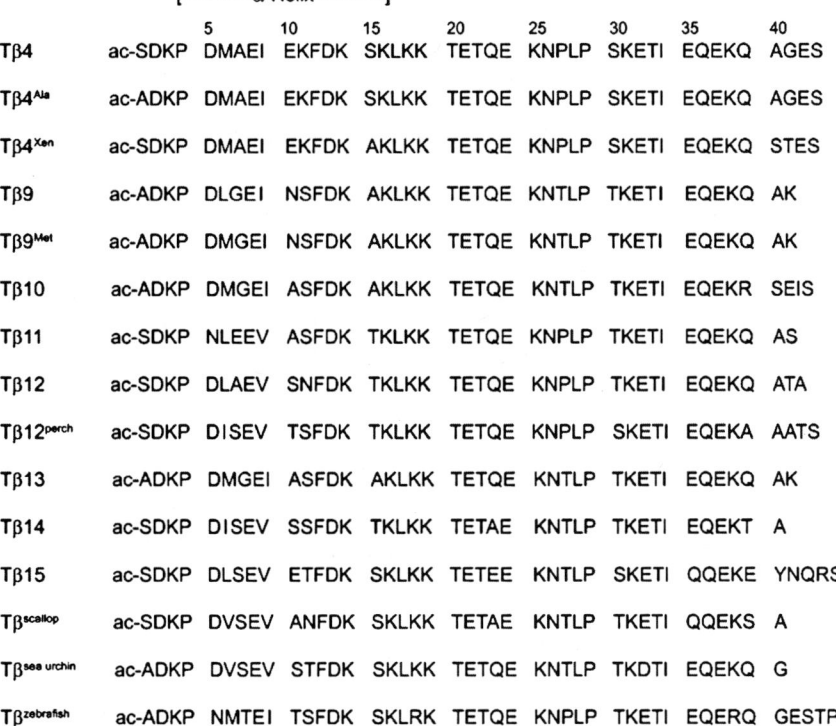

FIG. 1. Amino acid sequence of β-thymosins. Figure adapted from Huff et al. (2001), copyright 2001 with permission from Elsevier

Goldstein 1981). Tβ4 is usually accompanied in mammalian tissues by a second β-thymosin, thymosin β10, which is present at lower concentrations. In some species, other isoforms with minor sequence differences substitute for Tβ4 and/or Tβ10. Tβ4 localizes to both the cytoplasm and nucleus of vertebrate cells (Huff et al. 2004).

β-thymosin family members are found in metazoan species, from mammals to echinoderms (Fig. 1), and are absent from prokaryotes, plants, protists, and fungi. The mammalian gene encoding Tβ4 localizes to the X chromosome (Li et al. 1996). A number of Tβ4 homologues that have been isolated from invertebrates show >70% sequence identity to vertebrate Tβ4 (Safer and Chowrashi 1997). The invertebrate homologues are 1–3 amino acid residues shorter than the vertebrate thymosins, but 24 residues are invariant among the β-thymosins from different phyla. All β-thymosins that have been studied bind actin monomers, though some differences in affinity have been reported. A recent search for β-thymosins in *Dictyostelium* identified only a homolog of *Acanthamoeba* actobindin (Vancompernolle et al. 1991).

Tβ₄–Actin Monomer Binding Activity

Despite the ubiquitous presence of β-thymosins in cells and tissues, their intracellular function remained a mystery until 1991, when Safer and colleagues discovered that Tβ₄ was a G-actin sequestering peptide in platelets, where it forms a stable 1:1 complex with actin monomers (Safer, Elzinga and Nachmias 1991). Tβ₄ is present in high concentrations (560 μM in platelets) and sequesters a large pool of unpolymerized actin monomers (Weber et al. 1992). Because Tβ₄ binds ATP-actin more strongly than ADP-actin (Jean et al. 1994; De La Cruz et al. 2000), and the intracellular ATP concentration is much greater than that of ADP, the Tβ₄-sequestered actin monomer pool is comprised of ATP-actin. The rapid association (~2 μM^{-1} s^{-1}) and dissociation (~5 s^{-1}) rate constants (De La Cruz et al. 2000) allow the actin subunits to readily exchange between thymosin and free filament barbed ends on a time scale compatible with motile cellular events. In this regard, Tβ₄ acts as a cytoplasmic actin "buffer," preventing spontaneous polymerization and maintaining a large, concentrated pool of free actin monomers available for rapid polymerization upon stimulation by extracellular cues and factors.

Tβ₄ is predominantly unfolded in solution, although residues 4–16, and to a lesser extent residues 30–40, have some tendency toward an α-helical conformation (Czisch et al. 1993). Binding to actin favors partial Tβ₄ folding (Safer, Sosnick and Elzinga 1997; Domanski et al. 2004), but it is still predominantly in an unfolded conformation. A binding-linked folding reaction is likely to account for the thermodynamics of the actin–Tβ₄ interaction, including the large change in heat capacity (De La Cruz et al. 2000). Solution studies (De La Cruz et al. 2000; Dedova et al. 2006) suggest that actin also undergoes significant conformational rearrangement when Tβ₄ binds. Tβ₄ binding changes the distance between FRET probes on actin, indicating a rotation of actin subdomain 2 toward the nucleotide-binding site and away from subdomain 1 (Dedova et al. 2006). This change in conformation may account for the inhibition of nucleotide exchange by Tβ₄ (Goldschmidt-Clermont et al. 1992).

Site-specific mutagenesis indicates that amino acid residues throughout the Tβ₄ sequence contribute to high affinity actin monomer binding activity, consistent with a large actin–Tβ₄ binding interface. Truncation of the first 6 or 12 Tβ₄ amino acids weakens the actin-binding affinity ~20-fold. Truncation of the first 23 or last 26 amino acid residues essentially abolishes the interaction with actin monomers (Huff, Zerzawy and Hannappel 1995), indicating that both the N- and C-termini of Tβ₄ are critical for high actin-binding affinity.

It has been proposed that the sequence motif [17]LKKTETQEK[25] of Tβ₄ may participate in actin binding given its homology with the actin-binding sequence of actobindin, which binds the actin monomer barbed end (Safer, Elzinga and Nachmias 1991). The [17]LKK[19] segment within this motif plays an essential electrostatic role in actin binding (Van Troys et al. 1996). The three hydrophobic residues, Met6, Ile9, and Phe12, also play integral roles in the interaction.

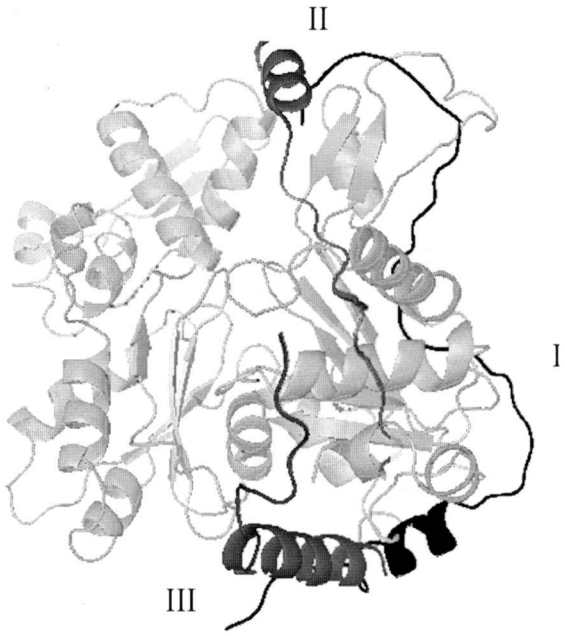

FIG. 2. Structural models for Tβ₄ bound to actin. Three different models for Tβ₄ are superimposed on actin structure 1ATN (shown in gray): (I) the model proposed for Tβ₄, based on its contacts with actin from crosslinking data, reproduced from Safer, Sosnick and Elzinga (1997), copyright 1997 The American Chemical Society; (II) Tβ₄ residues 21–43, taken from the crystal structure of a gelsolin-Tβ₄ chimera bound to actin (1T44; Irobi et al. 2004, copyright 2004 with permission from Macmillan Publishers Ltd); (III) Ciboulot residues 10–34, homologous with Tβ₄ residues 1–25, are taken from the crystal structure of that segment bound to actin–latrunculin (1SQK; Hertzog et al. 2004, copyright 2004 with permission from Elsevier)

Chemical crosslinking studies have permitted structural modeling of the actin–Tβ₄ complex. Tβ₄ wraps around an actin monomer contacting subdomains 1, 2, and 3 (Safer et al. 1997; De La Cruz et al. 2000) (Figure 2), thereby sterically inhibiting polymerization at both ends of the monomer. The N-terminal region of Tβ₄ makes direct contact with the barbed end of the actin monomer (Lys3 of Tβ₄ crosslinks to actin Glu167) while the C-terminal region of Tβ₄ contacts the pointed end (Lys38 of Tβ₄ to actin Gln41). The critical lysine residues at positions 18 and 19 of Tβ₄ are in contact with the cluster of acidic residues at the N-terminus of actin (Lys18 to actin residues 1–4) (Safer et al. 1997). Contact between the C-terminus of Tβ₄ and actin subdomain 2 was confirmed using a Tβ₄ mutant with cysteine substituted for serine (Ser43) at the C-terminus; Cys43 was crosslinked with high efficiency to His40 in subdomain 2 of actin (De La Cruz et al. 2000).

While the structure of the actin–Tβ_4 complex has not yet been solved directly, several models have been proposed on the basis of crystallographic structures of actin bound to Tβ_4 homologs, and on a spectroscopic study of the conformation of actin-bound Tβ_4 using isotope-edited NMR (Domanski et al. 2004). All three of these models show the N-terminal region of Tβ_4 in contact with the actin barbed end, and the C-terminus in contact with the pointed end, in general agreement with the contacts deduced from crosslinking. The various models differ in some details of the conformation and orientation of these segments of Tβ_4 (Fig. 2). Considering that actin-bound Tβ_4 shows considerable segmental mobility (Safer et al. 1997), it is possible that Tβ_4 can alternate between different sets of contacts with actin.

Competition with other actin monomer binding proteins is consistent with proposed actin–Tβ_4 models. Cofilin (Dedova et al. 2006) and gelsolin (Wriggers et al. 1998) bind actin subdomains 1 and 3, and compete with thymosin for actin monomer binding. Tβ_4 binds weakly to the actin–DNase I complex (Dedova et al. 2006), suggesting that interaction with actin subdomain 2 plays an important role in dictating high stability of the complex, although steric effects cannot be unequivocally eliminated. The fact that Tβ_4 binding reduces the solvent exposure of the fluorescent probes conjugated to Gln41 of actin (Dedova et al. 2006) and enhances proteolysis of actin at Gly46, is consistent with actin subdomain 2 forming part of the binding interface in addition to subdomains 1 and 3.

Although Tβ_4 functions as an actin monomer regulatory protein, weak binding to filaments (K_d = 5–10 mM) also occurs (Carlier et al. 1996) and may be significant, given the high physiological concentrations of Tβ_4 in cells, although other F-actin binding proteins, such as α-actinin and tropomyosin, are likely to compete for the same binding sites. Actin filaments in the presence of high concentrations of Tβ_4 appear irregular in structure. In the presence of phalloidin and exogenous nuclei, the crosslinked actin–Tβ_4 complex can be induced to polymerize (Ballweber et al. 2002). Image reconstruction of these filaments shows an increase in the crossover pitch, from 36.0 to 40.5 nm. Difference map analysis, in addition to showing the additional mass of Tβ_4, suggests that there may be subtle changes in actin conformation in the crosslinked filaments.

Tβ_4 in Wound Healing, Cancer Metastasis, and Angiogenesis

Tβ_4 occupies a unique place among actin-binding proteins in that it appears to perform important biological roles not only through its role in regulation of actin polymerization but also by functioning as a *bona fide* signaling molecule. In spite of the well-documented effects of Tβ_4 on many aspects of cell physiology, its mechanism of action in cell signaling remains to be firmly determined.

It is presently unknown whether Tβ₄ signaling proceeds through a set of Tβ₄-specific receptors (Goldstein, Hannappel and Kleinman 2005) or is mediated by changes in free G-actin concentration, for example, through nuclear actin-mediated transcriptional regulation (Huff et al. 2004; Pederson and Aebi 2005).

The experimental use of Tβ₄ as a potential therapeutic agent was initially motivated by its immunomodulatory activity (Low, Hu and Goldstein 1981), and was subsequently extended to other systems. Topical or systemic administration of Tβ₄ promotes wound healing in a variety of experimental models, including skin wounds in healthy, diabetic, and aged rodents (Malinda et al. 1999; Philp et al. 2003a) and rodent models of chemical cornea injury (Sosne et al. 2001, 2002b). Some of the possible mechanisms for Tβ₄ action in wound-healing (Fig. 3) include enhancement of epithelial and endothelial cell migration (Sosne et al. 2002a), modulation of production of proinflammatory cytokines (Sosne et al. 2002b), inhibition of neutrophil chemotaxis leading to reduced inflammation (Young et al. 1999), increased collagen deposition (Malinda et al. 1999; Philp et al. 2003b), and enhancement of angiogenesis (Grant et al. 1999). Wound-healing and proangiogenic activities of Tβ₄ have been mapped to the central seven amino acid peptide implicated in actin binding (Philp et al. 2003a,b).

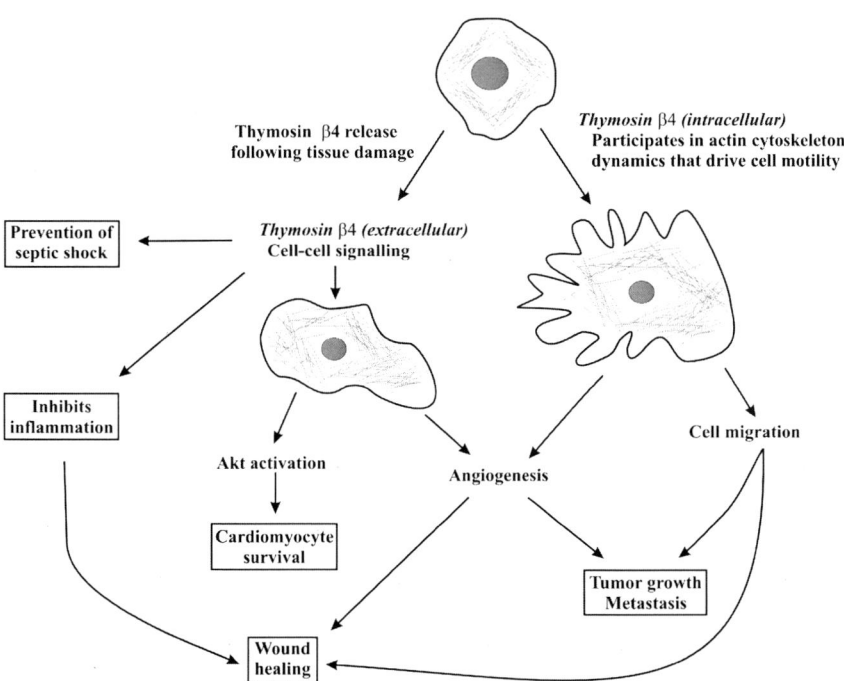

Fig. 3. Cellular functions attributed to Tβ₄ and Tβ₄SO. Adapted from Huff et al. (2001), copyright (2001) with permission from Elsevier

The role of Tβ$_4$ in angiogenesis may be highly physiologically relevant since Tβ$_4$ is present in large concentration in platelets, which are among the first cells to enter a wound and release factors that attract additional cells to the wound site. Tβ$_4$ can be chemically crosslinked to fibrin and collagen via a transglutaminase reaction (Huff et al. 2001). Thrombin-activated human platelets release large quantities of Tβ$_4$ and factor XIIIa, a transglutaminase enzyme, raising the possibility that factor XIIIa covalently crosslinks Tβ$_4$ to fibrin clots. This chemical coupling accumulates Tβ$_4$ in the wound and triggers migration of endothelial cells to facilitate angiogenesis (Huff et al. 2001).

In addition to promoting angiogenesis to improve wound-healing, Tβ$_4$ may also contribute to tumor-induced angiogenesis. Tumors from Tβ$_4$-overexpressing B16–F10 cells exhibit a marked increase in the number of blood vessels (Cha, Jeong and Kleinman 2003). Angiogenesis is required for aggressive tumor growth and metastasis (Hanahan and Folkman 1996), suggesting that Tβ$_4$ expression in tumors may be linked to poor prognosis.

Tβ$_4$ may also directly contribute to tumor cell metastasis (Fig. 3) since it has been observed that B16–F10 cells that have metastasized to the lung express high levels of Tβ$_4$. In addition, B16–F10 cells infected with Tβ$_4$-expressing adenovirus display more tumor growth and metastasis *in vivo* than B16–F10 cells infected with control adenovirus (Cha, Jeong and Kleinman 2003). Similarly, melanoma cells that overexpress Tβ$_4$ produce larger tumors and more metastatic lung nodules than injected melanoma cells that do not (Cha, Jeong and Kleinman 2003).

A recent study examining the role of Tβ$_4$ in cardiac injury repair found that extracellular application of Tβ$_4$ promoted cardiomyocyte survival and migration and resulted in improved cardiac function and reduced scar size following myocardial infarction (Bock-Marquette et al. 2004). This protective effect of Tβ$_4$ involved formation of Tβ$_4$ complex with PINCH (LIM-domain protein) and ILK (integrin-linked kinase). Formation of this signaling complex led to activation of ILK and phosphorylation of Akt, which is known to promote survival and growth of cardiomyocytes (Matsui and Rosenzweig 2005). In addition to activating ILK though a direct interaction, Tβ$_4$ may also affect ILK expression through actin-dependent regulation of transcription factors since ILK expression in the heart tissue is enhanced in the presence of Tβ$_4$ (Bock-Marquette et al. 2004). A common feature of these reports is that Tβ$_4$ promotes cell migration, which may result from down-regulation of cell adhesion through the ILK–Akt pathway.

Down-regulation of cell adhesion may also account for the anti-inflammatory activity of Tβ$_4$ (see chapter by Nunoi). Monocytes produce an oxidized form of Tβ$_4$ (Tβ$_4$ sulfoxide – referred to as Tβ$_{4SO}$) in response to glucocorticoids. Although oxidation attenuates the intracellular G-actin sequestering activity of Tβ$_4$, it greatly enhances its extracellular signaling properties (Young et al. 1999). Tβ$_{4SO}$ inhibits neutrophil chemotaxis *in vitro* and *in vivo*. While down-regulation of adhesion promotes the migration of nonmotile cells, e.g., in wound healing, cells that are already motile, such as neutrophils, cannot move without sufficient traction. In an animal model, the oxidized peptide inhibits carrageenin-induced edema in the BALB/c mouse paw. Significant suppression of swelling was evident after 6 h and was sustained up to 48 h. The suppression was dose-responsive and specific, as the nonoxidized peptide was inactive. $T\beta_{4SO}$ at a concentration of 20 µg mL^{-1} induced suppression equivalent to the suppression induced by 0.5 mg kg^{-1} dexamethasone, a potent anti-inflammatory steroid (Young et al. 1999). The possibility that $T\beta_{4SO}$ acts as an anti-inflammatory agent is an exciting prospect for therapy in inflammatory disease, because these molecules could give the therapeutic benefits of steroids without toxic side effects.

These complex observations may reflect more than one activity of $T\beta_4$: Cellular levels of $T\beta_4$ correlate with the size of the G-actin pool, and are higher in more motile cells, including metastatic cells. At the same time, high intracellular levels of $T\beta_4$ may result in locally higher levels outside the cell, which could promote the migration of endothelial cells by down-regulating adhesion, as well as other potential signaling effects.

As a major actin-sequestering protein in eukaryotic cells and a protein highly expressed in circulating blood cells (platelets and leukocytes) and present in serum, $T\beta_4$ may play an important role in alleviating the consequences of a release of large quantities of actin into the bloodstream. Significant quantities of actin are released into the blood and extracellular fluid following bacterial, fungal, and viral infections, and after tissue injury and trauma. The presence of large quantities of F-actin filaments contributes to cell necrosis and shock, and septic shock remains an important cause of mortality in infected patients admitted to intensive care units.

Under the ionic conditions of blood plasma, G-actin released from damaged tissues and dying cells polymerizes into F-actin, causing endothelial cell damage (Haddad et al. 1990; Erukhimov et al. 2000). G-actin in blood plasma is normally prevented from exerting its harmful effects by the plasma actin scavenging system consisting of vitamin D-binding protein and gelsolin. However, the actin sequestration system is overwhelmed under the conditions of mass cell lysis, and vascular damage can result (Haddad et al. 1990). Sequestration of G-actin by $T\beta_4$ may reduce toxic effects of actin release and prove useful for treatment of sepsis. In this situation, the relatively low affinity of $T\beta_4$ for actin may be outweighed by its high solubility and low immunogenicity. Indeed, injection of $T\beta_4$ following administration of an LD50 dose of LPS significantly decreased inflammatory cytokine production and increased the survival rate of mice injected with lethal doses of endotoxin with no detectable adverse side effects (Badamchian et al. 2003). Thus, administration of $T\beta_4$ to patients suffering from septic shock may be a useful therapeutic strategy.

Acknowledgments. EMDLC acknowledges support from the NSF (MCB-0546353), NIH (1 R01 GM071688), and the American Heart Association (0655849T).

References

Badamchian, M., Fagarasan, M. O., Danner, R. L., Suffredini, A. F., Damavandy, H. and Goldstein, A. L. 2003. Thymosin beta(4) reduces lethality and down-regulates inflammatory mediators in endotoxin-induced septic shock. Int. Immunopharmacol. 3, 1225–1233.

Ballweber, E., Hannappel, E., Huff, T., Stephan, H., Haener, M., Taschner, N., Stoffler, D., Aebi, U. and Mannherz, H. G. 2002. Polymerisation of chemically cross-linked actin:thymosin beta(4) complex to filamentous actin: Alteration in helical parameters and visualisation of thymosin beta(4) binding on F-actin. J. Mol. Biol. 315, 613–625.

Bock-Marquette, I., Saxena, A., White, M. D., Dimaio, J. M. and Srivastava, D. 2004. Thymosin beta 4 activates integrin-linked kinase and promotes cardiac cell migration, survival and cardiac repair. Nature 432, 466–472.

Carlier, M. F., Didry, D., Erk, I., Lepault, J., Van Troys, M. L., Vandekerckhove, J. H., Perelroizen, I., Yin, H., Doi, Y. and Pantaloni, D. 1996. Tbeta 4 is not a simple G-actin sequestering protein and interacts with F-actin at high concentration. J. Biol. Chem. 271, 9231–9239.

Cha, H., Jeong, M. and Kleinman, H. K. 2003. Role of thymosin beta 4 in tumor metastasis and angiogenesis. J. Natl Cancer Inst. 95, 1674–1680.

Czisch, M., Schleicher, M., Horger, S., Voelter, W. and Holak, T. A. 1993. Conformation of thymosin beta 4 in water determined by NMR spectroscopy. Eur. J. Biochem. 218, 335–344.

Dedova, I. V., Nikolaeva, O. P., Safer, D., De La Cruz, E. M. and dos Remedios, C. G. 2006. Thymosin beta 4 induces a conformational change in actin monomers. Biophys. J. 90, 985–992.

De La Cruz, E. M., Ostap, E. M., Brundage, R. A., Reddy, K. S., Sweeney, H. L. and Safer, D. 2000. Thymosin beta 4 Changes the conformation and dynamics of actin monomers. Biophys. J. 78, 2516–2527.

Domanski, M., Hertzog, M., Coutant, J., Gutsche-Perelroizen, I., Bontems, F., Carlier, M. F., Guittet, E. and van Heijenoort, C. 2004. Coupling of folding and binding of thymosin beta 4 upon interaction with monomeric actin monitored by nuclear magnetic resonance. J. Biol. Chem. 279, 23637–23645.

Erukhimov, J. A., Tang, Z. L., Johnson, B. A., Donahoe, M. P., Razzack, J. A., Gibson, K. F., Lee, W. M., Wasserloos, K. J., Watkins, S. A. and Pitt, B. R. 2000. Actin-containing sera from patients with adult respiratory distress syndrome are toxic to sheep pulmonary endothelial cells. Am. J. Respir. Crit. Care Med. 162, 288–294.

Goldschmidt-Clermont, P. J., Furman, M. I., Wachsstock, D., Safer, D., Nachmias, V. T. and Pollard, T. D. 1992. The control of actin nucleotide exchange by thymosin beta 4 and profilin. A potential regulatory mechanism for actin polymerization in cells. Mol. Biol. Cell 3, 1015–1024.

Goldstein, A. L., Hannappel, E. and Kleinman, H. K. 2005. Thymosin beta 4: Actin-sequestering protein moonlights to repair injured tissues. Trends Mol. Med. 11, 421–429.

Goldstein, A. L., Low, T. L., McAdoo, M., McClure, J., Thurman, G. B., Rossio, J., Lai, C. Y., Chang, D., Wang, S. S., Harvey, C., Ramel, A. H. and Meienhofer, J. 1977. Thymosin alpha 1: Isolation and sequence analysis of an immunologically active thymic polypeptide. Proc. Natl Acad. Sci. USA 74, 725–729.

Goldstein, A. L., Slater, F. D. and White, A. 1966. Preparation, assay, and partial purification of a thymic lymphocytopoietic factor (thymosin). Proc. Natl Acad. Sci. USA 56, 1010–1017.

Grant, D. S., Rose, W., Yaen, C., Goldstein, A., Martinez, J. and Kleinman, H. 1999. Thymosin beta 4 enhances endothelial cell differentiation and angiogenesis. Angiogenesis 3, 125–135.

Haddad, J. G., Harper, K. D., Guoth, M., Pietra, G. G. and Sanger, J. W. 1990. Angiopathic consequences of saturating the plasma scavenger system for actin. Proc. Natl Acad. Sci. USA 87, 1381–1385.

Hanahan, D. and Folkman, J. 1996. Patterns and emerging mechanisms of the angiogenic switch during tumorigenesis. Cell 86, 353–364.

Hertzog, M., van Heijenoort, C., Didry, D., Gaudier, M., Coutant, J., Gigant, B., Didelot, G., Preat, T., Knossow, M., Guittet, E. and Carlier, M. F. 2004. The beta-thymosin/WH2 domain: Structural basis for the switch from inhibition to promotion of actin assembly. Cell 117, 611–623.

Huff, T., Muller, C. S. G., Otto, A. M., Netzker, R. and Hannappel, E. 2001. β-Thymosins, small acidic peptides with multiple functions. Int. J. Biochem. Cell Biol. 33, 205–220.

Huff, T., Rosorius, O., Otto, A. M., Muller, C. S., Ballweber, E., Hannappel, E. and Mannherz, H. G. 2004. Nuclear localization of the G-actin sequestering peptide thymosin beta 4. J. Cell Sci. 117, 5333–5341.

Huff, T., Zerzawy, D. and Hannappel, E. 1995. Interactions of beta-thymosins, thymosin beta 4-sulfoxide, and N-terminally truncated thymosin beta 4 with actin studied by equilibrium centrifugation, chemical cross-linking and viscometry. Eur. J. Biochem. 230, 650–657.

Irobi, E., Aguda, A. H., Larsson, M., Guerin, C., Yin, H. L., Burtnick, L. D., Blanchoin, L. and Robinson, R. C. 2004. Structural basis of actin sequestration by thymosin-beta 4: Implications for WH2 proteins. EMBO J. 23, 3599–3608.

Jean, C., Rieger, K., Blanchoin, L., Carlier, M. F., Lenfant, M. and Pantaloni, D. 1994. Interaction of G-actin with thymosin beta 4 and its variants thymosin beta 9 and thymosin beta met9. J. Muscle Res. Cell Motil. 15, 278–286.

Li, X., Zimmerman, A., Copeland, N. G., Gilbert, D. J., Jenkins, N. A. and Yin, H. L. 1996. The mouse thymosin beta 4 gene: Structure, promoter identification, and chromosome localization. Genomics 32, 388–394.

Low, T. L. K., Hu, S. K. and Goldstein, A. L. 1981. Complete amino acid sequence of bovine thymosin beta 4: A thymic hormone that induces terminal deoxynucleotidyl transferase activity in thymocyte populations. Proc. Natl Acad. Sci. USA 78, 1162–1166.

Malinda, K. M., Sidhu, G. S., Mani, H., Banaudha, K., Maheshwari, R. K., Goldstein, A. L. and Kleinman, H. K. 1999. Thymosin beta 4 accelerates wound healing. J. Invest. Dermatol. 99, 364–368.

Matsui, T. and Rosenzweig, A. 2005. Convergent signal transduction pathways controlling cardiomyocyte survival and function: The role of PI 3-kinase and Akt. J. Mol. Cell. Cardiol. 38, 63–71.

Pederson, T. and Aebi, U. 2005. Nuclear actin extends, with no contraction in sight. Mol. Biol. Cell 16, 5055–5060.

Philp, D., Badamchian, M., Scheremeta, B., Nguyen, M., Goldstein, A. L. and Kleinman, H. K. 2003a. Thymosin beta 4 and a synthetic peptide containing its actin-binding domain promote dermal wound repair in db/db diabetic mice and in aged mice. Wound Repair Regen. 11, 19–24.

Philp, D., Huff, T., Gho, Y. S., Hannappel, E. and Kleinman, H. K. 2003b. The actin binding site on thymosin beta 4 promotes angiogenesis. FASEB J. 17, 2103–2105.

Safer, D. and Chowrashi, P. K. 1997. β-thymosins from marine invertebrates: Primary structure and interaction with actin. Cell Motil. Cytoskeleton 38, 163–171.

Safer, D., Elzinga, M. and Nachmias, V. T. 1991. Thymosin beta 4 and Fx, an actin-sequestering peptide, are indistinguishable. J. Biol. Chem. 266, 4029–4032.

Safer, D., Sosnick, T. R. and Elzinga, M. 1997. Thymosin beta 4 binds actin in an extended conformation and contacts both the barbed and pointed ends. Biochemistry 36, 5806–5816.

Sosne, G., Chan, C. C., Thai, K., Kennedy, M., Szliter, E. A., Hazlett, L. D. and Kleinman, H. K. 2001. Thymosin beta 4 promotes corneal wound healing and modulates inflammatory mediators in vivo. Exp. Eye Res. 72, 605–608.

Sosne, G., Hafeez, S., Greenberry, A. L. 2nd and Kurpakus-Wheater, M. 2002a. Thymosin beta 4 promotes human conjunctival epithelial cell migration. Curr. Eye Res. 24, 268–273.

Sosne, G., Szliter, E. A., Barrett, R., Kernacki, K. A., Kleinman, H. and Hazlett, L. D. 2002b. Thymosin beta 4 promotes corneal wound healing and decreases inflammation in vivo following alkali injury. Exp. Eye Res. 74, 293–299.

Vancompernolle, K., Vandekerckhove, J., Bubb, M. R. and Korn, E. D. 1991. The interfaces of actin and Acanthamoeba actobindin – identification of a new actin-binding motif. J. Biol. Chem. 266, 15427–15431.

Van Troys, M., Dewitte, D., Goethals, M., Carlier, M. F., Vandekerckhove, J. and Ampe, C. 1996. The actin binding site of thymosin beta 4 mapped by mutational analysis. EMBO J. 15, 201–210.

Weber, A., Nachmias, V. T., Pennise, C. R., Pring, M. and Safer, D. 1992. Interaction of thymosin beta 4 with muscle and platelet actin: Implications for actin sequestration in resting platelets. Biochemistry 31, 6179–6185.

Wriggers, W., Tang, J. X., Azuma, T., Marks, P. W. and Janmey, P. A. 1998 Cofilin and gelsolin segment-1: Molecular dynamics simulation and biochemical analysis predict a similar actin binding mode. J. Mol. Biol. 282, 921–932.

Young, J. D., Lawrence, A. J., MacLean, A. G., Leung, B. P., McInnes, I. B., Canas, B., Pappin, D. J. C. and Stevenson, R. D. 1999. Thymosin beta 4 sulfoxide is an anti-inflammatory agent generated by monocytes in the presence of glucocorticoids. Nat. Med. 5, 1424–1427.

10
Actin and Actin-Binding Proteins in Cancer Progression and Metastasis

Marleen Van Troys, Joël Vandekerckhove, and Christophe Ampe

Introduction

With over 10 million cases and over 5.7 million deaths a year (GLOBOCAN data 2002; Ferlay et al. 2004), cancer presents a major health problem worldwide. Dealing with cancer, both from a clinical and from a fundamental scientific view, is complicated by the extensive diversity within this disease. Based on their origin, more than 100 different human cancer types have been described and, within one organ, distinct subtypes can occur. In addition, tumors may progress to a malignant state that manifests itself mainly when tumor cells spread out from the primary lesion (or neoplasm) and metastasize, i.e., colonize distant sites of the patient's body. Metastatic cancer, accounts for 90% of cancer-related lethality (Sporn 1996) and hence forms the primary determining factor in patient outcome. Therefore a detailed molecular understanding of tumor cell spread is required to render treatment of cancer more specific and efficient.

A cancer cell exploits defects in normal cellular regulatory circuits. As outlined below, tumor cell migration and adhesion and tumor cell interactions with host extracellular matrix (ECM) and with host cells are important features during the switch to the metastatic state. The actin cytoskeleton is a central player in these processes and consequently it is necessarily involved in many aspects of cancer and cancer progression (Lambrechts, Van Troys and Ampe 2004).

In this chapter we present an overview of current knowledge on deregulations within the actin cytoskeleton during motility events in this disease. In a cancer cell both actin itself and members within the large group of actin-binding proteins (ABPs) may be affected. The differential functioning of the actin system is based on (1) the presence of mutant proteins (actins, ABPs), (2) altered expression levels of their genes, and/or (3) their altered activation status as a consequence of altered upstream signaling.

In addition, ABPs are without doubt proving to surpass their original name. Next to their capacity to reorganize the actin cytoskeleton, many ABPs have additional partners and/or appear to be involved in functions not directly related to regulating the actin cytoskeleton. Many can also reside in the nucleus where they may affect gene expression. Therefore these novel properties of ABPs may additionally turn out relevant in the contributions of these proteins to cancer cell motility and cancer progression. Although we are far from a complete understanding at the molecular level, the current data reveal that components of the actin system hold significant potential as targets in future cancer therapies.

Cell Migration Depends on a Dynamic Actin System

Cell motility and migration are pivotal to eukaryotic life (Pollard and Borisy 2003). Many unicellular eukaryotes migrate to nutrient sources in a chemotactic way, i.e., attracted by factors secreted by the organism they feed on. In multicellular organisms, proper development is impossible without completion of multiple migratory steps during embryogenesis and morphogenesis to lay down the body plan (Keller 2005) and accomplish the wiring of the nervous, vascular, and lymphatic systems (Weinstein 2005). After development, cell migration is limited since differentiated cells are confined within tissues by cell–cell adhesion or by interaction with ECM proteins. At this stage, cells that do display locomotion mainly act in response to potential harm. Fibroblasts and epithelial cells locally migrate during wound repair, and locomotion of white blood cells across vessel walls and within tissues is essential in immune surveillance.

In a strikingly similar way, malignant cancer cells are able to move out of the primary tumor and beyond the boundaries of the tissue or organ where the tumor initially developed (Mareel and Leroy 2003). Given the essential contribution of cell migration in tumor cell malignancy, we first briefly discuss the functioning of the actin-based machinery during normal cell migration. This also serves as an introduction to the function of key players of the actin system putatively deregulated during aberrant motility of tumor cells.

The Actin Machinery Drives Cell Migration

A well-described example of migrating cells is that of fibroblasts moving in a random or (semi)directional fashion over a two-dimensional surface, a migratory mode that is termed *mesemchymal* (Fig. 1). Various aspects of this type of migration have been studied in detail (see also Vicente-Manzanares, Webb and Horwitz 2005). Mesemchymal cells adopt an elongated spindle-like shape, are internally polarized (Franca-Koh and Devreotes 2004; Gamba et al. 2005; Nishiya et al. 2005), and succeed in moving or sliding in

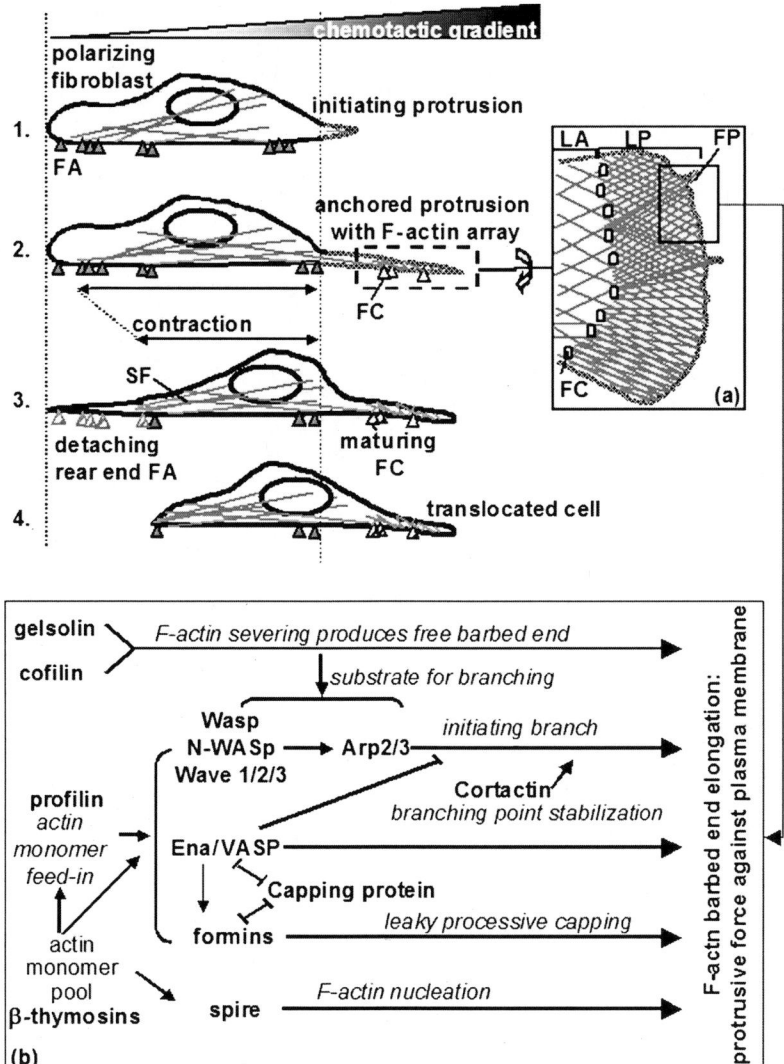

FIG. 1. Cell migration is dependent on actin organization and dynamics. The central scheme shows the four-step migration model for a polarized mesemchymal cell migrating up a chemotactic gradient, (1) protrusions form and extend depending on actin filament elongation downstream of receptor activation, (2) anchoring via integrin-based and actin array-associated focal contacts (FC). The FC mature into focal adhesions (FA), and (3) actomyosin-dependent contraction of stress fibers (SF) pulls the cell body forward, and (4) rear end focal adhesion sites (FA) release. *Inset* (**a**) shows a top view of the protruding membrane highlighting actin organization herein. In these protrusions the fast-growing barbed ends are oriented toward the membrane. *LP*, lamellipodium or leading edge consisting of a densely crosslinked actin network; *FP*, filopodium, needle-like protrusion extending beyond the edge containing F-actin bundles arising in or from the LP network. The lamella (*LA*) is a less dense actin network enriched in myosin II and tropomyosin (see also Gupton et al. 2005; Small and

(continued)

the "chosen" direction by repeating a four-step process (Fig. 1) of (1) membrane protrusion at the front of the cell, (2) substrate adhesion of this leading edge, (3) cell body contraction, and (4) detachment of the rear end of the cell (Lauffenburger and Horwitz 1996; Mitchison and Cramer 1996).

It is widely accepted and well documented that the actin cytoskeleton constitutes essential driving forces during all steps of this process as discussed next (Lambrechts, Van Troys and Ampe 2004). Within developing membrane protrusions at the front of a polarized cell (step 1, Fig. 1), an active molecular machinery tightly couples unidirectional actin filament elongation to forces pushing locally against the membrane (Fig. 1, inset a).

Unidirectional growth is an inherent property of actin filaments, related to the head-to-tail association of actin monomers and to the associated hydrolysis of ATP bound to the actin protomers. This results in a fast-growing (barbed) actin filament end and a slow-growing (pointed) end. Superimposed on this property and essential to the mechanistic coupling to produce force is the balanced activity of ABPs. These are recruited to and/or activated at the site of protrusion, for instance as a response to external chemotactic signals, where they control the growth of actin filaments and organize them in supramolecular structures.

Typical protrusions that form in a two-dimensional context are lamellipodia, thin veil-like structures that contain dense, branched actin filament networks, and filopodia, finger-like structures that contain parallel bundles of actin filaments (Svitkina and Borisy 1999; Small et al. 2002; Faix and Rottner 2006) (Fig. 1).

Figure 1 (inset b) shows ABPs that have been demonstrated to cooperate in locally generating protrusion. It is already worth noting that most of these proteins display altered expression in tumor cells (see below). These ABPs

---

FIG. 1. (continued) Resch 2005). *Inset* (**b**) lists actin-binding proteins that can induce or modulate barbed end elongation and that are present in LP or FP or both (for references see main text). Functional interaction between cofilin (DesMarais et al. 2004) or between gelsolin (Falet et al. 2002) and Arp2/3 has been demonstrated but these proteins may also produce F-actin ends compatible with Ena/Vasp activity. The connection drawn between Ena/VASP Arp2/3-activity is based upon the observation that VASP can reduce the frequency of Arp2/3-dependent branch formation *in vitro* (Skoble et al. 2001). The line drawn between capping protein and Ena/VASP-activity points at their competitive activity at the barbed end, i.e., Ena/VASP proteins exert anticapping activity as described by Barzik et al. (2005). A similar competitive effect between formin and capping protein is described in yeast (Kovar, Wu and Pollard 2005). The activity of the Spire protein is described by Schumacher et al. (2004) and Quinlan et al. (2005), the connection between formins and VASP by Grosse et al. (2003) and Schirenbeck et al. (2005)

display a wide range of activities on actin (for references see Van Troys, Vandekerckhove and Ampe 1999; Huff et al. 2001; Small et al. 2002; dos Remedios et al. 2003; Pollard and Borisy 2003; Lambrechts, Van Troys and Ampe 2004; Zigmond 2004; Polet et al. 2006).

β-Thymosins (see chapter by An et al.) and profilins (see chapter by Moens) are the most ubiquitous actin monomer-sequestering proteins and are important for feeding actin monomers to elongating ends. Cofilins (see chapter by Maloney et al.) and gelsolin (see chapter by Burtnick and Robinson) can act as filament-severing proteins. Their fragmenting capacities not only promote depolymerization but also provide free barbed ends that may form substrates for other ABPs (such as the WASP or Wave proteins, Ena/VASP proteins, formins, and Spire) resulting in barbed end elongation. In addition, Ena/VASP proteins (Lambrechts et al. 2000) and also the Arp2/3 complex have actin filament-nucleating properties with the feature that Arp2/3 generates branches on existing filaments. Finally, also barbed end capping activity (by capping protein, CapG, or gelsolin) is essential in actin dynamics as it competes with the elongation promoting ABPs (Barzik et al. 2005; Kovar, Wu and Pollard 2005).

On top of these activities, the formation of supramolecular actin filament-based structures in protrusions requires ABPs acting at interfilament connections. These include cortactin which stabilizes Arp2/3-induced branching points (Bryce et al. 2005); α-actinin, filamin, and EPLIN which are present in crosslinked networks (Flanagan et al. 2001; Maul et al. 2003); fascin and plastin/fimbrin which bundle filaments (Delanote, Vandekerckhove and Gettemans 2005; Giganti et al. 2005; Hashimoto, Skacel and Adams 2005); and tropomyosin which stabilizes them (Gunning et al. 2005).

For many of these ABPs some type of upstream signaling pathway has been documented. In particular, signaling cascades acting via small GTPases of the Rho-family and polyphosphoinositide (PPI)-modulating enzymes appear to be important (for details see Yin and Janmey 2003; Raftopoulou and Hall 2004; Stradal and Scita 2006).

The second step in the two-dimensional-migration model involves anchorage of newly formed protrusions to substrate molecules. In a physiological context, this will in most cases imply adhesion of β1 and/or β3 integrin transmembrane receptors to ECM proteins via the formation of transient focal contacts (FC) at the base of the dense lamellipodial actin network (Fig. 1). In these focal contacts, different anchor proteins are sequentially recruited and become involved in connecting the actin cytoskeleton to the membrane (Zaidel-Bar et al. 2003). Focal contacts are transient structures: as the lamellipod protrudes they either disassemble and are replaced by novel contacts positioned more anteriorly, or they mature into the more rigid focal adhesions (Ballestrem et al. 2001; Kaverina, Krylyshkina and Small 2002; Zaidel-Bar et al. 2003) (Fig. 1). It should be noted that too rigid adhesion negatively correlates with migratory speed (Huttenlocher, Ginsberg and Horwitz 1996).

Cell body contraction, the third essential step in cell locomotion, is also actin dependent. Stress fibers, that consist of myosin II decorated actin bundles and span the cell body of substrate adherent cells (Fig. 1), contract downstream of Rho and myosin light chain kinase. This shortening of stress fibers drags the trailing cell body forward and mechanically stresses rear cell attachments. Their detachment, e.g., by action of the intracellular protease calpain (Franco and Huttenlocher 2005), results in cell translocation (step 4).

Different Cell Types Migrate via Different Modes

Next to this mesemchymal- or fibroblast-like migration, other migratory modes or adaptations have been described that are less strongly dependent on integrin adhesion. Epidermal keratinocytes are rapidly moving cells in which the coupling between actin-based protrusion, adhesion, and retraction appears optimized: their cell body closely follows the protruding lamellipodium, they have no lagging tail and form less adhesion sites (Anderson, Wang and Small 1996). Leukocytes (lymphocytes, neutrophils) also display a faster crawling mode on a two-dimensional substrate. Their migration is termed "amoeboid" based on the similarity to migration by unicellular organisms like *Dictyostelium discoideum* (Condeelis 1993; Friedl, Borgmann and Brocker 2001). Lack of stress fibers, the presence of cortical actin and the establishment of weak transient contacts with the surface characterize this migration mode.

In general, cells adopting amoeboid movement display fast morphodynamics as they continuously change shape based on cortical actin filament polymerization and actin–myosin-based contraction. The differences between mesemchymal and amoeboid locomotion manifest themselves even more strongly in a three-dimensional context and are relevant to cancer cell migratory behavior.

In addition, it is well documented, though mechanistically less well understood, that specific cell types such as epithelial monolayers during gastrulation or wound healing move as one functional group. During this collective migration (reviewed by Friedl, Hegerfeldt and Tusch 2004) as sheets, strands, or clusters, cell–cell contacts are retained via cell adhesion molecules, a.o. cadherins. Motility aspects such as protrusion, traction force, and retraction are apparently executed by different subsets of cells within the motile group, however with a high degree of synchronization in each subset (Friedl, Hegerfeldt and Tusch 2004).

Cancer Cell Migration

Variations on the Theme of Normal Cell Migration

From histopathological images of human tumors of different degrees of progression, it has been evident for many decades that not all invasive tumors provide an identical picture. Indeed, some form protruding sheets or strands still attached to the primary tumor (collective migration mode), for example in melanoma (Friedl 2004) and colorectal carcinoma (Nabeshima et al. 1999) whereas in other tumors, cells detach as single cells of varying morphologies. These single cells can be fibroblast-like or have a rounded shape. The latter is usually correlated with a stronger level of dedifferentiation (Thiery 2002).

It can be inferred that, similar to normal cells, cancer cell migration modes are also the result of a balance between forward movement and cell–cell and/or cell–substrate adhesion. Obviously, in motile cancer cells these complex equilibria are deregulated in comparison to noncancerous cells in differentiated tissue. From the normal migration modes described above, it is evident that this balance is based upon the actions of integrins (cell–substrate adhesion), cadherins (cell–cell adhesion), and the actin system that are intricately linked by intracellular signaling pathways (Brunton, MacPherson and Frame 2004; Burridge and Wennerberg 2004; Huber, Kraut and Beug 2005; Nishiya et al. 2005).

The actin system functions downstream of these two adhesion systems but conversely can itself influence adhesion (Chu et al. 2004; Scott et al. 2005; Wiesner, Legate and Fassler 2005; Yamada et al. 2005b), and all three systems are themselves controlled by growth factor receptor signaling (references in Brunton, MacPherson and Frame 2004).

In a cancer cell, deregulation of either system can be "beneficial" to exert invasion as already evident from the identification of invasion-modulating proteins within these systems such as E-cadherin, focal adhesion components Src or focal adhesion kinase (FAK), the actin regulator RhoC, the ABP cortactin, etc. The alterations observed within the actin system in tumor cells are considered in detail below. Full understanding of cancer cell movement, however, requires to also take into account the microenvironments in which these cells move because these provide input for actin-based migration during tumor cell invasion.

Role of Matrix and Host Cells in Tumor Cell Invasion

Activating Signaling Pathways to the Actin Cytoskeleton

In vivo, migrating cells are confronted with a three-dimensional stromal lattice consisting of protein fibers (mainly collagen) and embedded cells. Consequently, migrating cancer cells are presented with a physical barrier. The density and composition of this ECM varies depending on the location. The most tightly packed ECM is the thin acellular layer formed by the basement membrane that, e.g., underlies all epithelia and endothelia. As the majority of tumors originate in epithelial lining (epidermal, endodermal, endothelial, glandular), disruption of the basement membrane is, in most cases, a prerequisite for tumor cell invasion. It appears, however, that escaping tumor cells are optimally equipped to face (and/or exploit) this degree of environmental variation and complexity during the various steps of the invasive process.

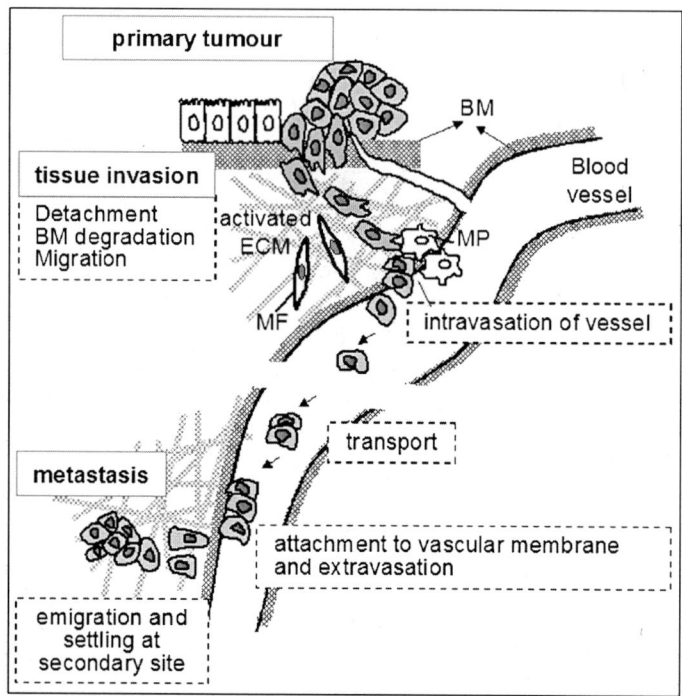

Fig. 2. Metastasis is a multistep process. The hallmarks of a grown tumor within differentiated tissue are self-sufficient growth and limitless replicative potential, escape from programmed cell death and sustained angiogenesis (Hanahan and Weinberg 2000). Malignancy is characterized by the capacity of cells within the tumor to initiate the invasive and metastatic process. By loss of homotypic cell–cell adhesion, malignant cells detach from the primary tumor mass, disrupt the basement membrane (BM), and migrate through surrounding tissue and along protein fibers toward blood or lymphatic vessels. Intracellular rearrangements in the actin cytoskeleton and extra-cellular matrix (ECM) degradation by proteolytic enzymes from the tumor cells (in the mesemchymal migration mode, see text) enable this migratory step. The tumor cells also activate or stimulate secretion of growth factors by host cells such as myofi-broblasts (MF) and macrophages (MP). This facilitates (chemotactic) tumor cell inva-sion and putatively also the intravasation through the basement membrane and endothelial layer of the vessel. At a distant site, upon extravasation, a reverse process occurs that again involves cytoskeletal rearrangements. Tumor cells adhere to the cap-illary endothelium, penetrate and migrate through the extravascular tissue where they again create a growth permissive microenvironment a.o. by attracting blood supply. This finally results in a secondary lesion or metastasis

Figure 2 gives a general overview of the consecutive steps of basement membrane disruption, local initial invasion, vessel intravasation, transport, vessel extravasation, and homing, that together form the metastatic process that malignant cells need to follow to ultimately give rise to a secondary tumor in a distant organ (see legend for details).

During the last decade, several research groups have focused on describing and understanding how tumor cells find their way through the matrix during the crucial initial migratory steps of this process. It is now evident that actin-based motility and adhesion are important features during this onset of malignancy. Important progress has been realized in this area via the development or optimization of novel real-time imaging techniques. Visualizing, at single cell resolution, the movements of individual normal and cancer cells (from established cell lines or cancer tissue explants) in reconstituted matrices of controlled composition has clarified fundamental aspects of migration in dense protein networks (Friedl et al. 1997; Maaser et al. 1999; Friedl and Brocker 2000; Hegerfeldt et al. 2002).

Using multiphoton-based intravital imaging, the behavior of invasive cells in and near a primary mammary tumor and the vessels surrounding it have been recorded *in situ* in living animals, providing fascinating high-resolution movies of malignant cells in action (Farina et al. 1998; Ahmed et al. 2002; Condeelis and Segall 2003; Wyckoff et al. 2004; Condeelis, Singer and Segall 2005; Yamaguchi, Wyckoff and Condeelis 2005). Simultaneously, these technologies allow dynamic imaging of ECM scaffolds and hence, visualize how migrating cancer cells interact with collagen fibers (Condeelis and Segall 2003; Wolf et al. 2003a). In addition, the interaction of tumor cells with host cells in the stromal tissue has also been recorded *in situ* either using dual fluorescence imaging in which host and tumor display different fluorescence (Yang et al. 2004b; Hoffman 2005), or employing specific transplantation chambers (Skobe et al. 1997; Bajou et al. 1998) or by employing a chemotactic *in vivo* invasion assay (Wyckoff et al. 2004; Condeelis, Singer and Segall 2005).

Figure 3 illustrates the different migratory modes observed for invasive cancer cells and some of the characteristic key effector molecules upstream of actin dynamics. Cells moving through stroma in a mesemchymal fashion utilize the multistep process described above but in addition remodel the matrix (Friedl and Wolf 2003). When their exploratory protrusions contact the matrix fibers, proteases such as matrix metalloproteinases (MMP and MT-MMPs) and serine proteases become enriched at the integrin adhesion sites. Concomitantly actin, ABPs, and signaling molecules are recruited to the intracellular side of the focal contacts (Wolf and Friedl 2005).

Hot spots of matrix degradation near the cell surface are located at the contact sites of invadopodia (Friedl 2004). The latter are actin rich membrane extensions that protrude deep into the matrix and are typically formed by aggressive tumor cells (Buccione, Orth and McNiven 2004; McNiven, Baldassarre and Buccione 2004). Invadopodia are reminiscent of podosomes. These are protrusions present on differentiated cells with high matrix penetrating or degrading capacity (macrophages, dendritic cells, osteoclast; reviewed in Linder and Kopp 2005). EGF-induced formation of podosomes requires the integrity of an activated actin

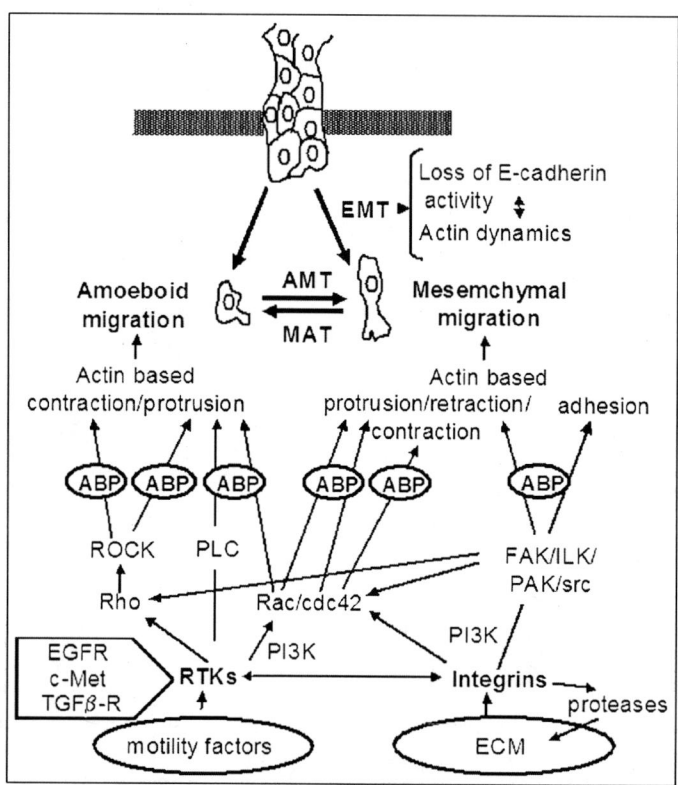

FIG. 3. Tumor cells use different migration modes during invasion. Depending on the cell type or on the extracellular conditions tumor cells either move as spindle-shaped mesemchymal cells that employ integrins and proteases to, respectively, adhere to and remodel the ECM or crawl and squeeze like an amoeboid through the matrix by adapting cell shape. Both migration types depend on actin dynamics in the front of the cell in response to extracellular input (ECM and secreted motility factors). Signaling proceeds via (lipid) kinases and small GTPases. In the amoeboid mode, Rho-based actin filament contraction is crucial. Downstream of these signaling molecules, specific actin-binding proteins, become activated (presented as ABP in this figure, for details see *inset* (**b**) of Fig. 1 and main text). Switches between migratory modes are discussed in the text. *EMT*, epithelial–mesemchymal transition is as shown dependent on loss of E-cadherin and involves extensive actin reorganization; *MAT*, mesemchymal–amoeboid transition; *AMT*, amoeboid–mesemchymal transition. This figure is adapted from a figure from Friedl (2004). *FAK*, focal adhesion kinase; *EGFR*, epidermal growth factor receptor; *ILK*, integrin-linked kinase; *PAK*, p21-activated kinase; *PI-3K*, phosphatidylinsitol-3-kinase; *PLC*, phospholipase C; *ROCK*, Rho-associated kinase; *RTKs*, receptor tyrosine kinases; *TGFβ-R*, transforming growth factor receptor

machinery containing in particular Arp2/3, N-WASp and its regulator Nck, cdc42, WIP, cortactin, and dynamin (McNiven, Baldassarre and Buccione 2004; Yamaguchi et al. 2005). Sustained podosome stability is required for efficient matrix degradation and depends on cofilin activity (Yamaguchi et al. 2005). Taken together, three-dimensional-mesemchymal movements depend on proteolysis, integrin adhesion, and actin dynamics (Maaser et al. 1999; Wolf et al. 2003a) (Fig. 3). In tumors this type of migration is observed in, e.g., fibrosarcomas, in gliomas, in progressed epithelial cancers (after epithelial–mesemchymal transition, see below), and in the cells at the invasive front of tumors protruding via collective movement (Friedl, Hegerfeldt and Tusch 2004).

However, pericellular proteolysis is not a prerequisite to cross ECM-barriers. Using an amoeboid-like, leukocyte type migration mode, certain tumor cells (e.g., lymphomas) display morphological adaptation to preformed matrices and are able to squeeze through preexisting holes in the matrix (Verschueren et al. 1994; Wolf et al. 2003b). In their escape from the primary tumor, amoeboid metastatic mammary carcinoma cells (MtLn3) use existing collagen fibers as tracks to move fast in the direction of blood vessels (Wang et al. 2002). *In vitro*, amoeboid tumor cell migration is protease independent and largely integrin independent (Fig. 3). It, however, requires functional Rho–ROCK signaling to control cortical actin filaments, a.o. for myosin II-based filament contraction (Sahai and Marshall 2003) and for cofilin-based filament remodeling (DesMarais et al. 2004).

Tumor cells not only display different morphologies and migratory modes, but they can also switch between these modes (Fig. 3). It is important to realize that this mimics (or mimics the reverse of) similar physiological transitions occurring during development and cell differentiation (reviewed in Thiery 2003; Friedl 2004). One such switch in morphology and motility is the epithelial to mesenchymal transition (EMT) that is considered as a point of no return in the progression of epithelial cancers (Thiery 2002). EMT is mainly characterized by loss of basal–apical polarity and of E-cadherin-based cell–cell adhesion (Fig. 3).

The various molecular mechanisms that can induce this switch are being elucidated. These include genetic and epigenetic changes and transcriptional control (reviewed by Berx and Van Roy 2001; Thiery 2003; Huber, Kraut and Beug 2005). Also, considerable cytoskeletal reorganization is involved during EMT and, interestingly, the expression levels of a number of actin polymerization-modulating ABPs (gelsolin, capping protein, etc.) are altered by activation of Snail – a transcriptional repressor-inducing epithelial dedifferentiation (De Craene et al. 2005).

Next to this EMT switch, it has recently been demonstrated that, upon altering environmental conditions counteracting one migratory mode, various cancer cells can keep moving by switching from mesemchymal to amoeboid movement (MTA) or vice versa (ATM; Fig. 3) (Sahai and Marshall 2003; Wolf et al. 2003a; Wolf and Friedl 2005). This was observed *in vitro* upon addition of either protease-, integrin-, or Rho/ROCK-blocking factors (Sahai and Marshall 2003; Wolf et al. 2003a). Marshall and colleagues recently showed that the Rho/ROCK-independent mesemchymal-like migration employs Cdc42/MRCK (myotonic dystrophy kinase-related Cdc42-binding kinase)-based signaling to control contractile processes (Wilkinson, Paterson and Marshall 2005). Since these escape mechanisms of malignant cells are relevant when applying specific therapeutics (e.g., MMP inhibitors), it may be more efficient to directly target the actin migration machinery in order to switch off migration.

Tumor cell–stroma interactions go beyond mere proteolytic digestion of the matrix to provide room to move or the capacity to trespass barriers. A cancer cell actively modulates the stroma into an environment supporting efficient tumor progression and invasion by the secretion of factors that promote tumor cell growth as well as migration. Next to autocrine signaling (e.g., tumor cell secreted EGF signals to the actin cytoskeleton of the same cells) there is considerable paracrine signaling. The tumor cells secrete growth factors and cytokines that attract and activate host cells such as fibroblast, myofibroblasts, macrophages, and endothelial cells (Fig. 2). In response these cells secrete (growth) factors facilitating tumor progression and/or migration. This is extensively reviewed in Mueller and Fusenig (2004). An early example of this is the recruitment of endothelial cells via vascular endothelial GF (VEGF) secretion by tumor cells during neoangiogenesis (reviewed in Carmeliet 2005) that is crucial both in early tumor growth and for invasion.

We will now expatiate on two recently reported examples that establish a link between host cell engagement and the functioning of the actin cytoskeleton in the tumor cell. De Wever and colleagues (2004a,b) describe a paracrine loop between colon cancer cells and transforming GF-β (TGF-β)-activated myofibroblasts. Upon activation, the latter produce hepatocyte GF (HGF) and the ECM component tenascin that, respectively, mediate activation of Rac and inactivation of Rho in the tumor cells, conditions favoring actin-based mesemchymal migration.

The Condeelis group (Wyckoff et al. 2004) discovered in an orthotopically grown mammary tumor in rat a promigratory paracrine loop between the carcinoma cells and macrophages accumulating near the blood vessel at the tissue side. The tumor cells produce colony-stimulating factor (CSF), a known promoter of tumor progression (references in Mueller and Fusenig 2004), that activates the infiltrating macrophages. In response, these cells produce EGF acting as chemoattractant for the tumor cells that then track with high directional persistence to the blood vessel. In an experimental set up consisting of macrophages and a needle containing EGF placed in the tumor microenvironment, the tumor cells migrate into the needle and can be collected (Wang et al. 2004a). The transcriptome of these collected invasive tumor cells extensively differs from that of the bulk tumor and provides insight in the requirement for the invasive switch: levels of mRNAs encoding for proteins that promote an antiproliferative, antiapoptotic, and highly motile state are altered. Accordingly, expression of many ABPs and their upstream regulators (Fig. 3) (see below) were found upregulated (Lorenz et al. 2004; Condeelis,

Singer and Segall 2005; Wang et al. 2005a; Yamaguchi, Wyckoff and Condeelis 2005) as a consequence of this host tumor interplay. These and other reported deregulations in the actin system in a tumor context are reviewed in detail below.

Deregulation Within the Actin System in tumor contexts

Given the tight link between actin dynamics and cell migration, a deregulated actin system obviously forms a principle element of the oncogenic and, even more likely, the invasive proteome of cancer cells. As causes for deregulation within the actin system, one can distinguish (1) mutations, translocations, or amplifications of coding genes (Futreal et al. 2004), (2) epigenetic changes affecting gene transcription, (3) changed mRNA levels, or (4) changed activity of the expressed protein. More recently an example of translational control has also been reported.

Actin Expression in Cancer Cells

Only few mutations in genes encoding components of the actin cytoskeleton have been reported in relation to cancer (see chapter by Sparrow and Laing). For β-actin two mutants have been described in tumor cells and their expression was associated with a neoplastic phenotype. Parental B16 mouse melanoma cells abundantly express a β-actin mutant with a single point mutation R28L (Sadano et al. 1988) of which the expression level in sublines negatively correlates with invasiveness and metastasis (Sadano, Taniguchi and Baba 1990).

β-Actin G244D (Vandekerckhove et al. 1980) was identified in immortalized human fibroblasts that produce tumors in athymic mice (Leavitt et al. 1987). *In vitro*, this mutant displays a polymerization defect and reduced binding to the chaperone CCT (Rommelaere et al. 2004). Interestingly, expression of subunits of the latter complex, which is largely dedicated to actin and tubulin folding, is upregulated in diverse (invasive) cancers (Yokota et al. 2001; Rhodes et al. 2004; Shen, Ghosh and Chinnaiyan 2004; Wang et al. 2004a) and this may enhance production levels of actin isoforms. The physiological role of CCT-subunit overexpression in cancer is, however, unexplored and could, next to actin, also be related to other CCT-substrates such as tubulins or the von Hippel–Landau protein (pVHL) – a tumor suppressor connected to response to hypoxia (Pugh and Ratcliffe 2003).

In cancer cells, differential transcriptional control of actin gene isoform expression is likely to occur since this is also an important feature of normal developmental programs and other pathologies (references in Chaponnier and Gabbiani 2004). Opposite expression level changes of β- and γ-actin have been reported in salivary gland adenocarcinoma as a function of metastatic capacity (Suzuki et al. 1998). More recently γ-actin-level changes were reported in a drug-resistant acute lymphoblastic leukemia (Verrills et al. 2006).

Microarray data and serial analysis of actin gene expression (SAGE) also reveal altered expression in cancer tissues compared to normal tissue, for example β- and γ-nonmuscle or α-smooth muscle actin. Similar observations have been made for tumor material at different stages of progression although consistent trends have been hard to define. In colon adenocarcinoma cell lines however, β-actin level and localization correlate with metastatic capacity (Nowak et al. 2005).

The zipcode-binding protein 1 (ZBP1) is an interesting modulator of actin expression. It binds the 3′ UTR of β-actin mRNA and assures its transport to sites of high actin dynamics in the cell periphery (Ross et al. 1997; Zhang et al. 2001; Oleynikov and Singer 2003). It was recently shown that the association of ZBP1 with β-actin mRNA represses its translation (Huttelmaier et al. 2005). However upon phosphorylation of ZBP1 by the exclusively peripheral Src kinase, ZBP1 binding is inhibited and β-actin protein is synthesized (Huttelmaier et al. 2005).

Interestingly, ZBP1-expression is downregulated in metastatic vs. non-metastatic breast tumors (Wang et al. 2002, 2004a). This is in line with observations that metastatic breast cancer cell lines have less peripherally located β-actin than their nonmetastatic counterparts (Shestakova et al. 1999). Restoring ZBP1-levels reduces invasiveness and the metastatic ability of the tumor cells (Wang et al. 2004a), indicating that polarization in β-actin synthesis by translational control negatively correlates with efficient invasion. This mechanism also provides a possible additional pathway from the Src oncogene to actin dynamics and cell migration but this has not been explored.

Gene Mutation and Deregulated Gene Transcription of ABPs in Tumors

Actin-Binding Proteins as Oncogenes?

Table 1 lists known genes coding for ABPs that have frequently been observed to display (a) genetic defect(s) in cancer cells suggesting causal contribution to oncogenesis or malignant tumor progression. All but two are extracted from a recently published "cancer census" that contains 291 "cancer genes" and is built using strict criteria. A cancer gene is assigned as such if it carries at least one germ line mutation and/or more than five unambiguous somatic mutations in at least two independently analyzed primary tumors (Futreal et al. 2004).

For most ABP-genes in Table 1 a dominant somatic mutation resulting from a chromosomal translocation is the mutational defect that underlies their identification as a potential cancer gene. These rearrangements result in chimeras (two genes are involved) in which the activity of the gene product is altered, favoring oncogenesis. Intriguingly, all of these ABPs bind filamentous actin and in a number of cases the fusion protein retains F-actin-binding capacity.

TABLE 1. Actin-binding protein genes mutated in cancer[a].

ABP	Gene	Mutated in[b]	Type[c]	Actin-binding
Abl	c-abl	CML, ALL	T	FA-binding
LASP1	LASP1	AML	T	FA-binding
L-plastin	LCP1	nHL	T	FA-bundling
Moesin	MSN	ALCL	T	FA-binding
Myosin heavy chain 11	MYH11	AML	T	FA-binding
Myosin heavy chain 9	MYH9	ALCL	T	FA-binding
Tropomyosin 3	TMP3	Papillary thyroid, ALCL, IMT	T	FA-binding
Tropomyosin 4	TMP4	ALCL	T	FA-binding
NF2	NF2	NF type 2, mesothelioma, melanoma	T, O	FA-binding
WASp	WAS	Lymphoma, Wiskott–Aldrich syndrome	O	G/FA-binding
Cortactin	CCTN[a]	Breast cancer, HNSCC	A	FA-crosslinking
Myopodin	SYNPO2	Prostate cancer	D	FA-bundling

[a]Obtained from the Wellcome Trust Sanger Institute Cancer Genome Project web site (http://www.sanger.ac.uk/genetics/CGP) and from Futreal et al. (2004), except for CCTN (the cortactin gene also termed EMS1) and SYNPO2.

[b]*AML* acute myelogenous leukemia, *ALCL* anaplastic large cell lymphoma, *ALL* acute lymphocytic leukemia, *CML* chronic myeloid leukemia, *IMT* inflammatory myofibroblastic tumor, *NF* neurofibromatosis, *nHL* non-Hodgkin lymphoma.

[c]*A* amplification, *D* deletion, *O* deletions, missense and nonsense mutations, splice site mutations, and others, *T* translocation (see http://www.sanger.ac.uk/genetics/CGP/ cancer census).

This is the case for the actin-binding tyrosine kinase (c-Abl), a well-recognized protooncogene. In a large set of human leukemias, the c-Abl tyrosine kinase is constitutively activated by fusion to sequences encoded by the breakpoint cluster region (bcr) gene (Gotoh and Broxmeyer 1997) (Table 1). In primary chronic myelogenous leukemia (CML) cells this causes, next to abnormal proliferation, motility defects such as increased spontaneous and persistent movement, decreased adhesion and decreased chemotaxis to stroma-derived factor-1α (Salgia et al. 1997). The F-actin-binding capacity of the Abl-kinase in the BCR/Abl-fusion has been implicated in the altered motile phenotype of these cancer cells (Ramaraj et al. 2004). In addition, many studies point to a regulatory role for Abl in actin remodeling both by small GTPase activation or by interacting with and modulating the activity of ABPs important in actin dynamics (e.g., Abl-kinase interactor 1, Ena/VASP proteins, WASp/Wave proteins; reviewed in Hernandez et al. 2004). Also in the myosin heavy chain (MHC) 11 fusion to the beta subunit of the transcription factor core-binding factor (Table 1) the F-actin-binding capacity of MHC11 is important. The dominant negative effect of this oncogenic fusion results from the cytosolic sequestering on actin filaments of the heterodimeric transcription factor via the fusion (Adya et al. 1998; Lukasik et al. 2002).

For the other ABP-genes that are involved in chromosomal translocations (Table 1) it is not yet clear whether their normal function in the actin system is a determining property of the fusion proteins in oncogenesis. They are fused to genes that are frequently rearranged and/or mutated in specific tumors and that are by themselves already associated with malignancy. The ABP LASP1 (LIM and SH3 protein) is only one of more than 30 fusion partners of the histon methyl transferase mixed lineage leukemia gene in acute myeloid leukemia (Strehl et al. 2003). L-plastin (or L-fimbrin) is fused to the nuclear transcriptional repressor B-cell lymphoma 6 (BCL6) in non-Hodgkin lymphoma (Galiegue-Zouitina et al. 1999).

Moesin, nonmuscle MHC9, and tropomyosins 3 and 4 form an activating fusion with the oncogenic receptor tyrosine kinase–anaplastic lymphoma kinase (ALK–RTK) in anaplastic large cell lymphoma and/or inflammatory myofibroblastic tumors (Pulford et al. 2004). For the F-actin-binding tropomyosins, it is hypothesized that their coiled-coil self-associated domain enables ligand-independent oligomerization and hence activation of the ALK receptor in the oncogenic fusions (Lawrence et al. 2000).

Based on these last examples one might assume that the function of the ABP-counterpart in observed oncogenic fusions is based on more general features of these proteins or of their genes (such as high gene promoter activity, cytosolic gene expression, or protein dimerization capacity). However it is intriguing that for most ABP listed in Table 1, altered expression levels of the unfused ABP-genes are also reported in one or more cancer types (see also Table 2). This may indicate that their capacity for organizing or regulating the actin cytoskeleton is an additional important feature of the fused oncogene or, conversely, the fusion may interfere with the action of the nonfused ABP or of other ABPs. It is conceivable that the ALK–RTK-fused tropomyosins interfere by competition with normal oligomerization of tropomyosins (Lawrence et al. 2000) resulting in aberrant functioning of the latter. This could induce a similar effect as the reported downregulation of specific tropomyosin isoforms in transformed cells (Novy et al. 1993; Varga et al. 2005) and their tumor suppression effect (Prasad et al. 1999).

The genes for the ABPs neurofibromatosis type 2 (NF2) and WAS display a range of inheritable mutations that result in loss of function of the ABP (Table 1). The WAS gene codes for the ABP WASp that is exclusively expressed in hematopoietic cells. Loss of WASp-expression is the cause of the immunodeficiency diseases Wiskott–Aldrich syndrome and X-linked thrombocytopenia. Lymphocytes from these patients show impaired actin remodeling and motility in response to stimuli since WASp is an important modulator of actin dynamics (Fig. 1) (Zicha et al. 1998).

The WAS gene is assigned as cancer gene because WAS patients have an increased risk of malignancy, most notably non-Hodgkins lymphoma, but the underlying mechanism is still unclear.

TABLE 2. Actin-binding proteins showing deregulated expression in tumor cells.

ABP[a] (actin function)	Cancer[b]	Up/down[b]	R/P F/M[b]	OE or I[b]: invasion	Regulators/ partners[b]
α-actinin1 (FA-cross-linking)	Melanoma	U	R, M	–	Integrins, vinculin, PPI-Mod. Enz.,[b] Zyxin, PI-3K
α-actinin3 (FA-cross-linking)	Breast	U[e,c]	R		
	Prostate (C/N)[b]	D[d]	R		
α-actinin4 (FA-cross-linking)	Colon (C/N)	U	P/F	–	
	Neuroblastoma	D	R/F		PAI-1
Arp2 (FA-binding Y-branching)	Colon (C/A)[b]	U	P	–	
	Gastric	D	R		
	Breast	U[e]	R		
	Prostate (C/N)	U[d]	R		Cortactin, WASp, WAVEs
Arp3 (FA-binding Y-branching)	Colon (C/A)	U	P	–	
	Gastric	D	R		
	Breast	U[e,f]	R		
	Prostate (C/MC)[b]	D[g]	R		
Calponin1 (FA-binding)	Fibrosarcoma	D	P, F	Melanoma, adeno-carcinoma: OE: –	Src, PKC
	Renal angiomy olipomas	D	P		
	Melanoma adenocar-cinoma	D	P, M		
	Breast	U[c]			
	Prostate	D[d]	R		
CapG (FA-capping)	Melanoma	U	P	Epithelial cells: OE: +	
	Glioblastoma	U	R		PPI-Mod. Enz., Src
	Breast	U[c,e]	R		
	Prostate (C/MC)	D[d]	R		
CAPZA (FA-capping)	Ovary	D	P	–	
	Breast	U[c]	R		PPI-Mod. Enz.
	Prostate (C/MC)	U[d]	R		
Cofilin1 (FA-dynamizing)	Ovary	U/D	R	Lung cancer: OE: – melanoma: OE: +	ROCK–LIMK, Rac1–PAK1, Cdc42, PPI-Mod. Enz., SSH1, 14-3-3ζ
	Breast	U[c,e]	R		
	RCC[b]	U	P		
	Glioma	U	R		
Coronin1a (FA-binding)	Breast	D[f]	R	–	PKC, Arp2/3
	Prostate	D[d]	R		

(continued)

TABLE 2. (continued).

ABP[a] (actin function)	Cancer[b]	Up/down[b]	R/P F/M[b]	OE or I[b]: invasion	Regulators/ partners[b]
Cortactin (FA-crosslinking)	Hepatocellular carcinoma	U	R, P, M	Fibroblasts: OE: + breast cancer cells:	Src, Arp2/3
	Breast	U[f], D	R		
	Prostate (C/MC)	U[d]	R	OE: +	
Ezrin (FA-binding)	Rhabdomy-osarcoma	U	R, M	Glioma, endometrial cancer cell:	
	Gastrointestinal	U	R	OE of a	
	Osteosarcoma	U	R, M	dominant	
	Medulloblastoma	U	R	negative	
	Esophagus	U	R, P, F	version:	CD44, ROCK,
	Bladder	U	P	– breast	FAS, HGF-R
	Breast	U[f]	R	cancer	
	Prostate (C/MC)	U[d]	R	cells: OE: +	
Eplin-α (FA-crosslinking)	Oral, prostate, breast cancer cells	D	R, P	Transformed – fibroblasts:	–
	Breast	D[e]	R	OE: –	
Fascin1 (FA-bundling)	Breast, colon, esophagus, lung ovary, pancreas, cervix, bladder	de novo	P/M	Colon epithelial cells: OE: + esophageal	PKCα p75 Neurotrophin receptor
	Breast	U[e]	R	squamous cell	
	Prostate (C/MC)	U[d]	R	carcinoma: I: –	
FilaminA/C (FA-crosslinking)	Melanoma	D	P	–	PSMA, PAK1, SHIP 1, β-integrin, S6 kinase, TRAF2
	Prostate (C/MC)	D[d]	R		
FilaminC	Gastric	D	R		
Gelsolin (FA-capping and severing)	Breast	D, U	P, M	Epithelial cells: OE: +	PPI-Mod. Enz., Rac, Ras, Src
	NSLC[b] stage I, II	D, U	P, M	bladder cancer cells,	
	Bladder	D, U	P	NSCLC,	
	Ovary	D	P	lung cancer	
	RCC	D	R	cells,	
	Prostate	D, D[d,g]		melanoma: OE: –	
LASP1 (FA-binding)	Breast	U	R	–	PKA, PKG, c-Abl
MIMA/B (GA-binding)	Bladder	D	R,P	–	
		D			PTPδ
	Prostate	C/N: D[d]	R		
		C/MC: U[d]			
Mena (G/FA-binding)	Breast	U	R, P	–	Zyxin, PPI-Mod. Enz., c-Abl

(continued)

TABLE 2. (continued).

ABP[a] (actin function)	Cancer[b]	Up/down[b] R/P F/M[b]		OE or I[b]: invasion	Regulators/ partners[b]
Moesin (FA-binding)	Bladder	U	R	–	
	Breast	U, U[c]	R, P		CD44
	Ovary	U	R		
L-Plastin (FA-bundling)	Colon, leukemia, cervix, fibrosarcoma, mammary, ovary cancer	*de novo*	R, P	Colon cancer cells: OE: + prostate carcinoma cell: I: –	PKA, vimentin
	Prostate	U[d]	R		
Profilin1 (GA-binding)	Breast	D	R, P	Breast cancer cells: OE: –	PPI-Mod. Enz., polypro-containing proteins
	Pancreas	D	P		
	Prostate	D[d]	R		
Profilin2 (GA-binding)	Breast	D	R	–	
	Prostate	U[d]	R		
Talin (FA-binding)	Prostate	U, D[d]	P	–	PPI-Mod. Enz., β-integrin, vinculin
	Breast	D, D[e,f]	P, R		
Thymosin β4 (GA-sequestering)	Fibrosarcoma	U		Colon cancer cells: OE: + fibrosarcoma: OE: +, I: –	ILK/PINCH
	Colon	U, D	P, M		
	OSCC[b]	U	R		
	NSCLC	U	R		
	Melanoma	U	R, M		
Thymosin β10 (GA-sequestering)	Ovary	D	R	–	
	Gastric	U	R, P		
	Thyroid, pancreas, breast, colon	U	P		Ras, tropomodulin E
	RCC, lung, melanoma, bladder, neuroblastoma, esophagus, ovary	U	R		
	Breast	U[c]	R		
Thymosin β15 (GA-sequestering)	Prostate, breast, lung	U	R, P, M	–	–
NB-thymosin β (GA-sequestering)	Neuroblastoma	U	R, P	–	–
	Breast	U[e,f]	R		
	Prostate (C/N)	U[h], U[d]	R		
	Prostate (C/MC)	D[d]	R		
Transgelin (FA-crosslinking)	Colon	D	P	–	–
	Esophagus	U	P		
	Prostate	D[d]	R		

(continued)

TABLE 2. (continued).

ABPᵃ (actin function)	Cancerᵇ	Up/downᵇ	R/P F/Mᵇ	OE or Iᵇ: invasion	Regulators/ partnersᵇ
Tropomyosin 1 (FA-binding)	Breast	D, Uᶜ	R	Breast cancer cells, transformed fibroblasts: OE: −	−
	Colon	D	R		
	Bladder, esophagus, neuroblastoma	D	P		
	Prostate	Dᵈ·ᵍ	R	−	
Tropomyosin 4/5 (FA-binding)	Bladder(5), Esophagus(4)	U	P	−	−
	Breast (4,5)	Uᶜ	R		
VASP (G/FA-binding)	Lung (C/N)	U	P	−	Zyxin, PPI-Mod. Enz., PKA, PKG
Vinculin (FA-binding)	Melanoma, breast, salivary gland	D	P, M	−	
	Breast, rhabdomyo sarcoma	U, Uᶜ	P, Rⁱ		PPI-Mod. Enz., α-actinin, talin
	Prostate	Dᵈ	R		
N-WASp (G/FA-binding)	Liver, prostate	U, Uᵈ	R	−	
WAVE2 (G/FA-binding)	Melanoma	U	P, M	Melanoma cells: OE: +, I: −	Arp2/3, PPI-Mod. Enz., SH3-containing proteins, Rho-GTPases
	Prostate	Dᵈ	R		
WAVE3 (G/FA-binding)	Breast	Uᶜ	R	Breast cancer cells: I: −	
	Prostate	Dᵈ	R		

ᵃα-actinin (Honda et al. 1998, 2005; Clark et al. 2000; Nikolopoulos et al. 2000); Arp2 and Arp3 (Kaneda et al. 2004; Otsubo et al. 2004); Calponin (Takeoka et al. 2002; Islam et al. 2004; Lener, Burgstaller and Gimona 2004); CapG (De Corte, Gettemans and Vandekerckhove 1997; Van Ginkel et al. 1998; Lal et al. 1999; De Corte et al. 2004); capping protein (Smith-Beckerman et al. 2005); cofilin (Gunnersen et al. 2000; Martoglio et al. 2000; Unwin et al. 2003; Ding et al. 2004; Lee et al. 2005b; Smith-Beckerman et al. 2005; Yap et al. 2005; Dang, Bamburg and Ramos 2006); cortactin (Patel et al. 1998; Wang et al. 2002; Chuma et al. 2004); ezrin (Ohtani et al. 1999; Wick et al. 2001; Park et al. 2003; Shen et al. 2003; Khanna et al. 2004; Koon et al. 2004; Yu et al. 2004a; Elliott et al. 2005; Langbein et al. 2006); eplin (Song et al. 2002; Maul et al. 2003); fascin (Jawhari et al. 2003; Hashimoto, Skacel and Adams 2005; Kabukcuoglu et al. 2005; Roma and Prayson 2005; Tong et al. 2005; Xie et al. 2005; Yoder et al. 2005); filamin C (Flanagan et al. 2001; Kaneda et al. 2002; Anilkumar et al. 2003); gelsolin (Chaponnier and Gabbiani 1989; Tanaka et al. 1995; Dosaka-Akita et al. 1998; Asch et al. 1999; Lee et al. 1999; Shieh et al. 1999; Fujita et al. 2001; Thor et al. 2001; De Corte et al. 2002; Sazawa et al. 2002; Yang et al. 2004a; Noske et al. 2005; Langbein et al. 2006); LASP1 (Tomasetto et al. 1995; Lin et al. 2004);

(continued)

NF2 is a germ line-based disorder caused by loss of the tumor suppressor NF2 (merlin, schwannomin). These patients develop slow growing and usually nonmalignant schwannomas, meningiomas, and medulloepithelioma (Gutmann 2001). Somatic mutations of the NF2 gene have also been reported in neoplasms of nonneuroectodermal origin, such as malignant mesothelioma and melanoma (Robinson, Musk and Lake 2005). Overexpression of NF2 in rat schwannoma cells inhibits their growth and impairs cell motility, adhesion, and spreading (Gutmann 2001). NF2 belongs to the ezrin/moesin/radixin proteins that bind actin filaments, interact with membrane lipids, with the hyaluronic acid transmembrane receptor CD44 and with HGF-regulated tyrosine kinase substrate (Gautreau et al. 2002; Scoles et al. 2002). Its activity is regulated by phosphorylation downstream of rac1 and cdc42 (Xiao et al. 2002). Consequently, NF2 is centrally positioned to influence cell adhesion, motility, and growth factor controlled cell growth as discussed in Gutmann (2001) and McClatchey and Giovannini (2005).

Amplification of the cortactin gene (CCTN) is not included in the cancer census because the chromosomal aberration at 11q13 affects several genes (Futreal et al. 2004). It occurs in 13–36% of human carcinomas mainly in breast cancer,

---

TABLE 2. (continued).

Mena (Di Modugno et al. 2004; Wang et al. 2004a); MIM (Lee et al. 2002; Nixdorf et al. 2004; Loberg et al. 2005); moesin (Carmeci et al. 1998; Martoglio et al. 2000; Langbein et al. 2006); L-Plastin (Lin et al. 1993; Delanote, Vandekerckhove and Gettemans 2005); profilin (Janke et al. 2000; Wang et al. 2002; Wittenmayer et al. 2004; Gronborg et al. 2006); talin (Glukhova et al. 1995; Everley et al. 2004); thymosin β4 (Yamamoto et al. 1993; Clark et al. 2000; Kobayashi et al. 2002; Cha, Jeong and Kleinman 2003; Muller-Tidow et al. 2004; Wang et al. 2004b; Vigneswaran et al. 2005); thymosin β10 (Hall 1994; Santelli et al. 1999; Lee et al. 2001; Takano et al. 2002; Oien et al. 2003; Chiappetta et al. 2004; Alldinger et al. 2005); thymosin β15 (Bao, Loda and Zetter 1998; Chakravatri et al. 2000); NB thymosin β (Yokoyama et al. 1996); transgelin (Qi et al. 2005; Yeo et al. 2006); tropomyosin 1 (Prasad, Fuldner and Cooper 1993; Gimona, Kazzaz and Helfman 1996; Prasad et al. 1999; Yager et al. 2003; Pawlak et al. 2004; Qi et al. 2005; Varga et al. 2005); tropomyosin 5 (Pawlak et al. 2004); VASP (Dertsiz et al. 2005); vinculin (Sadano, Inoue and Taniguchi 1992; Glukhova et al. 1995; Meyer and Brinck 1997); WAVE2 (Kurisu et al. 2005).
[b]*Cancer* cancer type in which upregulated (*U*) or downregulated (*D*) cancer level is observed on RNA (*R*) or protein (*P*) level, *F* the tumor suppression of promoting function has been validated in a functional assay (*in vitro* or *in vivo* invasion), *C/A* carcinoma vs. adenoma, *C/MC* carcinoma vs. metastatic carcinoma, *C/N* cancer vs. normal, *FA* F-actin, *I* forced downregulation of the ABP (obtained by RNAi or antisense technology) in these cells leads to inhibition (−) or to induction (+) of invasion, *M* correlated with metastasis, *NSCLC* nonsmall cell lung cancer, *OSCC* oral squamous cell carcinoma, *OE* overexpression of the ABP in these cells leads to inhibition (−) or to induction (+) of invasion, *PPI-Mod. Enz.* polyphosphoinositide-modulating enzymes, *RCC* renal cell carcinoma, *Regulators/partners* limited to those already implicated in cancer.
[c]Wang et al. (2004a).
[d]Yu et al. (2004b) only significant up- or downregulations are listed: significance criterion is default *p*-value < 0.05 (http://www.Oncomine.org).
[e]van 't Veer et al. (2002), as in footnote d.
[f]Wang et al. (2005b) as in footnote d.
[g]Rhodes et al. (2002).
[h]Dhaese et al. (in preparation)

oral squamous cell carcinoma, and head and neck squamous cell carcinomas. This gene encodes the actin filament crosslinking protein cortactin (Table 1). Its overexpression enhances cell migration, invasion, and metastasis *in vitro* (Patel et al. 1998) and *in vivo* (Rodrigo et al. 2000; Li et al. 2001; Chuma et al. 2004). This is in line with the role of cortactin as an F-actin organizer and Arp2/3 activator in the leading edge and in invadopodia of motile and invasive cells.

Li and coworkers (2001) demonstrated that the effect of cortactin overexpression on metastatic capacity *in vivo* is dependent on its phosphorylation by the protooncogene Src. In addition, a recent study demonstrated that cortactin overexpression attenuated ligand-induced EGFR downregulation (Timpson et al. 2005).

We also included in Table 1 the synaptopodin 2 gene (SYNPO2) coding for the actin-bundling protein myopodin or synaptopodin 2, because it is completely or partially deleted in 80% of malignant prostate cancers and strongly correlated with an invasive phenotype (Lin et al. 2001; Jing et al. 2004). Overexpression of myopodin in invasive prostate cancer cells suppresses tumor growth and invasion in mice (Jing et al. 2004) supporting its role as tumor suppressor in prostate cancer.

It is as yet not definitively determined if actin-binding is needed for the suppressive activity. Myopodin can reside both in nucleus and cytoplasm via the presence of a nuclear location signal (De Ganck et al. 2005) and its cytoplasmic (or nonnuclear) localization has been correlated with invasiveness in bladder cancer (Sanchez-Carbayo et al. 2003).

Altered Transcription of Actin Binding Proteins in Cancer Progression: On the Road to Revealing their Causal Roles

Intensified sequencing of cancer genomes will probably lead to the discovery of increased mutations in ABP-genes in cancer cells. However, it is already evident that many genes coding for ABPs display altered transcription and/or translation in specific cancers. The underlying mechanisms may be either genetic or epigenetic changes. Epigenetic changes have, e.g., been shown for gelsolin where a decreased transcription level in breast cancer cells is dependent on either histon deacetylase activity (Mielnicki et al. 1999) or ATF1 transcription repressor binding to the gelsolin promoter (Dong et al. 2002).

Below we report on the presence of ABPs in invasion or metastasis-associated expression profiles obtained through differential transcriptome or proteome profiling of normal and/or tumor samples (e.g., Clark et al. 2000; Rhodes et al. 2002). Deregulation on the level of upstream regulators of these proteins will also affect ABP activities (see below).

Based upon fascinating technological developments, in the last decade the scientific community has been provided with a massive amount of differential expression data. Several of these profiling studies (van 't Veer et al. 2002; Budhu et al. 2005; Weigelt et al. 2005) have recently contributed to the opinion that the metastatic capacity of a primary tumor may already be present and detectable in the (bulk) primary tumor in early phases of many cancer types.

This opposes the classical view that metastasis is an endpoint of tumor progression. In contrast, this suggests either that metastasis occurs at all stages in some tumors as evidenced from studies focusing on micrometastases (for example Coello et al. 2004; Klein 2004; Pantel and Woelfle 2004) or that some tumors are inherently metastatic (Weigelt et al. 2005).

Since actin-dependent migration and invasion are already required at the onset of the metastatic process relevant data on deregulation of ABPs may therefore be distilled from profiling data comparing primary tumors of different metastatic capacity. This optimistic view is, however, slightly attenuated by a more recent hypothesis, arising from diverse approaches. Several authors (Welch 2004; Brabletz et al. 2005; Condeelis, Singer and Segall 2005) converge on the fact that the "effector invasive proteome" manifests itself only in subsets of cancer cells of the primary tumor that are confronted with specific cues from the permissive and activated tumor microenvironment. Invasive and metastatic activity is putatively a transient and reversible feature of a cancer cell and can only be optimally derived upon mimicking the necessary input as has been performed by Condeelis, Singer and Segall (2005). They collected the invasive subpopulation of a mammary tumor using chemotaxis to an EGF-source, hereby mimicking the attraction by EGF-producing tumor-associated macrophages. This hypothesis of a transiently altered actin profile obviously has important implications for defining and evaluating metastatic transcriptome/proteome signatures as a means of determining prognosis. Indeed the optimal experimental set up is required to pinpoint key modulators of the actin system in invasive tumors and define what we might term the actual "effector actinome for invasion."

Apart from profiling data, we can gain important insights into the causal roles of ABPs in tumor cell migration and invasion from directed functional studies where up- or downregulation of expression of a specific ABP is induced in a specific (cancer) cell and the effect on invasion is monitored.

In Table 2 we compile data on altered ABP-transcription and expression levels in tumor cells with data from functional studies. This list is primarily based on a survey of the literature and analyses on protein level and on mRNA level are both included. In addition, for each of these ABPs we inspected the changes in expression level reported in a selected set of microarray-based analyses involving prostate and breast cancer. Yu and colleagues (2004b) compared benign prostate, prostate carcinoma, and metastatic prostate cancer samples. Rhodes et al. (2002) presented a meta-analysis of four microarray-based studies of prostate progression. van 't Veer et al. (2002) and Wang et al. (2005b) studied profiles of breast cancer metastasis and prognosis, and Wang et al. (2004a) determined the differential transcriptome of the invasive subpopulation of a mammary tumor in rat, collected via EGF-based chemotaxis. If the role of altered expression has been studied or verified in an invasion assays, the outcome is also indicated.

ABPs within Three Functional Groups Display Altered Expression Levels

Table 2 contains 32 different ABPs from 24 ABP families. The fact that changed expression of ABPs occurs in several cancer types suggests that alterations in the actin system are a general feature of a tumor cell. This deregulated expression of ABPs in cancers is in agreement with the earliest observations that the overall actin organization in cancer cells is strongly altered upon transformation (reviewed in Rao and Li 2004). It falls beyond the scope of this text to discuss in detail all cancer-related data on the ABPs in Table 2.

Several reviews have recently addressed individual ABPs, e.g., cofilins and Arp2/3 (Condeelis, Singer and Segall 2005; DesMarais et al. 2005), ezrin (Curto and McClatchey 2004; Hunter 2004), fascin (Hashimoto, Skacel and Adams 2005), plastins (Delanote, Vandekerckhove and Gettemans 2005), profilins (Polet et al. 2006), tropomyosins (Gunning et al. 2005), talin and vinculin (Critchley 2004), WASp and WAVEs (Chandrasekar et al. 2005), and some of these ABPs are addressed in this volume.

The ABPs listed in Table 2 largely fall into three functional groups. A first small set contains molecules (for example ezrin/moesin, talin, vinculin, α-actinins) that provide linkages between actin filaments and membrane-localized proteins or protein complexes, suggesting they affect cell–cell or cell–substrate adhesion and adhesive responses in cancer cells. As a second functional set of ABPs in Table 2, we distinguish proteins involved in supramolecular organization of actin filaments, i.e., F-actin crosslinking, bundling, and stabilization. We here also include proteins such as Arp2, Arp3, cortactin that promote formation and stabilization of branches in dendritic actin filaments arrays in lamellipodial protrusions in migrating cells (Svitkina and Borisy 1999). A recent analysis of microarray data, focusing on the identification of functional protein modules important in primary breast cancers destined to metastasize (Rhodes and Chinnaiyan 2005), revealed the "Y-branching pathway of actin filaments" as the most strongly enriched of all biological pathways (here a pathway as defined by BioCarta, http://www.biocarta.com). This is consistent with the knowledge that the higher-order organization of actin filaments into arrays, bundles, or stress fibers is essential for protrusive and contractile processes of a migrating cell.

A third set of ABPs with altered expression in human tumors is formed by ABPs important in the dynamics of the actin polymerization cycle includes Arp2 and 3 of the Arp2/3 complex, cortactin, CapG, CapZ (capping protein), cofilin1, gelsolin, profilins, β-thymosins, Ena/VASP proteins, N-WASp, and WAVEs.

Parallel Effector Paths for Protrusion Formation are Simultaneously Activated

Comparing this set with Fig. 1b illustrates that components of three different paths leading to filament elongation and to protrusion formation are strongly represented in Table 2 and thus deregulated in cancer. Within one tumor type, deregulated expression is frequently apparent for proteins of more than one of these paths.

From Table 2 it follows that expression of gelsolin, cofilin, CapZ, and thymosin β10 are altered in ovarian cancer whereas CapZ, CapG, profilin1, cortactin, gelsolin, N-WASp, and WAVE expression levels are changed in prostate cancer. Many of these protrusion-promoting factors are also part of the "invasion signature" of invasive chemotactic primary breast cancer cells (Wang et al.

2004a; see also Table 2). As discussed in Condeelis, Singer and Segall (2005), their coordinately upregulated expression can in breast cancer cells be expected to enhance the effects on protrusive activity and thus on cancer cell motility.

For the nonbreast cancer types in Table 2 this coordinated deregulation of specific ABPs is not readily extractable. In order to obtain independent insight within an additional tumor type, we exploited an advanced analysis tool combined with filtering for a specific functional module (Rhodes and Chinnaiyan 2005; Segal et al. 2005) to provide screening across multiple studies (see http://www.oncomine.org). We compared significant signals from a set of seven profiling studies of progressive prostate cancer (Dhanasekaran et al. (2001), Welsh et al. (2001), La Tulippe et al. (2002), Luo et al. (2002), Singh et al. (2002), Jingh et al. (2004), Lapointe et al. (2004)). As relevant functional modules we selected the BioCarta pathways "Y-branching of actin filaments" (12 genes) and "Rac1 cell motility signaling pathway" (21 genes; http://www.biocarta.com) and on the KEGG-pathway-162 "Regulation of the actin cytoskeleton" (213 genes; http://www.genome.jp/kegg/) (Kanehisa et al. 2006).

The results are presented in Table 3 and suggest that similar as in invasive breast cancer (Wang et al. 2004a; Rhodes and Chinnaiyan 2005) significant upregulation exists within the WASp/Wave–Arp2/3 path in conjunction with

TABLE 3. Actin functional module activity in prostate cancer.

Functional module[a]	Upregulated ABP-gene/protein[b]	Downregulated[b] ABP-gene/protein
Y-branching of FA	ARPC3/p21[c]	n.s.
	ARPC1A/p41 A[c]	
	WASL/N-WASp	
	WASFl/wave 1	
Rac1 cell motility	CFL1/cofilin 1	n.s.
	WASFl/wave1	
KEGG-162	ARPC3/p21[c]	VCL/vinculin
	DIAPH1/formin1	GSN/gelsolin
	CFL1/cofilin1	ACTN4/α-actinin 4
	ARPC1A/p41 A[c]	TMSB4X/thymosinβ4
	WASL/N-WASp	ACTNA/α-actinin 1
	MYH10/myosin heavy chain 10	DIAPH2/formin2
	PFN2/profilin2	WAS/WASp
	WASFl/wave 1	MYH10/myosin heavy chain10
	ARPC1B/p41B[c]	TMSB4Y/thymosinβ4
	VIL2/ezrin	PFN1/profilin1
	CFL1/cofilin1[d]	VCL/vinculin
	DIAPH1/formin1[d]	GSN/gelsolin[d]
	ARPC3/p21[c,d]	ACTNA/α-actinin 1[d]
		WAS/WASp[d]

[a]http://www.biocarta.com; KEGG-pathway-162: "Regulation of the actin cytoskeleton" (213 genes; http://www.genome.jp/kegg/) (Kanehisa et al. 2006).
[b]Listing is made based upon the *p*-value obtained after advanced analysis (http://www.oncomine.org) as measure for significance; genes underlined: *p*-value < 0.005; genes in bold: *p*-value < 0.05; others: $0.05 < p\text{-value} < 0.12$; *n.s.* no significant signals within these criteria.
[c]Not ABPs but components of the F-actin-binding Arp2/3 complex (Machesky et al. 1994).
[d]The expression of these genes are measured and significant in 6/7 studies; all others in 4/7 studies.

the activities of cofilin1 and formin1. Additionally isoforms of formins, profilins, and also WASp proteins are found to be either up- or downregulated, albeit only at lower significance. Downregulation in prostate cancer is evident for the actin filament-severing protein gelsolin, and for the focal adhesion component vinculin (Table 3). This largely follows the global picture derived from multiple cancers (Table 2).

Context Dependency and Isoform Specificities of Changed ABP Levels

A novel aspect of Table 3 is that it suggests that thymosin β4 and α-actinin1 and 4 are downregulated in prostate cancer. From Table 2, it is evident that for these proteins both up- and downregulation of expression are observed depending on the tumor type. Both alterations correlated with an invasive phenotype. In mouse melanoma and fibrosarcoma cells that yield lung metastasis, upregulation of thymosin β4 was observed (Kobayashi et al. 2002), whereas in human colorectal carcinoma cells showing metastasis in the liver, this protein is downregulated (Yamamoto et al. 1993). Likewise, the mRNA level of the homologous thymosin β10 is upregulated in many cancer types (Santelli et al. 1999), but decreased in ovarian cancers (Lee et al. 2001).

This example is not unique. It is somewhat frustrating that up- and down-regulation in different tumor types is observed for 21 of the 34 proteins in Table 2. An obvious – but none the less important – explanation is that the altered expression level of a particular ABP needs to be evaluated in light of the cancer cell context in which this change occurs. With regard to cancer cell motility and invasive capacity, this context in first instance refers to the accompanying changes in the actin cytoskeleton and, additionally, to other systems involved in migration, adhesion, and survival.

This underscores the need for multidisciplinary approaches combining profiling studies and functional assays of invasion and metastasis (preferentially *in vivo*) within a same experimental system, an approach already taken by a number of research groups. Especially for highly regulated ABPs, one needs to take into account both expression levels of the ABP and regulation by signaling events (see below) to evaluate causal contributions to invasive capacity.

The observed differential (either up or down) expression levels of the actin-severing protein gelsolin in breast and nonsmall cell lung carcinoma have led to an interesting additional hypothesis that an ABP may have different roles during different stages of tumor progression (Yang et al. 2004a). In both cancer types, gelsolin is generally downregulated (Dosaka-Akita et al. 1998; Asch et al. 1999) but in a small subpopulation of these cancers gelsolin upregulation is observed (Thor et al. 2001; Yang et al. 2004a) and the latter is correlated with bad prognosis.

From functional assays it is extractable that downregulation of gelsolin may be linked to tumor cell proliferation and antiapoptosis (Sazawa et al. 2002; Boccellino et al. 2004), which is thus favorable in early tumor stages. In contrast, gelsolin upregulation, as observed in later phases, contributes to enhanced cancer cell motility and invasion (Cunningham, Stossel and Kwiatkowski 1991;

Witke et al. 1995; De Corte et al. 2002). It will be interesting to investigate whether also other ABPs similarly display tumor stage-specific roles.

Table 2 also illustrates that different isoforms of a particular ABP that usually have similar *in vitro* actin-binding properties do not always display similar changes in expression level in the same cancer type. Parallel functional studies of isoform activities in specific cancer cells are therefore needed to clearly demonstrate isoform-specific effects in cancer.

For tropomyosins (TM), isoform switching is already apparent in normal and transformed cells. Extensive evidence exists that tropomyosin1 has tumor suppression activities. The phenotype of ras-, Kirsten virus-, and Src-transformed fibroblasts can be rescued by overexpression of tropomyosin1 (Braverman et al. 1996; Gimona, Kazzaz and Helfman 1996; Prasad et al. 1999). Similarly in breast cancer cells, malignant growth is reduced by restoring TM1 levels (Mahadev et al. 2002). Thus, it may be surprising that the analogous protein TM5 is upregulated in urinary bladder carcinoma cells concomitant with TM1 and 2 downregulation (Pawlak et al. 2004). This strongly suggests there are subtle differences in F-actin-binding function or in the interaction with other tropomyosin partners or actin-modulating proteins (Gunning et al. 2005).

By analogy, expression of the four human β-thymosin isoforms, β4, β10, β15, and *neuro*blastoma (NB) display intriguing differences in cancer cells. Overexpression of these actin monomer-sequestering proteins in cultured cells for all isoforms leads to decreased in stress fibers and positively affects cancer cell migration (Bao et al. 1996; Van Troys et al. 1996; Huff et al. 2001; Kobayashi et al. 2002; Cha, Jeong and Kleinman 2003; Dhaese et al. in preparation).

Thymosin β4 and 10 are ubiquitously coexpressed in normal cells and frequently in deregulated in human cancer cells (Table 2). However, thymosin β15 expression in adults appears to be tumor-specific and may be a urinary biomarker for prostate cancer (Bao et al. 1996; Hutchinson et al. 2005). NB-β-thymosin, originally claimed to be neuroblastoma-specific (Yokoyama et al. 1996), is also expressed in different cancer types (Dhaese et al. in preparation). The *de novo* expression of the β15 and NB isoforms in tumors may be due to their differential transcriptional regulation. It is currently unclear whether they also contribute differential properties to cancer cells. In this regard, several studies suggest that upregulation of the thymosin β4 and β10 in cancer cells (Table 2) is implicated in different aspects of tumor progression such as angiogenesis and antiapoptosis, respectively (Lee et al. 2001; Cha, Jeong and Kleinman 2003).

A major clue in understanding these isoform-specific activities in a tumor context may lie in the fact that differential nonactin partners have recently been described. Thymosin β4 interacts with integrin-linked kinase (ILK)/PINCH complex that acts upstream of the protein kinase act involved in apoptosis (Bock-Marquette et al. 2004). Thymosin β10 can bind to Ras (Lee et al. 2005a) and to the ABP, tropomodulin E (Rho et al. 2004). The Ras–thymosin β10 interaction inhibits the MAPK/Erk signaling pathway leading to decreased production of vascular endothelial growth factor

(Lee et al. 2005a). Accordingly, overexpressing thymosin β10 in tumor cells results in decreased tumor growth and vascularization in a mouse xenograft experiment (Lee et al. 2005a). Future studies are needed to fully elucidate the isoform-specific effects of this highly homologous ABP-family that is frequently deregulated in human tumors.

Altered ABP Levels in Function of their Nuclear and Non-actin Related Activities?

An additional aspect of ABP function is that many of them display both cytoplasmic and nuclear localization. For example, thymosin β4 has been observed in the nucleus of different cell types and an active transport into the nucleus has been suggested (McCormack et al. 1999; Huff et al. 2004). Gettemans and coworkers (2005) list 11 ABPs with known nuclear localization. Most of these are also present in Table 1 or 2 (profilin, thymosin β4, cofilin, CapG, gelsolin, α-actinins, myopodin, plastin, filaminA) and consequently display altered expression in cancer.

The nuclear actin field has seen great advances in the last years (reviewed by Pederson and Aebi 2005). Actin is involved in eukaryotic transcription based on RNA polymerase data and the observation of polymeric actin structures in the nucleus. The ABPs present in the nucleus may affect the dynamics of the "nuclear actin system," may associate with nuclear polyphosphoinositides, or affect the role of actin in transcription. In addition, a role in transcription is proposed for a number of the nuclear ABPs (e.g., by binding to steroid hormone receptors; references in Gettemans et al. 2005). Thus, in addition to the function of ABPs in cytosolic actin rearrangements during motility, ABPs effect the expression levels in the nuclei of cancer cells.

Numerous studies have demonstrated that by altering the level of a given ABP there are additional changes in the expression profile in the cell. Overexpression of thymosin β4 in normal and/or neoplastic cells induces increased expression of plasminogen inhibitor 1 (Al-Nedawi et al. 2004) and the matrix component laminin 5 (Sosne et al. 2004), and results in increased production and activation of MMPs (Blain, Mason and Duance 2002; Wang et al. 2004b; Sosne et al. 2005). It remains to be investigated whether the actin interaction of thymosin β4, its interactions with other cytosolic proteins, or its nuclear activities play a determining role in the properties of tumor cells. Interestingly, analogous questions are rising for other ABPs.

Altered Regulation of ABP Activity in Cancer Cells

As discussed above, ABPs can be seen as the final effectors of complex signaling pathways that result in actin cytoskeleton reorganization. Tables 2 and 3 demonstrate that changes in concentrations of ABPs are present in cancer vs. normal cells. A change in the overall cellular concentration of one or more ABPs will obviously result in aberrant local concentrations at sites of actin dynamics and in imbalanced relative concentrations of the other ABPs that act in concert.

We now go further into upstream signaling events in a cell that act on ABP activity. These may exert similar effects as altering the effective concentration of the ABP. Consequently there is a need to take these upstream regulators into account in studies aimed at understanding the role of a specific ABP in tumor invasion. It falls beyond the scope of this review to discuss this in great detail. However, we have selected a few examples to illustrate the wide variety of different types of regulation, including protein adaptor molecules, polyphosphoinositides, kinase, and phosphatase activity and small GTPases. Table 2 (last column) lists signaling molecules that may act upstream of ABPs or partners of the listed ABP for which a role in tumorigenesis or tumor progression has been suggested.

Altered Expression Levels of Scaffolding Proteins

These acts to recruit ABPs to subcellular sites of action. For example, the focal adhesions components zyxin (binding partner of α-actinin and Ena/VASP proteins) and paxillin (a vinculin partner) display altered expression levels in cancer cells (Sattler et al. 2000; Amsellem et al. 2005).

Membrane Lipids and Upstream Enzymes

Also, active lipid components ($PI[4,5]P_2$, $PI[3,4,5]P_3$) of cellular membranes can recruit ABPs, for example cofilin (see chapter by Maloney et al.), gelsolin (see chapter by Burtnick and Robinson), profilin (see chapter by Moens), vinculin, α-actinin, and talin to sites of high actin dynamics. ABP–lipid interactions also determine the activity of these ABPs (reviewed in Hilpela, Vartiainen and Lappalainen 2004) and recently, novel roles have been suggested for cell adhesion (e.g., for vinculin, Chandrasekar et al. 2005) and for migration (e.g., for cofilin, our unpublished data by Leyman et al.).

Upstream of these phospholipids are the many enzymes (lipid kinases, phosphatases, and phospholipases, generally termed PPI-modulating enzymes in Table 2) that modify the levels of these lipid signaling molecules (Yin and Janmey 2003; Niggli 2005). Many of these (e.g., PI3K, PTEN, PLCγ, SHIP) have disturbed expression or activity in tumor cells (e.g., Kassis et al. 1999; Steelman, Bertrand and McCubrey 2004; Wymann and Marone 2005).

Other Master Regulators

The Rho-GTPases are upstream master regulators of the actin cytoskeleton (Raftopoulou and Hall 2004), and are mutated or deregulated in tumor cells. The same holds for many of their regulators (GEFs, GAPs, or GDIs) or their direct effectors (e.g., ROCK and PAK isoforms) (see Futreal et al. 2004; also see microarray studies used in Table 2). It is noteworthy that PPI-modulating enzymes, small GTPases, and their effectors are strongly represented in the profile of the invasive subpopulation of breast cancer cells for which motility is a main cellular property (Wang et al. 2004a).

Phosphorylation

The activity of a number of ABPs is directly regulated by phosphorylation. Not surprisingly the kinases or phosphatases involved have been implicated in tumorigenesis or tumor progression. Vasodilator-stimulated phosphoprotein (VASP), that localizes to areas of focal contacts and the leading edge of lamellipodia, is a substrate for the cyclic adenosine monophosphate/cyclic guanosine monophosphate (cAMP/cGMP)-dependent protein kinases (PKA, PKG) (Chitaley et al. 2004). cAMP-dependent protein kinase phosphorylation of EVL, a Mena/VASP relative, regulates its interaction with actin and SH3 domains (Lambrechts et al. 2000). Both PKA (mainly its regulatory subunit RIA) (references in Bossis and Stratakis 2004) and recently also PKG (Hou et al. 2006) have been implicated in tumor proliferation and invasion in several cancer types. Gelsolin and cortactin are substrates of the Src-proto-oncogene and phosphorylation affects their effect on invasive capacity of cells *in vivo* or *in vitro* (De Corte, Gettemans and Vandekerckhove 1997; Daly 2004).

LIM Kinase 1 or 2 expression and activities have been correlated with invasive capacity of tumor cells by a.o. (Davila et al. 2003; Suyama et al. 2004; Wang et al. 2004a; Bagheri-Yarmand et al. 2006). The LIM kinases inactivate cofilins and are themselves activated by ROCK and PAK isoforms (Edwards et al. 1999; Ohashi et al. 2000). An increasing number of studies point at a role of these kinases in tumor malignancy and all are considered as potential therapeutic targets.

Many more examples demonstrate that the complex signaling pathways that drive actin reorganizations form a rich soil for potential deregulations in cancer cells affecting the ABP-mediated activities that generate cell migration.

Conclusions

Actin-based cell migration relies on the balanced activity of specific ABPs that drive the dynamics of the actin system and govern its spatial organization. These activities are highly regulated downstream of growth factor receptors, cell–cell and cell–substrate adhesion receptors.

Profiling of patient tumor samples and functional studies in malignant cells demonstrate that cancer cells that migrate in the tumor microenvironment in the initial phase of invasion or home to distant sites during metastasis display a wide range of changes in their actin cytoskeleton. These changes are genetic or epigenetic in nature and may even only be transient at crucial time points during the invasive process.

The alterations within the actin cytoskeleton vary widely from cancer type to cancer type and we have only identified key genes of the actin system that underlie tumor cell motility. It is highly probable that different combinations of changes can meet the demands of motile tumor cell.

Elucidating the motility requirements of a tumor cell on the molecular level is complicated by required insight in (1) alterations in upstream regulators (at various levels), (2) the activity state of actin-modulating proteins, and (3) the tight connection of factors involved in the actin motility system with other cellular functions such as cell–cell adhesion. In addition, we have described examples of ABPs that interact with other partners that have other cellular properties that also contribute to tumorigenesis and progression. This versatile activity of ABPs is even more evident in recent studies suggesting an active (in)direct role for specific ABPs in nuclear processes including the regulation of gene expression.

Much experimental work is consequently needed to elucidate the relationship between alterations in the actin system and the unique properties of tumor cells. Multidisciplinary approaches and careful experimental design will be crucial in deriving signatures that represent "actinomes" of invasive tumor cells and in identifying the common or cancer-specific features of these signatures.

Novel technological advances are needed such as intravital imaging of tumor cell activity within the complex host environment and "caught in the act" biosensor-based approaches to follow ABP activity in cells (Lorenz et al. 2004; Nalbant et al. 2004; Yamada et al. 2005a).

The actin and ABPs will most certainly contribute to providing inhibitors of the invasive steps of cancer that can be validated in a clinical setting.

Acknowledgments. This work was supported by grants FWO-G.0157.05 (to M.V.T. and C.A.), FWO-G0133.06 (to C.A.), BOF 01-J04806 (to M.V.T.), and IUAP-5/31 (to C.A.).

References

Adya, N., Stacy, T., Speck, N. A. and Liu, P. P. 1998. The leukemic protein core binding factor beta (CBFbeta)-smooth-muscle myosin heavy chain sequesters CBFalpha2 into cytoskeletal filaments and aggregates. Mol. Cell. Biol. 18, 7432–7443.

Ahmed, F., Wyckoff, J., Lin, E. Y., Wang, W., Wang, Y., Hennighausen, L., Miyazaki, J., Jones, J., Pollard, J. W., Condeelis, J. S. and Segall, J. E. 2002. GFP expression in the mammary gland for imaging of mammary tumor cells in transgenic mice. Cancer Res. 62, 7166–7169.

Al-Nedawi, K. N., Czyz, M., Bednarek, R., Szemraj, J., Swiatkowska, M., Cierniewska-Cieslak, A., Wyczolkowska, J. and Cierniewski, C. S. 2004. Thymosin beta 4 induces the synthesis of plasminogen activator inhibitor 1 in cultured endothelial cells and increases its extracellular expression. Blood 103, 1319–1324.

Alldinger, I., Dittert, D., Peiper, M., Fusco, A., Chiappetta, G., Staub, E., Lohr, M., Jesnowski, R., Baretton, G., Ockert, D., Saeger, H. D., Grutzmann, R. and Pilarsky, C. 2005. Gene expression analysis of pancreatic cell lines reveals genes overexpressed in pancreatic cancer. Pancreatology 5, 370–379.

Amsellem, V., Kryszke, M. H., Hervy, M., Subra, F., Athman, R., Leh, H., Brachet-Ducos, C. and Auclair, C. 2005. The actin cytoskeleton-associated protein zyxin acts as a tumor suppressor in Ewing tumor cells. Exp. Cell Res. 304, 443–456.

Anderson, K. I., Wang, Y. L. and Small, J. V. 1996. Coordination of protrusion and translocation of the keratocyte involves rolling of the cell body. J. Cell Biol. 134, 1209–1218.

Anilkumar, G., Rajasekaran, S. A., Wang, S., Hankinson, O., Bander, N. H. and Rajasekaran, A. K. 2003. Prostate-specific membrane antigen association with filamin A modulates its internalization and NAALADase activity. Cancer Res. 63, 2645–2648.

Asch, H. L., Winston, J. S., Edge, S. B., Stomper, P. C. and Asch, B. B. 1999. Down-regulation of gelsolin expression in human breast ductal carcinoma in situ with and without invasion. Breast Cancer Res. Treat. 55, 179–188.

Bagheri-Yarmand, R., Mazumdar, A., Sahin, A. A. and Kumar, R. 2006. LIM kinase 1 increases tumor metastasis of human breast cancer cells via regulation of the urokinase-type plasminogen activator system. Int. J. Cancer 118, 2703–2710.

Bajou, K., Noel, A., Gerard, R. D., Masson, V., Brunner, N., Holst-Hansen, C., Skobe, M., Fusenig, N. E., Carmeliet, P., Collen, D. and Foidart, J. M. 1998. Absence of host plasminogen activator inhibitor 1 prevents cancer invasion and vascularization. Nat. Med. 4, 923–928.

Ballestrem, C., Hinz, B., Imhof, B. A. and Wehrle-Haller, B. 2001. Marching at the front and dragging behind: Differential alphaVbeta3-integrin turnover regulates focal adhesion behavior. J. Cell Biol. 155, 1319–1332.

Bao, L., Loda, M., Janmey, P. A., Stewart, R., Anand-Apte, B. and Zetter, B. R. 1996. Thymosin beta 15: A novel regulator of tumor cell motility upregulated in metastatic prostate cancer. Nat. Med. 2, 1322–1328.

Bao, L., Loda, M. and Zetter, B. R. 1998. Thymosin beta15 expression in tumor cell lines with varying metastatic potential. Clin. Exp. Metastasis 16, 227–233.

Barzik, M., Kotova, T. I., Higgs, H. N., Hazelwood, L., Hanein, D., Gertler, F. B. and Schafer, D. A. 2005. Ena/VASP proteins enhance actin polymerization in the presence of barbed end capping proteins. J. Biol. Chem. 280, 28653–28662.

Berx, G. and Van Roy, F. 2001. The E-cadherin/catenin complex: An important gate-keeper in breast cancer tumorigenesis and malignant progression. Breast Cancer Res. 3, 289–293.

Blain, E. J., Mason, D. J. and Duance, V. C. 2002. The effect of thymosin beta4 on articular cartilage chondrocyte matrix metalloproteinase expression. Biochem. Soc. Trans. 30, 879–882.

Boccellino, M., Giuberti, G., Quagliuolo, L., Marra, M., D'Alessandro, A. M., Fujita, H., Giovane, A., Abbruzzese, A. and Caraglia, M. 2004. Apoptosis induced by interferon-alpha and antagonized by EGF is regulated by caspase-3-mediated cleavage of gelsolin in human epidermoid cancer cells. J. Cell. Physiol. 201, 71–83.

Bock-Marquette, I., Saxena, A., White, M. D., Dimaio, J. M. and Srivastava, D. 2004. Thymosin beta4 activates integrin-linked kinase and promotes cardiac cell migration, survival and cardiac repair. Nature 432, 466–472.

Bossis, I. and Stratakis, C. A. 2004. PRKAR1A: Normal and abnormal functions. Endocrinology 145, 5452–5458.

Brabletz, T., Jung, A., Spaderna, S., Hlubek, F. and Kirchner, T. 2005. Opinion: Migrating cancer stem cells – An integrated concept of malignant tumour progression. Nat. Rev. Cancer 5, 744–749.

Braverman, R. H., Cooper, H. L., Lee, H. S. and Prasad, G. L. 1996. Anti-oncogenic effects of tropomyosin: Isoform specificity and importance of protein coding sequences. Oncogene 13, 537–545.

Brunton, V. G., MacPherson, I. R. and Frame, M. C. 2004. Cell adhesion receptors, tyrosine kinases and actin modulators: A complex three-way circuitry. Biochim. Biophys. Acta 1692, 121–144.

Bryce, N. S., Clark, E. S., Leysath, J. L., Currie, J. D., Webb, D. J. and Weaver, A. M. 2005. Cortactin promotes cell motility by enhancing lamellipodial persistence. Curr. Biol. 15, 1276–1285.

Buccione, R., Orth, J. D. and McNiven, M. A. 2004. Foot and mouth: Podosomes, invadopodia and circular dorsal ruffles. Nat. Rev. Mol. Cell. Biol. 5, 647–657.

Budhu, A. S., Zipser, B., Forgues, M., Ye, Q. H., Sun, Z. and Wang, X. W. 2005. The molecular signature of metastases of human hepatocellular carcinoma. Oncology 69(Suppl. 1), 23–27.

Burridge, K. and Wennerberg, K. 2004. Rho and Rac take center stage. Cell 116, 167–179.

Carmeci, C., Thompson, D. A., Kuang, W. W., Lightdale, N., Furthmayr, H. and Weigel, R. J. 1998. Moesin expression is associated with the estrogen receptor-negative breast cancer phenotype. Surgery 124, 211–217.

Carmeliet, P. 2005. VEGF as a key mediator of angiogenesis in cancer. Oncology 69(Suppl. 3), 4–10.

Cha, H. J., Jeong, M. J. and Kleinman, H. K. 2003. Role of thymosin beta4 in tumor metastasis and angiogenesis. J. Natl Cancer Inst. 95, 1674–1680.

Chakravatri, A., Zehr, E. M., Zietman, A. L., Shipley, W. U., Goggins, W. B., Finkelstein, D. M., Young, R. H., Chang, E. L. and Wu, C. L. 2000. Thymosin beta-15 predicts for distant failure in patients with clinically localized prostate cancer – Results from a pilot study. Urology 55, 635–638.

Chandrasekar, I., Stradal, T. E., Holt, M. R., Entschladen, F., Jockusch, B. M. and Ziegler, W. H. 2005. Vinculin acts as a sensor in lipid regulation of adhesion-site turnover. J. Cell Sci. 118, 1461–1472.

Chaponnier, C. and Gabbiani, G. 1989. Gelsolin modulation in epithelial and stromal cells of mammary carcinoma. Am. J. Pathol. 134, 597–603.

Chaponnier, C. and Gabbiani, G. 2004. Pathological situations characterized by altered actin isoform expression. J. Pathol. 204, 386–395.

Chiappetta, G., Pentimalli, F., Monaco, M., Fedele, M., Pasquinelli, R., Pierantoni, G. M., Ribecco, M. T., Santelli, G., Califano, D., Pezzullo, L. and Fusco, A. 2004. Thymosin beta-10 gene expression as a possible tool in diagnosis of thyroid neoplasias. Oncol. Rep. 12, 239–243.

Chitaley, K., Chen, L., Galler, A., Walter, U., Daum, G. and Clowes, A. W. 2004. Vasodilator-stimulated phosphoprotein is a substrate for protein kinase C. FEBS Lett. 556, 211–215.

Chu, Y. S., Thomas, W. A., Eder, O., Pincet, F., Perez, E., Thiery, J. P. and Dufour, S. 2004. Force measurements in E-cadherin-mediated cell doublets reveal rapid adhesion strengthened by actin cytoskeleton remodeling through Rac and Cdc42. J. Cell Biol. 167, 1183–1194.

Chuma, M., Sakamoto, M., Yasuda, J., Fujii, G., Nakanishi, K., Tsuchiya, A., Ohta, T., Asaka, M. and Hirohashi, S. 2004. Overexpression of cortactin is involved in motility and metastasis of hepatocellular carcinoma. J. Hepatol. 41, 629–636.

Clark, E. A., Golub, T. R., Lander, E. S. and Hynes, R. O. 2000. Genomic analysis of metastasis reveals an essential role for RhoC. Nature 406, 532–535.

Coello, M. C., Luketich, J. D., Litle, V. R. and Godfrey, T. E. 2004. Prognostic significance of micrometastasis in non-small-cell lung cancer. Clin. Lung Cancer 5, 214–225.

Condeelis, J. 1993. Understanding the cortex of crawling cells: Insights from Dictyostelium. Trends Cell Biol. 3, 371–376.

Condeelis, J. and Segall, J. E. 2003. Intravital imaging of cell movement in tumours. Nat. Rev. Cancer 3, 921–930.

Condeelis, J., Singer, R. H. and Segall, J. E. 2005. The great escape: When cancer cells hijack the genes for chemotaxis and motility. Annu. Rev. Cell Dev. Biol. 21, 695–718.

Critchley, D. R. 2004. Cytoskeletal proteins talin and vinculin in integrin-mediated adhesion. Biochem. Soc. Trans. 32, 831–836.

Cunningham, C. C., Stossel, T. P. and Kwiatkowski, D. J. 1991. Enhanced motility in NIH 3T3 fibroblasts that overexpress gelsolin. Science 251, 1233–1236.

Curto, M. and McClatchey, A. I. 2004. Ezrin, a metastatic detERMinant? Cancer Cell 5, 113–114.

Daly, R. J. 2004. Cortactin signalling and dynamic actin networks. Biochem. J. 382, 13–25.

Dang, D., Bamburg, J. R. and Ramos, D. M. 2006. Alphavbeta3 integrin and cofilin modulate K1735 melanoma cell invasion. Exp. Cell Res. 312, 468–477.

Davila, M., Frost, A. R., Grizzle, W. E. and Chakrabarti, R. 2003. LIM kinase 1 is essential for the invasive growth of prostate epithelial cells: Implications in prostate cancer. J. Biol. Chem. 278, 36868–36875.

De Corte, V., Bruyneel, E., Boucherie, C., Mareel, M., Vandekerckhove, J. and Gettemans, J. 2002. Gelsolin-induced epithelial cell invasion is dependent on Ras–Rac signaling. EMBO J. 21, 6781–6790.

De Corte, V., Gettemans, J. and Vandekerckhove, J. 1997. Phosphatidylinositol 4, 5-bisphosphate specifically stimulates PP60(c-src) catalyzed phosphorylation of gelsolin and related actin-binding proteins. FEBS Lett. 401, 191–196.

De Corte, V., Van Impe, K., Bruyneel, E., Boucherie, C., Mareel, M., Vandekerckhove, J. and Gettemans, J. 2004. Increased importin-beta-dependent nuclear import of the actin modulating protein CapG promotes cell invasion. J. Cell Sci. 117, 5283–5292.

De Craene, B., Gilbert, B., Stove, C., Bruyneel, E., van Roy, F. and Berx, G. 2005. The transcription factor snail induces tumor cell invasion through modulation of the epithelial cell differentiation program. Cancer Res. 65, 6237–6244.

De Ganck, A., Hubert, T., Van Impe, K., Geelen, D., Vandekerckhove, J., De Corte, V. and Gettemans, J. 2005. A monopartite nuclear localization sequence regulates nuclear targeting of the actin binding protein myopodin. FEBS Lett. 579, 6673–6680.

De Wever, O., Nguyen, Q. D., Van Hoorde, L., Bracke, M., Bruyneel, E., Gespach, C. and Mareel, M. 2004a. Tenascin-C and SF/HGF produced by myofibroblasts in vitro provide convergent pro-invasive signals to human colon cancer cells through RhoA and Rac. FASEB J. 18, 1016–1018.

De Wever, O., Westbroek, W., Verloes, A., Bloemen, N., Bracke, M., Gespach, C., Bruyneel, E. and Mareel, M. 2004b. Critical role of N-cadherin in myofibroblast invasion and migration in vitro stimulated by colon-cancer-cell-derived TGF-beta or wounding. J. Cell Sci. 117, 4691–4703.

Delanote, V., Vandekerckhove, J. and Gettemans, J. 2005. Plastins: Versatile modulators of actin organization in (patho)physiological cellular processes. Acta Pharmacol. Sin. 26, 769–779.

Dertsiz, L., Ozbilim, G., Kayisli, Y., Gokhan, G. A., Demircan, A. and Kayisli, U. A. 2005. Differential expression of VASP in normal lung tissue and lung adenocarcinomas. Thorax 60, 576–581.

DesMarais, V., Ghosh, M., Eddy, R. and Condeelis, J. 2005. Cofilin takes the lead. J. Cell Sci. 118, 19–26.

DesMarais, V., Macaluso, F., Condeelis, J. and Bailly, M. 2004. Synergistic interaction between the Arp2/3 complex and cofilin drives stimulated lamellipod extension. J. Cell Sci. 117, 3499–3510.

Dhanasekaran, S. M., Barrette, T. R., Ghosh, D., Shah, R., Varambally, S., Kurachi, K., Pienta, K. J., Rubin, M. A. and Chinnaiyan, A. M. 2001. Delineation of prognostic biomarkers in prostate cancer. Nature 412, 822–826.

Di Modugno, F., Bronzi, G., Scanlan, M. J., Del Bello, D., Cascioli, S., Venturo, I., Botti, C., Nicotra, M. R., Mottolese, M., Natali, P. G., Santoni, A., Jager, E. and Nistico, P. 2004. Human Mena protein, a serex-defined antigen overexpressed in breast cancer eliciting both humoral and CD8+ T-cell immune response. Int. J. Cancer 109, 909–918.

Ding, S. J., Li, Y., Shao, X. X., Zhou, H., Zeng, R., Tang, Z. Y. and Xia, Q. C. 2004. Proteome analysis of hepatocellular carcinoma cell strains, MHCC97-H and MHCC97-L, with different metastasis potentials. Proteomics 4, 982–994.

Dong, Y., Asch, H. L., Ying, A. and Asch, B. B. 2002. Molecular mechanism of transcriptional repression of gelsolin in human breast cancer cells. Exp. Cell Res. 276, 328–336.

Dosaka-Akita, H., Hommura, F., Fujita, H., Kinoshita, I., Nishi, M., Morikawa, T., Katoh, H., Kawakami, Y. and Kuzumak, N. 1998. Frequent loss of gelsolin expression in non-small cell lung cancers of heavy smokers. Cancer Res. 58, 322–327.

Edwards, D. C., Sanders, L. C., Bokoch, G. M. and Gill, G. N. 1999. Activation of LIM-kinase by Pak1 couples Rac/Cdc42 GTPase signalling to actin cytoskeletal dynamics. Nat. Cell Biol. 1, 253–259.

Elliott, B. E., Meens, J. A., SenGupta, S. K., Louvard, D. and Arpin, M. 2005. The membrane cytoskeletal crosslinker ezrin is required for metastasis of breast carcinoma cells. Breast Cancer Res. 7, 365–373.

Everley, P. A., Krijgsveld, J., Zetter, B. R. and Gygi, S. P. 2004. Quantitative cancer proteomics: Stable isotope labeling with amino acids in cell culture (SILAC) as a tool for prostate cancer research. Mol. Cell Proteomics 3, 729–735.

Faix, J. and Rottner, K. 2006. The making of filopodia. Curr. Opin. Cell Biol. 18, 18–25.

Falet, H., Hoffmeister, K. M., Neujahr, R., Italiano, J. E., Jr., Stossel, T. P., Southwick, F. S. and Hartwig, J. H. 2002. Importance of free actin filament barbed ends for Arp2/3 complex function in platelets and fibroblasts. Proc. Natl Acad. Sci. USA 99, 16782–16787.

Farina, K. L., Wyckoff, J. B., Rivera, J., Lee, H., Segall, J. E., Condeelis, J. S. and Jones, J. G. 1998. Cell motility of tumor cells visualized in living intact primary tumors using green fluorescent protein. Cancer Res. 58, 2528–2532.

Ferlay, J., Bray, F., Pisani, P. and Parkin, D. M. 2004. GLOBOCAN 2002: Cancer Incidence, Mortality and Prevalence Worldwide. IARCPress, Lyon.

Flanagan, L. A., Chou, J., Falet, H., Neujahr, R., Hartwig, J. H. and Stossel, T. P. 2001. Filamin A, the Arp2/3 complex and the morphology and function of cortical actin filaments in human melanoma cells. J. Cell Biol. 155, 511–517.

Franca-Koh, J. and Devreotes, P. N. 2004. Moving forward: Mechanisms of chemoattractant gradient sensing. Physiology 19, 300–308.

Franco, S. J. and Huttenlocher, A. 2005. Regulating cell migration: Calpains make the cut. J. Cell Sci. 118, 3829–3838.

Friedl, P. 2004. Prespecification and plasticity: Shifting mechanisms of cell migration. Curr. Opin. Cell Biol. 16, 14–23.

Friedl, P., Borgmann, S. and Brocker, E. B. 2001. Amoeboid leukocyte crawling through extracellular matrix: Lessons from the Dictyostelium paradigm of cell movement. J. Leukoc. Biol. 70, 491–509.

Friedl, P. and Brocker, E. B. 2000. The biology of cell locomotion within three-dimensional extracellular matrix. Cell. Mol. Life Sci. 57, 41–64.

Friedl, P., Hegerfeldt, Y. and Tusch, M. 2004. Collective cell migration in morphogenesis and cancer. Int. J. Dev. Biol. 48, 441–449.

Friedl, P., Maaser, K., Klein, C. E., Niggemann, B., Krohne, G. and Zanker, K. S. 1997. Migration of highly aggressive MV3 melanoma cells in 3-dimensional collagen lattices results in local matrix reorganization and shedding of alpha2 and beta1 integrins and CD44. Cancer Res. 57, 2061–2070.

Friedl, P. and Wolf, K. 2003. Proteolytic and non-proteolytic migration of tumour cells and leucocytes. Biochem. Soc. Symp. 70, 277–285.

Fujita, H., Okada, F., Hamada, J., Hosokawa, M., Moriuchi, T., Koya, R. C. and Kuzumaki, N. 2001. Gelsolin functions as a metastasis suppressor in B16–BL6 mouse melanoma cells and requirement of the carboxyl-terminus for its effect. Int. J. Cancer 93, 773–780.

Futreal, P. A., Coin, L., Marshall, M., Down, T., Hubbard, T., Wooster, R., Rahman, N. and Stratton, M. R. 2004. A census of human cancer genes. Nat. Rev. Cancer 4, 177–183.

Galiegue-Zouitina, S., Quief, S., Hildebrand, M. P., Denis, C., Detourmignies, L., Lai, J. L. and Kerckaert, J. P. 1999. Nonrandom fusion of L-plastin(LCP1) and LAZ3(BCL6) genes by t(3;13)(q27;q14) chromosome translocation in two cases of B-cell non-Hodgkin lymphoma. Genes Chromosomes Cancer 26, 97–105.

Gamba, A., de Candia, A., Di Talia, S., Coniglio, A., Bussolino, F. and Serini, G. 2005. Diffusion-limited phase separation in eukaryotic chemotaxis. Proc. Natl Acad. Sci. USA 102, 16927–16932.

Gautreau, A., Manent, J., Fievet, B., Louvard, D., Giovannini, M. and Arpin, M. 2002. Mutant products of the NF2 tumor suppressor gene are degraded by the ubiquitin–proteasome pathway. J. Biol. Chem. 277, 31279–31282.

Gettemans, J., Van Impe, K., Delanote, V., Hubert, T., Vandekerckhove, J. and De Corte, V. 2005. Nuclear actin-binding proteins as modulators of gene transcription. Traffic 6, 847–857.

Giganti, A., Plastino, J., Janji, B., Van Troys, M., Lentz, D., Ampe, C., Sykes, C., and Friederich, E. 2005. Actin-filament cross-linking protein T-plastin increases Arp2/3-mediated actin-based movement. J. Cell Sci. 118, 1255–1265.

Gimona, M., Kazzaz, J. A. and Helfman, D. M. 1996. Forced expression of tropomyosin 2 or 3 in v-Ki-ras-transformed fibroblasts results in distinct phenotypic effects. Proc. Natl Acad. Sci. USA 93, 9618–9623.

Glukhova, M., Koteliansky, V., Sastre, X. and Thiery, J. P. 1995. Adhesion systems in normal breast and in invasive breast carcinoma. Am. J. Pathol. 146, 706–716.

Gotoh, A. and Broxmeyer, H. E. 1997. The function of BCR/ABL and related proto-oncogenes. Curr. Opin. Hematol. 4, 3–11.

Gronborg, M., Kristiansen, T. Z., Iwahori, A., Chang, R., Reddy, R., Sato, N., Molina, H., Jensen, O. N., Hruban, R. H., Goggins, M. G., Maitra, A. and Pandey, A. 2006.

Biomarker discovery from pancreatic cancer secretome using a differential proteomic approach. Mol. Cell. Proteomics 5, 157–171.

Grosse, R., Copeland, J. W., Newsome, T. P., Way, M. and Treisman, R. 2003. A role for VASP in RhoA-Diaphanous signalling to actin dynamics and SRF activity. EMBO J. 22, 3050–3061.

Gunnersen, J. M., Spirkoska, V., Smith, P. E., Danks, R. A. and Tan, S. S. 2000. Growth and migration markers of rat C6 glioma cells identified by serial analysis of gene expression. Glia 32, 146–154.

Gunning, P. W., Schevzov, G., Kee, A. J. and Hardeman, E. C. 2005. Tropomyosin isoforms: Divining rods for actin cytoskeleton function. Trends Cell Biol. 15, 333–341.

Gupton, S. L., Anderson, K. L., Kole, T. P., Fischer, R. S., Ponti, A., Hitchcock-DeGregori, S. E., Danuser, G., Fowler, V. M., Wirtz, D., Hanein, D. and Waterman-Storer, C. M. 2005. Cell migration without a lamellipodium: Translation of actin dynamics into cell movement mediated by tropomyosin. J. Cell Biol. 168, 619–631.

Gutmann, D. H. 2001. The neurofibromatoses: When less is more. Hum. Mol. Genet. 10, 747–755.

Hall, A. K. 1994. Amplification-independent overexpression of thymosin beta-10 mRNA in human renal cell carcinoma. Ren. Fail. 16, 243–254.

Hanahan, D. and Weinberg, R. A. 2000. The hallmarks of cancer. Cell 100, 57–70.

Hashimoto, Y., Skacel, M. and Adams, J. C. 2005. Roles of fascin in human carcinoma motility and signaling: Prospects for a novel biomarker? Int. J. Biochem. Cell Biol. 37, 1787–1804.

Hegerfeldt, Y., Tusch, M., Brocker, E. B. and Friedl, P. 2002. Collective cell movement in primary melanoma explants: Plasticity of cell–cell interaction, beta1-integrin function and migration strategies. Cancer Res. 62, 2125–2130.

Hernandez, S. E., Krishnaswami, M., Miller, A. L. and Koleske, A. J. 2004. How do Abl family kinases regulate cell shape and movement? Trends Cell Biol. 14, 36–44.

Hilpela, P., Vartiainen, M. K. and Lappalainen, P. 2004. Regulation of the actin cytoskeleton by PI(4,5)P2 and PI(3,4,5)P3. Curr. Top. Microbiol. Immunol. 282, 117–163.

Hoffman, R. M. 2005. In vivo cell biology of cancer cells visualized with fluorescent proteins. Curr. Top. Dev. Biol. 70, 121–144.

Honda, K., Yamada, T., Endo, R., Ino, Y., Gotoh, M., Tsuda, H., Yamada, Y., Chiba, H. and Hirohashi, S. 1998. Actinin-4, a novel actin-bundling protein associated with cell motility and cancer invasion. J. Cell Biol. 140, 1383–1393.

Honda, K., Yamada, T., Hayashida, Y., Idogawa, M., Sato, S., Hasegawa, F., Ino, Y., Ono, M. and Hirohashi, S. 2005. Actinin-4 increases cell motility and promotes lymph node metastasis of colorectal cancer. Gastroenterology 128, 51–62.

Hou, Y., Gupta, N., Shoenlein, P., Wong, E., Martindale, R., Ganapathy, V. and Browning, D. 2006. An anti-tumor role for cGMP-dependent protein kinase. Cancer Lett. 240, 60–68.

Huber, M. A., Kraut, N. and Beug, H. 2005. Molecular requirements for epithelial–mesenchymal transition during tumor progression. Curr. Opin. Cell Biol. 17, 548–558.

Huff, T., Muller, C. S., Otto, A. M., Netzker, R. and Hannappel, E. 2001. beta-thymosins, small acidic peptides with multiple functions. Int. J. Biochem. Cell Biol. 33, 205–220.

Huff, T., Rosorius, O., Otto, A. M., Muller, C. S., Ballweber, E., Hannappel, E. and Mannherz, H. G. 2004. Nuclear localisation of the G-actin sequestering peptide thymosin beta4. J. Cell Sci. 117, 5333–5341.

Hunter, K. W. 2004. Ezrin, a key component in tumor metastasis. Trends Mol. Med. 10, 201–204.

Hutchinson, L. M., Chang, E. L., Becker, C. M., Shih, M. C., Brice, M., DeWolf, W. C., Gaston, S. M. and Zetter, B. R. 2005. Use of thymosin beta15 as a urinary biomarker in human prostate cancer. Prostate 64, 116–127.

Huttelmaier, S., Zenklusen, D., Lederer, M., Dictenberg, J., Lorenz, M., Meng, X., Bassell, G. J., Condeelis, J. and Singer, R. H. 2005. Spatial regulation of beta-actin translation by Src-dependent phosphorylation of ZBP1. Nature 438, 512–515.

Huttenlocher, A., Ginsberg, M. H. and Horwitz, A. F. 1996. Modulation of cell migration by integrin-mediated cytoskeletal linkages and ligand-binding affinity. J. Cell Biol. 134, 1551–1562.

Islam, A. H., Ehara, T., Kato, H., Hayama, M. and Nishizawa, O. 2004. Loss of calponin h1 in renal angiomyolipoma correlates with aggressive clinical behavior. Urology 64, 468–473.

Janke, J., Schluter, K., Jandrig, B., Theile, M., Kolble, K., Arnold, W., Grinstein, E., Schwartz, A., Estevez-Schwarz, L., Schlag, P. M., Jockusch, B. M. and Scherneck, S. 2000. Suppression of tumorigenicity in breast cancer cells by the microfilament protein profilin 1. J. Exp. Med. 191, 1675–1686.

Jawhari, A. U., Buda, A., Jenkins, M., Shehzad, K., Sarraf, C., Noda, M., Farthing, M. J., Pignatelli, M. and Adams, J. C. 2003. Fascin, an actin-bundling protein, modulates colonic epithelial cell invasiveness and differentiation in vitro. Am. J. Pathol. 162, 69–80.

Jing, L., Liu, L., Yu, Y. P., Dhir, R., Acquafondada, M., Landsittel, D., Cieply, K., Wells, A. and Luo, J. H. 2004. Expression of myopodin induces suppression of tumor growth and metastasis. Am. J. Pathol. 164, 1799–1806.

Kabukcuoglu, S., Ozalp, S. S., Oner, U., Acikalin, M. F., Yalcin, O. T. and Colak, E. 2005. Fascin, an actin-bundling protein expression in cervical neoplasms. Eur. J. Gynaecol. Oncol. 26, 636–641.

Kaneda, A., Kaminishi, M., Sugimura, T. and Ushijima, T. 2004. Decreased expression of the seven ARP2/3 complex genes in human gastric cancers. Cancer Lett. 212, 203–210.

Kaneda, A., Kaminishi, M., Yanagihara, K., Sugimura, T. and Ushijima, T. 2002. Identification of silencing of nine genes in human gastric cancers. Cancer Res. 62, 6645–6650.

Kanehisa, M., Goto, S., Hattori, M., Aoki-Kinoshita, K. F., Itoh, M., Kawashima, S., Katayama, T., Araki, M. and Hirakawa, M. 2006. From genomics to chemical genomics: New developments in KEGG. Nucleic Acids Res. 34, 354–357.

Kassis, J., Moellinger, J., Lo, H., Greenberg, N. M., Kim, H. G. and Wells, A. 1999. A role for phospholipase C-gamma-mediated signaling in tumor cell invasion. Clin. Cancer Res. 5, 2251–2260.

Kaverina, I., Krylyshkina, O. and Small, J. V. 2002. Regulation of substrate adhesion dynamics during cell motility. Int. J. Biochem. Cell Biol. 34, 746–761.

Keller, R. 2005. Cell migration during gastrulation. Curr. Opin. Cell Biol. 17, 533–541.

Khanna, C., Wan, X., Bose, S., Cassaday, R., Olomu, O., Mendoza, A., Yeung, C., Gorlick, R., Hewitt, S. M. and Helman, L. J. 2004. The membrane-cytoskeleton linker ezrin is necessary for osteosarcoma metastasis. Nat. Med. 10, 182–186.

Klein, C. A. 2004. Gene expression sigantures, cancer cell evolution and metastatic progression. Cell Cycle 3, 29–31.

Kobayashi, T., Okada, F., Fujii, N., Tomita, N., Ito, S., Tazawa, H., Aoyama, T., Choi, S. K., Shibata, T., Fujita, H. and Hosokawa, M. 2002. Thymosin-beta4 regulates motility and metastasis of malignant mouse fibrosarcoma cells. Am. J. Pathol. 160, 869–882.

Koon, N., Schneider-Stock, R., Sarlomo-Rikala, M., Lasota, J., Smolkin, M., Petroni, G., Zaika, A., Boltze, C., Meyer, F., Andersson, L., Knuutila, S., Miettinen, M. and El-Rifai, W. 2004. Molecular targets for tumour progression in gastrointestinal stromal tumours. Gut 53, 235–240.

Kovar, D. R., Wu, J. Q. and Pollard, T. D. 2005. Profilin-mediated competition between capping protein and formin Cdc12p during cytokinesis in fission yeast. Mol. Biol. Cell 16, 2313–2324.

Kurisu, S., Suetsugu, S., Yamazaki, D., Yamaguchi, H. and Takenawa, T. 2005. Rac-WAVE2 signaling is involved in the invasive and metastatic phenotypes of murine melanoma cells. Oncogene 24, 1309–1319.

Lal, A., Lash, A. E., Altschul, S. F., Velculescu, V., Zhang, L., McLendon, R. E., Marra, M. A., Prange, C., Morin, P. J., Polyak, K., Papadopoulos, N., Vogelstein, B., Kinzler, K. W., Strausberg, R. L. and Riggins, G. J. 1999. A public database for gene expression in human cancers. Cancer Res. 59, 5403–5407.

Lambrechts, A., Kwiatkowski, A. V., Lanier, L. M., Bear, J. E., Vandekerckhove, J., Ampe, C. and Gertler, F. B. 2000. cAMP-dependent protein kinase phosphorylation of EVL, a Mena/VASP relative, regulates its interaction with actin and SH3 domains. J. Biol. Chem. 275, 36143–36151.

Lambrechts, A., Van Troys, M. and Ampe, C. 2004. The actin cytoskeleton in normal and pathological cell motility. Int. J. Biochem. Cell Biol. 6, 1890–1909.

Langbein, S., Lehmann, J., Harder, A., Steidler, A., Michel, M. S., Alken, P. and Badawi, J. K. 2006. Protein profiling of bladder cancer using the 2D-PAGE and SELDI-TOF-MS technique. Technol. Cancer Res. Treat. 5, 67–72.

Lapointe, J., Li, C., Higgins, J. P., van de Rijn, M., Bair, E., Montgomery, K., Ferrari, M., Egevad, L., Rayford, W., Bergerheim, U., Ekman, P., DeMarzo, A. M., Tibshirani, R., Botstein, D., Brown, P. O., Brooks, J. D. and Pollack, J. R. 2004. Gene expression profiling identifies clinically relevant subtypes of prostate cancer. Proc. Natl Acad. Sci. USA 101, 811–816.

LaTulippe, E., Satagopan, J., Smith, A., Scher, H., Scardino, P., Reuter, V. and Gerald, W. L. 2002. Comprehensive gene expression analysis of prostate cancer reveals distinct transcriptional programs associated with metastatic disease. Cancer Res. 62, 4499–4506.

Lauffenburger, D. A. and Horwitz, A. F. 1996. Cell migration: A physically integrated molecular process. Cell 84, 359–369.

Lawrence, B., Perez-Atayde, A., Hibbard, M. K., Rubin, B. P., Dal, C. P., Pinkus, J. L., Pinkus, G. S., Xiao, S., Yi, E., Fletcher, C. D. and Fletcher, J. A. 2000. TPM3-ALK and TPM4-ALK oncogenes in inflammatory myofibroblastic tumors. Am. J. Pathol. 157, 377–384.

Leavitt, J., Ng, S. Y., Varma, M., Latter, G., Burbeck, S., Gunning, P. and Kedes, L. 1987. Expression of transfected mutant beta-actin genes: Transitions toward the stable tumorigenic state. Mol. Cell. Biol. 7, 2467–2476.

Lee, H. K., Driscoll, D., Asch, H., Asch, B. and Zhang, P. J. 1999. Downregulated gelsolin expression in hyperplastic and neoplastic lesions of the prostate. Prostate 40, 14–19.

Lee, Y. G., Macoska, J. A., Korenchuk, S. and Pienta, K. J. 2002. MIM, a potential metastasis suppressor gene in bladder cancer. Neoplasia 4, 291–294.

Lee, Y. J., Mazzatti, D. J., Yun, Z. and Keng, P. C. 2005b. Inhibition of invasiveness of human lung cancer cell line H1299 by over-expression of cofilin. Cell Biol. Int. 29, 877–883.

Lee, S. H., Son, M. J., Oh, S. H., Rho, S. B., Park, K., Kim, Y. J., Park, M. S. and Lee, J. H. 2005a. Thymosin {beta}(10) inhibits angiogenesis and tumor growth by interfering with Ras function. Cancer Res. 65, 137–148.

Lee, S. H., Zhang, W., Choi, J. J., Cho, Y. S., Oh, S. H., Kim, J. W., Hu, L., Xu, J., Liu, J. and Lee, J. H. 2001. Overexpression of the thymosin beta-10 gene in human ovarian cancer cells disrupts F-actin stress fiber and leads to apoptosis. Oncogene 20, 6700–6706.

Lener, T., Burgstaller, G. and Gimona, M. 2004. The role of calponin in the gene profile of metastatic cells: Inhibition of metastatic cell motility by multiple calponin repeats. FEBS Lett. 556, 221–226.

Li, Y., Tondravi, M., Liu, J., Smith, E., Haudenschild, C. C., Kaczmarek, M. and Zhan, X. 2001. Cortactin potentiates bone metastasis of breast cancer cells. Cancer Res. 61, 6906–6911.

Lin, C. S., Park, T., Chen, Z. P. and Leavitt, J. 1993. Human plastin genes. Comparative gene structure, chromosome location and differential expression in normal and neoplastic cells. J. Biol. Chem. 268, 2781–2792.

Lin, Y. H., Park, Z. Y., Lin, D., Brahmbhatt, A. A., Rio, M. C., Yates, J. R. and Klemke, R. L. 2004. Regulation of cell migration and survival by focal adhesion targeting of Lasp-1. J. Cell Biol. 165, 421–432.

Lin, F., Yu, Y. P., Woods, J., Cieply, K., Gooding, B., Finkelstein, P., Dhir, R., Krill, D., Becich, M. J., Michalopoulos, G., Finkelstein, S. and Luo, J. H. 2001. Myopodin, a synaptopodin homologue, is frequently deleted in invasive prostate cancers. Am. J. Pathol. 159, 1603–1612.

Linder, S. and Kopp, P. 2005. Podosomes at a glance. J. Cell Sci. 118, 2079–2082.

Loberg, R. D., Neeley, C. K., Adam-Day, L. L., Fridman, Y., St John, L. N., Nixdorf, S., Jackson, P., Kalikin, L. M. and Pienta, K. J. 2005. Differential expression analysis of MIM (MTSS1) splice variants and a functional role of MIM in prostate cancer cell biology. Int. J. Oncol. 26, 1699–1705.

Lorenz, M., Yamaguchi, H., Wang, Y., Singer, R. H. and Condeelis, J. 2004. Imaging sites of N-wasp activity in lamellipodia and invadopodia of carcinoma cells. Curr. Biol. 14, 697–703.

Lukasik, S. M., Zhang, L., Corpora, T., Tomanicek, S., Li, Y., Kundu, M., Hartman, K., Liu, P. P., Laue, T. M., Biltonen, R. L., Speck, N. A. and Bushweller, J. H. 2002. Altered affinity of CBF beta-SMMHC for Runx1 explains its role in leukemogenesis. Nat. Struct. Biol. 9, 674–679.

Luo, J. H., Yu, Y. P., Cieply, K., Lin, F., Deflavia, P., Dhir, R., Finkelstein, S., Michalopoulos, G. and Becich, M. 2002. Gene expression analysis of prostate cancers. Mol. Carcinog. 33, 25–35.

Maaser, K., Wolf, K., Klein, C. E., Niggemann, B., Zanker, K. S., Brocker, E. B. and Friedl, P. 1999. Functional hierarchy of simultaneously expressed adhesion receptors: Integrin alpha2beta1 but not CD44 mediates MV3 melanoma cell migration and matrix reorganization within three-dimensional hyaluronan-containing collagen matrices. Mol. Biol. Cell 10, 3067–3079.

Machesky, L. M., Atkinson, S. J., Ampe, C., Vandekerckhove, J. and Pollard, T. D. 1994. Purification of a cortical complex containing two unconventional actins from Acanthamoeba by affinity chromatography on profilin-agarose. J. Cell Biol. 127, 107–115.

Mahadev, K., Raval, G., Bharadwaj, S., Willingham, M. C., Lange, E. M., Vonderhaar, B., Salomon, D. and Prasad, G. L. 2002. Suppression of the transformed phenotype of breast cancer by tropomyosin-1. Exp. Cell Res. 279, 40–51.

Mareel, M. and Leroy, A. 2003. Clinical, cellular, and molecular aspects of cancer invasion. Physiol. Rev. 83, 337–376.

Martoglio, A. M., Tom, B. D., Starkey, M., Corps, A. N., Charnock-Jones, D. S. and Smith, S. K. 2000. Changes in tumorigenesis- and angiogenesis-related gene transcript abundance profiles in ovarian cancer detected by tailored high density cDNA arrays. Mol. Med. 6, 750–765.

Maul, R. S., Song, Y., Amann, K. J., Gerbin, S. C., Pollard, T. D. and Chang, D. D. 2003. EPLIN regulates actin dynamics by cross-linking and stabilizing filaments. J. Cell Biol. 160, 399–407.

McClatchey, A. I. and Giovannini, M. 2005. Membrane organization and tumorigenesis – The NF2 tumor suppressor, Merlin. Genes Dev. 19, 2265–2277.

McCormack, S. A., Ray, R. M., Blanner, P. M. and Johnson, L. R. 1999. Polyamine depletion alters the relationship of F-actin, G-actin and thymosin beta4 in migrating IEC-6 cells. Am. J. Physiol. 276, 459–468.

McNiven, M. A., Baldassarre, M. and Buccione, R. 2004. The role of dynamin in the assembly and function of podosomes and invadopodia. Front. Biosci. 9, 1944–1953.

Meyer, T. and Brink, U. 1997. Immunohistochemical detection of vinculin in human rhabdomyosarcomas. Gen. Diagn. Pathol. 142, 191–198.

Mielnicki, L. M., Ying, A. M., Head, K. L., Asch, H. L. and Asch, B. B. 1999. Epigenetic regulation of gelsolin expression in human breast cancer cells. Exp. Cell Res. 249, 161–176.

Mitchison, T. J. and Cramer, L. P. 1996. Actin-based cell motility and cell locomotion. Cell 84, 371–379.

Mueller, M. M. and Fusenig, N. E. 2004. Friends or foes – Bipolar effects of the tumour stroma in cancer. Nat. Rev. Cancer 4, 839–849.

Muller-Tidow, C., Diederichs, S., Thomas, M. and Serve, H. 2004. Genome-wide screening for prognosis-predicting genes in early-stage non-small-cell lung cancer. Lung Cancer 45, S145–S150.

Nabeshima, K., Inoue, T., Shimao, Y., Kataoka, H. and Koono, M. 1999. Cohort migration of carcinoma cells, differentiated colorectal carcinoma cells move as coherent cell clusters or sheets. Histol. Histopathol. 14, 1183–1197.

Nalbant, P., Hodgson, L., Kraynov, V., Toutchkine, A. and Hahn, K. M. 2004. Activation of endogenous Cdc42 visualized in living cells. Science 305, 1615–1619.

Niggli, V. 2005. Regulation of protein activities by phosphoinositide phosphates. Annu. Rev. Cell Dev. Biol. 21, 57–79.

Nikolopoulos, S. N., Spengler, B. A., Kisselbach, K., Evans, A. E., Biedler, J. L. and Ross, R. A. 2000. The human non-muscle alpha-actinin protein encoded by the ACTN4 gene suppresses tumorigenicity of human neuroblastoma cells. Oncogene 19, 380–386.

Nishiya, N., Kiosses, W., Han, J. and Ginsberg, M. 2005. An alpha4 integrin–paxillin–Arf–GAP complex restricts Rac activation to the leading edge of migrating cells. Nat. Cell Biol. 7, 343–352.

Nixdorf, S., Grimm, M. O., Loberg, R., Marreiros, A., Russell, P. J., Pienta, K. J. and Jackson, P. 2004. Expression and regulation of MIM (Missing In Metastasis), a novel putative metastasis suppressor gene, and MIM-B, in bladder cancer cell lines. Cancer Lett. 215, 209–220.

Noske, A., Denkert, C., Schober, H., Sers, C., Zhumabayeva, B., Weichert, W., Dietel, M. and Wiechen, K. 2005. Loss of gelsolin expression in human ovarian carcinomas. Eur. J. Cancer 41, 461–469.

Novy, R. E., Lin, J. L., Lin, C. S. and Lin, J. J. 1993. Human fibroblast tropomyosin isoforms: Characterization of cDNA clones and analysis of tropomyosin isoform expression in human tissues and in normal and transformed cells. Cell Motil. Cytoskeleton 25, 267–281.

Nowak, D., Skwarek-Maruszewska, A., Zemanek-Zboch, M. and Malicka-Blaszkiewicz, M. 2005. Beta-actin in human colon adenocarcinoma cell lines with different metastatic potential. Acta Biochim. Pol. 52, 461–468.

Ohashi, K., Nagata, K., Maekawa, M., Ishizaki, T., Narumiya, S. and Mizuno, K. 2000. Rho-associated kinase ROCK activates LIM-kinase 1 by phosphorylation at threonine 508 within the activation loop. J. Biol. Chem. 275, 3577–3582.

Ohtani, K., Sakamoto, H., Rutherford, T., Chen, Z., Satoh, K. and Naftolin, F. 1999. Ezrin, a membrane-cytoskeletal linking protein, is involved in the process of invasion of endometrial cancer cells. Cancer Lett. 147, 31–38.

Oien, K. A., Vass, J. K., Downie, I., Fullarton, G. and Keith, W. N. 2003. Profiling, comparison and validation of gene expression in gastric carcinoma and normal stomach. Oncogene 22, 4287–4300.

Oleynikov, Y. and Singer, R. H. 2003. Real-time visualization of ZBP1 association with beta-actin mRNA during transcription and localization. Curr. Biol. 13, 199–207.

Otsubo, T., Iwaya, K., Mukai, Y., Mizokami, Y., Serizawa, H., Matsuoka, T. and Mukai, K. 2004. Involvement of Arp2/3 complex in the process of colorectal carcinogenesis. Mod. Pathol. 17, 461–467.

Pantel, K. and Woelfle, U. 2004. Micrometastasis in breast cancer and other solid tumors. J. Biol. Regul. Homeost. Agents 18, 120–125.

Park, P. C., Taylor, M. D., Mainprize, T. G., Becker, L. E., Ho, M., Dura, W. T., Squire, J. and Rutka, J. T. 2003. Transcriptional profiling of medulloblastoma in children. J. Neurosurg. 99, 534–541.

Patel, A. S., Schechter, G. L., Wasilenko, W. J. and Somers, K. D. 1998. Overexpression of EMS1/cortactin in NIH3T3 fibroblasts causes increased cell motility and invasion in vitro. Oncogene 16, 3227–3232.

Pawlak, G., McGarvey, T. W., Nguyen, T. B., Tomaszewski, J. E., Puthiyaveettil, R., Malkowicz, S. B. and Helfman, D. M. 2004. Alterations in tropomyosin isoform expression in human transitional cell carcinoma of the urinary bladder. Int. J. Cancer 110, 368–373.

Pederson, T. and Aebi, U. 2005. Nuclear actin extends, with no contraction in sight. Mol. Biol. Cell 16, 5055–5060.

Polet, D., Vandekerckhove, J., Ampe, C. and Lambrechts, A. 2006. Putting the biochemistry of profilins in a cellular context. Curr. Top. Biochem. Res. 8, 11–28.

Pollard, T. D. and Borisy, G. G. 2003. Cellular motility driven by assembly and disassembly of actin filaments. Cell 112, 453–465.

Prasad, G. L., Fuldner, R. A. and Cooper, H. L. 1993. Expression of transduced tropomyosin 1 cDNA suppresses neoplastic growth of cells transformed by the ras oncogene. Proc. Natl Acad. Sci. USA 90, 7039–7043.

Prasad, G. L., Masuelli, L., Raj, M. H. and Harindranath, N. 1999. Suppression of src-induced transformed phenotype by expression of tropomyosin-1. Oncogene 18, 2027–2031.

Pugh, C. W. and Ratcliffe, P. J. 2003. The von Hippel–Lindau tumor suppressor, hypoxia-inducible factor-1 (HIF-1) degradation and cancer pathogenesis. Semin. Cancer Biol. 13, 83–89.

Pulford, K., Lamant, L., Espinos, E., Jiang, Q., Xue, L., Turturro, F., Delsol, G. and Morris, S. W. 2004. The emerging normal and disease-related roles of anaplastic lymphoma kinase. Cell. Mol. Life Sci. 61, 2939–2953.

Qi, Y., Chiu, J. F., Wang, L., Kwong, D. L. and He, Q. Y. 2005. Comparative proteomic analysis of esophageal squamous cell carcinoma. Proteomics 5, 2960–2971.

Quinlan, M. E., Heuser, J. E., Kerkhoff, E. and Mullins, R. D. 2005. Drosophila Spire is an actin nucleation factor. Nature 433, 382–388.

Raftopoulou, M. and Hall, A. 2004. Cell migration: Rho GTPases lead the way. Dev. Biol. 265, 23–32.

Ramaraj, P., Singh, H., Niu, N., Chu, S., Holtz, M., Yee, J. K. and Bhatia, R. 2004. Effect of mutational inactivation of tyrosine kinase activity on BCR/ABL-induced abnormalities in cell growth and adhesion in human hematopoietic progenitors. Cancer Res. 64, 5322–5331.

Rao, J. and Li, N. 2004. Microfilament actin remodeling as a potential target for cancer drug development. Curr. Cancer Drug Targets 4, 345–354.

dos Remedios, C. G., Chhabra, D., Kekic, M., Dedova, I. V., Tsubakihara, M., Berry, D. A. and Nosworthy, N. J. 2003. Actin binding proteins: Regulation of cytoskeletal microfilaments. Physiol. Rev. 83, 433–473.

Rho, S. B., Chun, T., Lee, S. H., Park, K. and Lee, J. H. 2004. The interaction between E-tropomodulin and thymosin beta-10 rescues tumor cells from thymosin beta-10 mediated apoptosis by restoring actin architecture. FEBS Lett. 557, 57–63.

Rhodes, D. R., Barrette, T. R., Rubin, M. A., Ghosh, D. and Chinnaiyan, A. M. 2002. Meta-analysis of microarrays: Interstudy validation of gene expression profiles reveals pathway dysregulation in prostate cancer. Cancer Res. 62, 4427–4433.

Rhodes, D. R. and Chinnaiyan, A. M. 2005. Integrative analysis of the cancer transcriptome. Nat. Genet. 37(Suppl.), 31–37.

Rhodes, D. R., Yu, J., Shanker, K., Deshpande, N., Varambally, R., Ghosh, D., Barrette, T., Pandey, A. and Chinnaiyan, A. M. 2004. Large-scale meta-analysis of cancer microarray data identifies common transcriptional profiles of neoplastic transformation and progression. Proc. Natl Acad. Sci. USA 101, 9309–9314.

Robinson, B. W., Musk, A. W. and Lake, R. A. 2005. Malignant mesothelioma. Lancet 366, 397–408.

Rodrigo, J. P., Garcia, L. A., Ramos, S., Lazo, P. S. and Suarez, C. 2000. EMS1 gene amplification correlates with poor prognosis in squamous cell carcinomas of the head and neck. Clin. Cancer Res. 6, 3177–3182.

Roma, A. A. and Prayson, R. A. 2005. Fascin expression in 90 patients with glioblastoma multiforme. Ann. Diagn. Pathol. 9, 307–311.

Rommelaere, H., Waterschoot, D., Neirynck, K., Vandekerckhove, J. and Ampe, C. 2004. A method for rapidly screening functionality of actin mutants and tagged actins. Biol. Proced. Online 6, 235–249.

Ross, A. F., Oleynikov, Y., Kislauskis, E. H., Taneja, K. L. and Singer, R. H. 1997. Characterization of a beta-actin mRNA zipcode-binding protein. Mol. Cell. Biol. 17, 2158–2165.

Sadano, H., Inoue, M. and Taniguchi, S. 1992. Differential expression of vinculin between weakly and highly metastatic B16-melanoma cell lines. Jpn. J. Cancer Res. 83, 625–630.

Sadano, H., Taniguchi, S. and Baba, T. 1990. Newly identified type of beta actin reduces invasiveness of mouse B16-melanoma. FEBS Lett. 271, 23–27.

Sadano, H., Taniguchi, S., Kakunaga, T. and Baba, T. 1988. cDNA cloning and sequence of a new type of actin in mouse B16 melanoma. J. Biol. Chem. 263, 15868–15871.

Sahai, E. and Marshall, C. J. 2003. Differing modes of tumour cell invasion have distinct requirements for Rho/ROCK signalling and extracellular proteolysis. Nat. Cell Biol. 5, 711–719.

Salgia, R., Li, J. L., Ewaniuk, D. S., Pear, W., Pisick, E., Burky, S. A., Ernst, T., Sattler, M., Chen, L. B. and Griffin, J. D. 1997. BCR/ABL induces multiple abnormalities of cytoskeletal function. J. Clin. Invest. 100, 46–57.

Sanchez-Carbayo, M., Schwarz, K., Charytonowicz, E., Cordon-Cardo, C. and Mundel, P. 2003. Tumor suppressor role for myopodin in bladder cancer: Loss of nuclear expression of myopodin is cell-cycle dependent and predicts clinical outcome. Oncogene 22, 5298–5305.

Santelli, G., Califano, D., Chiappetta, G., Vento, M. T., Bartoli, P. C., Zullo, F., Trapasso, F., Viglietto, G. and Fusco, A. 1999. Thymosin beta-10 gene overexpression is a general event in human carcinogenesis. Am. J. Pathol. 155, 799–804.

Sattler, M., Pisick, E., Morrison, P. T. and Salgia, R. 2000. Role of the cytoskeletal protein paxillin in oncogenesis. Crit. Rev. Oncog. 11, 63–76.

Sazawa, A., Watanabe, T., Tanaka, M., Haga, K., Fujita, H., Harabayashi, T., Shinohara, N., Koyanagi, T. and Kuzumaki, N. 2002. Adenovirus mediated gelsolin gene therapy for orthotopic human bladder cancer in nude mice. J. Urol. 168, 1182–1187.

Schirenbeck, A., Arasada, R., Bretschneider, T., Schleicher, M. and Faix, J. 2005. Formins and VASPs may co-operate in the formation of filopodia. Biochem. Soc. Trans. 33, 1256–1259.

Schumacher, N., Borawski, J. M., Leberfinger, C. B., Gessler, M. and Kerkhoff, E. 2004. Overlapping expression pattern of the actin organizers Spir-1 and formin-2 in the developing mouse nervous system and the adult brain. Gene Expr. Patterns 4, 249–255.

Scoles, D. R., Nguyen, V. D., Qin, Y., Sun, C. X., Morrison, H., Gutmann, D. H. and Pulst, S. M. 2002. Neurofibromatosis 2 (NF2) tumor suppressor schwannomin and its interacting protein HRS regulate STAT signaling. Hum. Mol. Genet. 11, 3179–3189.

Scott, J. A., Shewan, A. M., den Elzen, N. R., Louriero, J. J., Gertler, F. B. and Yap, A. S. 2005. Ena/VASP proteins can regulate distinct modes of actin organization at cadherin-adhesive contacts. Mol. Biol. Cell 17, 1085–1095.

Segal, E., Friedman, N., Kaminski, N., Regev, A. and Koller, D. 2005. From signatures to models: Understanding cancer using microarrays. Nat. Genet. 37, S38–S45.

Shen, R., Ghosh, D. and Chinnaiyan, A. 2004. Prognostic meta-signature of breast cancer developed by two-stage mixture modeling of microarray data. BMC Genomics 5, 94.

Shen, Z., Xu, L., Chen, M., Li, E., Li, J., Wu, X. and Zeng, Y. 2003. Upregulated expression of Ezrin and invasive phenotype in malignantly transformed esophageal epithelial cells. World J. Gastroenterol. 9, 1182–1186.

Shestakova, E. A., Wyckoff, J., Jones, J., Singer, R. H. and Condeelis, J. 1999. Correlation of beta-actin messenger RNA localization with metastatic potential in rat adenocarcinoma cell lines. Cancer Res. 59, 1202–1205.

Shieh, D., Godleski, J., Herndon, J., Azuma, T., Mercer, H., Sugarbaker, D. and Kwiatkowski, D. 1999. Cell motility as a prognostic factor in Stage I nonsmall cell lung carcinoma: The role of gelsolin expression. Cancer 85, 47–57.

Singh, D., Febbo, P. G., Ross, K., Jackson, D. G., Manola, J., Ladd, C., Tamayo, P., Renshaw, A. A., D'Amico, A. V., Richie, J. P., Lander, E. S., Loda, M., Kantoff, P. W., Golub, T. R. and Sellers, W. R. 2002. Gene expression correlates of clinical prostate cancer behavior. Cancer Cell 1, 203–209.

Skobe, M., Rockwell, P., Goldstein, N., Vosseler, S. and Fusenig, N. E. 1997. Halting angiogenesis suppresses carcinoma cell invasion. Nat. Med. 3, 1222–1227.

Skoble, J., Auerbuch, V., Goley, E. D., Welch, M. D. and Portnoy, D. A. 2001. Pivotal role of VASP in Arp2/3 complex-mediated actin nucleation, actin branch-formation and Listeria monocytogenes motility. J. Cell Biol. 155, 89–100.

Small, J. V. and Resch, G. P. 2005. The comings and goings of actin: Coupling protrusion and retraction in cell motility. Curr. Opin. Cell Biol. 17, 517–523.

Small, J., Stradal, T., Vignal, E. and Rottner, K. 2002. The lamellipodium: Where motility begins. Trends Cell Biol. 12, 112–120.

Smith-Beckerman, D. M., Fung, K. W., Williams, K. E., Auersperg, N., Godwin, A. K. and Burlingame, A. L. 2005. Proteome changes in ovarian epithelial cells derived from women with BRCA1 mutations and family histories of cancer. Mol. Cell. Proteomics 4, 156–168.

Song, Y., Maul, R. S., Gerbin, C. S. and Chang, D. D. 2002. Inhibition of anchorage-independent growth of transformed NIH3T3 cells by epithelial protein lost in neoplasm (EPLIN) requires localization of EPLIN to actin cytoskeleton. Mol. Biol. Cell 13, 1408–1416.

Sosne, G., Christopherson, P. L., Barrett, R. P. and Fridman, R. 2005. Thymosin-beta4 modulates corneal matrix metalloproteinase levels and polymorphonuclear cell infiltration after alkali injury. Invest. Ophthalmol. Vis. Sci. 46, 2388–2395.

Sosne, G., Xu, L., Prach, L., Mrock, L. K., Kleinman, H. K., Letterio, J. J., Hazlett, L. D. and Kurpakus-Wheater, M. 2004. Thymosin beta 4 stimulates laminin-5 production independent of TGF-beta. Exp. Cell Res. 293, 175–183.

Sporn, M. B. 1996. The war on cancer. Lancet 347, 1377–1381.

Steelman, L. S., Bertrand, F. E. and McCubrey, J. A. 2004. The complexity of PTEN: Mutation, marker and potential target for therapeutic intervention. Expert Opin. Ther. Targets 8, 537–550.

Stradal, T. E. and Scita, G. 2006. Protein complexes regulating Arp2/3-mediated actin assembly. Curr. Opin. Cell Biol. 18, 4–10.

Strehl, S., Borkhardt, A., Slany, R., Fuchs, U. E., Konig, M. and Haas, O. A. 2003. The human LASP1 gene is fused to MLL in an acute myeloid leukemia with t(11;17)(q23;q21). Oncogene 22, 157–160.

Suyama, E., Wadhwa, R., Kawasaki, H., Yaguchi, T., Kaul, S. C., Nakajima, M. and Taira, K. 2004. LIM kinase-2 targeting as a possible anti-metastasis therapy. J. Gene Med. 6, 357–363.

Suzuki, H., Nagata, H., Shimada, Y. and Konno, A. 1998. Decrease in gamma-actin expression, disruption of actin microfilaments and alterations in cell adhesion systems associated with acquisition of metastatic capacity in human salivary gland adenocarcinoma cell clones. Int. J. Oncol. 12, 1079–1084.

Svitkina, T. M. and Borisy, G. G. 1999. Arp2/3 complex and actin depolymerizing factor/cofilin in dendritic organization and treadmilling of actin filament array in lamellipodia. J. Cell Biol. 145, 1009–1026.

Takano, T., Hasegawa, Y., Miyauchi, A., Matsuzuka, F., Yoshida, H., Kuma, K. and Amino, N. 2002. Quantitative analysis of thymosin beta-10 messenger RNA in thyroid carcinomas. Jpn. J. Clin. Oncol. 32, 229–232.

Takeoka, M., Ehara, T., Sagara, J., Hashimoto, S. and Taniguchi, S. 2002. Calponin h1 induced a flattened morphology and suppressed the growth of human fibrosarcoma HT1080 cells. Eur. J. Cancer 38, 436–442.

Tanaka, M., Mullauer, L., Ogiso, Y., Fujita, H., Moriya, S., Furuuchi, K., Harabayashi, T., Shinohara, N., Koyanagi, T. and Kuzumaki, N. 1995. Gelsolin: A candidate for suppressor of human bladder cancer. Cancer Res. 55, 3228–3232.

Thiery, J. P. 2002. Epithelial–mesenchymal transitions in tumour progression. Nat. Rev. Cancer 2, 442–454.

Thiery, J. P. 2003. Epithelial–mesenchymal transitions in development and pathologies. Curr. Opin. Cell Biol. 15, 740–746.

Thor, A. D., Edgerton, S. M., Liu, S., Moore, D. H., II. and Kwiatkowski, D. J. 2001. Gelsolin as a negative prognostic factor and effector of motility in erbB-2-positive epidermal growth factor receptor-positive breast cancers. Clin. Cancer Res. 7, 2415–2424.

Timpson, P., Lynch, D. K., Schramek, D., Walker, F. and Daly, R. J. 2005. Cortactin overexpression inhibits ligand-induced down-regulation of the epidermal growth factor receptor. Cancer Res. 65, 3273–3280.

Tomasetto, C., Moog-Lutz, C., Regnier, C. H., Schreiber, V., Basset, P. and Rio, M. C. 1995. Lasp-1 (MLN 50) defines a new LIM protein subfamily characterized by the association of LIM and SH3 domains. FEBS Lett. 373, 245–249.

Tong, G. X., Yee, H., Chiriboga, L., Hernandez, O. and Waisman, J. 2005. Fascin-1 expression in papillary and invasive urothelial carcinomas of the urinary bladder. Hum. Pathol. 36, 741–746.

Unwin, R. D., Craven, R. A., Harnden, P., Hanrahan, S., Totty, N., Knowles, M., Eardley, I., Selby, P. and Banks, R. E. 2003. Proteomic changes in renal cancer and co-ordinate demonstration of both the glycolytic and mitochondrial aspects of the Warburg effect. Proteomics 3, 1620–1632.

Van Ginkel, P. R., Gee, R. L., Walker, T. M., Hu, D. N., Heizmann, C. W. and Polans, A. S. 1998. The identification and differential expression of calcium-binding proteins associated with ocular melanoma. Biochim. Biophys. Acta 1448, 290–297.

Van Troys, M., Dewitte, D., Goethals, M., Carlier, M. F., Vandekerckhove, J. and Ampe, C. 1996. The actin binding site of thymosin beta 4 mapped by mutational analysis. EMBO J. 15, 201–210.

Van Troys, M., Vandekerckhove, J. and Ampe, C. 1999. Structural modules in actin-binding proteins: Towards a new classification. Biochim. Biophys. Acta 1448, 323–348.

Vandekerckhove, J., Leavitt, J., Kakunaga, T. and Weber, K. 1980. Coexpression of a mutant beta-actin and the two normal beta- and gamma-cytoplasmic actins in a stably transformed human cell line. Cell 22, 893–899.

Varga, A. E., Stourman, N. V., Zheng, Q., Safina, A. F., Quan, L., Li, X., Sossey-Alaoui, K. and Bakin, A. V. 2005. Silencing of the tropomyosin-1 gene by DNA methylation alters tumor suppressor function of TGF-beta. Oncogene 24, 5043–5052.

van 't Veer, L. J., Dai, H., van de Vijver, M. J., He, Y. D., Hart, A. A., Mao, M., Peterse, H. L., van der Kooy, K., Marton, M. J., Witteveen, A. T., Schreiber, G. J., Kerkhoven, R. M., Roberts, C., Linsley, P. S., Bernards, R. and Friend, S. H. 2002. Gene expression profiling predicts clinical outcome of breast cancer. Nature 415, 530–536.

Verrills, N. M., Liem, N. L., Liaw, T. Y., Hood, B. D., Lock, R. B. and Kavallaris, M. 2006. Proteomic analysis reveals a novel role for the actin cytoskeleton in vincristine resistant childhood leukemia – An in vivo study. Proteomics 6, 1681–1694.

Verschueren, H., Dewit, J., De Braekeleer, J., Schirrmacher, V. and De Baetselier, P. 1994. Motility and invasive potency of murine T-lymphoma cells: Effect of microtubule inhibitors. Cell Biol. Int. 18, 11–19.

Vicente-Manzanares, M., Webb, D. J. and Horwitz, A. R. 2005. Cell migration at a glance. J. Cell Sci. 118, 4917–4919.

Vigneswaran, N., Wu, J., Sacks, P., Gilcrease, M. and Zacharias, W. 2005. Microarray gene expression profiling of cell lines from primary and metastatic tongue squamous cell carcinoma: Possible insights from emerging technology. J. Oral Pathol. Med. 34, 77–86.

Wang, W. S., Chen, P. M., Hsiao, H. L., Wang, H. S., Liang, W. Y. and Su, Y. 2004b. Overexpression of the thymosin beta-4 gene is associated with increased invasion of SW480 colon carcinoma cells and the distant metastasis of human colorectal carcinoma. Oncogene 23, 6666–6671.

Wang, W., Goswami, S., Lapidus, K., Wells, A. L., Wyckoff, J. B., Sahai, E., Singer, R. H., Segall, J. E. and Condeelis, J. S. 2004a. Identification and testing of a gene expression signature of invasive carcinoma cells within primary mammary tumors. Cancer Res. 64, 8585–8594.

Wang, W., Goswami, S., Sahai, E., Wyckoff, J. B., Segall, J. E. and Condeelis, J. S. 2005a. Tumor cells caught in the act of invading: Their strategy for enhanced cell motility. Trends Cell Biol. 15, 138–145.

Wang, Y., Klijn, J. G., Zhang, Y., Sieuwerts, A. M., Look, M. P., Yang, F., Talantov, D., Timmermans, M., Meijer-van Gelder, M. E., Yu, J., Jatkoe, T., Berns, E. M., Atkins, D. and Foekens, J. A. 2005b. Gene-expression profiles to predict distant metastasis of lymph-node-negative primary breast cancer. Lancet 365, 671–679.

Wang, W., Wyckoff, J. B., Frohlich, V. C., Oleynikov, Y., Huttelmaier, S., Zavadil, J., Cermak, L., Bottinger, E. P., Singer, R. H., White, J. G., Segall, J. E. and Condeelis, J. S. 2002. Single cell behavior in metastatic primary mammary tumors correlated with gene expression patterns revealed by molecular profiling. Cancer Res. 62, 6278–6288.

Weigelt, B., Hu, Z., He, X., Livasy, C., Carey, L. A., Ewend, M. G., Glas, A. M., Perou, C. M. and van 't Veer, L. J. 2005. Molecular portraits and 70-gene prognosis signature are preserved throughout the metastatic process of breast cancer. Cancer Res. 65, 9155–9158.

Weinstein, B. M. 2005. Vessels and nerves: Marching to the same tune. Cell 120, 299–302.

Welch, D. R. 2004. Microarrays bring new insights into understanding of breast cancer metastasis to bone. Breast Cancer Res. 6, 61–64.

Welsh, J. B., Sapinoso, L. M., Su, A. I., Kern, S. G., Wang-Rodriguez, J., Moskaluk, C. A., Frierson, H. F., Jr. and Hampton, G. M. 2001. Analysis of gene expression identifies candidate markers and pharmacological targets in prostate cancer. Cancer Res. 61, 5974–5978.

Wick, W., Grimmel, C., Wild-Bode, C., Platten, M., Arpin, M. and Weller, M. 2001. Ezrin-dependent promotion of glioma cell clonogenicity, motility and invasion mediated by BCL-2 and transforming growth factor-beta2. J. Neurosci. 21, 3360–3368.

Wiesner, S., Legate, K. R. and Fassler, R. 2005. Integrin–actin interactions. Cell. Mol. Life Sci. 62, 1081–1099.

Wilkinson, S., Paterson, H. F. and Marshall, C. J. 2005. Cdc42-MRCK and Rho–ROCK signalling cooperate in myosin phosphorylation and cell invasion. Nat. Cell Biol. 7, 255–261.

Witke, W., Sharpe, A. H., Hartwig, J. H., Azuma, T., Stossel, T. P. and Kwiatkowski, D. J. 1995. Hemostatic, inflammatory and fibroblast responses are blunted in mice lacking gelsolin. Cell 81, 41–51.

Wittenmayer, N., Jandrig, B., Rothkegel, M., Schluter, K., Arnold, W., Haensch, W., Scherneck, S. and Jockusch, B. M. 2004. Tumor suppressor activity of profilin requires a functional actin binding site. Mol. Biol. Cell 15, 1600–1608.

Wolf, K. and Friedl, P. 2005. Functional imaging of pericellular proteolysis in cancer cell invasion. Biochimie 87, 315–320.

Wolf, K., Mazo, I., Leung, H., Engelke, K., von Andrian, U. H., Deryugina, E. I., Strongin, A. Y., Brocker, E. B. and Friedl, P. 2003a. Compensation mechanism in tumor cell migration: Mesenchymal–amoeboid transition after blocking of pericellular proteolysis. J. Cell Biol. 160, 267–277.

Wolf, K., Muller, R., Borgmann, S., Brocker, E. B. and Friedl, P. 2003b. Amoeboid shape change and contact guidance: T-lymphocyte crawling through fibrillar collagen is independent of matrix remodeling by MMPs and other proteases. Blood 102, 3262–3269.

Wyckoff, J., Wang, W., Lin, E. Y., Wang, Y., Pixley, F., Stanley, E. R., Graf, T., Pollard, J. W., Segall, J. and Condeelis, J. 2004. A paracrine loop between tumor cells and macrophages is required for tumor cell migration in mammary tumors. Cancer Res. 64, 7022–7029.

Wymann, M. and Marone, R. 2005. Phosphoinositide 3-kinase in disease: Timing, location and scaffolding. Curr. Opin. Cell Biol. 17, 141–149.

Xiao, G. H., Beeser, A., Chernoff, J. and Testa, J. R. 2002. p21-activated kinase links Rac/Cdc42 signaling to merlin. J. Biol. Chem. 277, 883–886.

Xie, J. J., Xu, L. Y., Zhang, H. H., Cai, W. J., Mai, R. Q., Xie, Y. M., Yang, Z. M., Niu, Y. D., Shen, Z. Y. and Li, E. M. 2005. Role of fascin in the proliferation and invasiveness of esophageal carcinoma cells. Biochem. Biophys. Res. Commun. 337, 355–362.

Yager, M. L., Hughes, J. A., Lovicu, F. J., Gunning, P. W., Weinberger, R. P. and O'Neill, G. M. 2003. Functional analysis of the actin-binding protein, tropomyosin 1, in neuroblastoma. Br. J. Cancer 89, 860–863.

Yamada, A., Hirose, K., Hashimoto, A. and Iino, M. 2005a. Real-time imaging of myosin II regulatory light-chain phosphorylation using a new protein biosensor. Biochem. J. 385, 589–594.

Yamada, S., Pokutta, S., Drees, F., Weis, W. I. and Nelson, W. J. 2005b. Deconstructing the cadherin–catenin–actin complex. Cell 123, 889–901.

Yamaguchi, H., Lorenz, M., Kempiak, S., Sarmiento, C., Coniglio, S., Symons, M., Segall, J., Eddy, R., Miki, H., Takenawa, T. and Condeelis, J. 2005. Molecular mechanisms of invadopodium formation: The role of the N-WASP–Arp2/3 complex pathway and cofilin. J. Cell Biol. 168, 441–452.

Yamaguchi, H., Wyckoff, J. and Condeelis, J. 2005. Cell migration in tumors. Curr. Opin. Cell Biol. 17, 559–564.

Yamamoto, T., Gotoh, M., Kitajima, M. and Hirohashi, S. 1993. Thymosin beta-4 expression is correlated with metastatic capacity of colorectal carcinomas. Biochem. Biophys. Res. Commun. 193, 706–710.

Yang, M., Reynoso, J., Jiang, P., Li, L., Moossa, A. R. and Hoffman, R. M. 2004b. Transgenic nude mouse with ubiquitous green fluorescent protein expression as a host for human tumors. Cancer Res. 64, 8651–8656.

Yang, J., Tan, D., Asch, H. L., Swede, H., Bepler, G., Geradts, J. and Moysich, K. B. 2004a. Prognostic significance of gelsolin expression level and variability in non-small cell lung cancer. Lung Cancer 46, 29–42.

Yap, C. T., Simpson, T. I., Pratt, T., Price, D. J. and Maciver, S. K. 2005. The motility of glioblastoma tumour cells is modulated by intracellular cofilin expression in a concentration-dependent manner. Cell Motil. Cytoskeleton 60, 153–165.

Yeo, M., Kim, D. K., Park, H. J., Oh, T. Y., Kim, J. H., Cho, S. W., Paik, Y. K. and Hahm, K. B. 2006. Loss of transgelin in repeated bouts of ulcerative colitis-induced colon carcinogenesis. Proteomics 6, 1158–1165.

Yin, H. L. and Janmey, P. A. 2003. Phosphoinositide regulation of the actin cytoskeleton. Annu. Rev. Physiol. 65, 761–789.

Yoder, B. J., Tso, E., Skacel, M., Pettay, J., Tarr, S., Budd, T., Tubbs, R. R., Adams, J. C. and Hicks, D. G. 2005. The expression of fascin, an actin-bundling motility protein, correlates with hormone receptor-negative breast cancer and a more aggressive clinical course. Clin. Cancer Res. 11, 186–192.

Yokota, S., Yamamoto, Y., Shimizu, K., Momoi, H., Kamikawa, T., Yamaoka, Y., Yanagi, H., Yura, T. and Kubota, H. 2001. Increased expression of cytosolic chaperonin CCT in human hepatocellular and colonic carcinoma. Cell Stress Chaperones 6, 345–350.

Yokoyama, M., Nishi, Y., Yoshii, J., Okubo, K. and Matsubara, K. 1996. Identification and cloning of neuroblastoma-specific and nerve tissue-specific genes through compiled expression profiles. DNA Res. 3, 311–320.

Yu, Y., Khan, J., Khanna, C., Helman, L., Meltzer, P. S. and Merlino, G. 2004a. Expression profiling identifies the cytoskeletal organizer ezrin and the developmental homeoprotein Six-1 as key metastatic regulators. Nat. Med. 10, 175–181.

Yu, Y. P., Landsittel, D., Jing, L., Nelson, J., Ren, B., Liu, L., McDonald, C., Thomas, R., Dhir, R., Finkelstein, S., Michalopoulos, G., Becich, M. and Luo, J. H. 2004b. Gene expression alterations in prostate cancer predicting tumor aggression and preceding development of malignancy. J. Clin. Oncol. 22, 2790–2799.

Zaidel-Bar, R., Ballestrem, C., Kam, Z. and Geiger, B. 2003. Early molecular events in the assembly of matrix adhesions at the leading edge of migrating cells. J. Cell Sci. 116, 4605–4613.

Zhang, H. L., Eom, T., Oleynikov, Y., Shenoy, S. M., Liebelt, D. A., Dictenberg, J. B., Singer, R. H. and Bassell, G. J. 2001. Neurotrophin-induced transport of a beta-actin mRNP complex increases beta-actin levels and stimulates growth cone motility. Neuron 31, 261–275.

Zicha, D., Allen, W. E., Brickell, P. M., Kinnon, C., Dunn, G. A., Jones, G. E. and Thrasher, A. J. 1998. Chemotaxis of macrophages is abolished in the Wiskott–Aldrich syndrome. Br. J. Haematol. 101, 659–665.

Zigmond, S. H. 2004. Beginning and ending an actin filament: Control at the barbed end. Curr. Top. Dev. Biol. 63, 145–188.

11
Diseases with Abnormal Actin and Actin-Binding Proteins in Leukocyte and Nonmuscle Cells

Hiroyuki Nunoi, MD. PhD.

Introduction

Actin and actin-binding proteins (ABPs) in nonmuscle cells play important functional roles in defense against microbial infection, in phagocytes, and in the alignment of epithelial cells. Phagocytes migrate to sites of infection to ingest and destroy pathogens by generating reactive oxygen species and releasing the contents of their granules into phagosomes and the extracellular medium. Epithelial cells maintain their shape and use villi to remove pathogens and waste materials. The underlying mechanisms for these activities are closely associated with reorganization of the cytoskeleton. This reorganization is a cyclic process that includes polymerization of G-actin to filaments, crosslinking of filaments to form supramolecular assemblies anchored to membranes, and depolymerization of F-actin to G-actin (Moraczewska et al. 1996) (Fig. 1).

Classes of ABPs

Various ABPs are known to regulate the dynamics of the actin cytoskeleton. Schleicher and colleagues (1995) have classified these proteins into three categories. The first group of proteins includes those that bind to monomeric actin and thus reversibly remove polymerizable actin from equilibrium with F-actin. The second group includes end-binding or capping proteins that inhibit further addition of monomers to actin polymers, and keep filaments short. The third group includes those proteins that bind along the side of actin filaments and either stabilize filaments, crosslink filaments to form three-dimensional networks, anchor filaments to membranes, or work as motors (Fig. 1b).

Various functional studies have been performed on actin and associated proteins. Spontaneous or genetically engineered mutants of actin or associated proteins have been developed in yeast, the amoebae form of the slime mold *Dictyostelium*, the protozoan *Tetrahymena*, the indirect flight muscle of

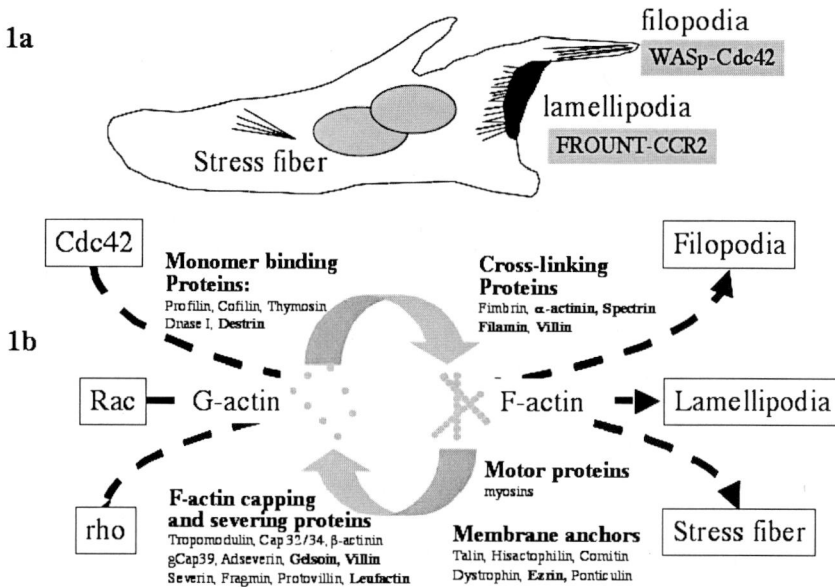

FIG. 1. The structural organization of the microfilament network on cell membranes (**a**) as well as the equilibrium between monomeric G-actin and filamentous F-actin (**b**) are regulated by actin-binding proteins (ABPs) and small GTP binding proteins. They either bind to monomeric actin thus inhibiting polymerization, or they cap, sever, anchor, crosslink or move actin filaments by binding to the ends or along the filaments. Active forms of each small GTP-binding protein display specific interactions with the microfilament network and form cell filopodia, lamelliopodia, and stress-fiber formations (**a, b**) (*See Color Plates*)

Drosophila fruit flies (Drummond, Hennessey and Sparrow 1991), and mouse gelsolin (Witke et al. 1995).

In phagocyte disorders due to cytoskeletal abnormality, motility or chemotactic defects of the cells are indicated. A small number of patients have been identified with neutrophils showing motility defects *in vivo* and *in vitro*, such as neutrophil actin dysfunction (NAD) (Boxer, Hedley-Whyte and Stossel 1974), neutrophil actin dysfunction due to 47 and 89 kDa protein abnormalities (NAD 47/89), and leufactin disease (Coates et al. 1991). Abnormal β-actin (Nunoi et al. 2000) and mutant Rac2 (8–10) have recently been reported. This paper provides an overview of diseases involving defects of actin and ABPs in human neutrophils and nonmuscle cells (Tables 1 and 2).

Reduced Chemotactic Responsiveness of Neonatal Neutrophils

Chemotaxis is deficient in neonates, particularly those delivered prematurely, and this is likely to contribute to increased vulnerability to severe infections

TABLE 1. Clinical and laboratory findings of patients with neutrophil cytoskeletal diseases.

	Actin dysfunction	Leufactin disease	β-actin mutant	Rac 2 mutation
M/F	Male	Male	Female	3 Males
Onset age	3 days	2 months	12 years	1 year, 5 weeks, 5 weeks
Symptoms	Blepharitis	Mucosal ulceration	Tuberculous pneu.	Multiple recurrent life-threatening infections
	No pus	Hepatosplenomegaly	Furunculosis	
		Petechiae	Thrombocytopenia	Perirectal abscess
			Hyperphotosensitivity	Absence of pus
			MR, short stature	
Pathogen	*Candida albicans Staphylococcus*	*Aspergillus nigrans*		*E. Coli, Entero sp.*
WBC per µl	17,600–133,000	16,000	3,000–4,000	21,300
			Stab > segmented PMN	Leukocytosis
Hb g dl^{-1}	?	9	9.5	?
Plat per µl	567,000	64,000	64,000	?
Bleeding tendency	Normal	–	–	?
Chemotaxis	Decrease	Decrease	Decrease	Absent
Adhesion	Not observed	Decrease	Not observed	Decrease
Phagocytosis	Decrease	ND	Decrease	Decrease
Degranulation	Enchanced	Not observed	Not observed	Decrease MPO
Superpoxide	Enchanced	Normal (NBT)	Decrease	Absent (fMLP+cyto-B) Present (PMA)
Actin polymerization	Decrease	Decrease	Normal	Not observed
Pathogenesis	Actin-binding protein?	Leufactin increase	Mutant β-actin (E364K)	Mutant Rac 2 (D57N)
Interactive substrate	Actin	Actin	Profilin, etc.	GTP binding position
Inheritance	AR	AR	AD	AD
Reference	[5, 15, 16]	[6, 17]	[7]	[8, 9, 10]

such as septicemia. Although potential mechanisms underlying impaired chemotactic responsiveness of neonatal neutrophils have been investigated from various perspectives such as signaling (Weinberger et al. 2001), response pattern to chemoattractants (Fox et al. 2005) and serum chemokines levels (Sullivan et al. 2002), the precise mechanisms have yet to be elucidated.

TABLE 2. Diseases involving defects of actin and actin-binding proteins in nonmuscle cells.

Categoria	Proteins	Abribation	Gene locus	Distribution	Related disorders
Monomer binding	Destrin	DSTN	20	Brain	Irregular thickening of the corneal epithelium (mouse)
Crosslinking	Actinin 4	ACTN4	19q13	Kidney podocytes	Focal and segmental glomerulosclerosis
	Spectrin	SPTA1	1q22–q25	Erythroid cell	Elliptocytosis, spherocytosis
		SPTB	14q22–q23.2	Erythroid cell	Pyropoikilocytosis
	Filamin	FLNA	Xq28	All brain cortical layers	Periventricular heterotopia
		FLNB	3p14.3	Endochondral ossification	Spondylocarpotarsal syn, Larsen syn, AOI, AOIII
	Villin	VIL1	2q35–q36	The brush border cytoskeleton	Progressive cholestasis and hepatic failure
Motor	Myosins	MYH9	22q11.2	Ubiquitous	May-Hegglin S, Sebastian S., Fechtner S., Epstein S.
Membrane anchors	Ezrin	VIL2	6q25–q26	Intestine, kidney, and lung	Brush border microvilli intestinal epithelial cell (mice)
F-actin capping and severing	Gelsolin	GSN	9q34	Leukocytes, platelets, and other cells	Amyloidosis, familial Finnish type

Wolach, Gavrieli and Pomeranz et al. (2000) reported that preincubation with G- and GM-CSF (colony stimulating factor) improved chemotactic disability of cord blood neutrophils *in vitro*. The effects of G- and GM-CSF in neonates, however, have neither been confirmed *in vivo* nor contributed in medical care. More precise analysis of the mechanisms involved and exploration of potential treatments for neonates are required.

Neutrophil Actin Dysfunction

NAD was first reported in 1974 (Boxer, Hadley-Whyte and Stossel 1974). The patient was a male infant who suffered from staphylococcal blepharitis on day three after birth. At 5 months, the child had been infected with

Staphylococcus aureus on the back and abdomen. Peripheral neutrophil counts were elevated (17,600–133,000 cells per μl) and no infiltration into the focus of necrosis was observed. This raised the question of whether these two disorders represented the same disease. In fact, surface expression of two CR3 subunits on neutrophils of NAD family members was found to be depressed. However, actin molecules obtained from neutrophil extracts were poorly polymerized in NAD patients, but were normally assembled in all patients with CR3 deficiency. These findings indicate that CR3 deficiency is not associated with actin filament assembly defects and support the conclusion that NAD represents a unique disease entity (Southwick et al. 1989).

Laboratory analysis revealed severe neutrophil motility disorder in the patient and bone marrow transplantation from the brother was successfully performed. No marked structural defects in actin purified from neutrophils taken from the father were noted on high-performance liquid chromatography (HPLC) and two-dimensional thin layer chromatography after tryptic digestion. Clearly, therefore, NAD is the result of a defect in neutrophil actin assembly, quite possibly due to defects or dysfunction in an actin-associated protein (Southwick, Dabiri and Stossel 1988), and not in the actin molecule itself. The disease may be inherited in an autosomal recessive manner.

Neutrophil Actin Dysfunction with 47 and 89 kDa Protein Abnormalities: Leufactin Disease

The patient was a two-month-old boy who presented with fever, severe skin and mucosal infections, and hepatosplenomegaly. Thrombocytopenia was noted, but neutrophil counts and morphology were normal. Two siblings had died in infancy exhibiting similar clinical pictures. The patient was successfully treated with bone marrow transplantation from his brother at 7 months (Coates et al. 1991).

Neutrophils from the patient showed functional abnormalities, particularly for motility. Random migration, chemotaxis, and morphological changes were decreased, whereas superoxide generation and elastase from granules were markedly elevated. Actin polymerization in fMLP (*N*-formyl-met-leu-phe)-stimulated neutrophils determined by NBD-phallacidin (7-nitrobenz-2-oxa-1, 3-phallacidin)-stained F-actin content was lower than neutrophils of controls. Partial defects in actin polymerization and scattering were also observed in neutrophils from both parents of the patient.

Sodium dodecyl sulfate (SDS) electrophoresis of whole cell proteins from the patient's neutrophils displayed marked decreases in the 89 kDa protein (8% of controls) and a marked increase in the 47 kDa protein (4.2-fold). Specimens from both the mother and father exhibited similar but milder changes in these proteins, although neither had experienced recurrent infections or chemotactic defects. The 47 kDa protein (called leufactin, after leukocyte F-actin binding protein (Li, Guerrero and Howard 1995) is known to interact with F-actin mainly via the electrostatic C-terminal domain. This protein is also known to be important for adjustment of actin polymerization. These studies described a new inherited actin dysfunction syndrome and drew attention to leufactin, which may be important in the regulation of actin polymerization in human neutrophils.

Wiskott–Aldrich Syndrome

Wiskott–Aldrich syndrome (WAS) is an X-linked recessive disorder originally characterized by the clinical triad of eczema, thrombocytopenia, and severe immunodeficiency, with recurrent bacterial and viral infections. Chemotactic deficiency has previously been reported in WAS patients, but the precise mechanisms have not been elucidated. Recently, a study of chemotaxis in macrophages from patients with classic WAS suggested that Cdc42-WASp-mediated filopodial extension is a requirement for chemotaxis (Zicha et al. 1998).

Upstream of the WASp (WAS protein) family of proteins lies the receptor tyrosine kinases, G-protein-coupled receptors, phosphoinositide-3-OH kinase (PI3-kinase), and the rho family of GTPases. These receive and transduce signals that lead to actin nucleation through WASp-Arp2/3 action (Jones 2000). Conversely, lamellipodium protrusion is initiated by ligation of CCR2-coupled FROUNT on monocytes and macrophages with CCL2 through a cascade comprising PI3-kinase and the small G-protein Rac (Fig. 1a, b) (Terashima et al. 2005). For T-cell chemotaxis *in vitro* and homing *in vivo*, WASp and its partner, the WAS protein-interacting protein (WIP), reportedly play important roles in actin reorganization (Gallego et al. 2005).

β-Actin Dysfunction

We reported the first case of mutant β-actin disease at the molecular level. The patient was a 12-year-old girl with a history of recurrent pyrogenic infection, recurrent stomatitis, photosensitivity, moderate intellectual impairment, and short stature. By 15-years, she had developed cardiomegaly (see chapter by Stefani et al.), hepatomegaly, and hypothyroidism. At that time, she presented with persistent fevers and died from septicemia despite intensive therapy (Nunoi et al. 2000).

Laboratory analyses showed persistently high levels of C-reactive protein, thrombocytopenia, and leukopenia. Band-form leukocytes were consistently the most abundant type of leukocyte in both blood samples (30–35%) and bone marrow (45%), and the ratio to other cells did not change following steroid or epinephrine challenge. A poor influx of leukocytes to skin window test sites was noted. Subsets of lymphocytes were within normal range and their responses to mitogens were normal.

Neutrophils from this patient demonstrated depressed chemotactic responses, reduced superoxide generation upon stimulation, and weaker fMLP-induced depolarization/repolarization relative to control neutrophils.

Using 2DE (two-dimensional electrophoresis) protein profiles, a unique 42 kDa spot was found in her neutrophils and other cells. Using an expression cloning technique after construction of the full-length cDNA library by cap method, a single nucleotide substitution was found in three of eight clones coding the β-actin sequence. Since a *Hin*fI DNase restriction site in the normal sequence was eliminated in the mutant sequence, this mutation was confirmed in the actin gene of the patient by amplifying fragments predicted to contain the mutation site.

The mutation site in the actin molecule occurred at a binding site for profilin (see chapter by Moens). The mutant actin could polymerize and depolymerize similar to normal actins according to analysis using 2D-PAGE. However, binding activity of mutant β-actin with profilin was confirmed to be decreased compared with normal actins. This represented the first description of an actin molecule mutation.

Very recently, four papers have described a mutant beta actin (R183W) and six mutant gamma actins (T89I, K118M, P264L, T278I, P332A, V370A). The former showed developmental midline malformations, sensory hearing loss, and a delayed-onset generalized dystonia (Procaccio et al. 2006). The latter showed autosomal dominant, progressive, sensorineural hearing impairment (van Wijk et al. 2003; Zhu et al. 2003; Rendtorff et al. 2006).

Rac2 Mutation (Neutrophil Immunodeficiency Syndrome)

Three patients with Rac2 mutation were reported independently. The first case involved a 5-week-old boy who presented with leukocytosis, severe bacterial infections, and poor wound healing (Ambruso et al. 2000). Neutrophils from this patient showed decreased chemotaxis, polarization, azurophilic granule secretion, and superoxide anion production upon stimulation of cells with fMLP or platelet-activated factor (PAF). In contrast, when cells were stimulated using PMA (phorbol-12-myristate-13-acetate), no differences were observed in superoxide generation compared with control neutrophils, and surface expression of CD11b was somewhat elevated. Western blot analysis of lysate from the patient's neutrophils demonstrated decreased levels of Rac2 protein.

The second case involved a 1-year-old boy who showed recurrent infections and defective neutrophil cellular function (Williams et al. 2000) similar to those found in Rac2-deficient mice. The third case involved a newborn boy who displayed multiple rapidly progressive soft-tissue infections (Kurkchubasche et al. 2001).

In these cases, mutations were identified at the same site in one allele of the Rac2 gene (D57N). Asp57 is conserved in all defined GTP-binding proteins. Rac2 (D57N) binds GDP, but not GTP, and inhibits activation of superoxide-generating phagocyte oxidase *in vitro*.

In the case of the first patient, addition of recombinant Rac to extracts from the patient's neutrophils resulted in a recovery of superoxide-producing activity *in vitro*. In the case of the second patient, cloned Rac cDNA from the patient's cells was introduced into normal bone marrow cells using retroviral vectors. Neutrophils expressing mutant Rac showed decreased cell movement and reduced production of superoxide in response to fMLP. The expressed recombinant protein did not bind GTP. These functional studies demonstrated that the D57N mutant behaves in a dominant-negative way at the cellular level.

These cases showed that rho GTPases control various cellular processes, including actin polymerization, integrin complex formation, cell adhesion (Nobes and Hall 1999), gene transcription, cell cycle progression, and cell proliferation.

Other Cases with Defects of ABPs

Table 2 is a brief overview of diseases associated with defects of ABPs in humans and animals.

α-Actinin

α-actinin (ACTN4) is highly expressed in glomerular podocytes (see chapter by dos Remedios), and is important in nonmuscular cytoskeletal function. ACTN4 is also upregulated early in the course of some animal models of nephrotic syndrome (Kalpan et al. 2000). Missense mutations in ACTN4 RNA (LYS228GLU, THR232ILE, SER235PRO) have been confirmed in affected individuals from a family with focal segmental glomerulosclerosis mapping to 19q (603278).

Spectrin

Spectrin is a tetrameric cytoskeletal protein essential for determination of cell shape, resilience of membranes to mechanical stress, positioning of transmembrane proteins, and organization of organelles and molecular traffic. Antiparallel dimers that self-associate to create the spectrin tetramer are formed by α- and β-spectrin subunits. In hereditary spherocytosis, elliptocytosis, and pyropoikilocytosis, molecular defects exist in the spectrin α- and β-chains (SPTA1 and SPTB, respectively) (Gallagher and Forget 1998).

Filamin

Filamin is a widely expressed protein that regulates reorganization of the actin cytoskeleton by interacting with integrins, transmembrane receptor complexes, and second messengers. FLNA is required for neuronal migration to the cortex and is essential for embryogenesis. Mutations of FLNA and numerous allelic variants have been reported in patients with periventricular heterotopia as an X-linked trait (Huttenlocher, Taravath and Mojtahedi

1994). Mutations in the FLNB gene have been identified in four human skeletal disorders: spondylocarpotarsal syndrome (SCT); autosomal-dominant Larsen syndrome characterized by bilateral dislocation of the knees; type I atelosteogenesis characterized by lethal chondrodysplasia (AOI); and type III atelosteogenesis (AOIII) (Krakow et al. 2004).

Nonmuscle Myosin

Myosin 9 (MYH9) is the only class II nonmuscle myosin readily and highly detectable in mouse T-cell cDNA myosin family members (Jacobelli et al. 2004). Expression is enriched in the uropod during T-cell crawling. Inactivation of the myosin motor may represent a key step in the T-cell "stop" response during antigen recognition. However, the mutation phenotype of MYH9 did not show any immune deficiencies, instead displaying autosomal-dominant giant-platelet disorders such as May-Hegglin anomaly, Fechtner syndrome, and Sebastian syndrome, all of which share the triad of thrombocytopenia, large platelets, and characteristic leukocyte inclusions called Dohle-like bodies. The N93K mutation was also associated with a tendency for myosin to aggregate, which may explain the leukocyte inclusions associated with this mutation in humans (Seri et al. 2000). An attempt was made to detect any defect of chemotaxis in neutrophils, but no such defects were identified (data not shown).

Gelsolin

The amyloid protein in the Finnish type of hereditary amyloidosis represents a fragment of the actin-filament-binding region of a variant gelsolin molecule. Using PCR and allele-specific oligonucleotide hybridization analysis of genomic DNA, a single base mutation, 654G-A, was identified in all five unrelated patients with Finnish amyloidosis studied, but not in the 45 unrelated control subjects (Maury, Alli and Baumann 1990) (see chapter on gelsolin by Burtnick and Robinson).

Summary

Actin and ABPs in neutrophils not only act as fundamental mobile vehicles and maintain cell shape, but are also involved in physiological regulatory functions such as the production of superoxide by neutrophils and of platelets by cytoplasmic budding of megakaryocytes. Although the molecular mechanisms have not been elucidated, many chemotactic insufficiencies have been reported, such as neonatal neutrophil abnormality, NAD, Wiskott–Aldrich syndrome and neutrophil actin dysfunction with abnormal 47 and 89 kDa proteins. Recently, abnormal-β-actin disease and diseases involving Rac2 mutations have been analyzed at the molecular and physiological level. Neutrophils exhibit not only chemotactic disability, but also defects such as thrombocytopenia, photohypersensitivity, and mental retardation. The above represents an overview of the literature on diseases involving defects of actin and the actin-binding protein Wiscott–Aldrich Syndrome protein (WASp) in human neutrophils and nonmuscle cells.

Concluding Remarks

Based on the recent development of novel technologies such as FACS (fluorescence-activated cell sorter) analysis with NBD-phallacidin-staining, 2DE protein analysis and molecular cloning, and from the growing knowledge about cytoskeletal proteins, numerous diseases involving defects of actin and associated binding proteins have been reported. The mode of inheritance for these diseases is sometimes autosomal dominant. Based on the distribution of causative proteins, patients suffer not only from recurrent infections due to poor neutrophil infiltration, but also mental retardation, and other symptoms such as thrombocytopenia, tendency to bleed, and photohypersensitivity. Patients with blood diseases are currently managed symptomatically and prophylactically using antibiotics. It should be noted that bone marrow transplantation has been successfully performed in patients with defective NAD, leufactin, and inhibitory Rac2 mutant disease.

References

Ambruso, D. R., Knall, C., Abell, A. N., Panepinto, J., Kurkchubasche, A., Thuman, G., Gonzales-Aller, C., Hiester, A., deBoer, M., Harbeck, R. J., Oyer, R., Johnson, G. L. and Roos, D. 2000. Human neutrophil immunodeficiency syndrome is associated with an inhibitory Rac2 mutation. Proc. Natl Acad. Sci. USA 97, 4654–4659.

Boxer, L. A., Hedley-Whyte, E. T. and Stossel, T. P. 1974. Neutrophil actin dysfunction and abnormal neutrophil behavior. N. Engl. J. Med. 291, 1093–1099.

Coates, T. D., Torkildson, J. C., Torres, M., Church, J. A. and Howard, T. H. 1991. An inherited defect of neutrophil motility and microfilamentous cytoskeleton associated with abnormalities in 47 kD and 89 kD proteins. Blood 78, 1338–1346.

Drummond, D. R., Hennessey, E. S. and Sparrow, J. C. 1991. Characterisation of missense mutations in the Act88F gene of *Drosophila melanogaster*. Mol. Genet. 226, 70–80.

Fox, S. E., Lu, W., Maheshwari, A., Christensen, R. D. and Calhoun, D. A. 2005. The effects and comparative differences of neutrophil specific chemokines on neutrophil chemotaxis of the neonate. Cytokine 29, 135–140.

Gallagher, P. G. and Forget, B. G. 1998. Hematologically important mutations: spectrin and ankyrin variants in hereditary spherocytosis. Blood Cells Mol. Dis. 24, 539–543.

Gallego, M. D., de la Fuente, M. A., Anton, I. M., Snapper, S., Fuhlbrigge, R. and Geha, R. S. 2005. WIP and WASP play complementary roles in T cell homing and chemotaxis to SDF-1 α. Int. Immunol. 18, 221–232.

Huttenlocher, P. R., Taravath, S. and Mojtahedi, S. 1994. Periventricular heterotopia and epilepsy. Neurology 44, 51–55.

Jacobelli, J., Chmura, S. A., Buxton, D. B., Davis, M. M. and Krummel, M. F. 2004. A single class II myosin modulates T cell motility and stopping, but not synapse formation. Nat. Immunol. 5, 531–538.

Jones, G. E. 2000. Cellular signaling in macrophage migration and chemotaxis. J. Leukoc. Biol. 68, 593–602.

Kaplan, J. M., Kim, S. H., North, K. N., Rennke, H., Correia, L. A., Tong, H. Q., Mathis, B. J., Rodriguez-Perez, J. C., Allen, P. G., Beggs, A. H. and Pollak, M. R. 2000. Mutations in ACTN4, encoding alpha-actinin-4, cause familial focal segmental glomerulosclerosis. Nat. Genet. 24, 251–256.

Krakow, D., Robertson, S. P., King, L. M., Morgan, T., Sebald, E. T., Bertolotto, C., Wachsmann-Hogiu, S., Acuna, D., Shapiro, S. S., Takafuta, T., Aftimos, S., Kim, C. A., Firth, H., Steiner, C. E., Cormier-Daire, V., Superti-Furga, A., Bonafe, L., Graham, J. M. Jr., Grix, A., Bacino, C. A., Allanson, J., Bialer, M. G., Lachman, R. S., Rimoin, D. L. and Cohn, D. H. 2004. Mutations in the gene encoding filamin B disrupt vertebral segmentation, joint formation and skeletogenesis. Nat. Genet. 36, 405–410.

Kurkchubasche, A. G., Panepinto, J. A., Tracy, T. F. Jr., Thurman, G. W. and Ambruso, D. R. 2001. Clinical features of a human Rac2 mutation: A complex neutrophil dysfunction disease. J. Pediatr. 139, 141–147.

Li, Y., Guerrero, A. and Howard, T. H. 1995. The actin-binding protein, lymphocyte-specific protein 1, is expressed in human leukocytes and human myeloid and lymphoid cell lines. J. Immunol. 155, 3563–3569.

Maury, C. P. J., Alli, K. and Baumann, M. 1990. Finnish hereditary amyloidosis: Amino acid sequence homology between the amyloid fibril protein and human plasma gelsolin. FEBS Lett. 260, 85–87.

Moraczewska, J., Strzelecka-Golaszewska, H., Moens, P. D. J. and dos Remedios, C. G. 1996. Structural changes in the small domain of actin detected by fluorescence resonance energy transfer spectroscopy. Biochem. J. 317, 605–611.

Nobes, C. D. and Hall, A. 1999. Rho GTPases control polarity, protrusion, and adhesion during cell movement. J. Cell Biol. 144, 1235–1244.

Nunoi, H., Yamazaki, T., Tsuchiya, H., Kato, S., Malech, H. L., Matsuda, I. and Kanegasaki, S. 2000. A heterozygous mutation of β-actin associated with neutrophil dysfunction and recurrent infection. Proc. Natl Acad. Sci. USA 96, 8693–8698.

Procaccio, V., Salazar, G., Ono, S., Styers, M. L., Gearing, M., Davila, A., Jimenez, R., Juncos, J., Gutekunst, C. A., Meroni, G., Fontanella, B., Sontag, E., Sontag, J. M., Faunde, V. and Wainer, B. H. 2006. A mutation of beta-actin that alters depolymerization dynamics is associated with autosomal dominant developmental malformations, deafness, and dystonia. Am. J. Hum. Genet. 78, 947–960.

Rendtorff, N. D., Zhu, M., Fagerheim, T., Antal, T. L., Jones, M., Teslovich, T. M., Gillanders, E. M., Barmada, M., Teig, E., Trent, J. M., Friderici, K. H., Stephan, D. A. and Tranebjaerg, L. 2006. A novel missense mutation in ACTG1 causes dominant deafness in a Norwegian DFNA20/26 family, but ACTG1 mutations are not frequent among families with hereditary hearing impairment. Eur. J. Hum. Genet. 14, 1097–1105.

Schleicher, M., Andre, B., Andreoli, C., Eichinger, L., Haugwitz, M., Hofmann, A., Karakesisoglou, J., Stockelhuber, M. and Noegel, A. A. 1995. Structure/function studies on cytoskeletal proteins in *Dictyostelium* amoebae as a paradigm. FEBS Lett. 369, 38–42.

Seri, M., Cusano, R., Gangarossa, S., Caridi, G., Bordo, D., Lo Nigro, C., Ghiggeri, G. M., Ravazzolo, R., Savino, M., Del Vecchio, M., d'Apolito, M., Iolascon, A., Zelante, L. L., Savoia, A., Balduini, C. L., Noris, P., Magrini, U., Belletti, S., Heath, K. E., Babcock, M., Glucksman, M. J., Aliprandis, E., Bizzaro, N., Desnick, R. J. and Martignetti, J. A. 2000. Mutations in MYH9 result in the May-Hegglin anomaly, and Fechtner and Sebastian syndromes. The May-Heggllin/Fechtner Syndrome Consortium. Nat. Genet. 26, 103–105.

Southwick, F. S., Dabiri, G. A. and Stossel, T. P. 1988. Neutrophil actin dysfunction is a genetic disorder associated with partial impairment of neutrophil actin assembly in three family members. J. Clin. Invest. 2, 1525–1531.

Southwick, F. S., Howard, T. H., Holbrook, T., Anderson, D. C., Stossel, T. P. and Arnaout, M. A. 1989. The relationship between CR3 deficiency and neutrophil actin assembly. Blood 73, 1973–1979.

Sullivan, S. E., Staba, S. L., Gersting, J. A., Hutson, A. D., Theriaque, D., Christensen, R. D. and Calhoun, D. A. 2002. Circulating concentrations of chemokines in cord blood, neonates, and adults. Pediatr. Res. 51, 653–657.

Terashima, Y., Onai, N., Murai, M., Enomoto, M., Poonpiriya, V., Hamada, T., Motomura, K., Suwa, M., Ezaki, T., Haga, T., Kenegasaki, S. and Matsushima, K. 2005. Pivotal function for cytoplasmic protein FROUNT in CCR2-mediated monocyte chemotaxis. Nat. Immunol. 6, 827–835.

van Wijk, E., Krieger, E., Kemperman, M. H., De Leenheer, E. M., Huygen, P. L., Cremers, C. W., Cremers, F. P. and Kremer, H. 2003, A mutation in the gamma actin 1 (ACTG1) gene causes autosomal dominant hearing loss (DFNA20/26). J. Med. Genet. 40, 879–884.

Weinberger, B., Laskin, D. L., Mariano, T. M., Sunil, V. R., DeCoste, C. J., Heck, D. E., Gardner, C. R. and Laskin, J. D. 2001. Mechanisms underlying reduced responsiveness of neonatal neutrophils to distinct chemoattractants. J. Leukoc. Biol. 70, 969–976.

Williams, D. A., Tao, W., Yang, F., Kim, C., Gu, Y., Mansfield, P., Levine, J. E., Petryniak, B., Derrow, C. W., Harris, C., Jia, B., Zheng, Y., Ambruso, D. R., Lowe, J. B., Atkinson, S. J., Dinauer, M. C. and Boxer, L. 2000. Dominant negative mutation of the hematopoietic-specific rho GTPase, Rac2, is associated with a human phagocyte immunodeficiency. Blood 96, 1646–1654.

Witke, W., Sharpe, A. H., Hartwig, J. H., Azuma, T., Stossel, T. P. and Kwiatkowski, D. J. 1995. Hemostatic, inflammatory, and fibroblast responses are blunted in mice lacking gelsolin. Cell 81, 41–51.

Wolach, B., Gavrieli, R. and Pomeranz, A. 2000. Effect of granulocyte and granulocyte macrophage colony stimulating factors (G-CSF and GM-CSF) on neonatal neutrophil functions. Pediatr. Res. 48, 369–373.

Zhu, M., Yang, T., Wei, S., DeWan, A. T., Morell, R. J., Elfenbein, J. L., Fisher, R. A., Leal, S. M., Smith, R. J. and Friderici K. H. 2003. Mutations in the gamma-actin gene (ACTG1) are associated with dominant progressive deafness (DFNA20/26). Am. J. Hum. Genet. 73, 1082–1091.

Zicha, D., Allen, W. E., Brickell, P. M., Kinnon, C., Dunn, G. A., Jones, G. E. and Thrasher, A. J. 1998. Chemotaxis of macrophages is abolished in the Wiskott–Aldrich syndrome. Br. J. Haematol. 101, 659–665.

12
The Role of PIP$_2$ in Actin, Actin-Binding Proteins and Disease

C. G. dos Remedios and Neil J. Nosworthy

Introduction

What is PIP$_2$?

Phosphatidylinositol(4,5) bisphosphate (PI(4,5)P$_2$ or more simply, PIP$_2$) is only a minor component of the cell membrane lipids but it is the most abundant of the bisphorylated phosphoinositides. PIP$_2$ is a key ligand in regulating the activity of a number of actin-binding proteins that regulate the assembly of actin microfilaments (Logan and Mandato 2006; Yin and Janmey 2003). Because of its central role, abnormal accumulation of PIP$_2$ in the cell might be expected to produce defects in the assembly of cytoplasmic actin microfilaments. Similarly, a substantial reduction in cellular PIP$_2$ is also not likely to be tolerated. Thus, mutations in enzymes that control cellular PIP$_2$ content may be responsible, either directly or indirectly, for a number of human diseases/disorders.

Where is it Located in the Cell?

PIP$_2$ is synthesized in the cell by the phosphorylation of PI(4)P by PIP kinase type I or by the phosphorylation of PI(5)P by PIP kinase type II. The former is localized at the cell membrane and is consistent with the cellular abundance of PIP$_2$ which is also located on the intracellular surface of the cell membrane. PIP kinase type I is the predominantly active kinase. PIP kinase type II is located closer to the nucleus (Doughman, Firestone and Anderson 2003).

PIP$_2$ Metabolism

PIP$_2$ is hydrolyzed by the action of phospholipase C (PLC) and possibly some of its isoforms, producing diacylglycerol (DAG) and inositol 1,4,5-triphosphate (IP$_3$). The latter is an activator of calmodulin that in turn activates gelsolin (see chapter by Burtnick and Robinson) and cofilin/ADF (see chapter by Maloney et al.). Thus PIP$_2$ is a precursor of DAG and IP$_3$, both known second messengers. PIP$_2$ itself is a second messenger because of its role as regulator of the actin cytoskeleton. PIP$_2$ can be converted into phosphatidylinositol 3,4,5-triphosphate (PI(1,4,5)P$_3$) by phosphoinosityl 3-kinase (Stephens, Jackson and Hawkins 1993). And finally, there is a large family of PIP$_2$ 5-phosphatases that hydrolyze phosphate from the 5 position of PIP$_2$ (Godi, Di Campli and De Matteis 2004). This reference should be consulted for more details of the metabolism of PIP$_2$.

PIP$_2$ and Disease

Lowe Syndrome

The Lowe syndrome is also known as the oculocerebrorenal syndrome of Lowe (OCRL). As the name suggests, the syndrome involves the eyes, brain, and kidneys (Addis et al. 2004). In the eye, the problem is mainly the development of cataracts as early as in utero, early postnatal development of glaucoma as well as a number of associated dysfunctional features. The most serious cerebral defect is severe hypotonia and a lack of deep tendon reflex that can cause serious respiration difficulties at birth. In addition there is moderate to severe mental retardation in about 10% of cases. The third part of the nomenclature involves the kidneys. This renal malfunction develops after birth and results in a failure to thrive, finally resulting in renal failure requiring renal dialysis.

Lowe syndrome is a very rare (1:500,000) X-linked disease that has been the focus of several papers since it was first described (Lowe, Terrey and MacLachan 1952; Loi 2006). The genetic defect was later identified on the X chromosome in the region of locus Xq26 (Mueller et al. 1985), the gene was subsequently cloned by Nussbaum et al. (1997), and the gene product OCRL1 is a member of the PIP$_2$ 5-phosphatase type II family (Suchy, Olivos-Glander and Nussbaum 1995). In cells, it is localized in the trans-Golgi network (Olivos-Glander, Janne and Nussbaum 1995; Nussbaum and Suchy 2001).

The syndrome is characterized by the accumulation of PIP$_2$ caused by reduced or deficient activity of the protein product (OCRL1). The defective gene was identified in a report of two females who presented with the Lowe clinical phenotype and had balanced X-autosome translocations in the region Xq26. The mutations include truncation mutations (nonsense, splice-site, and frame-shift), and missense mutations both inside and outside the catalytic domains of this enzyme (reviewed by Loi 2006). There are several consequences of the presence of elevated PIP$_2$, all of which involve altered levels of activities of actin-binding proteins which in turn produce enhanced assembly of cytoplasmic actin (Suchy and Nussbaum 2002). PIP$_2$ does not directly affect the assembly of actin.

Fibroblasts from OCRL Patients

How does increased PIP$_2$ account for a complex syndrome that affects the eyes, the central nervous system, and the kidneys? Suchy and Nussbaum (2002) recently addressed this question by studying fibroblasts from patients with Lowe syndrome and comparing them to one control subject and from ATCC fibroblast cell lines. The expectation was that the defective gene would be evident in these cells and an observed abnormality may be informative for the molecular defects observed in the eye, brain, and kidneys of these patients.

Suchy and Nussbaum (2002) cultured Lowe patients' fibroblasts and found that the PIP_2 activity was <10% of the control activity. The cells exhibited significantly reduced stress fibers (actin microfilament) identified using Alexa488-labeled phalloidin that only binds to filaments, not actin monomers. Staining patterns revealed that the filaments were in thinner and shorter bundles, the microfilaments were less well packed or absent, and the staining was more punctate, especially around the nucleus suggesting that F-actin forms very short bundles in this region of the cell. Furthermore, the actin filaments were less stable, being easily depolymerized by agents such as cytochalasin D and latrunculin A (see chapter by Braet et al.), compared to control F-actin. This "punctate" appearance is consistent with the pattern of fluorescent-labeled gelsolin, which was distributed at the ends of short filaments in these cells. The authors suggested that the syndrome is associated with increased actin severing and capping activity by gelsolin associated with the endoplasmic reticulum. Other likely candidate actin-binding proteins (profilin, cofilin, vinculin, villin, and α-actinin) were tested using fluorescent-labeled monoclonal antibodies but the staining pattern was only abnormal for α-actinin. Since the primary defect is the chronic deficiency of PIP_2 5-phosphatase, the resulting accumulation of PIP_2 will inhibit microfilament assembly needed for lens development and reduce cell–cell adhesion needed for renal tubule function (Suchy and Nussbaum 2002).

Altered Extracellular Matrix

Podocytes are renal glomerular epithelial cells that have cytoplasmic processes that make close contact with the basal lamina that separates the podocytes from the endothelial cells of the capillary. The connection between the podocyte (actin cytoskeleton and laminin, collagen) and other molecular components of this basal lamina is essential. It is effected by integrin adhesion molecules, the major one being integrin α3β1 (Bijian et al. 2005). In a number of diseases, the foot-like connections of the podocytes become defective, resulting in the kind of phenotype described above for the OCRL syndrome. These defects involve the actin cytoskeleton, which in turn is regulated by the action of PIP_2 on actin-binding proteins. Thus, extracellular matrix proteins appear to play a key role in modulating the state of assembly of the actin cytoskeleton.

PIP_2 Uncaps the Barbed (Growing) Ends of Actin Filaments

PIP_2 can induce polymerization of actin in stimulated cells by increasing the rate of dissociation of these ABPs forcing them to leave this site. In platelets it has been calculated that approximately 500 new free barbed ends are created from about 2,000 capped filaments within 30–60 sec of stimulation (Schafer, Jennings and Cooper 1996). This process would be more efficient in the face of a low cellular content of PIP$_2$ if it was nonuniformly distributed.

Altered Cell Signaling

Extracellular stimuli can set in train the hydrolysis of membrane-bound lipids that then generate second messengers including DAG, IP$_3$, and PIP$_2$. How then does PIP$_2$ act as a second messenger? PIP$_2$-mediated signals are involved in channel gating, vesicle trafficking as well as controlling the regulators of actin assembly, namely the ABPs. Studies using antibodies to PIP$_2$- and GFP-labeled plextrin homology (PH) domains suggest that PIP$_2$ is uniformly distributed in both fixed and live cell membranes (Laux et al. 2000; Stauffer, Ahn and Meyer 1998). It is possible that small (~250 nm) lipid rafts exist and that PIP$_2$ is associated with them (Chamberlain 2004). Restriction of PIP$_2$ to rafts makes some sense because, for example, it would explain why there is a localization of the effects of PIP$_2$ breakdown by PLC. However, this idea was recently challenged by van Rheenen et al. (2005). These authors used methods that exceed the resolution of the light microscope. These include failure to observe FRET (here the donor and acceptors must be separated by at least 10 nm) and uniform labeling of PH domains by gold particles observed in ultrathin sections in electron microscopy. Rafts or not, the location of PIP$_2$ at the cell membrane explains the localization of PIP$_2$-dependent ABPs like cofilin in the cell.

Type 2 Diabetes

In type 2 diabetes, cellular uptake of insulin is impaired and insulin resistance (diabetes type 2) develops (Bedi et al. 2006). Chen et al. (2004) recently tested whether this insulin resistance affected PIP$_2$ content of a mouse 3T3 cell line. They noted that there was loss of this lipid from the plasma membrane, but more importantly they used a PIP$_2$ delivery system to restore the deficiency at the cell membrane. They found that when they treated 3T3 cells that were made insulin-insensitive by sustained insulin treatment, the loss of responsiveness was restored by simply replenishing PIP$_2$. This is another example of the key role of PIP$_2$ in cellular responses.

The Role of PIP$_2$ in Regulating ABPs

Altered Regulation of Cytoskeleton Assembly

The question of how ABPs sever actin microfilaments, promote their assembly, effect new branch points, and promote side-to-side associations is dealt with in other chapters in this volume (for cofilin, see chapter by Maloney et al.; for profilin, see chapter by Moens; for gelsolin, see chapter by Burtnick and Robinson). The roles of these ABPs can be particularly puzzling when one realizes that some can both sever and promote assembly. Gelsolin and cofilin can do both. PIP_2 does not bind to actin monomers or polymer and so it cannot directly regulate its assembly. This is achieved by binding to ABPs. For example, PIP_2 (in concert with high levels of Ca ions) binds to the N-terminal half of gelsolin and villin, and inhibits their ability to sever actin filaments by inhibiting their ability to nucleate new filaments or their ability to cap filaments.

Identifying the Cofilin PIP_2-Binding Site

A number of phosphoinositide-binding domains, such as the PH domain, have been identified (Niggli et al. 2001). There are no atomic structures of ABPs complexed to PIP_2, and, attempts to identify the lipid-binding region of ABPs have largely relied on indirect methods such as peptide-inhibition of PIP_2 binding and site-directed mutants in the region of putative PIP_2 binding.

There are no known atomic-resolution structures of an ABP complexed to PIP_2, presumably because the complex does not form good crystals. To solve this problem, we took a more direct approach. Using chick cofilin (PDB file 1TVJ), we produced base-line resolution NMR spectra that yielded a 1.8 Å resolution structure, the highest reported so far. We then used NMR chemical shift mapping supported by site-directed mutagenesis to identify the amino acid side chains whose resonances were shifted by PIP_2 binding. PIP_2 binding abrogates the interaction of cofilin with F-actin and there is a clear overlap (coincidence) of the F-actin-binding site and the binding site for PIP_2. Furthermore, it explained why phosphorylation at a distant site (at the N-terminus) did not affect the PIP_2–cofilin interaction (Gorbatyuk et al. 2006).

The PIP_2-binding site was mapped to the C-terminal region of cofilin which is also the F-actin-binding site. Amino acid side chains that we identified as being contact points and/or whose environment was affected due to the binding of PIP_2 are shown in Fig. 1a (Gorbatyuk et al. 2006). Direct contact points were determined using a Clean-SEA HSQC experiment. Two cationic residues K132 and H133 were shown to be direct contact points. We changed these residues to alanine and found that the double mutant (K132A and H133A) reduced but did not abrogate PIP_2 binding and these residues are shown spaced-filled. We found it interesting to contrast this region of the structure with two other cofilin structures.

Schutt and colleagues (2000) solved the structure of *Arabidopsis thaliana* ADF1, a close relative of cofilin. In their report they noted the presence of lauryldimethylamine oxide (LDAO), a lipid-like molecule, in the same location as the PIP_2 seen in our structure (Fig. 1b) (for original see PDB file 1F7S). LDAO

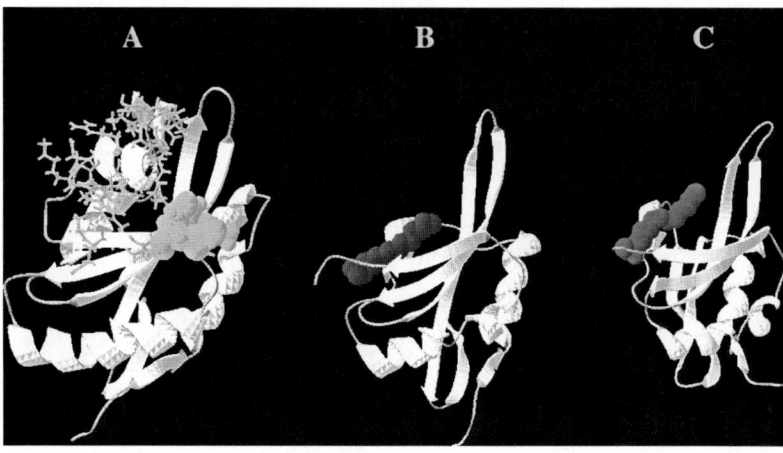

FIG. 1. (**a**) NMR structure of chick cofilin (PDB file 1TVJ) with residues perturbed by the binding of PIP$_2$. K132 and H133 are shown spaced-filled and were shown to be direct contact points. (**b**) Crystal structure of *Arabidopsis thaliana* ADF1 (PDB file 1F7S) showing single molecule of cocrystallized LDAO. (**c**) Crystal structure of fission yeast cofilin (PDB file 2I2Q) showing single molecule of cocrystallized LDAO (*See Color Plates*)

was present in the crystallizing solution conditions. More recently, the atomic coordinates of fission yeast cofilin-containing LDAO were deposited (see PDB 2I2Q) in the protein database (Fig. 1c). Once again the LDAO was present during crystal formation and once again, it was resolved in the same location as the *Arabidopsis* ADF. These structures of these three members of the cofilin family are compared in Fig. 1a–c. There is a remarkable similarity between the residues responsible for PIP$_2$ binding and the locations of the LDAO. All reside in a region of cofilin known to bind to F-actin.

It remains to be seen if the PIP$_2$-binding site on cofilin is unique. This site may well be structurally distinct from PIP$_2$-binding sites on gelsolin, villin, and related proteins.

References

Addis, M., Loi, M., Lepiani, C. and Melis, M. A. 2004. OCRL mutation analysis in Italian patients with Lowe syndrome. Hum. Mutat. 23, 524–525.

Bedi, D., Clarke, K. J., Dennis, J. C., Zhong, Q., Brunson, B. L., Morrison, E. E. and Judd, R. L. 2006. Endothelin-1 inhibits adiponectin secretion through a phosphatidylinositol 4,5-bisphosphate/actin-dependent mechanism. Biochem. Biophys. Res. Commun. 345, 332–339.

Bijian, K., Takano, T., Papillon, J., Le Berre, L., Michaud, J. L., Kennedy, C. R. and Cybulsky, A. V. 2005. Actin cytoskeleton regulates extracellular matrix-dependent survival signals in glomerular epithelial cells. Am. J. Physiol. Renal Physiol. 289, F1313–F1323.

Bowman, G. D., Nodelman, I. M., Hong, Y., Chua, N.-H., Lindberg, U. and Schutt, C. E. 2000. A comparative structural analysis of the ADF/cofilin family. Proteins 41, 374–384.

Chamberlain, L. H. 2004. Detergents as tools for the purification and classification of lipid rafts. FEBS Lett. 559, 1–5.

Doughman, R. L., Firestone, A. L. and Anderson, R. A. 2003. Phosphatidylinositol phosphate kinases put PI(4,5)P$_2$ in its place. J. Membr. Biol. 194, 77–89.

Godi, A., Di Campli, A. and De Matteis, A. 2004. Phosphoinositides and membrane traffic in health and disease. Top. Curr. Genet. 10, 171–192.

Gorbatyuk, V. Y., Nosworthy, N. J., Robson, S. A., Bains, N. P. S., Maciejewski, M. W., dos Remedios, C. G. and King, G. F. 2006. Mapping of a novel phosphoinositide binding site on chick cofilin explains how PIP$_2$ regulated the cofilin-actin interaction. Mol. Cell 24, 511–522.

Laux, T., Fukami, K., Thelen, M., Golub, T., Frey, D. and Caroni, P. 2000. GAP43, MARKS, and CAP23 modulate PI(4,5)P(2) at plasmalemmal rafts, and regulate cell cortex actin dynamics through a common mechanism. J. Cell Biol. 149, 1445–1472.

Logan, M. R. and Mandato, C. A. 2006. Regulation of the actin cytoskeleton by PIP2 in cytokinesis. Biol. Cell 98, 377–388.

Loi, M. 2006. Lowe syndrome. Orphanet J. Rare Dis. 1: 16.

Lowe, C. U., Terrey, M. and MacLachan, E. A. 1952. Organic aciduria, decreased renal ammonia production, hydrophthalmos, and mental retardation; a clinical entity. Am. J. Dis. Child. 83, 164–184.

Mueller, O. Y., Hartsfield, J. K., Hughes, E., Crolla, J. A., Dubowitz, V. and Bobrow, M. 1985. A balanced de-novo X/autosome translocation in a girl with manifestation of Lowe syndrome. Am. J. Med. Genet. 23, 837–847.

Nussbaum, R. L., Orrison, B. M., Janne, P. A., Charnas, L. and Chinault, A. C. 1997. Physical mapping and genomic structure of the Lowe syndrome gene OCRL1. Hum. Genet. 99, 145–150.

Nussbaum, R. L. and Suchy, S. F. 2001. The oculocerebrorenal syndrome of Lowe (Lowe syndrome). In Metabolic and Molecular Basis of Inherited Disease, vol. 252, 8th edition. C. R. Scriver, A. L. Beaudet, W. S. Sly and D. Valle (Editors). McGraw-Hill, New York. pp. 6257–6266.

Olivos-Glander, I. M., Janne, P. A. and Nussbaum, R. L. 1995. The oculocerebrorenal syndrome gene product is a 105-kD protein localized in the Golgi complex. Am. J. Hum. Genet. 57, 817–823.

van Rheenen, J., Achame, E. M., Janssen, H., Calafat, J. and Jalink, K. 2005. PIP$_2$ signaling in lipid domains: A critical re-evaluation. EMBO J. 24, 1664–1673.

Schafer, D. A., Jennings, P. B. and Cooper, J. A. 1996. Dynamics of capping proteins and actin assembly in vitro: Uncapping barbed ends by phosphoinositides. J. Cell Biol. 135, 169–179.

Stauffer, T. P., Ahn, S. and Meyer, T. 1998. Receptor-induced transient reduction in plasma membrane PtdIn(4,5)P2 concentration monitored in living cells. Curr. Biol. 8, 343–346.

Stephens, L. R., Jackson, T. R. and Hawkins, P. T. 1993. Agonist-stimulated synthesis of phosphatidylinositol(3,4,5)-trisphosphate: A new intracellular signalling system? Biochim. Biophys. Acta 1179, 27–75.

Suchy, S. F. and Nussbaum, R. L. 2002. The deficiency of PIP2 5-phosphatase in Lowe syndrome affects actin polymerization. Am. J. Hum. Genet. 71, 1420–1427.

Suchy, S. F., Olivos-Glander, I. M. and Nussbaum, R. L. 1995. Lowe syndrome, a deficiency of a phosphatidylinositol 4,5 bisphosphate 5 phosphatase in the Golgi apparatus. Hum. Mol. Genet. 4, 2245–2250.

Yin, H. L. and Janmey, P. A. 2003. Phosphoinositide regulation of the actin cytoskeleton. Annu. Rev. Physiol. 65, 761–789.

13
Intracellular Pathogens and the Actin Cytoskeleton

E. L. Bearer

Introduction

Intracellular pathogens co-opt cellular machinery in many ways. It has long been recognized that the life cycle of most viruses depends on host cell DNA replication enzymes. More recently, intracellular bacteria as well as viruses have been discovered to disrupt and redirect the host cell cytoskeleton to assist their survival and growth. Understanding pathogen–cytoskeleton interactions will provide fresh insights into targets for drug therapies or for design of immunogens for preventive vaccine development.

Intracellular Bacteria Assembles Host Cell Actin

The interplay between the microbe and the host cell's actin machinery has provided key insights into normal cell function as well as the life cycle of the microbe. For example, it was long known that eukaryotic cells must have an actin polymerization machine. But none could be found until the seminal discovery that the bacteria, *Listeria monocytogenes*, induced polymerization of actin during its intracellular life (Tilney, Connelly and Portnoy 1990). Extracellular *Listeria* was not capable of inducing filament formation from purified G-actin. Thus, a separate host cell factor was required. Using fluorescence microscopy of actin filament formation to assay for polymerization activity in fractionated cytoplasm, the Arp2/3 complex of proteins was identified as a host cell factor sufficient for nucleation of actin filament assembly by *Listeria* (Welch, Iwamatsu and Mitchison 1997). Arp2/3, a 7-subunit complex, had been independently identified previously as an actin-binding protein in platelets (Bearer 1991, 1992, 1995), and as a polyproline/profilin-binding complex from soil amoeba (Machesky et al. 1994). The importance of Arp2/3 for the formation of actin filaments in normal eukaryotic cells was first demonstrated in platelets, where inhibition of the Arp2 subunit blocked shape change and actin filament polymerization and reorganization (Li, Kim and Bearer 2002).

Arp2/3 and Intracellular Pathogens

It has since been shown that most intracellular bacteria as well as other microbes use the Arp2/3 machinery, either by direct recruitment or via upstream activators, to power intracellular movements. Such examples include Salmonella, Shigella and Ricketsia. Details of these interactions have been the subject of numerous reviews (Bearer and Satpute-Krishnan 2002; Galan and Cossart 2005; Gouin, Welch and Cossart 2005; Rottner, Stradal and Wehland 2005; Carlsson and Brown 2006), and hence will not be repeated here.

While similarities to the *Listeria* Arp2/3 recruitment among microbes from viruses to protozoa continue to be rapidly found, less common has been further exploitation of the strategy that identified Arp2/3. This strategy used the microbe as a tool to reach into the cytoplasm and manipulate the cell's own machinery, thereby identifying new molecular mechanisms for normal cellular behavior. Emerging work from my lab and others has begun to exploit herpesvirus for identification of molecular mechanisms governing cargo–motor interactions for actin-based and microtubule-based movements. This will be described in more detail below.

New insights into the cytoskeletal dependence of intracellular protozoan parasites are likely to provide novel targets for drug design from among cytoskeletal proteins, since protozoan infection appears to require both the parasitic actin–myosin machinery and the host cell cytoskeleton (Cowman and Crabb 2006; Schuler and Matuschewski 2006). Apicomplexan family of protozoan parasites, which include malaria and toxoplasma, use both of these actin-based systems for penetration, intracellular reproduction, and release from the host cell.

Here we will review recent developments in viral and protozoal interactions with actin and propose a unifying hypothesis of emergent behavior of actin cytoskeleton that could guide recognition and identification of additional, pathogen-based, actin-interacting networks. The simple rules governing interactions of multiple contributors to actin cytoskeleton formation serve as an entry point by parasitic invaders. These rules of protein–protein interactions that produce complex cellular structures, for which substitutions of players and redundancies of function are the norm, can result in similar outcomes from different players or different outcomes from the same team of players.

The Actin Cytoskeleton

Actin serves two principle roles inside cells (1) scaffolding of stable filaments for maintenance of cellular architecture, coordination of enzyme assemblies or tracks upon which other cellular components move; and (2) dynamic movements mediated by polymerization/depolymerization of actin monomers into filaments which produce force that can drive cellular surface contour changes and intracellular movements pushed by new filament formation. So far, pathogens have been found to destabilize the actin scaffold and then co-opt the monomers for incorporation into new filaments that can support intracellular movements of the invading microbe. Pathogen co-option of contractile processes, actin scaffolds, and actin-based motors is just beginning to be recognized.

Toxins produced by pathogens inside and outside the cell also affect the actin cytoskeleton. Actin-interacting toxins from poisonous noninvasive organisms are commonly used in studies of actin dynamics and structure, including phalloidin (from the mushroom *Amanita phalloides*), cytochalasin (from fungus), and jasplakinolide and latrunculin (see chapter by Braet et al.). Extracellular invasive pathogens also secrete toxins that affect the actin-based cytoskeleton, either directly or by affecting the regulatory network that influences actin filament dynamics, such as cholera toxins and the LE factor of *Anthrax*.

Cellular actin networks are most vulnerable to pathogenic effects when they are undergoing dynamic processes, such as surface protrusion or other shape change events. These vulnerable points render cells more susceptible to pathogenic insult or co-option of the cytoskeletal network. How pathogenic disturbance of the cytoskeleton leads to cell death is only recently being determined, and as yet little is known about the molecular mechanisms by which disrupted actin dynamics influence cell division or death pathways.

Viruses

Viral Effects on the Actin Cytoskeleton

To generalize about viruses is difficult, as this group of pathogens differs widely in structural, genetic, and molecular components, with each subtype having very different cellular effects. However, some common characteristics of viral–host cell interactions can be defined (see Fig. 1 for general principles). All viruses enter a cell to replicate. Some kill the cell directly, while in some other viral infections the cell dies due to host immune attack. Other viruses persist in cells for the lifetime of the organism by becoming latent. Some viruses, such as alpha herpesviruses, have both effects depending on which cell type they infect, epithelial or neuronal.

After replication, infectious viral particles are released or otherwise passed on to another cell, which propagates the virus. Release of infectious virus may occur by lysis of the cell or more gradually by transport and exocytosis of viral particles at the cell surface.

Experimental control of viral infections in cultured cells is challenging for a number of reasons. First, many viral particles in a given preparation of infectious particles are "duds." Hence, few achieve the goal of replication inside the next cell. For example, in highly purified preparations of herpes simplex virus, less than 10% of the particles are capable of creating a plaque when exposed to receptive epithelial monolayers (Honess and Roizman 1973). This may be in part because viral survival strategies rely on abundance rather than

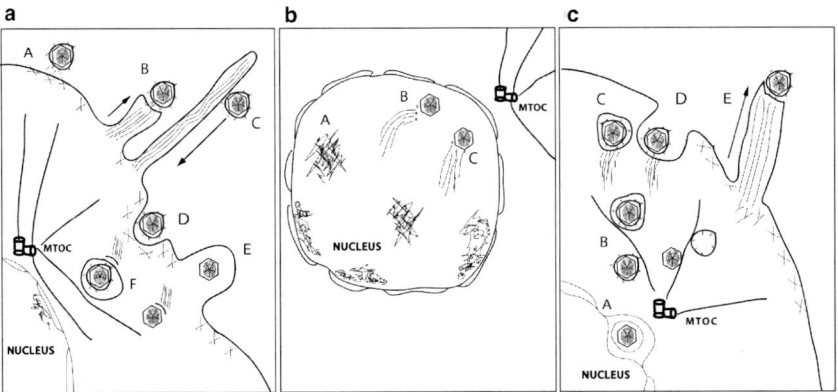

FIG. 1. Diagrams of the general principles of viral interactions with host cell actin filaments. (**a**) *Entry*. During viral entry, infecting particles bind to cell surface (*A*) where binding between viral surface and host cell receptors is stabilized by the actin filament-containing membrane skeleton. Some types of infecting pathogens, both viral and bacterial, induce actin polymerization beneath the surface-binding site, thereby creating "pedestals" (*B*). Several viruses have been observed to "surf" down existing filopodia toward the cell surface (*C*). After binding, virus is internalized either via uptake in endosomes, which requires actin nucleation and myosin reorganization of membrane filaments (*D*), or via fusion of the viral coat/envelope with the cell membrane, liberating the viral nucleocapsid containing the viral genome into the cellular cytoplasm (*E*). Once internalized, viral particles induce actin nucleation, usually via the Arp2/3 complex or its upstream activators, which drives viral movements through cortical cytoplasm. Once in the hyalomere, viral particles hop onto the microtubule system for transport to the microtubule-organizing center (MTOC) in the perinuclear region. (**b**) *Replication*. During intranuclear replication of at least two viruses, baculovirus and HSV, monomeric actin enters the nucleus and forms filaments around nascent capsids. These filaments may serve to organize transcriptional domains (*A*), and/or to drive virus toward the nuclear membrane for export (*B*). Viral movements in the nucleus may be polarized toward the side facing the MTOC. Arp2/3 may be responsible for these actin filament formations since a virally encoded WASP-like activator is required (*C*, *empty circles* at the viral–actin interface). Movements of nuclear viral particles on polymerized actin may also be driven by myosin V. See text for citations. (**c**) *Egress*. After export from the nucleus (*A*) and transport out of the perinuclear region on microtubules arising in the MTOC (*B*), viral particles leave the microtubule system and transit the cortical cytoplasm probably by nucleating actin filament polymerization (*C*). The surface components of the virus that mediate this transport may include amyloid precursor protein recruited from the cell during secondary envelopment. Nucleation of actin filaments on viral particles at the periphery of the cell likely uses the same machinery as for entry, including the Arp2/3 complex and/or its upstream activators. One question is whether the virus employs the same machinery during egress as for entry. Upon reaching the surface, the virus must first penetrate the membrane skeleton and then be released through the membrane (*D*). Once externalized, some viruses remain attached to the external cell surface, continuing to nucleate actin across the viral envelope and cell membrane to create filopodia that project the virus outward. This new actin-based projection facilitates viral spread to adjacent cells

precision. In addition, during propagation of virus *in vitro*, virus is released from cells both as mature virions from relatively healthy cells as well as by lysis of dying cells. After lysis, viral particles at all stages of maturation are released indiscriminately. Not only is variability high in any given viral preparation, but also intracellular viral particles do not behave consistently. Some stall at different time points in the infection process. Such infecting particles stalled on the way into the cell persist in the cytoplasm and can be confused with newly synthesized particles on the way out. These caveats have produced confusion in the field even for those viruses whose behavior is best characterized.

Virus Entry Depends on Actin

Viral entry begins with attachment to the cell surface. This is followed by a variety of processes of penetration through the membrane into the cellular cytoplasm (Fig. 1a). For enveloped viruses with a membrane, penetration occurs by two alternative pathways: fusion of the viral envelope with the cell plasma membrane which delivers capsids directly into the cytoplasm or endocytosis of the virus into a cellular endocytic compartment where it subsequently fuses with the internal membrane compartment for release into the cytoplasm.

Binding of the virus to the cell surface may result in destruction of the membrane skeleton, including its actin filaments. Rotavirus, responsible for diarrhea in children, enters by binding to microvilli of intestinal epithelial cells. Rotavirus binding acts like localized cytochalasin to depolymerize and reorganize the actin filaments in the microvilli (Gardet et al. 2006).

Actin dependence of retroviral infectivity during entry into cells models the steps followed by other enveloped viruses and has recently been reviewed in detail (Fackler and Krausslich 2006). Binding to the surface and either internalization via an endosomal process or surface penetration requires clustering of host cell surface receptors. For HIV, clustering on the host cell surface of receptors is actin-dependent. Furthermore, disruption of actin by the actin filament inhibitor, cytochalasin D, reduces infectivity (Bukrinskaya et al. 1998). Decreased infectivity could be due to the requirement of receptor clustering for cell entry, or to failure of cytoplasmic virus to be transported to the site of efficient replication deeper in the cell. This latter transport, involving short centripedal movements of intracellular infecting particles, is independent of microtubules and thus likely to be actin-based. Indeed, knock-down of the actin nucleation complex, Arp2/3, abolishes these short subcortical movements (Komano et al. 2004).

Attachment, Receptor Clustering, Pedestals, Surfing, Endocytosis, and Tails

At the surface, attachment of viral particles to the host cell receptors can result in "surfing" down filopodia on actin filaments via a myosin-mediated process (Lehmann et al. 2005). This surfing has been imaged for vesicular stomatitis virus (VSV), mouse leukemia virus (MLV), avian leukosis virus, and HIV (Lehmann et al. 2005; Lehmann and Frischknecht 2006). Surfing is actin- and myosin II-dependent. Myosin VI, a retrograde motor (Wells et al. 1999), may also be involved, but this has not yet been tested. How these many different viruses hook onto the retrograde transport machinery is not known – it could be by activating a cellular transmembrane receptor that secondarily attaches directly or indirectly to the motor(s), or via a direct interaction through penetration of the membrane of virally encoded proteins.

Once inside the cell, most invading viruses need to traverse the cortical cytoplasm and then travel to the perinuclear region for replication. For influenza, endocytosed virus travels to the perinuclear region before acidification and release (Lakadamyali et al. 2003). This centripedal movement may be driven in part by actin filament nucleation as the rate of movement is consistent with actin-based motility.

The Role of Nef

Many viruses appear to use cellular actin polymerization machines, Arp2/3 or formin, to power transition from cortical cytoplasm onto the microtubule transport system that transports them deeper into the cell interior. Which viral protein(s) mediate interaction with the host actin cytoskeleton during this phase of intracellular transport is not yet well understood, although some viral proteins have been implicated. For HIV, the virally encoded Nef protein appears to modulate the activity of the Arp2/3 activator, N-WASp (Haller et al. 2006), which is consistent with the requirement for Arp2/3-mediated actin polymerization in the early stages of intracellular transport. Nef is found in the detergent-insoluble cytoskeleton (Niederman, Hastings and Ratner 1993), and binds to Vav (Fackler et al. 1999). Depolymerization of actin complements the ability of Nef to enhance HIV infection (Campbell, Nunez and Hope 2004).

Nef's interaction with actin is complex. In contrast to lymphocytes where Nef depolymerizes actin, in dendritic cells Nef induces rearrangement of actin filaments and ruffling (Quaranta et al. 2003). Furthermore, the HIV Nef protein, which enhances HIV replication, plays an opposite role when expressed in recombinant vaccinia virus-infected cells. In HIV, Nef enhances pathogenicity and replication, whereas in vaccinia, Nef expression alters plaque morphology in culture and attenuates infection in mice (Chan et al. 2005). Arp2/3, one of the Nef targets, is required for the normal life cycle of both vaccinia and lentivirus infections as inhibition of Arp2/3 has similar effects on both types of virus (Komano et al. 2004). Thus, although an HIV-encoded protein has been identified that influences the actin effects of HIV, the biochemical basis of these effects remains obscure, and this protein does not perform a similar function for other viruses. Hence, viral proteins that mediate actin dynamics in infected cells are likely to be virus-type specific.

Vaccinia

Vaccinia virus, a member of the poxvirus family that includes smallpox, has its own method of activating the Arp2/3 actin polymerization machine – through tyrosine kinases such as Abl and Src (Frischknecht et al. 1999; Newsome et al. 2006) that feed into Nck, WIP, WASp, and Grb2 (Moreau et al. 2000; Scaplehorn et al. 2002; Newsome et al. 2006). This indirect upstream activation of cellular signaling pathways is highly efficient as it allows a single viral protein to stimulate a bifurcating cascade of events, one of which is the Arp2/3 actin polymerization machine. Like HIV, vaccinia also activates Vav (see Munter, Way and Frischknecht 2006 for recent review of such signaling pathway).

The vaccinia viral protein responsible for activation of the Arp2/3 polymerization machine has been suggested as A36R (Rottger et al. 1999) while A33R and B5R may also play a role (Katz et al. 2003). A36R is a type 1b membrane protein and component of the intracellular envelope of the virus during packaging and egress (van Eijl, Hollinshead and Smith 2000). Knockouts of A36R have reduced plaque size and do not form actin tails (Frischknecht et al. 1999). The actin tails induced by vaccinia form filopodia projecting from the surface of the cell, with the extracellular emerging viral particle at the tip. Such viral induction of cell surface projections is thought to facilitate transfer of the virus to adjacent uninfected cells. This process and its commonality among viruses will be discussed further below in the section on viral egress.

Dependence of Viral Replication on Actin: Actin Tails in the Nucleus

A role for actin in viral replication was not suspected until recently. For at least two very different viruses, baculovirus and the alpha herpesvirus, pseudorabies (PRV), actin filaments are induced in the nucleus during replication of viral DNA and capsid assembly (Charlton and Volkman 1991; Feierbach et al. 2006; Goley et al. 2006) (Fig. 1b). These two viruses replicate in the nucleus of the infected cell and the first stages in packaging of viral DNA into the viral capsid also occur in the nucleus. Both baculoviruses and herpesviruses are enveloped double-stranded DNA viruses. Baculovirus infects larvae of lepidopteran insects while alpha herpes virus primarily infects mammalian epithelial and neuronal cells.

Induction of actin filaments in the host cell nucleus by baculovirus is required for production of viral progeny (Kasman and Volkman 2000). This is mediated by a two-step process. G-actin is allowed to enter the nucleus where it is normally not found and accumulates there, a process mediated by six virally encoded proteins (Ohkawa, Rowe and Volkman 2002). Late viral gene products stimulate actin polymerization into filaments (Charlton and Volkman 1991; Ohkawa, Rowe and Volkman 2002). One of these is a viral-encoded WASp-like protein, p78/83 (Goley et al. 2006), that stimulates actin filament formation via the host cell Arp2/3 complex. Baculoviral p78/83 activates Arp2/3 *in vitro* and colocalizes with it in the nucleus of infected cells. Mutations in p78/83 cause defects in actin nucleation and in viral replication, but p78/83 is not necessary for monomeric actin accumulation in the nucleus. Arp2/3 also accumulates in the nucleus in infected cells, and this is not dependent on p78/83.

While the baculovirus story has been unfolding for over a decade, it was not recognized until very recently that other double-stranded DNA viruses that replicate in the nucleus might also co-opt cellular cytoplasmic actin to organize transcriptional or replication processes. Recent evidence on mammalian herpesviruses, PRV and HSV, demonstrates that nuclear actin filaments may be a common feature of double-stranded DNA virus replication (Simpson-Holley et al. 2005; Feierbach et al. 2006). The accumulation of monomeric actin may be a secondary effect of nuclear laminin disruption by viral proteins, UL31 and 34. Disruption of filaments does not seem to affect viral replication (Simpson-Holley et al. 2005). Filaments begin to appear with expression of capsid protein, VP26, and are physically associated with capsids by electron microscopy (Feierbach et al. 2006). Nuclear actin filaments are formed in both epithelial and neuronal cells by either PRV or HSV. Filament assembly does not require replication but is inhibited by cyclohexamide blockade of transcription.

Interestingly, HSV capsids are motile within the nucleus at velocities consistent with actin-based transport and this motility is sensitive to actin filament inhibitors, cyclohexamide, and latrunculin, as well as BDM, a general inhibitor of myosin motors (Forest, Barnard and Baines 2005). This nuclear motility of capsids may serve to aggregate emerging capsids into compartments where gene expression and replication occur. Alternatively, it may serve to carry capsids to the nuclear membrane for export to the cytoplasm. Colocalization studies suggest that the motor driving this actin-based movement is myosin V (Feierbach et al. 2006).

Nuclear motility of prepackaged capsids may be a common feature of viral replication in the nucleus. HIV, an RNA virus that also replicates in the nucleus of infected cells, also undergoes active transport in the nucleus (Arhel et al. 2006), although in this case actin filaments have not yet been observed and the machinery is unknown.

Viral Egress: Outbound Transport, Cortical Actin Penetration, Exocytosis and Filopodia Formation

Once newly synthesized viral particles are formed, they are shipped out of the cell for dissemination to adjacent cells or a new host (Fig. 1c). Active transport of virions is most dramatic in neurons, where nascent viral particles must travel long distances to reach the body surface for dissemination.

A general mechanism for movement of packaged viral DNA to the cell surface appears to involve two steps: active transport on microtubules, and directed actin polymerization at the periphery.

Herpesvirus

Studies in the neurotropic alpha herpesviruses, particularly PRV and HSV, have identified many of the players in microtubule-based transport, both to and from the nucleus, although the role of actin in herpesvirus cytoplasmic transport in egress has been little studied (Bearer and Satpute-Krishnan 2002). Despite several decades of intense investigation, the details of microtubule-based transport of virus remain unexplained (Bearer and Satpute-Krishnan 2002). Even for the normal cell, the molecular mechanism that connects cargo to motors for directed active transport is an area of vigorous investigation, debate, and much contradictory evidence. For herpesvirus, even the basic form of the virus that is transported is controversial. Much evidence suggests that herpes may travel as separate components to be assembled at the periphery (Saksena et al. 2006; Snyder, Wisner and Johnson 2006), whereas other studies show that herpes virions travel as enveloped virus (Antinone and Smith 2006), probably as enveloped virus inside a second membrane derived from the cellular Golgi apparatus (Satpute-Krishnan, DeGiorgis and Bearer 2003).

Herpesvirus appears to have at least two distinct modes of transport in neurons. Incoming virus probably loses its envelope when it fuses with the cell membrane at the cell surface (Bearer and Satpute-Krishnan 2002; Radtke, Dohner and Sodeik 2006). The nucleocapsid together with tegument proteins somehow passes through the cortical actin network and then are picked up by the host cell retrograde motors, probably dynein although a retrograde kinesin has not been ruled out (Bearer et al. 2000; Luxton et al. 2005; Dohner et al. 2006). This process is distinct from the viral recruitment of anterograde motors that, in cells with polarized microtubules, delivers nascent virus along with normal cellular organelles, to the cell periphery or outward within the axon to its termini.

Anterograde Transport

For anterograde transport, host cell membranes provide one mechanism for motor recruitment. By budding into Golgi cisternae or transport vesicles, viral particles can simultaneously obtain envelope and a second cellular membrane that normally recruits anterograde motors (Bearer and Satpute-Krishnan 2002; Satpute-Krishnan, DeGiorgis and Bearer 2003). Virus devoid of membranes goes uniquely retrograde in a reconstituted assay using the polarized microtubules in the mature squid giant axon. In contrast, viral particles inside host cell membranes travel in the anterograde direction. Thus in this reconstituted system, transport direction is determined by the surface of the virus: viral tegument/capsid is directed to the nucleus, virus inside membranes is directed outward to the surface or to nerve terminal (Satpute-Krishnan, DeGiorgis and Bearer 2003).

Amyloid Precursor Protein

In a recent exciting discovery, my laboratory has found that HSV virions undergoing transport in the anterograde direction are physically associated with large amounts (three times that of viral proteins) of amyloid precursor protein (APP) (Bearer and Satpute-Krishnan 2002; Satpute-Krishnan, DeGiorgis and Bearer 2003). When proteolyzed, APP produces the amyloid plaques that clog the brain in Alzheimer's disease. When the viral particle is replaced with a fluorescent bead of the same size as the virus and coupled only to APP, the bead is also transported anterograde (Satpute-Krishnan et al. 2006), demonstrating that APP is sufficient to recruit the cellular motor machinery to carry the virus to the terminal. Further reduction of the amino acid sequence of the cytoplasmic domain of APP identified a 15 amino acid peptide as sufficient for transport of viral-sized fluorescent beads. Thus, this strategy to use HSV as a tool to identify normal cellular machinery has paid off by identifying the first peptide zip-code sufficient to hitch cargo to motors for transport in cells.

Role of Actin in Long-Distance Virus Transport

It is likely that actin plays a role in this directed long-distance transport, although ancillary to the major role of the microtubule-based motor system. Two pieces of evidence support this. First, the APP beads travel slightly less rapidly and pause more frequently than membranated virus (Satpute-Krishnan et al. 2006). This behavior suggests that the virus recruits additional motors, possibly those for actin-based transport. Second, myosin V knockouts also transport organelles slightly less efficiently, with more pauses and shorter run lengths, than in wild type neurons. We have identified a novel unconventional myosin II that is associated with axoplasmic vesicles (DeGiorgis, Reese and Bearer 2002) and may contribute to viral transport as well.

Egress of Virus

Egress of the poxvirus, vaccinia, in epithelial cells has led the way in our understanding of the relative contributions of microtubule-based transport and actin polymerization in viral egress (Way 1998; Smith, Murphy and Law 2003). Vaccinia virus travels the longer distance from perinuclear region of replication near the nucleus to the cell periphery on microtubules (Rietdorf et al. 2001; Smith, Murphy and Law 2003; Newsome, Scaplehorn and Way 2004). Src activation appears to release the virus from microtubules and initiate interaction with cortical actin. It is not clear whether the virus then nucleates actin filaments to drive itself through the cortical web or manages to exocytose on its own. After exiting the cell, the enveloped virion remains attached to the cell surface where it drives actin polymerization within the cell beneath it, across both viral and cellular membranes. These new actin filaments form filopodial extensions with the infective viral particle at the tip. These extensions propel the virus outward from the infected cell surface toward adjacent cells.

Viruses and Filopodia

The signaling cascade involved in filopodia formation involves many of the players identified in other systems, including those mentioned above for entry: Abl and Src, Nck, Grb2, Vav, N-WASp, and the Arp2/3 actin nucleator. Although Src and Grb2 are known to be involved in adhesion plaques, these may also form as a consequence of the Ena/VASP actin nucleation system (Svitkina et al. 2003). It will be of interest to see what role these proteins play in vaccinia-induced filopodia.

The induction of filopodia by exiting virus may be a universal phenomenon for intracellular pathogen exit. Initially observed during the departure of *L. monocytogenes*, an intracellular bacterium, from cells and then of vaccinia, it now appears that other intracellular pathogens, including viruses in addition to vaccinia, also use this strategy. African swine fever virus stimulates filopodia during egress (Jouvenet et al. 2006), and influenza virus also requires actin filaments in the final phase of budding (Simpson-Holley et al. 2002). In each case, viral particles are found at the tips of actin filament-rich extensions from the infected cell surface. Thus, filopodia formation by nascent virus may be a common mechanism for directed delivery of infective virus across extracellular space to penetrate adjacent cells and spread infection.

Protozoan Intracellular Parasites

Parasite Actin–Myosin Contractility Powers Host Cell Penetration

Intracellular protozoan parasites causing malaria and toxoplasmosis have recently been shown to have their own actin-based cytoskeleton as well as to interact with the host actin machinery and scaffolding in important ways. Three major advances in this parasitology research are making considerable headway in identifying actin-related processes necessary for intracellular protozoan life – penetration of host cells, survival and replication inside the host cell, and release of infective parasites into the host blood stream for subsequence rounds of disease within the same host and spread to other individuals. These three advances are (1) the ability to maintain parasites in culture in the laboratory, (2) the genome projects, and (3) mutagenesis and construction of dominant negative or conditional mutations.

The genome projects for malaria (*Plasmodium falciparum*) is complete for one isolate, 3D7, and a close relative of *Plasmodium f*, *Toxoplasma gondi*, is in progress (Kronstad 2006). Mining of these genome databases reveals that the Apicomplexan family, of which plasmodium and toxoplasma are members, has a small set of genes with homology to other eukaryotic actin regulators (Schuler and Matuschewski 2006). These include two genes for actin in *P. falciparum* and one in *T. gondii*, and six myosin genes, encoding several myosin type XIV (Heintzelman and Schwartzman 2001; Chaparro-Olaya et al. 2005). Other genes with homology to actin regulators include two cofilins (see chapter by Maloney et al.), one each of 14-3-3, profilin (see chapter by Moens), chronophin, coronin, and possibly many actin-like proteins (ALP). These ALP may serve analogous functions of the dynactin subunit, Arp1, and substitute functionally for the Arp2/3 complex, although none are homologous to Arp2/3 (Gordon and Sibley 2005). There are also five types of putative nucleating/capping proteins with formin-like or capping protein homologies. A few of these proteins have been tested in biochemical assays for effects on actin dynamics, including the cofilin-like actin-depolymerizing factor (Allen et al. 1997; Schuler, Mueller and Matuschewski 2005) and coronin (Tardieux et al. 1998).

Actin and Parasite Invasion

Actin filaments are required for parasite invasion of host cells (Miller et al. 1979; Dobrowolski and Sibley 1996). Actin filaments in the parasites grown in blood culture are short both *in situ* (Schmitz et al. 2005) and when polymerized *in vitro* (Sahoo et al. 2006). Tachyzoites of *T. gondii* exhibit robust actin filament formation at the conoid when treated with jasplakinolide, an actin-stabilizing toxin (Shaw and Tilney 1999) (Fig. 2). S1 meromyosin decoration in thin section electron microscopy confirmed these filaments are composed of actin, but the dense packing did not allow determination of their polarity (Shaw and Tilney 1999). Since cytochalasin blocks polymerization and parasite penetration, and jasplakinolide stabilizes growing filaments, polymerization may also accompany myosin motility in host cell penetration. The molecules regulating such polymerization remain to be discovered.

The role of myosin–actin sliding in Apicomplexan penetration of host cells has been fairly well defined in the past year (Cowman and Crabb 2006; Schuler and Matuschewski 2006) (Fig. 3). After binding to the host cell, the merozoite reorients itself such that its apical end is directed perpendicular to the host cell membrane. As for viral entry, an initial phase involves clustering of receptors in both the erythrocyte and parasite membranes at their adhesion point. A "tight junction" forms and this zippers around the parasite as it is pulled into the host cell like a sock being pulled over the foot. In the

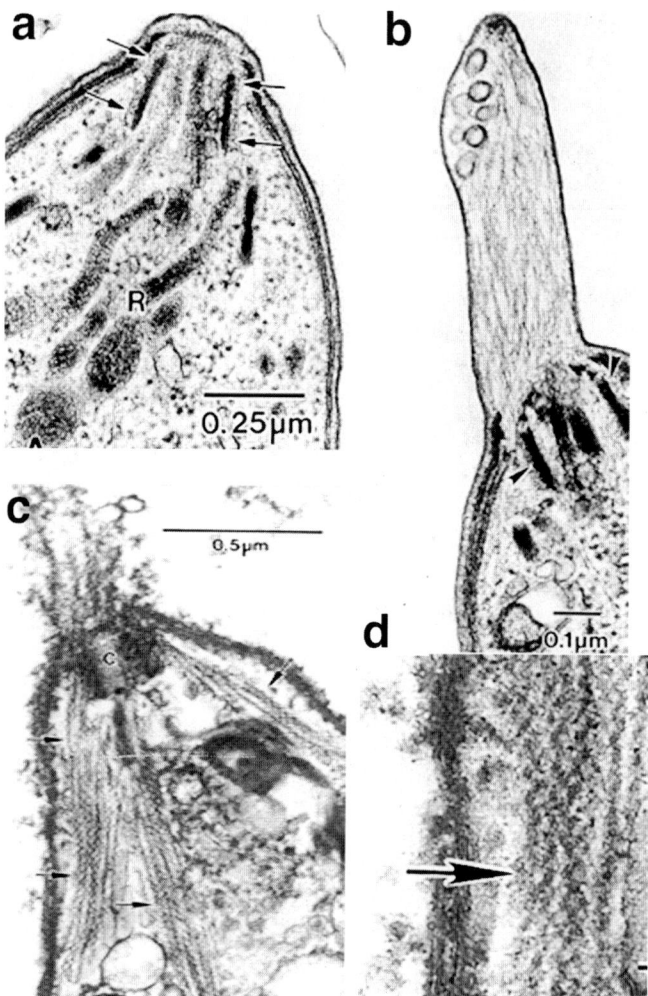

FIG. 2. Actin filaments in the conoid of Apicomplexan trophozoites. Protozoan parasites encode their own actomyosin machinery that drives penetration of host cells. In thin section electron micrographs, this is hard to detect (**a**) but after treatment with the actin-stabilizing drug, jasplakinolide, actin filaments form as in the acrosomal process of invertebrate sperm (**b**), suggesting that these filaments in the conoid of the parasite are constantly undergoing rapid turnover. Decoration with S1 fragment of myosin confirms that the filaments are composed of actin filaments (**c**), low magnification and (**d**), higher magnification of region indicated by *arrow* to the left in panel **c**). The polarity of these filaments could not be defined (reprinted with permission from Shaw and Tilney 1999)

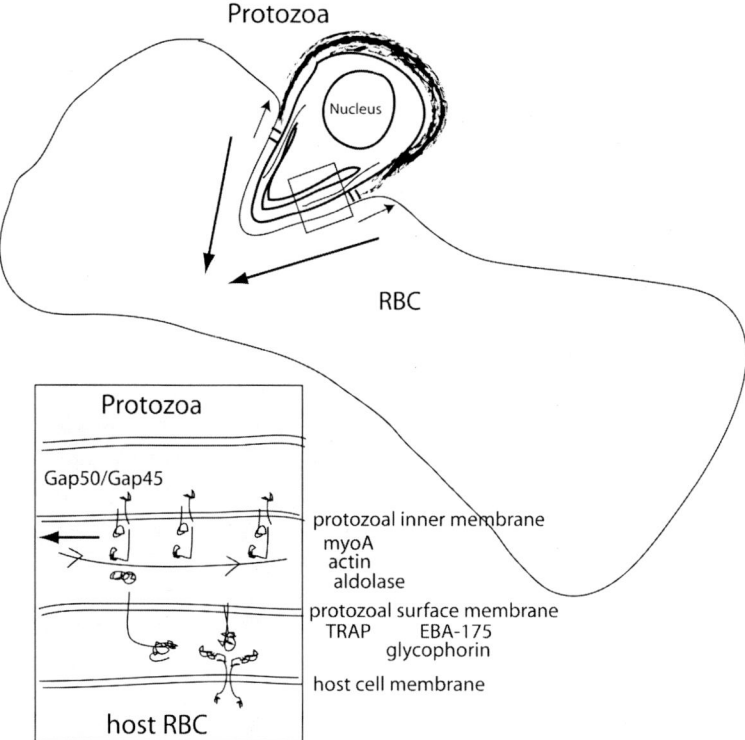

Fɪɢ. 3. Diagram of Apicomplexan protozoal host cell penetration. Features of both malarial merozoites and Toxoplasmi trophozoites are combined in this schematic drawing. The red blood cell, penetrated by the malarial parasite, is analogous to other host cells in principle. The protozoan adheres to the host cell surface and forms a "tight junction," diagrammed as double lines between parasite and host cell surfaces. This tight junction moves posterior as the parasite is thrust into the host cell via the parasite myosin–actin contractile process driven by myoA, the parasite type XIV myosin. If we suppose that the actin filaments are oriented with the barbed or fast-growing plus ends toward the host cell, then the myosin, attached to the protozoan inner membrane via GAP45/50, could move along the filaments to drive the parasite into the host cell. Attachment of the filaments to a membrane anchor, such as the EBA-175 transmembrane glycoprotein bound to red cell glycophorin, would stabilize the attachment site and pull the red cell membrane past the invading parasite. As yet no specific link between the host cell membrane receptor and the parasite actomyosin apparatus has been found, although it is speculated that "TRAP" a thrombospondin-type transmembrane glycoprotein of the parasite, may be the link. Attachment of TRAP to the actin filaments is proposed to occur via aldolase. The molecular machinery that induces the polarized formation of actin filaments is not known (not drawn to scale) (adapted from Cowman and Crabb 2006)

plasmodium sporozoite that invades liver cells in the first stage of infection, a thrombospondin-like protein on the surface of the parasite (TRAP) links the parasite surface to its actin filaments via aldolase, a protein that binds the cytoplasmic tail of TRAP to the actin filament (Jewett and Sibley 2003). Myosin A, a type XIV myosin, links the actin filaments to the inner membrane complex deeper in the merozoite via attachments of the myosin tail to inner membrane complex proteins GAP45 and GAP59 in *T. gondii* (Gaskins et al. 2004). This attachment appears to be mediated through a myosin light chain, which unlike other MLCs, binds the tail of myoA. This set of proteins is highly conserved, and thus a similar scenario appears likely in *P. falciparum* (Baum et al. 2006).

Yet to be discovered is how the short actin filaments in the parasite support long-range powerful movements as would be required for penetration into the host cell. Crosslinking proteins, which could provide more stability and rigidity to actin filaments, are apparently lacking in the parasite genome. Polarity of the filaments and their placement beneath the parasite surface are also a key to producing penetration movements. That inhibition of actin polymerization by cytochalasins blocks parasite penetration argues for intense efforts to identify the molecular components of this process to serve as targets for drug development and disease treatment. For a diagram of Apicomplexan protozoan-host cell penetration, see Fig. 3.

Summary

Intracellular pathogens use actin filaments to accomplish crucial events in their life cycle. Bacterial pathogens such as *L. monocytogenes* and herpesviruses are providing key insights into cellular mechanisms. Universal mechanisms common to many pathogens include co-option of the cellular actin nucleation machine, the Arp2/3 complex, either by direct interaction with pathogen-encoded proteins or by upstream activation through signaling molecules. Both actin polymerization and myosin-based actin motility contribute to pathogen entry, transport, replication, and egress. Thus, actin regulatory machinery and the pathogen-encoded proteins that recruit actin are prime targets for antimicrobial drugs or for immunotherapy strategies.

Acknowledgments. I am indebted to many years of discussion with Lew Tilney and with Thomas S. Reese in the summers at Marine Biological Laboratory at Woods Hole, MA, that inspired the work in my lab described here, and gratefully acknowledge NIH-NIGMS (RO1-GM47368), NIH-NINDS (RO1-NS046820), and NCCR (P3 G03109), as well as the Moore Foundation and Dart Neurosciences LLP for funding that makes this work possible.

References

Allen, M. L., Dobrowolski, J. M., Muller, H., Sibley, L. D. and Mansour, T. E. 1997. Cloning and characterization of actin depolymerizing factor from *Toxoplasma gondii*. Mol. Biochem. Parasitol. 88, 43–52.

Antinone, S. E. and Smith, G. A. 2006. Two modes of herpesvirus trafficking in neurons: Membrane acquisition directs motion. J. Virol. 80, 11235–11240.

Arhel, N., Genovesio, A., Kim, K. A., Miko, S., Perret, E., Olivo-Marin, J. C., Shorte, S. and Charneau, P. 2006. Quantitative four-dimensional tracking of cytoplasmic and nuclear HIV-1 complexes. Nat. Methods 3, 817–824.

Baum, J., Richard, D., Healer, J., Rug, M., Krnajski, Z., Gilberger, T. W., Green, J. L., Holder, A. A. and Cowman, A. F. 2006. A conserved molecular motor drives cell invasion and gliding motility across malaria life cycle stages and other apicomplexan parasites. J. Biol. Chem. 281, 5197–5208.

Bearer, E. L. 1991. Direct observation of actin filament severing by gelsolin and binding by gCap39 and CapZ. J. Cell Biol. 115, 1629–1638.

Bearer, E. L. 1992. An actin-associated protein present in the microtubule organizing center and the growth cones of PC-12 cells. J. Neurosci. 12, 750–761.

Bearer, E. L. 1995. Cytoskeletal domains in the activated platelet. Cell Motil. Cytoskeleton 30, 50–66.

Bearer, E. L., Breakefield, X. O., Schuback, D., Reese, T. S. and LaVail, J. H. 2000. Retrograde axonal transport of herpes simplex virus: Evidence for a single mechanism and a role for tegument. Proc. Natl Acad. Sci. USA 97, 8146–8150.

Bearer, E. L. and Satpute-Krishnan, P. 2002. The role of the cytoskeleton in the life cycle of viruses and intracellular bacteria: Tracks, motors, and polymerization machines. Curr. Drug Targets Infect. Disord. 2, 247–264.

Bukrinskaya, A., Brichacek, B., Mann, A. and Stevenson, M. 1998. Establishment of a functional human immunodeficiency virus type 1 (HIV-1) reverse transcription complex involves the cytoskeleton. J. Exp. Med. 188, 2113–2125.

Campbell, E. M., Nunez, R. and Hope, T. J. 2004. Disruption of the actin cytoskeleton can complement the ability of Nef to enhance human immunodeficiency virus type 1 infectivity. J. Virol. 78, 5745–5755.

Carlsson, F. and Brown, E. J. 2006. Actin-based motility of intracellular bacteria, and polarized surface distribution of the bacterial effector molecules. J. Cell. Physiol. 209, 288–296.

Chan, K. S., Verardi, P. H., Legrand, F. A. and Yilma, T. D. 2005. Nef from pathogenic simian immunodeficiency virus is a negative factor for vaccinia virus. Proc. Natl Acad. Sci. USA 102, 8734–8739.

Chaparro-Olaya, J., Margos, G., Coles, D. J., Dluzewski, A. R., Mitchell, G. H., Wasserman, M. M. and Pinder, J. C. 2005. *Plasmodium falciparum* myosins: Transcription and translation during asexual parasite development. Cell Motil. Cytoskeleton 60, 200–213.

Charlton, C. A. and Volkman, L. E. 1991. Sequential rearrangement and nuclear polymerization of actin in baculovirus-infected *Spodoptera frugiperda* cells. J. Virol. 65, 1219–1227.

Cowman, A. F. and Crabb, B. S. 2006. Invasion of red blood cells by malaria parasites. Cell 124, 755–766.

DeGiorgis, J. A., Reese, T. S. and Bearer, E. L. 2002. Association of a nonmuscle myosin II with axoplasmic organelles. Mol. Biol. Cell 13, 1046–1057.

Dobrowolski, J. M. and Sibley, L. D. 1996. Toxoplasma invasion of mammalian cells is powered by the actin cytoskeleton of the parasite. Cell 84, 933–939.

Dohner, K., Radtke, K., Schmidt, S. and Sodeik, B. 2006. Eclipse phase of herpes simplex virus type 1 infection: Efficient dynein-mediated capsid transport without the small capsid protein VP26. J. Virol. 80, 8211–8224.

van Eijl, H., Hollinshead, M. and Smith, G. L. 2000. The vaccinia virus A36R protein is a type Ib membrane protein present on intracellular but not extracellular enveloped virus particles. Virology 271, 26–36.

Fackler, O. T. and Krausslich, H. G. 2006. Interactions of human retroviruses with the host cell cytoskeleton. Curr. Opin. Microbiol. 9, 409–415.

Fackler, O. T., Luo, W., Geyer, M., Alberts, A. S. and Peterlin, B. M. 1999. Activation of Vav by Nef induces cytoskeletal rearrangements and downstream effector functions. Mol. Cell 3, 729–739.

Feierbach, B., Piccinotti, S., Bisher, M., Denk, W. and Enquist, L. W. 2006. Alpha-Herpesvirus infection induces the formation of nuclear actin filaments. PLoS Pathog. 2, e85.

Forest, T., Barnard, S. and Baines, J. D. 2005. Active intranuclear movement of herpesvirus capsids. Nat. Cell Biol. 7, 429–431.

Frischknecht, F., Moreau, V., Rottger, S., Gonfloni, S., Reckmann, I., Superti-Furga, G. and Way, M. 1999. Actin-based motility of vaccinia virus mimics receptor tyrosine kinase signalling. Nature 401, 926–929.

Galan, J. E. and Cossart, P. 2005. Host–pathogen interactions: A diversity of themes, a variety of molecular machines. Curr. Opin. Microbiol. 8, 1–3.

Gardet, A., Breton, M., Fontanges, P., Trugnan, G. and Chwetzoff, S. 2006. Rotavirus spike protein VP4 binds to and remodels actin bundles of the epithelial brush border into actin bodies. J. Virol. 80, 3947–3956.

Gaskins, E., Gilk, S., DeVore, N., Mann, T., Ward, G. and Beckers, C. 2004. Identification of the membrane receptor of a class XIV myosin in Toxoplasma gondii. J. Cell Biol. 165, 383–393.

Goley, E. D., Ohkawa, T., Mancuso, J., Woodruff, J. B., D'Alessio, J. A., Cande, W. Z., Volkman, L. E. and Welch, M. D. 2006. Dynamic nuclear actin assembly by Arp2/3 complex and a baculovirus WASP-like protein. Science 314, 464–467.

Gordon, J. L. and Sibley, L. D. 2005. Comparative genome analysis reveals a conserved family of actin-like proteins in apicomplexan parasites. BMC Genomics 6, 179.

Gouin, E., Welch, M. D. and Cossart, P. 2005. Actin-based motility of intracellular pathogens. Curr. Opin. Microbiol. 8, 35–45.

Haller, C., Rauch, S., Michel, N., Hannemann, S., Lehmann, M. J., Keppler, O. T. and Fackler, O. T. 2006. The HIV-1 pathogenicity factor Nef interferes with maturation of stimulatory T-lymphocyte contacts by modulation of N-Wasp activity. J. Biol. Chem. 281, 19618–19630.

Heintzelman, M. B. and Schwartzman, J. D. 2001. Myosin diversity in Apicomplexa. J. Parasitol. 87, 429–432.

Honess, R. W. and Roizman, B. 1973. Proteins specified by herpes simplex virus. XI. Identification and relative molar rates of synthesis of structural and nonstructural herpes virus polypeptides in the infected cell. J. Virol. 12, 1347–1365.

Jewett, T. J. and Sibley, L. D. 2003. Aldolase forms a bridge between cell surface adhesins and the actin cytoskeleton in apicomplexan parasites. Mol. Cell 11, 885–894.

Jouvenet, N., Windsor, M., Rietdorf, J., Hawes, P., Monaghan, P., Way, M. and Wileman, T. 2006. African swine fever virus induces filopodia-like projections at the plasma membrane. Cell. Microbiol. 8, 1803–1811.

Kasman, L. M. and Volkman, L. E. 2000. Filamentous actin is required for lepidopteran nucleopolyhedrovirus progeny production. J. Gen. Virol. 81, 1881–1888.

Katz, E., Ward, B. M., Weisberg, A. S. and Moss, B. 2003. Mutations in the vaccinia virus A33R and B5R envelope proteins that enhance release of extracellular virions and eliminate formation of actin-containing microvilli without preventing tyrosine phosphorylation of the A36R protein. J. Virol. 77, 12266–12275.

Komano, J., Miyauchi, K., Matsuda, Z. and Yamamoto, N. 2004. Inhibiting the Arp2/3 complex limits infection of both intracellular mature vaccinia virus and primate lentiviruses. Mol. Biol. Cell 15, 5197–5207.

Kronstad, J. W. 2006. Serial analysis of gene expression in eukaryotic pathogens. Infect. Disord. Drug Targets 6, 281–297.

Lakadamyali, M., Rust, M. J., Babcock, H. P. and Zhuang, X. 2003. Visualizing infection of individual influenza viruses. Proc. Natl Acad. Sci. USA 100, 9280–9285.

Lehmann, M. J. and Frischknecht, F. 2006. Surfing through a sea of sharks: Report on the British Society for Cell Biology meeting on 'Signaling and Cytoskeletal Dynamics During Infection', October 2–5, 2005, Edinburgh, Scotland. Traffic 7, 479–487.

Lehmann, M. J., Sherer, N. M., Marks, C. B., Pypaert, M. and Mothes, W. 2005. Actin- and myosin-driven movement of viruses along filopodia precedes their entry into cells. J. Cell Biol. 170, 317–325.

Li, Z., Kim, E. S. and Bearer, E. L. 2002. Arp2/3 complex is required for actin polymerization during platelet shape change. Blood 99, 4466–4474.

Luxton, G. W., Haverlock, S., Coller, K. E., Antinone, S. E., Pincetic, A. and Smith, G. A. 2005. Targeting of herpesvirus capsid transport in axons is coupled to association with specific sets of tegument proteins. Proc. Natl Acad. Sci. USA 102, 5832–5837.

Machesky, L. M., Atkinson, S. J., Ampe, C., Vandekerckhove, J. and Pollard, T. D. 1994. Purification of a cortical complex containing two unconventional actins from Acanthamoeba by affinity chromatography on profilin-agarose. J. Cell Biol. 127, 107–115.

Miller, L. H., Aikawa, M., Johnson, J. G. and Shiroishi, T. 1979. Interaction between cytochalasin B-treated malarial parasites and erythrocytes. Attachment and junction formation. J. Exp. Med. 149, 172–184.

Moreau, V., Frischknecht, F., Reckmann, I., Vincentelli, R., Rabut, G., Stewart, D. and Way, M. 2000. A complex of N-WASP and WIP integrates signalling cascades that lead to actin polymerization. Nat. Cell Biol. 2, 441–448.

Munter, S., Way, M. and Frischknecht, F. 2006. Signaling during pathogen infection. Sci. STKE 2006, re5.

Newsome, T. P., Scaplehorn, N. and Way, M. 2004. SRC mediates a switch from microtubule- to actin-based motility of vaccinia virus. Science 306, 124–129.

Newsome, T. P., Weisswange, I., Frischknecht, F. and Way, M. 2006. Abl collaborates with Src family kinases to stimulate actin-based motility of vaccinia virus. Cell. Microbiol. 8, 233–241.

Niederman, T. M., Hastings, W. R. and Ratner, L. 1993. Myristoylation-enhanced binding of the HIV-1 Nef protein to T cell skeletal matrix. Virology 197, 420–425.

Ohkawa, T., Rowe, A. R. and Volkman, L. E. 2002. Identification of six *Autographa californica* multicapsid nucleopolyhedrovirus early genes that mediate nuclear localization of G-actin. J. Virol. 76, 12281–12289.

Quaranta, M. G., Mattioli, B., Spadaro, F., Straface, E., Giordani, L., Ramoni, C., Malorni, W. and Viora, M. 2003. HIV-1 Nef triggers Vav-mediated signaling pathway leading to functional and morphological differentiation of dendritic cells. FASEB J. 17, 2025–2036.

Radtke, K., Dohner, K. and Sodeik, B. 2006. Viral interactions with the cytoskeleton: A hitchhiker's guide to the cell. Cell. Microbiol. 8, 387–400.

Rietdorf, J., Ploubidou, A., Reckmann, I., Holmstrom, A., Frischknecht, F., Zettl, M., Zimmermann, T. and Way, M. 2001. Kinesin-dependent movement on microtubules precedes actin-based motility of vaccinia virus. Nat. Cell Biol. 3, 992–1000.

Rottger, S., Frischknecht, F., Reckmann, I., Smith, G. L. and Way, M. 1999. Interactions between vaccinia virus IEV membrane proteins and their roles in IEV assembly and actin tail formation. J. Virol. 73, 2863–2875.

Rottner, K., Stradal, T. E. and Wehland, J. 2005. Bacteria–host-cell interactions at the plasma membrane: Stories on actin cytoskeleton subversion. Dev. Cell 9, 3–17.

Sahoo, N., Beatty, W., Heuser, J., Sept, D. and Sibley, L. D. 2006. Unusual kinetic and structural properties control rapid assembly and turnover of actin in the parasite *Toxoplasma gondii*. Mol. Biol. Cell 17, 895–906.

Saksena, M. M., Wakisaka, H., Tijono, B., Boadle, R. A., Rixon, F., Takahashi, H. and Cunningham, A. L. 2006. Herpes simplex virus type 1 accumulation, envelopment, and exit in growth cones and varicosities in mid-distal regions of axons. J. Virol. 80, 3592–3606.

Satpute-Krishnan, P., DeGiorgis, J. A. and Bearer, E. L. 2003. Fast anterograde transport of herpes simplex virus: Role for the amyloid precursor protein of Alzheimer's disease. Aging Cell 2, 305–318.

Satpute-Krishnan, P., DeGiorgis, J. A., Conley, M. P., Jang, M. and Bearer, E. L. 2006. A peptide zipcode sufficient for anterograde transport within amyloid precursor protein. Proc. Natl Acad. Sci. USA 103, 16532–16537.

Scaplehorn, N., Holmstrom, A., Moreau, V., Frischknecht, F., Reckmann, I. and Way, M. 2002. Grb2 and Nck act cooperatively to promote actin-based motility of vaccinia virus. Curr. Biol. 12, 740–745.

Schmitz, S., Grainger, M., Howell, S., Calder, L. J., Gaeb, M., Pinder, J. C., Holder, A. A. and Veigel, C. 2005. Malaria parasite actin filaments are very short. J. Mol. Biol. 349, 113–125.

Schuler, H. and Matuschewski, K. 2006. Regulation of apicomplexan microfilament dynamics by a minimal set of actin-binding proteins. Traffic 7, 1433–1439.

Schuler, H., Mueller, A. K. and Matuschewski, K. 2005. A *Plasmodium* actin-depolymerizing factor that binds exclusively to actin monomers. Mol. Biol. Cell 16, 4013–4023.

Shaw, M. K. and Tilney, L. G. 1999. Induction of an acrosomal process in *Toxoplasma gondii*: Visualization of actin filaments in a protozoan parasite. Proc. Natl Acad. Sci. USA 96, 9095–9099.

Simpson-Holley, M., Colgrove, R. C., Nalepa, G., Harper, J. W. and Knipe, D. M. 2005. Identification and functional evaluation of cellular and viral factors involved in the alteration of nuclear architecture during herpes simplex virus 1 infection. J. Virol. 79, 12840–12851.

Simpson-Holley, M., Ellis, D., Fisher, D., Elton, D., McCauley, J. and Digard, P. 2002. A functional link between the actin cytoskeleton and lipid rafts during budding of filamentous influenza virions. Virology 301, 212–225.

Smith, G. L., Murphy, B. J. and Law, M. 2003. Vaccinia virus motility. Annu. Rev. Microbiol. 57, 323–342.

Snyder, A., Wisner, T. W. and Johnson, D. C. 2006. Herpes simplex virus capsids are transported in neuronal axons without an envelope containing the viral glycoproteins. J. Virol. 80, 11165–11177.

Svitkina, T. M., Bulanova, E. A., Chaga, O. Y., Vignjevic, D. M., Kojima, S., Vasiliev, J. M. and Borisy, G. G. 2003. Mechanism of filopodia initiation by reorganization of a dendritic network. J. Cell Biol. 160, 409–421.

Tardieux, I., Liu, X., Poupel, O., Parzy, D., Dehoux, P. and Langsley, G. 1998. A *Plasmodium falciparum* novel gene encoding a coronin-like protein which associates with actin filaments. FEBS Lett. 441, 251–256.

Tilney, L. G., Connelly, P. S. and Portnoy, D. A. 1990. Actin filament nucleation by the bacterial pathogen, *Listeria monocytogenes*. J. Cell Biol. 111, 2979–2988.

Way, M. 1998. Interaction of vaccinia virus with the actin cytoskeleton. Folia Microbiol. (Praha) 43, 305–310.

Welch, M. D., Iwamatsu, A. and Mitchison, T. J. 1997. Actin polymerization is induced by Arp2/3 protein complex at the surface of *Listeria monocytogenes*. Nature 385, 265–269.

Wells, A. L., Lin, A. W., Chen, L. Q., Safer, D., Cain, S. M., Hasson, T., Carragher, B. O., Milligan, R. A. and Sweeney, H. L. 1999. Myosin VI is an actin-based motor that moves backwards. Nature 401, 505–508.

14
Actin and Its Binding Proteins in Heart Failure

Maurizio Stefani, Masako Tsubakihara, Brett D. Hambly,
Choon C. Liew, Paul D. Allen, Peter S. Macdonald,
and Cristobal G. dos Remedios

Pathophysiology of Heart Failure

Heart failure (HF) is one of the leading causes of combined morbidity and mortality among developed nations. It is the final clinical presentation of a variety of cardiovascular diseases and disorders, such as coronary artery disease, hypertension, valvular heart disease, myocarditis, diabetes, alcohol abuse, and familial cardiomyopathies (Narula et al. 1996). This pathophysiological state is characterized by progressive deterioration of ventricular function, usually in the left ventricle (LV).

Pathologic stressors and increased demands for cardiac work result in compensatory mechanisms, such as hypertrophy of cardiomyocytes, increased sympathetic activity, and activation of the rennin–angiotensin–aldosterone system (Curtiss et al. 1978; Levine et al. 1982; Pfeffer et al. 1988; Anversa, Olivetti and Capasso 1991). These short-term adaptive mechanisms eventually become maladaptive and lead to HF.

Understanding the mechanisms underlying HF has been difficult since it arises from the interaction between environmental factors and genetic susceptibility, and cannot be explained by a single gene or pathway (Seidman and Seidman 2001). Considerable research has been undertaken to explain why short-term compensatory mechanisms often become maladaptive in the long term. This chapter will focus on mutations in actin within the heart (both sarcomeric and cytoskeletal) and alterations in the actin binding proteins (ABPs) that interact with actin. (For a detailed discussion see the chapter by Sparrow and Laing.)

Cardiomyopathies Due to Cardiac α-Actin Mutations

Familial Hypertrophic Cardiomyopathy

Familial hypertrophic cardiomyopathy (FHC) is an autosomal dominant congenital disease, affecting 1 in 500 individuals, but with variable penetrance (Doolan, Nguyen and Semsarian 2004). Affected individuals may experience dyspnea, syncope, angina, diastolic dysfunction, heart failure, and sudden death due to spontaneous arrhythmias (Doolan, Nguyen and Semsarian 2004). FHC is the leading cause of sudden death in young people (Maron 2002) and is characterized by otherwise unexplained hypertrophy of the left ventricle, without dilation (Doolan, Nguyen and Semsarian 2004).

More than 100 mutations, predominantly in myofibrillar genes, including actin and ABPs, predispose an individual to developing this condition. Actin mutations include Ala295Ser (Mogensen et al. 1999), Tyr166Cys and Met305Leu (Mogensen et al. 2004a), Pro164Ala and Ala331Pro (Olson et al. 2000), Glu99Lys (Olson et al. 2000; Arad et al. 2005), and Ala230Val (Van Driest et al. 2003) (see Table 1 and Figs. 1 and 2). Given the high degree of conservation in the amino acid sequence of actin, any mutation is likely to have a significant effect on structure and function (Bookwalter and Trybus 2006).

In addition to having a deleterious effect due to the position of the substituted residues in the actin structure, Vang et al. (2005) found that the mutated actins implicated in FHC exhibit defects that affect folding and incorporation of monomers into the filament. Mutant actins Glu99Lys, Pro164Lys, and Met305Leu were found to fold slowly *in vitro* compared to the wild-type assays, and exhibited the lowest filament incorporation in transfected COS-7 and HEK293 cells (Vang et al. 2005). Given that certain actin mutants fold slowly, and exhibit reduced incorporation into filaments, the disease is likely to be associated with a reduction in functioning sarcomeric or cytoskeletal actin.

Thus, mutations in regions of cardiac α-actin involved in binding to a number of thick and thin filament components can result in FHC, although the mechanisms involved may include both direct interference in function and/or haploinsufficiency.

TABLE 1. Cardiac α-actin mutations that are etiologically linked to familial cardiomyopathies.

Function	Residue change	Phenotype	Reference
β-Myosin binding	Glu99Lys	FHC	Olson et al. (2000)
β-Myosin binding	Ala295Ser	FHC	Mogensen et al. (1999)
β-Myosin binding	Ala331Pro	FHC	Olson et al. (2000)
β-Myosin binding; α-actin binding	Pro164Ala	FHC	Olson et al. (2000)
β-Myosin binding; α-actin binding	Tyr166Cys	FHC	Mogensen et al. (2004a)
β-Myosin binding; ATP/ADP binding	Met305Leu	FHC	Mogensen et al. (2004a)
Tropomyosin 1 binding	Ala230Val	FHC	Van Driest et al. (2003)
Intercalated disc and Z-disc binding	Arg312His	FDC	Olson et al. (1998)
α-Actinin binding, dystrophin binding	Glu361Gly	FDC	Olson et al. (1998), Kuhlman et al. (1992), and Levine et al. (1992)

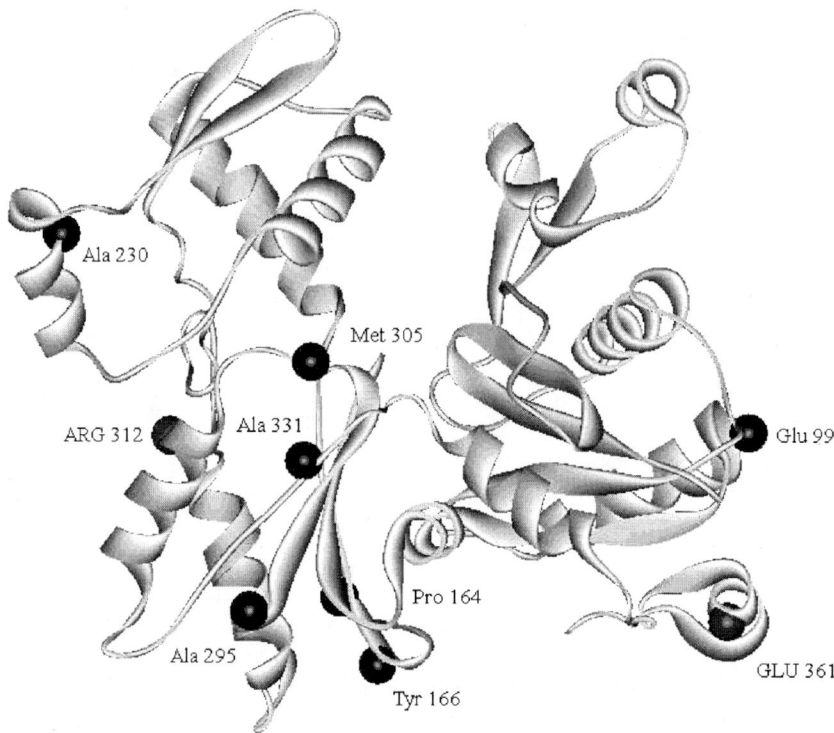

Fig. 1. Crystal structure of actin adapted from the Protein Data Base. Amino acid residues mutated in FHC are labeled in upper case, and residues mutated in FDC are entitled in capitals

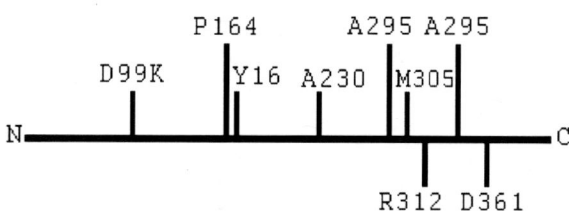

Fig. 2. Diagram indicating the positions of mutations in cardiac α-actin (scaled). Note, mutations predisposing to FHC are listed above and mutations predisposing to FDC are listed below. N refers to the N-terminus and C refers to the C-terminus

Familial Dilated Cardiomyopathy

Like FHC, familial dilated cardiomyopathy (FDC) is an autosomal dominant condition with variable penetrance. It is caused by missense mutations in a variety of sarcomeric and cytoskeletal genes. It is the leading cause of heart failure in youth. The phenotype is characterized by dilatation of the ventricles, systolic dysfunction, and eventually thinning of the heart wall.

Two mutations in cardiac actin have been shown to predispose an individual to FDC, namely Arg312His and Glu361Gly (Olson et al. 1998) (see Table 1 and Figs. 1 and 2). Both mutations are in residues that are invariant in all human actins and in lower organisms (Olson et al. 1998). These mutations occur in subdomains 1 and 3, which are in the region of actin that is located close to the Z-disc and intercalated disc. The Glu361Gly substitution occurs in an exposed region of cardiac actin that can bind to the actin-regulator protein α-actinin (Kuhlman, Hemmings and Critchley 1992). This binding domain also binds to the N-terminal region of dystrophin (Levine et al. 1992). Therefore, Olson et al. (1998) proposed that the FDC-related actin mutations cause defects in force transmission that predisposes myocytes to mechanical injury and cell death that in turn lead to interstitial fibrosis and cardiac dilation.

Vang et al. (2005) have shown that these actins exhibit a defective ability to fold and incorporate into filaments. The Arg312His mutant actin is particularly affected. There is a possibility that a reduction in the incorporation of cardiac-actin into filaments either has an effect on contractile function, or that the accumulation of misfolded actins in the cytosol disrupts cell function, or both (Vang et al. 2005).

Cardiac and Skeletal α-Actin Abundance in Heart Failure

Normal adult human myocardium expresses both α-skeletal and α-cardiac actins (Vandekerckhove, Bugiasky and Buckingham 1986; Bennetts, Burnett and dos Remedios, 1986). Skeletal actin mRNA increases in the human heart during ontogenic development and is the major isoform of control and failing adult hearts (Boheler et al. 1991). In normal adults, cardiomyocytes which stain positive for α-skeletal actin protein show a transmural gradient of abundance in the ventricles, with the highest proportion being in the subendocardium (Suurmeijer et al. 2003). In contrast, all cardiomyocytes, in the LV at least, exhibit positive staining for α-cardiac actin protein (Suurmeijer et al. 2003).

In patients with hypertrophy associated with ischemic cardiomyopathy and other causes, there are more cardiomyocytes staining positive for α-skeletal actin protein, and they are present throughout the myocardium (Suurmeijer et al. 2003). This increase in α-skeletal actin may be related to an increase in myocardial contractility in states of hypertrophy (Hewett et al. 1994). Interestingly, in most patients with HF due to idiopathic dilated cardiomyopathy, there was a marked reduction in α-skeletal actin levels by Suurmeijer et al. (2003). This reduction, however, may be explained by the fact that all these patients were treated with ACE inhibitors which have been shown to reduce α-skeletal actin expression in rats (Dalton et al. 2000).

The prevalence of DCM is 1:2,500 and an incidence of 7/100,000 per year (but it may be under diagnosed). FDC may account for 20–48% of DCM (Taylor, Carniel and Mestroni 2006). Note that since genetic testing of dilated cardiomyopathy patients is not routine in clinical practices, a diagnosis of FDC is usually made on the basis of a family history, and if not, a diagnosis of "idiopathic"is made. Therefore, many cases of idiopathic dilated cardiomyopathy might actually be FDC. In failing hearts due to idiopathic dilated cardiomyopathy, an increase in α-cardiac actin was observed by Western blotting while in failing hearts from patients exhibiting hypertrophic phenotypes, the same authors reported a slight decline in α-cardiac actin expression (Suurmeijer et al. 2003). There seems to be no significant difference in the 5-year survival rate for familial vs. non-FDC (Michels et al. 2003).

Transgenic mice overexpressing serum response factor (SRF) exhibited cardiac hypertrophy with associated collagen deposition and interstitial fibrosis (similar to FHC in humans) (Zhang et al. 2001). SRF controls the expression of cardiac-specific genes, and in these transgenic mice, skeletal actin was upregulated and cardiac actin was down-regulated (Zhang et al. 2001), mirroring the changes found in humans with cardiac hypertrophy.

These results indicate that changes in actin isoform expression are associated with all types of HF and the determinant of isoform abundance is related to the particular disease mechanism that results in HF.

LIM Proteins and Heart Failure

LIM domain-containing proteins are a structurally and functionally diverse class that are emerging as a crucial link between the actin cytoskeleton and transcriptional regulation (Kadrmas and Beckerle 2004). The LIM domain is approximately 55 residues in length and contains two zinc-binding regions. The consensus sequence is: C-X2-C-X16-23-H-X2-C-X2-C-X16-23-C-X2-C, where the H is histidine, the C is cysteine, and the X is any amino acid. The LIM domain has been found to function as a protein–protein interaction domain with the variable amino acids conferring substrate specificity (Kadrmas and Beckerle 2004).

LIM domains can bind to a diverse array of targets, found in the cytoplasm as well as in the nucleus. The presence of LIM domains, in conjunction with other domains, allows these proteins to transmit signals between the sarcolemma, the cytoskeleton, and the nucleus, and may therefore couple changes in contractile tension with appropriate changes in transcriptional activity.

Muscle LIM Protein

A subfamily of the LIM domain containing proteins is the cysteine-rich protein family (CRP), which exhibit a common structure of two tandemly arranged LIM domains (Weiskirchen et al. 1995). Muscle LIM protein

(MLP) (also known as CRP3) is a member of this family, and each LIM domain is followed by a glycine-rich repeat region (Weiskirchen et al. 1995). In primary human cardiac cultures, MLP initially localizes to the nucleus and later is localized primarily along cytoskeletal actin filaments in the cytoplasm (Arber and Caroni 1996). *In vitro*, MLP binds to zyxin (Louis et al. 1997), a phosphoprotein involved in the spatial control of actin filament assembly at the Z-disc (Beckerle 1997). MLP also binds α-actinin in both *in vitro* binding assays (Louis et al. 1997) and yeast-2-hybrid assays (Knoll et al. 2002). α-actinin is localized to the Z-disc (Dubreuil 1991) and thus the mechanism for MLP localization to the Z-disc includes its interaction with α-actinin.

Disease-Causing Mutations in MLP

Missense mutations in MLP have been etiologically linked to the development of HF in human clinical studies. A Lys69Arg substitution mutation in MLP has been shown to predispose patients to the development of FDC, for example HF was identified in a single family by Mohapatra et al. (2003). This mutation occurs in the glycine-rich region near the N-terminal LIM1 domain. Lys69Arg is highly conserved among LIM1 domains, and this mutation disrupts LIM1 domain binding to α-actinin in co-immunoprecipitation studies (Mohapatra et al. 2003). A Trp4Arg substitution in MLP has also been linked to development of FDC and HF (Knoll et al. 2002). On the other hand, in three families with Leu44Pro, Cys58Gly, and Ser54Arg/Asp55Gly mutations, individuals developed FHC (Geier et al. 2003). The aforementioned mutations changed highly conserved amino acids in the LIM1 domain of MLP. As mentioned above, the LIM1 domain comprises the surface that binds to α-actinin, and, at least for the Cys58Gly mutation, binding was heavily impaired in yeast-2-hybird and dot blot overlay assays (Geier et al. 2003). Interestingly, Gln9Arg α-actinin, which disrupts α-actinin binding to MLP was found to predispose patients to FDC (Mohapatra et al. 2003). Therefore, a defect in MLP structure/function can lead to the development of HF (Fig. 3).

FIG. 3. Diagram indicating the position of mutations in MLP (scaled). Note, mutations predisposing to FHC are listed above and mutations predisposing to FDC are listed below. N refers to the N-terminus and C refers to the C-terminus

Altered Expression of MLP in Disease

Our group has found that left ventricular MLP mRNA expression is down-regulated in idiopathic, doxorubicin-induced, and viral-induced dilated cardiomyopathy compared to nonfailing donor hearts (Tsubakihara 2005). Zolk, Caroni and Bohm (2000) found that left ventricular MLP protein levels are down by approximately 50% in HF due to ischemia and dilated cardiomyopathy compared with nonfailing donor hearts. Therefore, in addition to MLP mutations, a decrease in MLP abundance also appears to be associated with HF.

Pathogenic Mechanism Underlying MLP Abnormalities

The Z-disc is a key component of a mechanical stress sensor pathway, and down-regulation or a mutational defect in MLP may result in destabilization of the cardiac Z-disc and a deficit in primary tension generation, and therefore a loss of contractility (Knoll et al. 2002). This loss is a hallmark of dilated cardiomyopathies and HF. MLP interacts with and potentiates the effects of nuclear MyoD (an inducer of myocyte terminal differentiation) on myogenesis, and this interaction is mediated by the LIM1 domain of MLP (Kong et al. 1997). This pathway may promote hypertrophy in the heart, given that these cells are terminally differentiated. It is possible that a loss of binding to α-actinin leads to a net movement of MLP to the nucleus to stimulate a hypertrophy-inducing program of gene transcription. This might be crucial in the development of FHC. This hypothesis is strengthened by the finding that Lys69Arg MLP in FHC is predominantly perinuclear in C2C12 myoblast cells, whereas the wild-type protein is predominantly cytoplasmic (Mohapatra et al. 2003).

MLP has been implicated further in other regulatory activities within the cardiac myocyte, including the intercalated disk and in subsarcolemmal structures (Hoshijima 2006). The pathological implications of these mechanisms warrant further investigation.

LMCD1

This is strongly expressed in human heart muscle (Bespalova and Burmeister, 2000). It binds to and represses GATA-6 transcription by blocking DNA binding to GATA-6-specific promoters (Rath et al. 2005). GATA-6 is a transcription factor that regulates heart-specific gene expression (Morrisey et al. 1996). LMCD1 is possibly regulated by nucleus-cytoplasmic shuttling, e.g. in the nucleus it represses GATA-6 activity (Bespalova and Burmeister 2000). We found LMCD1 mRNA to be downregulated in end-stage HF of a variety of etiologies including FHC and FDC, as well as peripartum and idiopathic dilated cardiomyopathy (Tsubakihara 2005). GATA-6 has been shown to induce the transcription of ANF and BNP (Charron et al. 1999), therefore a reduction of LMCD1 repression of GATA-6 activity accords with experimental findings in the literature.

CRP2

Our group found that left ventricular CRP2 (also known as SmLIM) mRNA is downregulated in idiopathic dilated cardiomyopathy (Tsubakihara 2005). CRP2 is highly expressed in murine embryonic cardiac tissues, and drops to low levels in adults (Jain et al. 1998), indicating it has a role in normal cardiac development. CRP2 can bind directly to F-actin (Grubinger and Gimona 2004), as well as to zyxin and α-actinin (Louis et al. 1997), indicating it has a role in regulation or maintenance of the actin cytoskeleton and sarcomere. Grubinger and Gimona (2004) suggest that CRP2 has a role in stabilizing actin filaments, however, functional studies in human hearts or cardiac cell lines are necessary to dissect out the contribution of this protein to HF.

CLP36

CLP36 (also known as hCLIM1) mRNA has been shown by our group to be downregulated in both idiopathic and doxorubicin-induced dilated cardiomyopathic failing hearts compared with nonfailing hearts (Tsubakihara 2005). Northern blotting has shown that CLP36 is most highly expressed in human heart (Kotaka et al. 1999), indicating that it has an important, yet so far unknown role in human cardiac function. CLP36 has been shown to bind to α-actinin 2 in yeast-2-hybrid and co-immunoprecipitation assays, and localizes at Z-discs in human myocardium using immunofluorescence (Kotaka et al. 2000). While functional studies in human myocardium are lacking, in nonmuscle fibres, CLP36 recruits LIM kinase to actin stress fibres via its interaction with α-actinin (Bauer et al. 2000). LIM kinase in turn can phosphorylate and inactivate cofilin (Arber et al. 1998), an ABP that depolymerizes F-actin (see chapter by Maloney et al.). In the heart, a reduction of CLP36 might lead to increased cofilin activity and as a result, destabilization of sarcomeric and cytoskeletal actin filaments.

Four and a Half LIM Protein 2

Four and a half LIM protein 2 (FHL2) is also known as SLIM3. It has been shown to bind to α-skeletal actin as well as a variety of other cytoskeletal proteins (Coghill et al. 2003). FHL2 is highly expressed in human heart and depending on the human cell line probed, endogenous FHL2 has been found both in the nucleus and the cytoplasm (Johannessen et al. 2006). A variety of stimuli can induce nuclear localization (Johannessen et al. 2006), suggesting a signal transduction role for FHL2 in cardiomyocytes where it has been shown to localize at the Z-disc (Purcell et al. 2004).

We found that FHL2 is downregulated in idiopathic and viral dilated cardiomyopathy compared with nonfailing hearts (Tsubakihara 2005). However, in a smaller study, Grzeskowiak et al. (2003) found that FHL2 was upregulated in dilated cardiomyopathic human failing hearts.

No cardiac abnormalities were found in a FHL2 null mouse (Chu et al. 2000), however, these FHL2 null mice were found to exhibit a greater hypertrophic response to β-adrenergic stimulation compared to wild-type mice (Kong et al. 2001). The underlying reason for the increased hypertrophic response to β-adrenergic stimulation may be that FHL2 inhibits the hypertrophy-inducing MEK1-ERK1/2 signaling pathway (Purcell et al. 2004). Changes in FHL2, coupled with the increased sympathetic activity that is characteristic of HF, may directly influence the remodeling that occurs during HF.

Sarcomeric Actin-Binding Protein Mutations and Heart Failure

Troponin T

Approximately 5–10% of patients with FHC have mutations in TNNT2 (cardiac TnT) located on chromosome 1q32 (Townsend et al. 1994). Most of the 31 different FHC-causing TNNT2 mutations are missense mutations, but cause TnT truncation (Richard et al. 2003). The phenotype associated with mutated cardiac TnT is characterized by mild hypertrophy and a poor prognosis with a high incidence of sudden death (Song et al. 2005). On the other hand, TNNT2 mutations causing FDC are associated with severe disease and complete penetrance (Mogensen et al. 2004b). The functional consequence of the mutations is believed to be a decrease in calcium sensitivity of the actomyosin interaction.

Troponin I

The gene coding for TnI (TNNI3) is located on chromosome 19p13.2–q13.2 (Vallins et al. 1990). Most of the 27 mutations found in TNNI3 that cause FHC are missense mutations, with a prevalence of about 3% (Doolan et al. 2005; Mogensen et al. 2004c). There is no clear clinical genotype–phenotype pattern in FHC. The mutation Arg145Gly interferes with TnI binding to TnC (Lindhout et al. 2002), while the Arg21Cys mutation renders TnI more readily degraded by calpain, affects the Ca^{2+} sensitizing effect of the Tn complex and the ability of TnI to be phosphorylated (Gomes, Harada and Potter 2005). Only one recessive mutation in TNNI3 causing FDC has been found, Ala2Val, which results in impaired interaction with TnT (Murphy et al. 2004).

Troponin C

TNNC1, the gene that codes for TnC, is located on chromosome 3p21.3–3p14.3 (Song et al. 1996). Only one FHC-causing mutation in TNNC1, Leu29Glu (Hoffmann et al. 2001), has been found, which causes septal hypertrophy and atrial fibrillation. The mutation hinders transduction of the phosphorylation signal from TnI to TnC (Schmidtmann et al. 2005). Similarly, only one mutation in TNNC1 has been found to cause FDC, Gly159Asp (Mogensen et al. 2004a), which results in a decrease in the rate of force production (Preston et al. 2006).

α-Tropomyosin

The TPM1 gene is located on 15q22.1 (Lees-Miller and Helfman 1991). Eight missense mutations account for ~5% of FHC cases. While hypertrophy has been found to be mild in most cases studied, disease prognosis varies with the specific mutation. Ala63Val, Lys70Thr, and Val95Ala mutations are associated with a poor prognosis (Yamauchi-Takihara et al. 1996; Karibe et al. 2001) and both Leu185Arg and Glu62Gln mutations are malignant at an early age (Van Driest et al. 2002; Earing, Ackerman and O'Leary 2003; Jongbloed et al. 2003).

Histories of sudden cardiac death have also been reported in patients with a Glu180Gly mutation and this mutation in transgenic mice resulted in an increase in the Ca^{2+} sensitivity of force production (Michele et al. 2002). Asp175Asn resulted in a variable phenotype but was the unconventional TPM1 mutation in that it showed a good prognosis (Coviello et al. 1997). Finally, in a case with severe systolic dysfunction and the rare progression of hypertrophic to dilated cardiomyopathy, the Glu180Val mutation in TPM1 has been identified (Yamauchi-Takihara et al. 1996; Regitz-Zagrosek et al. 2000). Two mutations have been identified that are specifically associated with FDC, Glu40Lys and Glu54Lys (Olson et al. 2001). Both alter the surface charge of tropomyosin.

α-Actinin

Only one mutation causing FDC has been identified in α-actinin-2, Gln9Arg (Mohapatra et al. 2003). It lies close to the N-terminus and is predicted to extend an α-helix. This region is involved in actin binding, and may also interfere with the induction of differentiation by interfering with intranuclear localization.

Nonsarcomeric Actin-Binding Protein Mutations and Heart Failure

DNase I

Evidence of extensive apoptosis has been found in idiopathic dilated cardiomyopathic failing hearts (Narula et al. 1996). Elevation of DNase I is indicative of apoptosis in aging Fisher 344 rats (Nitahara et al. 1998). Increased levels of DNase I protein have been found in end-stage failing hearts from patients with idiopathic dilated cardiomyopathy (Yao et al. 1996), which indicates that apoptosis mediated by increased DNase I levels is involved in the pathophysiology of this disease.

It has been suggested that DNase I may have a major role in sequestering actin monomers intracellularly (dos Remedios et al. 2003). The very high affinity of DNase I for G-actin may be responsible for decreased filament incorporation. This might expose the cardiomyocytes to increased mechanical damage and thus be involved in the pathophysiology of DCM.

Gelsolin

Increased expression of gelsolin mRNA (3.4-fold) and protein (1.6–2.3-fold) has been reported in failing DCM hearts compared to nonfailing hearts (Yang et al. 2000). A similar increase in gelsolin mRNA (2.7-fold) was also found in failing hearts with ischemic cardiomyopathy (Yang et al. 2000), however, protein levels of gelsolin were not tested. Gelsolin is involved in the regulation of actin filament assembly/disassembly (see chapter by Burtnick and Robinson). Therefore, increased levels of gelsolin may undermine the stability of the cardiomyocyte cytoskeleton (Kuhne et al. 1993), allowing the "slippage" that is characteristic of DCM.

Gelsolin has been implicated in regulation of L-type cardiac Ca^{2+} channel regulation, with increased gelsolin resulting in decreased channel activity (decreased Ca^{2+} current) in neonatal mouse cardiomyocytes (Lader, Kwiatkowski and Caniello 1999). In heart failure, there is evidence of abnormal Ca^{2+} kinetics, and this increase in gelsolin, coupled with a probable decrease in L-type Ca^{2+} abundance (Yang et al. 2000), may contribute. A caveat is that Yang et al. (2000) did not have age-matched control hearts, and given that gelsolin has been shown to increase with age in rats (Ahn et al. 2003), their finding of increased gelsolin in HF may be an artifact of the sample population used in the study.

Summary

Actin and its binding proteins are directly involved in the disease mechanism of heart failure from a variety of causes. Mutations in cardiac α-actin which alter binding to other sarcomeric elements, alter folding kinetics and filament incorporation, predispose to the development of HF. In addition, with respect to HF from both congenital and non-congenital causes, actin isoform abundance changes are observed, which may mediate changes in contractility. Sarcomeric and cytoskeletal actins are bound by many associated proteins, including the class of LIM domain containing proteins. Mutations in these actin-binding proteins, e.g. MLP and α-actinin also predispose to HF. In addition, the abundance of many ABPs, a few of which have been mentioned in this review, increase or decrease, and underlie the cytoskeletal and transcriptional changes that are etiologically responsible for this lethal disease.

References

Ahn, J. S., Jang, I.-S., Kim, D.-I., Cho, K. A., Park, Y. H., Kim, K., Kwak, C. S. and Chul Park, S. 2003. Aging-associated increase of gelsolin for apoptosis resistance. Biochem. Biophys. Res. Commun. 312, 1335–1341.

Anversa, P., Olivetti, G. and Capasso, J. M. 1991. Cellular basis of ventricular remodeling after myocardial infarction. Am. J. Cardiol. 68, 7D–16D.

Arad, M., Penas-Lado, M., Monserrat, L., Maron, B. J., Sherrid, M., Ho, C. Y., Barr, S., Karim, A., Olson, T. M., Kamisago, M., Seidman, J. G. and Seidman, C. E. 2005. Gene mutations in apical hypertrophic cardiomyopathy. Circulation 112, 2805–2811.

Arber, S., Barbayannis, F. A., Hanser, H., Schneider, C., Stanyon, C. A., Bernard, O. and Caroni, P. 1998. Regulation of actin dynamics through phosphorylation of cofilin by LIM-kinase. Nature 393, 805–809.

Arber, S. and Caroni, P. 1996. Specificity of single LIM motifs in targeting and LIM/LIM interactions in situ. Genes Dev. 10, 289–300.

Bauer, K., Kratzer, M., Otte, M., de Quintana, K. L., Hagmann, J., Arnold, G. J., Eckerskorn, C., Lottspeich, F. and Siess, W. 2000. Human CLP36, a PDZ-domain and LIM-domain protein, binds to alpha-actinin-1 and associates with actin filaments and stress fibers in activated platelets and endothelial cells. Blood 96, 4236–4245.

Beckerle, M. C. 1997. Zyxin: Zinc fingers at sites of cell adhesion. BioEssays 19, 949–957.

Bennetts B. H., Burnett, L. and dos Remedios, C. G. 1986. Differential co-expression of α-actin genes within the human heart. J. Mol. Cell. Cardiol. 18, 993–996.

Bespalova, I. N. and Burmeister, M. 2000. Identification of a novel LIM domain gene, LMCD1, and chromosomal localization in human and mouse. Genomics 63, 69–74.

Boheler, K. R., Carrier, L., de la Bastie, D., Allen, P. D., Komajda, M., Mercadier, J. J. and Schwartz, K. 1991. Skeletal actin mRNA increases in the human heart during ontogenic development and is the major isoform of control and failing adult hearts. J. Clin. Invest. 88, 323–330.

Bookwalter, C. S. and Trybus, K. M. 2006. Functional consequences of a mutation in an expressed human α-cardiac actin at a site implicated in familial hypertrophic cardiomyopathy. J. Biol. Chem. 281, 16777–16784.

Charron, F., Paradis, P., Bronchain, O., Nemer, G. and Nemer, M. 1999. Cooperative interaction between GATA-4 and GATA-6 regulates myocardial gene expression. Mol. Cell. Biol. 19, 4355–4365.

Chu, P.-H., Bardwell, W. M., Gu, Y., Ross, J. Jr. and Chen, J. 2000. FHL2 (SLIM3) is not essential for cardiac development and function. Mol. Cell. Biol. 20, 7460–7462.

Coghill, I. D., Brown, S., Cottle, D. L., McGrath, M. J., Robinson, P. A., Nandurkar, H. H., Dyson, J. M. and Mitchell, C. A. 2003. FHL3 is an actin-binding protein that regulates α-actinin-mediated actin bundling: FHL3 localizes to actin stress fibers and enhances cell spreading and stress fiber disassembly. J. Biol. Chem. 278, 24139–24152.

Coviello, D. A., Maron, B. J., Spirito, P., Watkins, H., Vosberg, H. P., Thierfelder, L., Schoen, F. J., Seidman, J. G. and Seidman, C. E. 1997. Clinical features of hypertrophic cardiomyopathy caused by mutation of a "hot spot" in the α-tropomyosin gene. J. Am. Coll. Cardiol. 29, 635–640.

Curtiss, C., Cohn, J. N., Vrobel, T. and Franciosa, J. A. 1978. Role of the renin-angiotensin system in the systemic vasoconstriction of chronic congestive heart failure. Circulation 58, 763–770.

Dalton, G. R., Jones, J. V., Levi, A. J. and Levy, A. 2000. Changes in contractile protein gene expression with ageing and with captopril-induced regression of hypertrophy in the spontaneously hypertensive rats. J. Hypertens. 18, 1297–1306.

Doolan, A., Nguyen, L. and Semsarian, C. 2004. Hypertrophic hardiomyopathy: From "Heart tumour" to a complex molecular genetic disorder. Heart Lung Circ. 13, 15–25.

Doolan, A., Tebo, M., Ingles, J., Nguyen, L., Tsoutsman, T., Lam, L., Chiu, C., Chung, J., Weintraub, R. G. and Semsarian, C. 2005. Cardiac troponin I mutations in Australian families with hypertrophic cardiomyopathy: Clinical, genetic and functional consequences. J. Mol. Cell. Cardiol. 38, 387–393.

Dubreuil, R. R. 1991. Structure and evolution of the actin crosslinking proteins. BioEssays 13, 219–226.

Earing, M. G., Ackerman, M. J. and O'Leary, P. W. 2003. Diastolic ventricular dysfunction as a marker for hypertrophic cardiomyopathy in a family with a novel α-tropomyosin mutation. J. Am. Soc. Echocardiogr. 16, 698–702.

Geier, C., Perrot, A., Ozcelik, C., Binner, P., Counsell, D., Hoffmann, K., Pilz, B., Martiniak, Y., Gehmlich, K., van der Ven, P. F. M., Furst, D. O., Vornwald, A., von Hodenberg, E., Nurnberg, P., Scheffold, T., Dietz, R. and Osterziel, K. J. 2003. Mutations in the human muscle LIM protein gene in families with hypertrophic cardiomyopathy. Circulation 107, 1390–1395.

Gomes, A. V., Harada, K. and Potter, J. D. 2005. A mutation in the N-terminus of troponin I that is associated with hypertrophic cardiomyopathy affects the Ca^{2+}-sensitivity, phosphorylation kinetics and proteolytic susceptibility of troponin. J. Mol. Cell. Cardiol. 39, 754–765.

Grubinger, M. and Gimona, M. 2004. CRP2 is an autonomous actin-binding protein. FEBS Lett. 557, 88–92.

Grzeskowiak, R., Witt, H., Drungowski, M., Thermann, R., Hennig, S., Perrot, A., Osterziel, K. J., Klingbiel, D., Scheid, S. and Spang, R. 2003. Expression profiling of human idiopathic dilated cardiomyopathy. Cardiovasc. Res. 59, 400–411.

Hewett, T. E., Grupp, I. L., Grupp, G. and Robbins, J. 1994. α-skeletal actin is associated with increased contractility in the mouse heart. Circ. Res. 74, 740–746.

Hoffmann, B., Schmidt-Traub, H., Perrot, A., Osterziel, K. J. and Gessner, R. 2001. First mutation in cardiac troponin C, L29Q, in a patient with hypertrophic cardiomyopathy. Hum. Mutat. 17, 524.

Hoshijima, M. 2006. Mechanical stress–strain sensors embedded in cardiac cytoskeleton: Z disk, titin, and associated structures. Am. J. Physiol. Heart Circ. Physiol. 290, H1313–H1325.

Jain, M. K., Kashiki, S., Hsieh, C. M., Layne, M. D., Yet, S. F., Sibinga, N. E., Chin, M. T., Feinberg, M. W., Woo, I., Maas, R. L., Haber, E. and Lee, M. E. 1998. Embryonic expression suggests an important role for CRP2/SmLIM in the developing cardiovascular system. Circ. Res. 83, 980–985.

Johannessen, M., Moller, S., Hansen, T., Moens, U. and Van Ghelue, M. 2006. The multifunctional roles of the four-and-a-half-LIM only protein FHL2. Cell. Mol. Life Sci. 63, 268–284.

Jongbloed, R. J., Marcelis, C. L., Doevendans, P. A., Schmeitz-Mulkens, J. M., Van Dockum, W. G., Geraedts, J. P. and Smeets, H. J. 2003. Variable clinical manifestation of a novel missense mutation in the α-tropomyosin (TPM1) gene in familial hypertrophic cardiomyopathy. J. Am. Coll. Cardiol. 41, 981–986.

Kadrmas, J. L. and Beckerle, M. C. 2004. The LIM domain: From the cytoskeleton to the nucleus. Nat. Rev. Mol. Cell Biol. 5, 920–931.

Karibe, A., Tobacman, L. S., Strand, J., Butters, C., Back, N., Bachinski, L. L., Arai, A. E., Ortiz, A., Roberts, R., Homsher, E. and Fananapazir, L. 2001. Hypertrophic cardiomyopathy caused by a novel α-tropomyosin mutation (V95A) is associated with mild cardiac phenotype, abnormal calcium binding to troponin, abnormal myosin cycling, and poor prognosis. Circulation 103, 65–71.

Knoll, R., Hoshijima, M., Hoffman, H. M., Person, V., Lorenzen-Schmidt, I., Bang, M., Hayashi, T., Shiga, N., Yasukawa, H. and Schaper, W. 2002. The cardiac mechanical stretch sensor machinery involves a Z disc complex that is defective in a subset of human dilated cardiomyopathy. Cell 111, 943–955.

Kong, Y., Flick, M. J., Kudla, A. J. and Konieczny, S. F. 1997. Muscle LIM protein promotes myogenesis by enhancing the activity of MyoD. Mol. Cell. Biol. 17, 4750–4760.

Kong, Y., Shelton, J. M., Rothermel, B., Li, X., Richardson, J. A., Bassel-Duby, R. and Williams, R. S. 2001. Cardiac-specific LIM protein FHL2 modifies the hypertrophic response to β-adrenergic stimulation. Circulation 103, 2731–2738.

Kotaka, M., Kostin, S., Ngai, S., Chan, K., Lau, Y., Lee, S. M., Li, H., Ng, E. K., Schaper, J., Tsui, S. K., Fung, K., Lee, C. and Waye, M. M. 2000. Interaction of hCLIM1, an enigma family protein, with alpha-actinin 2. J. Cell. Biochem. 78, 558–565.

Kotaka, M., Ngai, S. M., Garcia-Barcelo, M., Tsui, S. K., Fung, K. P., Lee, C. Y. and Waye, M. M. 1999. Characterization of the human 36 kDa carboxyl terminal LIM domain protein (hCLIM1). J. Cell. Biochem. 72, 279–285.

Kuhlman, P. A., Hemmings, L. and Critchley, D. R. 1992. The identification and characterisation of an actin-binding site in α-actinin by mutagenesis. FEBS Lett. 304, 201–206.

Kuhne, W., Besselmann, M., Noll, T., Muhs, A., Watanabe, H. and Piper, H. M. 1993. Disintegration of cytoskeletal structure of actin filaments in energy-depleted endothelial cells. Am. J. Physiol. – Heart Circ. Physiol. 264, H1599–H1608.

Lader, A. S., Kwiatkowski, D. J. and Cantiello, H. F. 1999. Role of gelsolin in the actin filament regulation of cardiac L-type calcium channels. Am. J. Physiol. Cell Physiol. 277, C1277–C1283.

Lees-Miller, J. P. and Helfman, D. M. 1991. The molecular basis for tropomyosin isoform diversity. BioEssays 13, 429–437.

Levine, T. B., Francis, G. S., Goldsmith, S. R., Simon, A. B. and Cohn, J. N. 1982. Activity of the sympathetic nervous system and renin–angiotensin system assessed by plasma hormone levels and their relation to hemodynamic abnormalities in congestive heart failure. Am. J. Cardiol. 49, 1659–1666.

Levine, B. A., Moir, A. J. G., Patchell, V. B. and Perry, S. V. 1992. Binding sites involved in the interaction of actin with the N-terminal region of dystrophin. FEBS Lett. 298, 44–48.

Lindhout, D. A., Li, M. X., Schieve, D. and Sykes, B. D. 2002. Effects of T142 phosphorylation and mutation R145G on the interaction of the inhibitory region of human cardiac troponin I with the C-domain of human cardiac troponin C. Biochemistry 41, 7267–7274.

Louis, H. A., Pino, J. D., Schmeichel, K. L., Pomies, P. and Beckerle, M. C. 1997. Comparison of three members of the cysteine-rich protein family reveals functional conservation and divergent patterns of gene expression. J. Biol. Chem. 272, 27484–27491.

Maron, B. J. 2002. Hypertrophic cardiomyopathy: A systematic review. J. Am. Med. Assoc. 287, 1308–1320.

Michele, D. E., Gomez, C. A., Hong, K. E., Westfall, M. V. and Metzger, J. M. 2002. Cardiac dysfunction in hypertrophic cardiomyopathy mutant tropomyosin mice is transgene-dependent, hypertrophy-independent, and improved by β-blockade. Circ. Res. 91, 255–262.

Michels, V. V., Driscoll, D. J., Miller, F. A., Olson, T. M., Atkinson, E. J., Olswold, C. L. and Schaid, D. J. 2003. Progression of familial and non-familial dilated cardiomyopathy: Long term follow up. Heart 89, 757–761.

Mogensen, J., Klausen, I. C., Pedersen, A. K., Egeblad, H., Bross, P., Kruse, T. A., Gregersen, N., Hansen, P. S., Baandrup, U. and Borglum, A. D. 1999. α-cardiac actin is a novel disease gene in familial hypertrophic cardiomyopathy. J. Clin. Invest. 103, R39–R43.

Mogensen, J., Murphy, R. T., Kubo, T., Bahl, A., Moon, J. C., Klausen, I. C., Elliott, P. M. and McKenna, W. J. 2004c. Frequency and clinical expression of cardiac troponin I mutations in 748 consecutive families with hypertrophic cardiomyopathy. J. Am. Coll. Cardiol. 44, 2315–2325.

Mogensen, J., Murphy, R. T., Shaw, T., Bahl, A., Redwood, C., Watkins, H., Burke, M., Elliott, P. M. and McKenna, W. J. 2004b. Severe disease expression of cardiac troponin C and T mutations in patients with idiopathic dilated cardiomyopathy. J. Am. Coll. Cardiol. 44, 2033–2040.

Mogensen, J., Perrot, A., Andersen, P. S., Havndrup, O., Klausen, I. C., Christiansen, M., Bross, P., Egeblad, H., Bundgaard, H., Osterziel, K. J., Haltern, G., Lapp, H., Reinecke, P., Gregersen, N. and Borglum, A. D. 2004a. Clinical and genetic characteristics of α-cardiac actin gene mutations in hypertrophic cardiomyopathy. J. Med. Genet. 41, 10e.

Mohapatra, B., Jimenez, S., Lin, J. H., Bowles, K. R., Coveler, K. J., Marx, J. G., Chrisco, M. A., Murphy, R. T., Lurie, P. R. and Schwartz, R. J. 2003. Mutations in the muscle LIM protein and α-actinin-2 genes in dilated cardiomyopathy and endocardial fibroelastosis. Mol. Genet. Metab. 80, 207–215.

Morrisey, E. E., Ip, H. S., Lu, M. M. and Parmacek, M. S. 1996. GATA-6: A zinc finger transcription factor that is expressed in multiple cell lineages derived from lateral mesoderm. Dev. Biol. 177, 309–322.

Murphy, R. T., Kubo, T., Bahl, A., Moon, J. C., Klausen, I. C., Elliott, P. M. and McKenna, W. J. 2004. Frequency and clinical expression of cardiac troponin I mutations in 748 consecutive families with hypertrophic cardiomyopathy. J. Am. Coll. Cardiol. 44, 2315–2325.

Narula, J., Haider, N., Virmani, R., DiSalvo, T. G., Kolodgie, F. D., Hajjar, R. J., Schmidt, U., Semigran, M. J., Dec, G. W. and Khaw, B. 1996. Apoptosis in myocytes in end-stage heart failure. N. Engl. J. Med. 335, 1182–1189.

Nitahara, J. A., Cheng, W., Liu, Y., Li, B., Leri, A., Li, P., Mogul, D., Gambert, S. R., Kajstura, J. and Anversa, P. 1998. Intracellular calcium, DNase activity and myocyte apoptosis in aging Fischer 344 rats. J. Mol. Cell. Cardiol. 30, 519–535.

Olson, T. M., Doan, T. P., Kishimoto, N. Y., Whitby, F. G., Ackerman, M. J. and Fananapazir, L. 2000. Inherited and de novo mutations in the cardiac actin gene cause hypertrophic cardiomyopathy. J. Mol. Cell. Cardiol. 32, 1687–1694.

Olson, T. M., Kishimoto, N. Y., Whitby, F. G. and Michels, V. V. 2001. Mutations that alter the surface charge of α-tropomyosin are associated with dilated cardiomyopathy. J. Mol. Cell. Cardiol. 33, 723–732.

Olson, T. M., Michels, V. V., Thibodeau, S. N., Tai, Y.-S. and Keating, M. T. 1998. Actin mutations in dilated cardiomyopathy, a heritable form of heart failure. Science 280, 750–752.

Pfeffer, M. A., Lamas, G. A., Vaughan, D. E., Parisi, A. F. and Braunwald E. 1988. Effect of captopril on progressive ventricular dilatation after anterior myocardial infarction. N. Engl. J. Med. 319, 80–86.

Preston, L. C., Lipscomb, S., Robinson, P., Mogensen, J., McKenna, W. J., Watkins, H., Ashley, C. C. and Redwood, C. S. 2006. Functional effects of the DCM mutant Gly159Asp troponin C in skinned muscle fibres. Eur. J. Physiol. 453, 771–776.

Purcell, N. H., Darwis, D., Bueno, O. F., Muller, J. M., Schule, R. and Molkentin, J. D. 2004. Extracellular signal-regulated kinase 2 interacts with and is negatively regulated by the LIM-only protein FHL2 in cardiomyocytes. Mol. Cell. Biol. 24, 1081–1095.

Rath, N., Wang, Z., Lu, M. M. and Morrisey, E. E. 2005. LMCD1/Dyxin is a novel transcriptional cofactor that restricts GATA6 function by inhibiting DNA binding. Mol. Cell. Biol. 25, 8864–8873.

Regitz-Zagrosek, V., Erdmann, J., Wellnhofer, E., Raible, J. and Fleck, E. 2000. Novel mutation in the α-tropomyosin gene and transition from hypertrophic to hypocontractile dilated cardiomyopathy. Circulation 102, E112–E116.

dos Remedios, C. G., Chhabra, D., Kekic, M., Dedova, I. V., Tsubakihara, M., Berry, D. A. and Nosworthy, N. J. 2003. Actin binding proteins: Regulation of cytoskeletal microfilaments. Physiol. Rev. 83, 433–473.

Richard, P., Charron, P., Carrier, L., Ledeuil, C., Cheav, T., Pichereau, C., Benaiche, A., Isnard, R., Dubourg, O., Burban, M., Gueffet, J. P., Millaire, A., Desnos, M., Schwartz, K., Hainque, B., Komajda, M. and Eurogene Heart Failure Project. 2003. Hypertrophic cardiomyopathy: Distribution of disease genes, spectrum of mutations, and implications for a molecular diagnosis strategy. Circulation 107, 2227–2232.

Schmidtmann, A., Lindow, C., Villard, S., Heuser, A., Mugge, A., Gessner, R., Granier, C. and Jaquet, K. 2005. Cardiac troponin C-L29Q, related to hypertrophic cardiomyopathy, hinders the transduction of the protein kinase A dependent phosphorylation signal from cardiac troponin I to C. FEBS J. 272, 6087–6097.

Seidman, J. G. and Seidman, C. 2001. The genetic basis for cardiomyopathy, from mutation identification to mechanistic paradigms. Cell 104, 557–567.

Song, W. J., Van Keuren, M. L., Drabkin, H. A., Cypser, J. R., Gemmill, R. M. and Kurnit, D. M. 1996. Assignment of the human slow twitch skeletal muscle/cardiac troponin C gene (TNNC1) to human chromosome 3p21.3→3p14.3 using somatic cell hybrids. Cytogenet. Cell Genet. 75, 36–37.

Song, L., Zou, Y., Wang, J., Wang, Z., Zhen, Y., Lou, K., Zhang, Q., Wang, X., Wang, H., Li, J. and Hui, R. 2005. Mutations profile in Chinese patients with hypertrophic cardiomyopathy. Clin. Chim. Acta 351, 209–216.

Suurmeijer, A. J., Clement, S., Francesconi, A., Bocchi, L., Angelini, A., van Veldhuisen, D. J., Spagnoli, L. G., Gabbiani, G. and Orlandi, A. 2003. α-actin isoform distribution in normal and failing human heart: a morphological, morphometric, and biochemical study. J. Pathol. 199, 387–397.

Taylor, M. R., Carniel, E. and Mestroni, L. 2006. Cardiomyopathy, familial dilated. Orphanet J. Rare Dis. 1, 27.

Townsend, P. J., Farza, H., MacGeoch, C., Spurr, N. K., Wade, R., Gahlmann, R., Yacoub, M. H. and Barton, P. J. 1994. Human cardiac troponin T: Identification of fetal isoforms and assignment of the TNNT2 locus to chromosome 1q. Genomics 21, 311–316.

Tsubakihara, M. 2005. Transcription profiling study of the human heart. Ph.D. Thesis, The University of Sydney.

Vallins, W. J., Brand, N. J., Dabhade, N., Butler-Browne, G., Yacoub, M. H. and Barton, P. J. 1990. Molecular cloning of human cardiac troponin I using polymerase chain reaction. FEBS Lett. 270, 57–61.

Van Driest, S. L., Ellsworth, E. G., Ommen, S. R., Tajik, A. J., Gersh, B. J. and Ackerman, M. J. 2003. Prevalence and spectrum of thin filament mutations in an outpatient referral population with hypertrophic cardiomyopathy. Circulation 108, 445–451.

Van Driest, S. L., Will, M. L., Atkins, D. L. and Ackerman, M. J. 2002. A novel TPM1 mutation in a family with hypertrophic cardiomyopathy and sudden cardiac death in childhood. Am. J. Cardiol. 90, 1123–1127.

Vandekerckhove, J., Bugaisky, G. and Buckingham, M. 1986. Simultaneous expression of skeletal muscle and heart actin proteins in various striated muscle tissues and cells. A quantitative determination of the two actin isoforms. J. Biol. Chem. 261, 1838–1843.

Vang, S., Corydon, T. J., Borglum, A. D., Scott, M. D., Frydman, J., Mogensen, J., Gregersen, N. and Bross, P. 2005. Actin mutations in hypertrophic and dilated cardiomyopathy cause inefficient protein folding and perturbed filament formation. FEBS Lett. 272, 2037–2049.

Weiskirchen, R., Pino, J. D., Macalma, T., Bister, K. and Beckerle, M. C. 1995. The cysteine-rich protein family of highly related LIM domain proteins. J. Biol. Chem. 270, 28946–28954.

Yamauchi-Takihara, K., Nakajima-Taniguchi, C., Matsui, H., Fujio, Y., Kunisada, K., Nagata, S. and Kishimoto, T. 1996. Clinical implications of hypertrophic cardiomyopathy associated with mutations in the α-tropomyosin gene. Heart 76, 63–65.

Yang, J., Moravec, C. S., Sussman, M. A., DiPaola, N. R., Fu, D., Hawthorn, L., Mitchell, C. A., Young, J. B., Francis, G. S., McCarthy, P. M. and Bond, M. 2000. Decreased SLIM1 expression and increased gelsolin expression in failing human hearts measured by high-density oligonucleotide arrays. Circulation 102, 3046–3052.

Yao, M., Keogh, A., Spratt, P., dos Remedios, C. G. and Kiessling, P. C. 1996. Elevated DNase I levels in human idiopathic dilated cardiomyopathy: An indicator of apoptosis? J. Mol. Cell. Cardiol. 28, 95–101.

Zhang, X., Azhar, G., Chai, J., Sheridan, P., Nagano, K., Brown, T., Yang, G., Khrapko, K., Borras, A. M., Lawitts, J., Misra, R. P. and Wei, J. Y. 2001. Cardiomyopathy in transgenic mice with cardiac-specific overexpression of serum response factor. Am. J. Physiol. Heart Circ. Physiol. 280, H1782–H1792.

Zolk, O., Caroni, P. and Bohm, M. 2000. Decreased expression of the cardiac LIM domain protein MLP in chronic human heart failure. Circulation 101, 2674–2677.

Index

Printed in the United States of America